Precalculus

An Investigation of Functions

Edition 2.1

David Lippman
Melonie Rasmussen

This book is also available to read free online at
http://www.opentextbookstore.com/precalc/
If you want a printed copy, buying from the bookstore is cheaper than printing yourself.

ii

In addition to these rights, we give explicit permission to remix small portions of this book (less than 10% cumulative) into works that are CC-BY, CC-BY-SA-NC, or GFDL licensed.

Selected exercises were remixed from *Precalculus* by D.H. Collingwood and K.D. Prince, originally licensed under the GNU Free Document License, with permission from the authors. These are marked in the book as [UW].

Portions of chapter 3 were remixed from *College Algebra* by Carl Stitz and Jeff Zeager, originally licensed under a Creative Commons Attribution ShareAlike Non-Commercial license, used with permission from the authors.

Portions of chapter 9 were remixed from work by Lara Michaels, and contains content remixed from *Precalculus* by OpenStax, originally licensed under a Creative Commons Attribution license. The original version is available online for free at OpenStax.org.

Cover photos by Ralph Morasch and David Lippman, of artwork by John Rogers.
Lituus, 2010. Dichromatic glass and aluminum.
Washington State Arts Commission in partnership with Pierce College

About the Authors

David Lippman received his master's degree in mathematics from Western Washington University and has been teaching at Pierce College since Fall 2000.

Melonie Rasmussen also received her master's degree in mathematics from Western Washington University and has been teaching at Pierce College since Fall 2002. Prior to this Melonie taught for the Puyallup School district for 6 years after receiving her teaching credentials from Pacific Lutheran University.

We have both been long time advocates of open learning, open materials, and basically any idea that will reduce the cost of education for students. It started by supporting the college's calculator rental program, and running a book loan scholarship program. Eventually the frustration with the escalating costs of commercial text books and the online homework systems that charged for access led them to take action.

First, David developed IMathAS, open source online math homework software that runs WAMAP.org and MyOpenMath.com. Through this platform, we became integral parts of a vibrant sharing and learning community of teachers from around Washington State that support and contribute to WAMAP. Our pioneering efforts, supported by dozens of other dedicated faculty and financial support from the WA-SBCTC, have led to a system used by thousands of students every quarter, saving hundreds of thousands of dollars over comparable commercial offerings.

David continued further and wrote his first open textbook, *Math in Society*, a math for liberal arts majors book, after being frustrated by students having to pay $100+ for a textbook for a terminal course. Together, frustrated by both cost and the style of commercial texts, we began writing *PreCalculus: An Investigation of Functions* in 2010. Since then, David has contributed to several other open texts.

Acknowledgements

We would like to thank the following for their generous support and feedback.

- The community of WAMAP users and developers for creating a majority of the homework content used in our online homework sets.

- Pierce College students in our Fall 2010 - Summer 2011 Math 141 and Math 142 classes for helping correct typos, identifying videos related to the homework, and being our willing test subjects.

- The Open Course Library Project for providing the support needed to produce a full course package for these courses.

- Mike Kenyon, Chris Willett, Tophe Anderson, and Vauhn Foster-Grahler for reviewing the course and giving feedback and suggestions.

- Our Pierce College colleagues for providing their suggestions.

- Tophe Anderson, James Gray, and Lawrence Morales for their feedback and suggestions in content and examples.

- Jeff Eldridge for extensive proofreading and suggestions for clarification.

- James Sousa for developing videos associated with the online homework.

- Kevin Dimond for his work on indexing the book and creating PowerPoint slides.

- Faculty at Green River Community College and the Maricopa College District for their feedback and suggestions.

- Lara Michaels for contributing the basis for a conics chapter.

- The dozens of instructors who have sent us typos or suggestions over the years.

Preface

Over the years, when reviewing books we found that many had been mainstreamed by the publishers in an effort to appeal to everyone, leaving them with very little character. There were only a handful of books that had the conceptual and application driven focus we liked, and most of those were lacking in other aspects we cared about, like providing students sufficient examples and practice of basic skills. The largest frustration, however, was the never ending escalation of cost and being forced into new editions every three years. We began researching open textbooks, however the ability for those books to be adapted, remixed, or printed were often limited by the types of licenses, or didn't approach the material the way we wanted.

This book is available online for free, in both Word and PDF format. You are free to change the wording, add materials and sections or take them away. We welcome feedback, comments and suggestions for future development at precalc@opentextbookstore.com. Additionally, if you add a section, chapter or problems, we would love to hear from you and possibly add your materials so everyone can benefit.

In writing this book, our focus was on the story of functions. We begin with function notation, a basic toolkit of functions, and the basic operation with functions: composition and transformation. Building from these basic functions, as each new family of functions is introduced we explore the important features of the function: its graph, domain and range, intercepts, and asymptotes. The exploration then moves to evaluating and solving equations involving the function, finding inverses, and culminates with modeling using the function.

The "rule of four" is integrated throughout - looking at the functions verbally, graphically, numerically, as well as algebraically. We feel that using the "rule of four" gives students the tools they need to approach new problems from various angles. Often the "story problems of life" do not always come packaged in a neat equation. Being able to think critically, see the parts and build a table or graph a trend, helps us change the words into meaningful and measurable functions that model the world around us.

There is nothing we hate more than a chapter on exponential equations that begins "Exponential functions are functions that have the form $f(x)=a^x$." As each family of functions is introduced, we motivate the topic by looking at how the function arises from life scenarios or from modeling. Also, we feel it is important that precalculus be the bridge in level of thinking between algebra and calculus. In algebra, it is common to see numerous examples with very similar homework exercises, encouraging the student to mimic the examples. Precalculus provides a link that takes students from the basic plug & chug of formulaic calculations towards building an understanding that equations and formulas have deeper meaning and purpose. While you will find examples and similar exercises for the basic skills in this book, you will also find examples of multistep problem solving along with exercises in multistep problem solving. Often times these exercises will not exactly mimic the exercises, forcing the students to employ their critical thinking skills and apply the skills they've learned to new situations. By developing students' critical thinking and problem solving skills this course prepares students for the rigors of Calculus.

While we followed a fairly standard ordering of material in the first half of the book, we took some liberties in the trig portion of the book. It is our opinion that there is no need to separate unit circle trig from triangle trig, and instead integrated them in the first chapter. Identities are introduced in the first chapter, and revisited throughout. Likewise, solving is introduced in the second chapter and revisited more extensively in the third chapter. As with the first part of the book, an emphasis is placed on motivating the concepts and on modeling and interpretation.

About the Second Edition

About 4 years and several minor typo revisions after the original release of this book, we started contemplating creating a second edition. We didn't want to change much; we've always found it very annoying when new editions change things just for the sake of making it seem different. However, in talking with instructors from around the country, we knew there were a few topics that we had left out that other schools need. We didn't want to suffer the same "content bloat" that many commercial books do, but we also wanted to make it easier for more schools to adopt open resources.

We put our plans for a new revision on hold after OpenStax started working on a precalculus book, using the first edition of this text as a base. After the final product came out, though, we felt it had strayed a bit far from our original vision. We had written this text, not to be an encyclopedic reference text, but to be a concise, easy-to-read, student-friendly approach to precalculus. We valued contextual motivation and conceptual understanding over procedural skills. Our book took, in places, a non-traditional approach to topics and content ordering. Ultimately, we decided to go ahead with this second edition.

The primary changes in the second edition are:
- New, higher resolution graphs throughout
- New sections added to Chapter 3:
  - 3.4 Factor theorem (includes long division of polynomials)
  - 3.5 Real zeros of polynomials (using rational roots theorem)
  - 3.6 Complex zeros of polynomials
- Coverage of oblique asymptotes added to the rational equations section (now 3.7)
- A new section 8.5 on dot product of vectors
- A new chapter 9 on conic sections

There were many additional refinements, some new examples added, and Try it Now answers expanded, but most of the book remains unchanged.

Instructor Resources

As part of the Washington Open Course Library project, we developed a full course package to accompany this text. The course shell was built for the IMathAS online homework platform, and is available for Washington State faculty at www.wamap.org and mirrored for others at www.myopenmath.com. It contains:

- Online homework for each section (algorithmically generated, free response), most with video help associated.
- Video lessons for each section. The videos were mostly created and selected by James Sousa, of Mathispower4u.
- A selection of printable class worksheets, activities, and handouts
- Support materials for an example course (does not include all sections):
  - Suggested syllabus and Day by day course guide
  - Instructor guide with lecture outlines and examples
  - Discussion forums
  - Diagnostic review
  - Chapter review problems
  - Sample quizzes and sample chapter exams

The course shell was designed to follow Quality Matters (QM) guidelines, but has not yet been formally reviewed.

Getting Started

To get started using this textbook and the online supplementary materials,
- Request an instructor account on WAMAP (in Washington) or MyOpenMath (outside Washington).
- Review the table of contents of the text, and compare it to your course outcomes or student learning objectives. Determine which sections you will need to cover, and which to omit. If there are topics in your outcomes that are not in the text, explore other sources like the Stitz/Zeager Precalc or OpenStax Precalc to supplement from. Also check the book's website, as we may offer additional online-only topics.
- Once your instructor account is approved, log in, and click **Add New Course**
- From the "Use content from a template course", select "Precalculus – Lippman/Rasmussen 2nd Ed". Note that you might also see two half-book templates, one covering chapters $1 - 4$, and the other covering chapters $5 - 9$.
- Once you have copied the course, go through and remove any sections you don't need for your course. Refer to the Training Course Quickstart videos in MyOpenMath and WAMAP for more details on how to make those changes.

How To Be Successful In This Course

This is not a high school math course, although for some of you the content may seem familiar. There are key differences to what you will learn here, how quickly you will be required to learn it and how much work will be required of you.

You will no longer be shown a technique and be asked to mimic it repetitively as the only way to prove learning. Not only will you be required to master the technique, but you will also be required to extend that knowledge to new situations and build bridges between the material at hand and the next topic, making the course highly cumulative.

As a rule of thumb, for each hour you spend in class, you should expect this course will require an average of 2 hours of out-of-class focused study. This means that some of you with a stronger background in mathematics may take less, but if you have a weaker background or any math anxiety it will take you more.

Notice how this is the equivalent of having a part time job, and if you are taking a fulltime load of courses as many college students do, this equates to more than a full time job. If you must work, raise a family and take a full load of courses all at the same time, we recommend that you get a head start & get organized as soon as possible. We also recommend that you spread out your learning into daily chunks and avoid trying to cram or learn material quickly before an exam.

To be prepared, read through the material before it is covered in class and note or highlight the material that is new or confusing. The instructor's lecture and activities should not be the first exposure to the material. As you read, test your understanding with the Try it Now problems in the book. If you can't figure one out, try again after class, and ask for help if you still can't get it.

As soon as possible after the class session recap the day's lecture or activities into a meaningful format to provide a third exposure to the material. You could summarize your notes into a list of key points, or reread your notes and try to work examples done in class without referring back to your notes. Next, begin any assigned homework. The next day, if the instructor provides the opportunity to clarify topics or ask questions, do not be afraid to ask. If you are afraid to ask, then you are not getting your money's worth! If the instructor does not provide this opportunity, be prepared to go to a tutoring center or build a peer study group. Put in quality effort and time and you can get quality results.

Lastly, if you feel like you do not understand a topic. Don't wait, ASK FOR HELP!

ASK: Ask a teacher or tutor, **S**earch for ancillaries, **K**eep a detailed list of questions
FOR: Find additional resources, **O**rganize the material, **R**esearch other learning options
HELP: Have a support network, **E**xamine your weaknesses, **L**ist specific examples & **P**ractice

Best of luck learning! We hope you like the course & love the price.
David & Melonie

Table of Contents

Chapter 1: Functions

Section 1.1 Functions and Function Notation

What is a Function?

The natural world is full of relationships between quantities that change. When we see these relationships, it is natural for us to ask "If I know one quantity, can I then determine the other?" This establishes the idea of an input quantity, or independent variable, and a corresponding output quantity, or dependent variable. From this we get the notion of a functional relationship in which the output can be determined from the input.

For some quantities, like height and age, there are certainly relationships between these quantities. Given a specific person and any age, it is easy enough to determine their height, but if we tried to reverse that relationship and determine age from a given height, that would be problematic, since most people maintain the same height for many years.

Function
Function: A rule for a relationship between an input, or independent, quantity and an output, or dependent, quantity in which each input value uniquely determines one output value. We say "the output is a function of the input."

Example 1
In the height and age example above, is height a function of age? Is age a function of height?
In the height and age example above, it would be correct to say that height is a function of age, since each age uniquely determines a height. For example, on my 18th birthday, I had exactly one height of 69 inches.
However, age is not a function of height, since one height input might correspond with more than one output age. For example, for an input height of 70 inches, there is more than one output of age since I was 70 inches at the age of 20 and 21.

Example 2

At a coffee shop, the menu consists of items and their prices. Is price a function of the item? Is the item a function of the price?

We could say that price is a function of the item, since each input of an item has one output of a price corresponding to it. We could not say that item is a function of price, since two items might have the same price.

Example 3

In many classes the overall percentage you earn in the course corresponds to a decimal grade point. Is decimal grade a function of percentage? Is percentage a function of decimal grade?

For any percentage earned, there would be a decimal grade associated, so we could say that the decimal grade is a function of percentage. That is, if you input the percentage, your output would be a decimal grade. Percentage may or may not be a function of decimal grade, depending upon the teacher's grading scheme. With some grading systems, there are a range of percentages that correspond to the same decimal grade.

One-to-One Function

Sometimes in a relationship each input corresponds to exactly one output, and every output corresponds to exactly one input. We call this kind of relationship a **one-to-one function**.

From Example 3, *if* each unique percentage corresponds to one unique decimal grade point and each unique decimal grade point corresponds to one unique percentage then it is a one-to-one function.

Try it Now
Let's consider bank account information.
1. Is your balance a function of your bank account number?
   (if you input a bank account number does it make sense that the output is your balance?)

2. Is your bank account number a function of your balance?
   (if you input a balance does it make sense that the output is your bank account number?)

Function Notation

To simplify writing out expressions and equations involving functions, a simplified notation is often used. We also use descriptive variables to help us remember the meaning of the quantities in the problem.

Rather than write "height is a function of age", we could use the descriptive variable h to represent height and we could use the descriptive variable a to represent age.

"height is a function of age"   if we name the function f we write
"h is f of a"   or more simply
$h = f(a)$ we could instead name the function h and write
$h(a)$ which is read "h of a"

Remember we can use any variable to name the function; the notation $h(a)$ shows us that h depends on a. The value "a" must be put into the function "h" to get a result. Be careful - the parentheses indicate that age is input into the function (Note: do not confuse these parentheses with multiplication!).

Function Notation

The notation output = f(input) defines a function named f. This would be read "output is f of input"

Example 4

Introduce function notation to represent a function that takes as input the name of a month, and gives as output the number of days in that month.

The number of days in a month is a function of the name of the month, so if we name the function f, we could write "days = f(month)" or $d = f(m)$. If we simply name the function d, we could write $d(m)$

For example, d(March) = 31, since March has 31 days. The notation $d(m)$ reminds us that the number of days, d (the output) is dependent on the name of the month, m (the input)

Example 5

A function $N = f(y)$ gives the number of police officers, N, in a town in year y. What does $f(2005) = 300$ tell us?

When we read $f(2005) = 300$, we see the input quantity is 2005, which is a value for the input quantity of the function, the year (y). The output value is 300, the number of police officers (N), a value for the output quantity. Remember $N=f(y)$. This tells us that in the year 2005 there were 300 police officers in the town.

Tables as Functions

Functions can be represented in many ways: Words (as we did in the last few examples), tables of values, graphs, or formulas. Represented as a table, we are presented with a list of input and output values.
In some cases, these values represent everything we know about the relationship, while in other cases the table is simply providing us a few select values from a more complete relationship.

Table 1: This table represents the input, number of the month (January = 1, February = 2, and so on) while the output is the number of days in that month. This represents everything we know about the months & days for a given year (that is not a leap year)

(input) Month number, m	1	2	3	4	5	6	7	8	9	10	11	12
(output) Days in month, D	31	28	31	30	31	30	31	31	30	31	30	31

Table 2: The table below defines a function $Q = g(n)$. Remember this notation tells us g is the name of the function that takes the input n and gives the output Q.

n	1	2	3	4	5
Q	8	6	7	6	8

Table 3: This table represents the age of children in years and their corresponding heights. This represents just some of the data available for height and ages of children.

(input) a, age in years	5	5	6	7	8	9	10
(output) h, height inches	40	42	44	47	50	52	54

Example 6

Which of these tables define a function (if any)? Are any of them one-to-one?

Input	Output
2	1
5	3
8	6

Input	Output
-3	5
0	1
4	5

Input	Output
1	0
5	2
5	4

The first and second tables define functions. In both, each input corresponds to exactly one output. The third table does not define a function since the input value of 5 corresponds with two different output values.

Only the first table is one-to-one; it is both a function, and each output corresponds to exactly one input. Although table 2 is a function, because each input corresponds to exactly one output, each output does not correspond to exactly one input so this function is not one-to-one. Table 3 is not even a function and so we don't even need to consider if it is a one-to-one function.

Try it Now

3. If each percentage earned translated to one letter grade, would this be a function? Is it one-to-one?

Solving and Evaluating Functions:

When we work with functions, there are two typical things we do: evaluate and solve. Evaluating a function is what we do when we know an input, and use the function to determine the corresponding output. Evaluating will always produce one result, since each input of a function corresponds to exactly one output.

Solving equations involving a function is what we do when we know an output, and use the function to determine the inputs that would produce that output. Solving a function could produce more than one solution, since different inputs can produce the same output.

Example 7

Using the table shown, where $Q=g(n)$

a) Evaluate $g(3)$

n	1	2	3	4	5
Q	8	6	7	6	8

Evaluating $g(3)$ (read: "g of 3")
means that we need to determine the output value, Q, of the function g given the input value of $n=3$. Looking at the table, we see the output corresponding to $n=3$ is $Q=7$, allowing us to conclude $g(3) = 7$.

b) Solve $g(n) = 6$

Solving $g(n) = 6$ means we need to determine what input values, n, produce an output value of 6. Looking at the table we see there are two solutions: $n = 2$ and $n = 4$.

When we input 2 into the function g, our output is $Q = 6$

When we input 4 into the function g, our output is also $Q = 6$

Try it Now

4. Using the function in Example 7, evaluate $g(4)$

Graphs as Functions

Oftentimes a graph of a relationship can be used to define a function. By convention, graphs are typically created with the input quantity along the horizontal axis and the output quantity along the vertical.

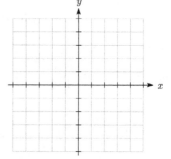

The most common graph has y on the vertical axis and x on the horizontal axis, and we say y is a function of x, or $y = f(x)$ when the function is named f.

Example 8

Which of these graphs defines a function $y=f(x)$? Which of these graphs defines a one-to-one function?

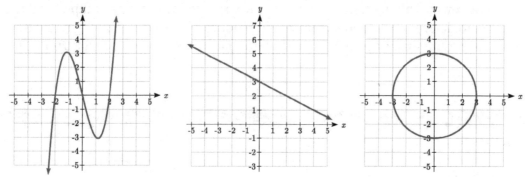

Looking at the three graphs above, the first two define a function $y=f(x)$, since for each input value along the horizontal axis there is exactly one output value corresponding, determined by the y-value of the graph. The 3rd graph does not define a function $y=f(x)$ since some input values, such as $x=2$, correspond with more than one output value.

Graph 1 is not a one-to-one function. For example, the output value 3 has two corresponding input values, -1 and 2.3

Graph 2 is a one-to-one function; each input corresponds to exactly one output, and every output corresponds to exactly one input.

Graph 3 is not even a function so there is no reason to even check to see if it is a one-to-one function.

Vertical Line Test

The **vertical line test** is a handy way to think about whether a graph defines the vertical output as a function of the horizontal input. Imagine drawing vertical lines through the graph. If any vertical line would cross the graph more than once, then the graph does not define only one vertical output for each horizontal input.

Horizontal Line Test

Once you have determined that a graph defines a function, an easy way to determine if it is a one-to-one function is to use the **horizontal line test**. Draw horizontal lines through the graph. If any horizontal line crosses the graph more than once, then the graph does not define a one-to-one function.

Evaluating a function using a graph requires taking the given input and using the graph to look up the corresponding output. Solving a function equation using a graph requires taking the given output and looking on the graph to determine the corresponding input.

Example 9

Given the graph of $f(x)$
a) Evaluate $f(2)$
b) Solve $f(x) = 4$

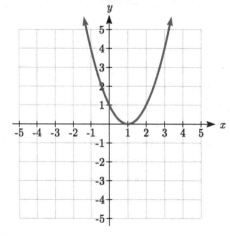

a) To evaluate $f(2)$, we find the input of $x=2$ on the horizontal axis. Moving up to the graph gives the point (2, 1), giving an output of $y=1$. $f(2) = 1$.

b) To solve $f(x) = 4$, we find the value 4 on the vertical axis because if $f(x) = 4$ then 4 is the output. Moving horizontally across the graph gives two points with the output of 4: (-1,4) and (3,4). These give the two solutions to $f(x) = 4$: $x = -1$ or $x = 3$
This means $f(-1)=4$ and $f(3)=4$, or when the input is -1 or 3, the output is 4.

Notice that while the graph in the previous example is a function, getting two input values for the output value of 4 shows us that this function is not one-to-one.

Try it Now
5. Using the graph from example 9, solve $f(x)=1$.

Formulas as Functions

When possible, it is very convenient to define relationships using formulas. If it is possible to express the output as a formula involving the input quantity, then we can define a function.

Example 10

Express the relationship $2n + 6p = 12$ as a function $p = f(n)$ if possible.

To express the relationship in this form, we need to be able to write the relationship where p is a function of n, which means writing it as $p =$ [something involving n].

$2n + 6p = 12$ subtract $2n$ from both sides
$6p = 12 - 2n$ divide both sides by 6 and simplify

$$p = \frac{12 - 2n}{6} = \frac{12}{6} - \frac{2n}{6} = 2 - \frac{1}{3}n$$

Having rewritten the formula as $p=$, we can now express p as a function:
$$p = f(n) = 2 - \frac{1}{3}n$$

It is important to note that not every relationship can be expressed as a function with a formula.

Note the important feature of an equation written as a function is that the output value can be determined directly from the input by doing evaluations - no further solving is required. This allows the relationship to act as a magic box that takes an input, processes it, and returns an output. Modern technology and computers rely on these functional relationships, since the evaluation of the function can be programmed into machines, whereas solving things is much more challenging.

Example 11

Express the relationship $x^2 + y^2 = 1$ as a function $y = f(x)$ if possible.

If we try to solve for y in this equation:
$$y^2 = 1 - x^2$$
$$y = \pm\sqrt{1 - x^2}$$

We end up with two outputs corresponding to the same input, so this relationship cannot be represented as a single function $y = f(x)$.

As with tables and graphs, it is common to evaluate and solve functions involving formulas. Evaluating will require replacing the input variable in the formula with the value provided and calculating. Solving will require replacing the output variable in the formula with the value provided, and solving for the input(s) that would produce that output.

Example 12

Given the function $k(t) = t^3 + 2$

a) Evaluate $k(2)$
b) Solve $k(t) = 1$

a) To evaluate $k(2)$, we plug in the input value 2 into the formula wherever we see the input variable t, then simplify

$k(2) = 2^3 + 2$
$k(2) = 8 + 2$
So $k(2) = 10$

b) To solve $k(t) = 1$, we set the formula for $k(t)$ equal to 1, and solve for the input value that will produce that output

$k(t) = 1$ substitute the original formula $k(t) = t^3 + 2$

$t^3 + 2 = 1$ subtract 2 from each side

$t^3 = -1$ take the cube root of each side

$t = -1$

When solving an equation using formulas, you can check your answer by using your solution in the original equation to see if your calculated answer is correct.

We want to know if $k(t) = 1$ is true when $t = -1$.

$k(-1) = (-1)^3 + 2$
$\quad\quad = -1 + 2$
$\quad\quad = 1$ which was the desired result.

Example 13

Given the function $h(p) = p^2 + 2p$

a) Evaluate $h(4)$
b) Solve $h(p) = 3$

To evaluate $h(4)$ we substitute the value 4 for the input variable p in the given function.

a) $h(4) = (4)^2 + 2(4)$
$\quad\quad = 16 + 8$
$\quad\quad = 24$

b) $h(p) = 3$ Substitute the original function $h(p) = p^2 + 2p$

$p^2 + 2p = 3$ This is quadratic, so we can rearrange the equation to get it $= 0$

$p^2 + 2p - 3 = 0$ subtract 3 from each side

$p^2 + 2p - 3 = 0$ this is factorable, so we factor it

$(p+3)(p-1) = 0$

By the zero factor theorem since $(p+3)(p-1) = 0$, either $(p+3) = 0$ or $(p-1) = 0$ (or both of them equal 0) and so we solve both equations for p, finding $p = -3$ from the first equation and $p = 1$ from the second equation.

This gives us the solution: $h(p) = 3$ when $p = 1$ or $p = -3$

We found two solutions in this case, which tells us this function is not one-to-one.

Try it Now

6. Given the function $g(m) = \sqrt{m-4}$
    a. Evaluate $g(5)$
    b. Solve $g(m) = 2$

Basic Toolkit Functions

In this text, we will be exploring functions – the shapes of their graphs, their unique features, their equations, and how to solve problems with them. When learning to read, we start with the alphabet. When learning to do arithmetic, we start with numbers. When working with functions, it is similarly helpful to have a base set of elements to build from. We call these our "toolkit of functions" – a set of basic named functions for which we know the graph, equation, and special features.

For these definitions we will use x as the input variable and $f(x)$ as the output variable.

Toolkit Functions
Linear

Constant:	$f(x) = c$, where c is a constant (number)
Identity:	$f(x) = x$

Absolute Value:	$f(x) = \|x\|$

Power

Quadratic:	$f(x) = x^2$
Cubic:	$f(x) = x^3$
Reciprocal:	$f(x) = \dfrac{1}{x}$
Reciprocal squared:	$f(x) = \dfrac{1}{x^2}$
Square root:	$f(x) = \sqrt[2]{x} = \sqrt{x}$
Cube root:	$f(x) = \sqrt[3]{x}$

You will see these toolkit functions, combinations of toolkit functions, their graphs and their transformations frequently throughout this book. In order to successfully follow along later in the book, it will be very helpful if you can recognize these toolkit functions and their features quickly by name, equation, graph and basic table values.

Not every important equation can be written as $y = f(x)$. An example of this is the equation of a circle. Recall the distance formula for the distance between two points:

$$dist = \sqrt{(x_2 - x_1)^2 + (y_2 - y_1)^2}$$

A circle with radius r with center at (h, k) can be described as all points (x, y) a distance of r from the center, so using the distance formula, $r = \sqrt{(x - h)^2 + (y - k)^2}$, giving

Equation of a circle
A circle with radius r with center (h, k) has equation $r^2 = (x - h)^2 + (y - k)^2$

Graphs of the Toolkit Functions

Constant Function: $f(x) = 2$ Identity: $f(x) = x$ Absolute Value: $f(x) = |x|$

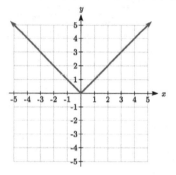

Quadratic: $f(x) = x^2$ Cubic: $f(x) = x^3$ Square root: $f(x) = \sqrt{x}$

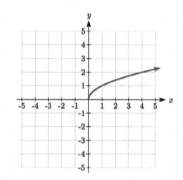

Cube root: $f(x) = \sqrt[3]{x}$ Reciprocal: $f(x) = \dfrac{1}{x}$ Reciprocal squared: $f(x) = \dfrac{1}{x^2}$

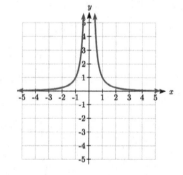

Important Topics of this Section
Definition of a function
Input (independent variable)
Output (dependent variable)
Definition of a one-to-one function
Function notation
Descriptive variables
Functions in words, tables, graphs & formulas
Vertical line test
Horizontal line test
Evaluating a function at a specific input value
Solving a function given a specific output value
Toolkit Functions

Try it Now Answers

1. Yes: for each bank account, there would be one balance associated

2. No: there could be several bank accounts with the same balance

3. Yes it's a function; No, it's not one-to-one (several percents give the same letter grade)

4. When $n=4$, $Q=g(4)=6$

5. There are two points where the output is 1: $x = 0$ or $x = 2$

6. a. $g(5) = \sqrt{5-4} = 1$

   b. $\sqrt{m-4} = 2$. Square both sides to get $m-4 = 4$. $m = 8$

Section 1.1 Exercises

1. The amount of garbage, G, produced by a city with population p is given by
    $G = f(p)$. G is measured in tons per week, and p is measured in thousands of people.
    a. The town of Tola has a population of 40,000 and produces 13 tons of garbage
        each week. Express this information in terms of the function f.
    b. Explain the meaning of the statement $f(5) = 2$.

2. The number of cubic yards of dirt, D, needed to cover a garden with area a square
    feet is given by $D = g(a)$.
    a. A garden with area 5000 ft^2 requires 50 cubic yards of dirt. Express this
        information in terms of the function g.
    b. Explain the meaning of the statement $g(100) = 1$.

3. Let $f(t)$ be the number of ducks in a lake t years after 1990. Explain the meaning of
    each statement:
    a. $f(5) = 30$ b. $f(10) = 40$

4. Let $h(t)$ be the height above ground, in feet, of a rocket t seconds after launching.
    Explain the meaning of each statement:
    a. $h(1) = 200$ b. $h(2) = 350$

5. Select all of the following graphs which represent y as a function of x.

a   b   c

d   e   f

6. Select all of the following graphs which represent y as a function of x.

a

b

c

d

e

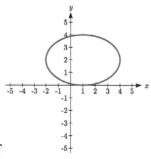

f

7. Select all of the following tables which represent y as a function of x.

a.
x	5	10	15
y	3	8	14

b.
x	5	10	15
y	3	8	8

c.
x	5	10	10
y	3	8	14

8. Select all of the following tables which represent y as a function of x.

a.
x	2	6	13
y	3	10	10

b.
x	2	6	6
y	3	10	14

c.
x	2	6	13
y	3	10	14

9. Select all of the following tables which represent y as a function of x.

a.
x	y
0	-2
3	1
4	6
8	9
3	1

b.
x	y
-1	-4
2	3
5	4
8	7
12	11

c.
x	y
0	-5
3	1
3	4
9	8
16	13

d.
x	y
-1	-4
1	2
4	2
9	7
12	13

10. Select all of the following tables which represent y as a function of x.

a.
x	y
-4	-2
3	2
6	4
9	7
12	16

b.
x	y
-5	-3
2	1
2	4
7	9
11	10

c.
x	y
-1	-3
1	2
5	4
9	8
1	2

d.
x	y
-1	-5
3	1
5	1
8	7
14	12

11. Select all of the following tables which represent y as a function of x **and** are one-to-one.

a.
x	3	8	12
y	4	7	7

b.
x	3	8	12
y	4	7	13

c.
x	3	8	8
y	4	7	13

12. Select all of the following tables which represent y as a function of x **and** are one-to-one.

a.
x	2	8	8
y	5	6	13

b.
x	2	8	14
y	5	6	6

c.
x	2	8	14
y	5	6	13

13. Select all of the following graphs which are **one-to-one functions**.

a.

b.

c.

d.

e.

f.

14. Select all of the following graphs which are **one-to-one functions**.

a

b

c

d

e

f

Given each function $f(x)$ graphed, evaluate $f(1)$ and $f(3)$

15.

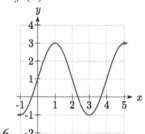

16.

17. Given the function $g(x)$ graphed here,
   a.   Evaluate $g(2)$
   b.   Solve $g(x) = 2$

18. Given the function $f(x)$ graphed here.
   a.   Evaluate $f(4)$
   b.   Solve $f(x) = 4$

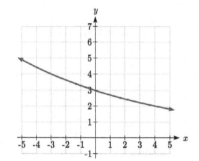

19. Based on the table below,

   a. Evaluate $f(3)$ b. Solve $f(x) = 1$

x	0	1	2	3	4	5	6	7	8	9
$f(x)$	74	28	1	53	56	3	36	45	14	47

20. Based on the table below,

   a. Evaluate $f(8)$ b. Solve $f(x) = 7$

x	0	1	2	3	4	5	6	7	8	9
$f(x)$	62	8	7	38	86	73	70	39	75	34

For each of the following functions, evaluate: $f(-2)$, $f(-1)$, $f(0)$, $f(1)$, and $f(2)$

21. $f(x) = 4 - 2x$

22. $f(x) = 8 - 3x$

23. $f(x) = 8x^2 - 7x + 3$

24. $f(x) = 6x^2 - 7x + 4$

25. $f(x) = -x^3 + 2x$

26. $f(x) = 5x^4 + x^2$

27. $f(x) = 3 + \sqrt{x+3}$

28. $f(x) = 4 - \sqrt[3]{x-2}$

29. $f(x) = (x-2)(x+3)$

30. $f(x) = (x+3)(x-1)^2$

31. $f(x) = \dfrac{x-3}{x+1}$

32. $f(x) = \dfrac{x-2}{x+2}$

33. $f(x) = 2^x$

34. $f(x) = 3^x$

35. Suppose $f(x) = x^2 + 8x - 4$. Compute the following:

    a. $f(-1) + f(1)$ b. $f(-1) - f(1)$

36. Suppose $f(x) = x^2 + x + 3$. Compute the following:

    a. $f(-2) + f(4)$ b. $f(-2) - f(4)$

37. Let $f(t) = 3t + 5$

    a. Evaluate $f(0)$ b. Solve $f(t) = 0$

38. Let $g(p) = 6 - 2p$

    a. Evaluate $g(0)$ b. Solve $g(p) = 0$

39. Match each function name with its equation.

    a. $y = x$ i.   Cube root
    b. $y = x^3$ ii.   Reciprocal
    c. $y = \sqrt[3]{x}$ iii. Linear
                               iv. Square Root
    d. $y = \dfrac{1}{x}$ v.   Absolute Value
                               vi. Quadratic
    e. $y = x^2$ vii. Reciprocal Squared
    f. $y = \sqrt{x}$ viii. Cubic
    g. $y = |x|$
    h. $y = \dfrac{1}{x^2}$

40. Match each graph with its equation.

    a. $y = x$ i.   ii.   iii.   iv.

    b. $y = x^3$

    c. $y = \sqrt[3]{x}$

    d. $y = \dfrac{1}{x}$

    e. $y = x^2$

    f. $y = \sqrt{x}$ v.   vi.   vii.   viii.

    g. $y = |x|$

    h. $y = \dfrac{1}{x^2}$

41. Match each table with its equation.

a. $y = x^2$

b. $y = x$

c. $y = \sqrt{x}$

d. $y = 1/x$

e. $y = |x|$

f. $y = x^3$

i.

In	Out
-2	-0.5
-1	-1
0	_
1	1
2	0.5
3	0.33

ii.

In	Out
-2	-2
-1	-1
0	0
1	1
2	2
3	3

iii.

In	Out
-2	-8
-1	-1
0	0
1	1
2	8
3	27

iv.

In	Out
-2	4
-1	1
0	0
1	1
2	4
3	9

v.

In	Out
-2	_
-1	_
0	0
1	1
4	2
9	3

vi.

In	Out
-2	2
-1	1
0	0
1	1
2	2
3	3

42. Match each equation with its table

   a. Quadratic
   b. Absolute Value
   c. Square Root
   d. Linear
   e. Cubic
   f. Reciprocal

i.

In	Out
-2	-0.5
-1	-1
0	_
1	1
2	0.5
3	0.33

ii.

In	Out
-2	-2
-1	-1
0	0
1	1
2	2
3	3

iii.

In	Out
-2	-8
-1	-1
0	0
1	1
2	8
3	27

iv.

In	Out
-2	4
-1	1
0	0
1	1
2	4
3	9

v.

In	Out
-2	_
-1	_
0	0
1	1
4	2
9	3

vi.

In	Out
-2	2
-1	1
0	0
1	1
2	2
3	3

43. Write the equation of the circle centered at $(3, -9)$ with radius 6.

44. Write the equation of the circle centered at $(9, -8)$ with radius 11.

45. Sketch a reasonable graph for each of the following functions. [UW]
   a. Height of a person depending on age.
   b. Height of the top of your head as you jump on a pogo stick for 5 seconds.
   c. The amount of postage you must put on a first class letter, depending on the weight of the letter.

46. Sketch a reasonable graph for each of the following functions. [UW]

a. Distance of your big toe from the ground as you ride your bike for 10 seconds.

b. Your height above the water level in a swimming pool after you dive off the high board.

c. The percentage of dates and names you'll remember for a history test, depending on the time you study.

47. Using the graph shown,

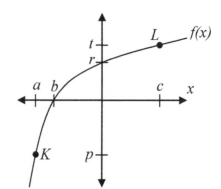

a. Evaluate $f(c)$

b. Solve $f(x) = p$

c. Suppose $f(b) = z$. Find $f(z)$

d. What are the coordinates of points L and K?

48. Dave leaves his office in Padelford Hall on his way to teach in Gould Hall. Below are several different scenarios. In each case, sketch a plausible (reasonable) graph of the function $s = d(t)$ which keeps track of Dave's distance s from Padelford Hall at time t. Take distance units to be "feet" and time units to be "minutes." Assume Dave's path to Gould Hall is long a straight line which is 2400 feet long. [UW]

gould

padelford

a. Dave leaves Padelford Hall and walks at a constant spend until he reaches Gould Hall 10 minutes later.

b. Dave leaves Padelford Hall and walks at a constant speed. It takes him 6 minutes to reach the half-way point. Then he gets confused and stops for 1 minute. He then continues on to Gould Hall at the same constant speed he had when he originally left Padelford Hall.

c. Dave leaves Padelford Hall and walks at a constant speed. It takes him 6 minutes to reach the half-way point. Then he gets confused and stops for 1 minute to figure out where he is. Dave then continues on to Gould Hall at twice the constant speed he had when he originally left Padelford Hall.

d. Dave leaves Padelford Hall and walks at a constant speed. It takes him 6 minutes to reach the half-way point. Then he gets confused and stops for 1 minute to figure out where he is. Dave is totally lost, so he simply heads back to his office, walking the same constant speed he had when he originally left Padelford Hall.

e. Dave leaves Padelford heading for Gould Hall at the same instant Angela leaves Gould Hall heading for Padelford Hall. Both walk at a constant speed, but Angela walks twice as fast as Dave. Indicate a plot of "distance from Padelford" vs. "time" for the both Angela and Dave.

f. Suppose you want to sketch the graph of a new function $s = g(t)$ that keeps track of Dave's distance s from Gould Hall at time t. How would your graphs change in (a)-(e)?

Section 1.2 Domain and Range

One of our main goals in mathematics is to model the real world with mathematical functions. In doing so, it is important to keep in mind the limitations of those models we create.

This table shows a relationship between circumference and height of a tree as it grows.

Circumference, c	1.7	2.5	5.5	8.2	13.7
Height, h	24.5	31	45.2	54.6	92.1

While there is a strong relationship between the two, it would certainly be ridiculous to talk about a tree with a circumference of -3 feet, or a height of 3000 feet. When we identify limitations on the inputs and outputs of a function, we are determining the domain and range of the function.

Domain and Range
Domain: The set of possible input values to a function
Range: The set of possible output values of a function

Example 1

Using the tree table above, determine a reasonable domain and range.

We could combine the data provided with our own experiences and reason to approximate the domain and range of the function $h = f(c)$. For the domain, possible values for the input circumference c, it doesn't make sense to have negative values, so $c > 0$. We could make an educated guess at a maximum reasonable value, or look up that the maximum circumference measured is about 119 feet[1]. With this information, we would say a reasonable domain is $0 < c \leq 119$ feet.

Similarly for the range, it doesn't make sense to have negative heights, and the maximum height of a tree could be looked up to be 379 feet, so a reasonable range is $0 < h \leq 379$ feet.

---

[1] http://en.wikipedia.org/wiki/Tree, retrieved July 19, 2010

Example 2

When sending a letter through the United States Postal Service, the price depends upon the weight of the letter[2], as shown in the table below. Determine the domain and range.

Letters	
Weight not Over	Price
1 ounce	$0.50
2 ounces	$0.71
3 ounces	$0.92
3.5 ounces	$1.13

Suppose we notate Weight by w and Price by p, and set up a function named P, where Price, p is a function of Weight, w. $p = P(w)$.

Since acceptable weights are 3.5 ounces or less, and negative weights don't make sense, the domain would be $0 < w \le 3.5$. Technically 0 could be included in the domain, but logically it would mean we are mailing nothing, so it doesn't hurt to leave it out.

Since possible prices are from a limited set of values, we can only define the range of this function by listing the possible values. The range is p = $0.50, $0.71, $0.92, or $1.13.

Try it Now

1. The population of a small town in the year 1960 was 100 people. Since then the population has grown to 1400 people reported during the 2010 census. Choose descriptive variables for your input and output and use interval notation to write the domain and range.

Notation

In the previous examples, we used inequalities to describe the domain and range of the functions. This is one way to describe intervals of input and output values, but is not the only way. Let us take a moment to discuss notation for domain and range.

Using inequalities, such as $0 < c \le 163$, $0 < w \le 3.5$, and $0 < h \le 379$ imply that we are interested in all values between the low and high values, including the high values in these examples.

However, occasionally we are interested in a specific list of numbers like the range for the price to send letters, p = $0.44, $0.61, $0.78, or $0.95. These numbers represent a set of specific values: {0.44, 0.61, 0.78, 0.95}

---

[2] https://pe.usps.com/text/dmm300/Notice123.htm, retrieved August 6, 2018

Representing values as a set, or giving instructions on how a set is built, leads us to another type of notation to describe the domain and range.

Suppose we want to describe the values for a variable x that are 10 or greater, but less than 30. In inequalities, we would write $10 \leq x < 30$.

When describing domains and ranges, we sometimes extend this into **set-builder notation**, which would look like this: $\{x \mid 10 \leq x < 30\}$. The curly brackets {} are read as "the set of", and the vertical bar | is read as "such that", so altogether we would read $\{x \mid 10 \leq x < 30\}$ as "the set of x-values such that 10 is less than or equal to x and x is less than 30."

When describing ranges in set-builder notation, we could similarly write something like $\{f(x) \mid 0 < f(x) < 100\}$, or if the output had its own variable, we could use it. So for our tree height example above, we could write for the range $\{h \mid 0 < h \leq 379\}$. In set-builder notation, if a domain or range is not limited, we could write $\{t \mid t \text{ is a real number}\}$, or $\{t \mid t \in \mathbb{R}\}$, read as "the set of t-values such that t is an element of the set of real numbers.

A more compact alternative to set-builder notation is **interval notation**, in which intervals of values are referred to by the starting and ending values. Curved parentheses are used for "strictly less than," and square brackets are used for "less than or equal to." Since infinity is not a number, we can't include it in the interval, so we always use curved parentheses with ∞ and -∞. The table below will help you see how inequalities correspond to set-builder notation and interval notation:

Inequality	Set Builder Notation	Interval notation
$5 < h \leq 10$	$\{h \mid 5 < h \leq 10\}$	(5, 10]
$5 \leq h < 10$	$\{h \mid 5 \leq h < 10\}$	[5, 10)
$5 < h < 10$	$\{h \mid 5 < h < 10\}$	(5, 10)
$h < 10$	$\{h \mid h < 10\}$	$(-\infty, 10)$
$h \geq 10$	$\{h \mid h \geq 10\}$	$[10, \infty)$
all real numbers	$\{h \mid h \in \mathbb{R}\}$	$(-\infty, \infty)$

To combine two intervals together, using inequalities or set-builder notation we can use the word "or". In interval notation, we use the union symbol, $\cup$, to combine two unconnected intervals together.

Example 3

Describe the intervals of values shown on the line graph below using set builder and interval notations.

To describe the values, *x*, that lie in the intervals shown above we would say, "*x* is a real number greater than or equal to 1 and less than or equal to 3, or a real number greater than 5."

As an inequality it is: $1 \le x \le 3$ or $x > 5$

In set builder notation: $\{x \mid 1 \le x \le 3 \text{ or } x > 5\}$

In interval notation: $[1,3] \cup (5,\infty)$

Remember when writing or reading interval notation:
Using a square bracket [means the start value is included in the set
Using a parenthesis (means the start value is not included in the set

Try it Now

2. Given the following interval, write its meaning in words, set builder notation, and interval notation.

Domain and Range from Graphs

We can also talk about domain and range based on graphs. Since domain refers to the set of possible input values, the domain of a graph consists of all the input values shown on the graph. Remember that input values are almost always shown along the horizontal axis of the graph. Likewise, since range is the set of possible output values, the range of a graph we can see from the possible values along the vertical axis of the graph.

Be careful – if the graph continues beyond the window on which we can see the graph, the domain and range might be larger than the values we can see.

26 Chapter 1

Example 4

Determine the domain and range of the graph below.

In the graph above[3], the input quantity along the horizontal axis appears to be "year", which we could notate with the variable y. The output is "thousands of barrels of oil per day", which we might notate with the variable b, for barrels. The graph would likely continue to the left and right beyond what is shown, but based on the portion of the graph that is shown to us, we can determine the domain is $1975 \leq y \leq 2008$, and the range is approximately $180 \leq b \leq 2010$.

In interval notation, the domain would be [1975, 2008] and the range would be about [180, 2010]. For the range, we have to approximate the smallest and largest outputs since they don't fall exactly on the grid lines.

Remember that, as in the previous example, x and y are not always the input and output variables. Using descriptive variables is an important tool to remembering the context of the problem.

---

[3]

Try it Now

3. Given the graph below write the domain and range in interval notation

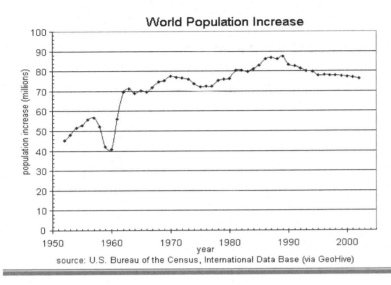

World Population Increase

source: U.S. Bureau of the Census, International Data Base (via GeoHive)

Domains and Ranges of the Toolkit functions

We will now return to our set of toolkit functions to note the domain and range of each.

Constant Function: $f(x) = c$
The domain here is not restricted; x can be anything. When this is the case we say the domain is all real numbers. The outputs are limited to the constant value of the function.
Domain: $(-\infty, \infty)$
Range: $[c]$
Since there is only one output value, we list it by itself in square brackets.

Identity Function: $f(x) = x$
Domain: $(-\infty, \infty)$
Range: $(-\infty, \infty)$

Quadratic Function: $f(x) = x^2$
Domain: $(-\infty, \infty)$
Range: $[0, \infty)$
Multiplying a negative or positive number by itself can only yield a positive output.

Cubic Function: $f(x) = x^3$
Domain: $(-\infty, \infty)$
Range: $(-\infty, \infty)$

Reciprocal: $f(x) = \dfrac{1}{x}$

Domain: $(-\infty, 0) \cup (0, \infty)$

Range: $(-\infty, 0) \cup (0, \infty)$

We cannot divide by 0 so we must exclude 0 from the domain.
One divide by any value can never be 0, so the range will not include 0.

Reciprocal squared: $f(x) = \dfrac{1}{x^2}$

Domain: $(-\infty, 0) \cup (0, \infty)$

Range: $(0, \infty)$

We cannot divide by 0 so we must exclude 0 from the domain.

Cube Root: $f(x) = \sqrt[3]{x}$

Domain: $(-\infty, \infty)$

Range: $(-\infty, \infty)$

Square Root: $f(x) = \sqrt[2]{x}$, commonly just written as, $f(x) = \sqrt{x}$

Domain: $[0, \infty)$

Range: $[0, \infty)$

When dealing with the set of real numbers we cannot take the square root of a negative number so the domain is limited to 0 or greater.

Absolute Value Function: $f(x) = |x|$

Domain: $(-\infty, \infty)$

Range: $[0, \infty)$

Since absolute value is defined as a distance from 0, the output can only be greater than or equal to 0.

Example 5

Find the domain of each function: a) $f(x) = 2\sqrt{x + 4}$ b) $g(x) = \dfrac{3}{6 - 3x}$

a) Since we cannot take the square root of a negative number, we need the inside of the square root to be non-negative.
$x + 4 \geq 0$ when $x \geq -4$.
The domain of *f(x)* is $[-4, \infty)$.

b) We cannot divide by zero, so we need the denominator to be non-zero.
$6 - 3x = 0$ when $x = 2$, so we must exclude 2 from the domain.
The domain of *g(x)* is $(-\infty, 2) \cup (2, \infty)$.

Piecewise Functions

In the toolkit functions we introduced the absolute value function $f(x) = |x|$.

With a domain of all real numbers and a range of values greater than or equal to 0, the absolute value can be defined as the magnitude or modulus of a number, a real number value regardless of sign, the size of the number, or the distance from 0 on the number line. All of these definitions require the output to be greater than or equal to 0.

If we input 0, or a positive value the output is unchanged
$$f(x) = x \quad \text{if} \quad x \geq 0$$

If we input a negative value the sign must change from negative to positive.
$$f(x) = -x \quad \text{if} \quad x < 0, \quad \text{since multiplying a negative value by -1 makes it positive.}$$

Since this requires two different processes or pieces, the absolute value function is often called the most basic piecewise defined function.

Piecewise Function

A **piecewise function** is a function in which the formula used depends upon the domain the input lies in. We notate this idea like:

$$f(x) = \begin{cases} \text{formula 1} & \text{if} & \text{domain to use formula 1} \\ \text{formula 2} & \text{if} & \text{domain to use formula 2} \\ \text{formula 3} & \text{if} & \text{domain to use formula 3} \end{cases}$$

Example 6

A museum charges $5 per person for a guided tour with a group of 1 to 9 people, or a fixed $50 fee for 10 or more people in the group. Set up a function relating the number of people, n, to the cost, C.

To set up this function, two different formulas would be needed. $C = 5n$ would work for n values under 10, and $C = 50$ would work for values of n ten or greater. Notating this:

$$C(n) = \begin{cases} 5n & \text{if} & 0 < n < 10 \\ 50 & \text{if} & n \geq 10 \end{cases}$$

Example 7

A cell phone company uses the function below to determine the cost, C, in dollars for g gigabytes of data transfer.

$$C(g) = \begin{cases} 25 & if \quad 0 < g < 2 \\ 25 + 10(g-2) & if \quad g \geq 2 \end{cases}$$

Find the cost of using 1.5 gigabytes of data, and the cost of using 4 gigabytes of data.

To find the cost of using 1.5 gigabytes of data, $C(1.5)$, we first look to see which piece of domain our input falls in. Since 1.5 is less than 2, we use the first formula, giving $C(1.5) = \$25$.

The find the cost of using 4 gigabytes of data, $C(4)$, we see that our input of 4 is greater than 2, so we'll use the second formula. $C(4) = 25 + 10(4-2) = \$45$.

Example 8

Sketch a graph of the function $f(x) = \begin{cases} x^2 & if \quad x \leq 1 \\ 3 & if \quad 1 < x \leq 2 \\ 6 - x & if \quad x > 2 \end{cases}$

The first two component functions are from our library of Toolkit functions, so we know their shapes. We can imagine graphing each function, then limiting the graph to the indicated domain. At the endpoints of the domain, we put open circles to indicate where the endpoint is not included, due to a strictly-less-than inequality, and a closed circle where the endpoint is included, due to a less-than-or-equal-to inequality.

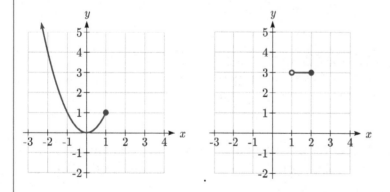

For the third function, you should recognize this as a linear equation from your previous coursework. If you remember how to graph a line using slope and intercept, you can do that. Otherwise, we could calculate a couple values, plot points, and connect them with a line.

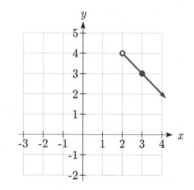

At $x = 2$, $f(2) = 6 - 2 = 4$. We place an open circle here.
At $x = 3$, $f(3) = 6 - 3 = 3$. Connect these points with a line.

Now that we have each piece individually, we combine them onto the same graph:

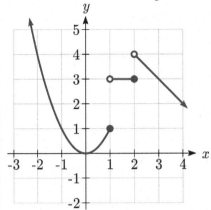

Try it Now

4. At Pierce College during the 2009-2010 school year tuition rates for in-state residents were \$89.50 per credit for the first 10 credits, \$33 per credit for credits 11-18, and for over 18 credits the rate is \$73 per credit[4]. Write a piecewise defined function for the total tuition, T, at Pierce College during 2009-2010 as a function of the number of credits taken, c. Be sure to consider a reasonable domain and range.

Important Topics of this Section
Definition of domain
Definition of range
Inequalities
Interval notation
Set builder notation
Domain and Range from graphs
Domain and Range of toolkit functions
Piecewise defined functions

---

[4] https://www.pierce.ctc.edu/dist/tuition/ref/files/0910_tuition_rate.pdf, retrieved August 6, 2010

Try it Now Answers

1. Domain; y = years $[1960, 2010]$; Range, p = population, $[100, 1400]$

2. a. Values that are less than or equal to -2, or values that are greater than or equal to -1 and less than 3
   b. $\{x \mid x \le -2 \ or \ -1 \le x < 3\}$
   c. $(-\infty, -2] \cup [-1, 3)$

3. Domain; y=years, $[1952, 2002]$; Range, p=population in millions, $[40, 88]$

4. $T(c) = \begin{cases} 89.5c & if & c \le 10 \\ 895 + 33(c-10) & if & 10 < c \le 18 \\ 1159 + 73(c-18) & if & c > 18 \end{cases}$ Tuition, T, as a function of credits, c.

Reasonable domain should be whole numbers 0 to (answers may vary), e.g. $[0, 23]$
Reasonable range should be $0 – (answers may vary), e.g. $[0, 1524]$

Section 1.2 Exercises

Write the domain and range of the function using interval notation.

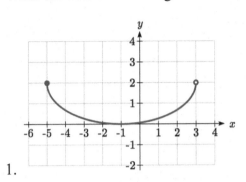

1.   2.

Write the domain and range of each graph as an inequality.

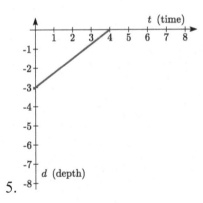

3.   4.

Suppose that you are holding your toy submarine under the water. You release it and it begins to ascend. The graph models the depth of the submarine as a function of time, stopping once the sub surfaces. What is the domain and range of the function in the graph?

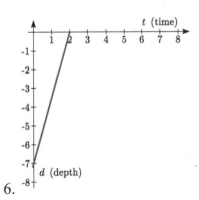

5.   6.

Find the domain of each function

7. $f(x) = 3\sqrt{x-2}$

8. $f(x) = 5\sqrt{x+3}$

9. $f(x) = 3 - \sqrt{6-2x}$

10. $f(x) = 5 - \sqrt{10-2x}$

11. $f(x) = \dfrac{9}{x-6}$

12. $f(x) = \dfrac{6}{x-8}$

13. $f(x) = \dfrac{3x+1}{4x+2}$

14. $f(x) = \dfrac{5x+3}{4x-1}$

15. $f(x) = \dfrac{\sqrt{x+4}}{x-4}$

16. $f(x) = \dfrac{\sqrt{x+5}}{x-6}$

17. $f(x) = \dfrac{x-3}{x^2+9x-22}$

18. $f(x) = \dfrac{x-8}{x^2+8x-9}$

Given each function, evaluate: $f(-1)$, $f(0)$, $f(2)$, $f(4)$

19. $f(x) = \begin{cases} 7x+3 & if \quad x < 0 \\ 7x+6 & if \quad x \geq 0 \end{cases}$

20. $f(x) = \begin{cases} 4x-9 & if \quad x < 0 \\ 4x-18 & if \quad x \geq 0 \end{cases}$

21. $f(x) = \begin{cases} x^2-2 & if \quad x < 2 \\ 4+|x-5| & if \quad x \geq 2 \end{cases}$

22. $f(x) = \begin{cases} 4-x^3 & if \quad x < 1 \\ \sqrt{x+1} & if \quad x \geq 1 \end{cases}$

23. $f(x) = \begin{cases} 5x & if \quad x < 0 \\ 3 & if \quad 0 \leq x \leq 3 \\ x^2 & if \quad x > 3 \end{cases}$

24. $f(x) = \begin{cases} x^3+1 & if \quad x < 0 \\ 4 & if \quad 0 \leq x \leq 3 \\ 3x+1 & if \quad x > 3 \end{cases}$

Write a formula for the piecewise function graphed below.

25.

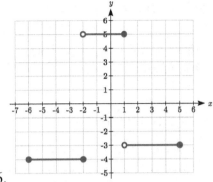

26.

27.

28.

29.

30.

Sketch a graph of each piecewise function

31. $f(x) = \begin{cases} |x| & if \quad x < 2 \\ 5 & if \quad x \geq 2 \end{cases}$

32. $f(x) = \begin{cases} 4 & if \quad x < 0 \\ \sqrt{x} & if \quad x \geq 0 \end{cases}$

33. $f(x) = \begin{cases} x^2 & if \quad x < 0 \\ x+2 & if \quad x \geq 0 \end{cases}$

34. $f(x) = \begin{cases} x+1 & if \quad x < 1 \\ x^3 & if \quad x \geq 1 \end{cases}$

35. $f(x) = \begin{cases} 3 & if \quad x \leq -2 \\ -x+1 & if \quad -2 < x \leq 1 \\ 3 & if \quad x > 1 \end{cases}$

36. $f(x) = \begin{cases} -3 & if \quad x \leq -2 \\ x-1 & if \quad -2 < x \leq 2 \\ 0 & if \quad x > 2 \end{cases}$

Section 1.3 Rates of Change and Behavior of Graphs

Since functions represent how an output quantity varies with an input quantity, it is natural to ask about the rate at which the values of the function are changing.

For example, the function $C(t)$ below gives the average cost, in dollars, of a gallon of gasoline t years after 2000.

t	2	3	4	5	6	7	8	9
$C(t)$	1.47	1.69	1.94	2.30	2.51	2.64	3.01	2.14

If we were interested in how the gas prices had changed between 2002 and 2009, we could compute that the cost per gallon had increased from $1.47 to $2.14, an increase of $0.67. While this is interesting, it might be more useful to look at how much the price changed *per year*. You are probably noticing that the price didn't change the same amount each year, so we would be finding the **average rate of change** over a specified amount of time.

The gas price increased by $0.67 from 2002 to 2009, over 7 years, for an average of $\frac{\$0.67}{7\,years} \approx 0.096$ dollars per year. On average, the price of gas increased by about 9.6 cents each year.

Rate of Change

A **rate of change** describes how the output quantity changes in relation to the input quantity. The units on a rate of change are "<u>output units</u> per <u>input units</u>"

Some other examples of rates of change would be quantities like:
- A population of rats increases by 40 rats per week
- A barista earns $9 per hour (dollars per hour)
- A farmer plants 60,000 onions per acre
- A car can drive 27 miles per gallon
- A population of grey whales decreases by 8 whales per year
- The amount of money in your college account decreases by $4,000 per quarter

---

Average Rate of Change

The **average rate of change** between two input values is the total change of the function values (output values) divided by the change in the input values.

$$\text{Average rate of change} = \frac{\text{Change of Output}}{\text{Change of Input}} = \frac{\Delta y}{\Delta x} = \frac{y_2 - y_1}{x_2 - x_1}$$

Example 1

Using the cost-of-gas function from earlier, find the average rate of change between 2007 and 2009

From the table, in 2007 the cost of gas was $2.64. In 2009 the cost was $2.14.

The input (years) has changed by 2. The output has changed by $2.14 - $2.64 = -0.50.

The average rate of change is then $\dfrac{-\$0.50}{2\,years}$ = -0.25 dollars per year

---

Try it Now

1. Using the same cost-of-gas function, find the average rate of change between 2003 and 2008

Notice that in the last example the change of output was *negative* since the output value of the function had decreased. Correspondingly, the average rate of change is negative.

Example 2

Given the function *g(t)* shown here, find the average rate of change on the interval [0, 3].

At *t* = 0, the graph shows $g(0) = 1$
At *t* = 3, the graph shows $g(3) = 4$

The output has changed by 3 while the input has changed by 3, giving an average rate of change of:
$$\frac{4-1}{3-0} = \frac{3}{3} = 1$$

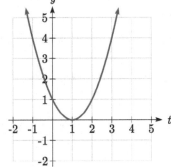

On a road trip, after picking up your friend who lives 10 miles away, you decide to record your distance from home over time. Find your average speed over the first 6 hours.

t (hours)	0	1	2	3	4	5	6	7
D(t) (miles)	10	55	90	153	214	240	292	300

Here, your average speed is the average rate of change.
You traveled 282 miles in 6 hours, for an average speed of
$$\frac{292-10}{6-0} = \frac{282}{6} = 47 \text{ miles per hour}$$

We can more formally state the average rate of change calculation using function notation.

Average Rate of Change using Function Notation

Given a function $f(x)$, the average rate of change on the interval $[a, b]$ is

$$\text{Average rate of change} = \frac{\text{Change of Output}}{\text{Change of Input}} = \frac{f(b) - f(a)}{b - a}$$

Compute the average rate of change of $f(x) = x^2 - \dfrac{1}{x}$ on the interval $[2, 4]$

We can start by computing the function values at each endpoint of the interval
$$f(2) = 2^2 - \frac{1}{2} = 4 - \frac{1}{2} = \frac{7}{2}$$
$$f(4) = 4^2 - \frac{1}{4} = 16 - \frac{1}{4} = \frac{63}{4}$$

Now computing the average rate of change

$$\text{Average rate of change} = \frac{f(4) - f(2)}{4 - 2} = \frac{\frac{63}{4} - \frac{7}{2}}{4 - 2} = \frac{\frac{49}{4}}{2} = \frac{49}{8}$$

Try it Now

2. Find the average rate of change of $f(x) = x - 2\sqrt{x}$ on the interval $[1, 9]$

Example 5

The magnetic force F, measured in Newtons, between two magnets is related to the distance between the magnets d, in centimeters, by the formula $F(d) = \dfrac{2}{d^2}$. Find the average rate of change of force if the distance between the magnets is increased from 2 cm to 6 cm.

We are computing the average rate of change of $F(d) = \dfrac{2}{d^2}$ on the interval [2, 6].

Average rate of change $= \dfrac{F(6) - F(2)}{6 - 2}$ Evaluating the function

$\dfrac{F(6) - F(2)}{6 - 2} =$

$\dfrac{\dfrac{2}{6^2} - \dfrac{2}{2^2}}{6 - 2}$ Simplifying

$\dfrac{\dfrac{2}{36} - \dfrac{2}{4}}{4}$ Combining the numerator terms

$\dfrac{\dfrac{-16}{36}}{4}$ Simplifying further

$\dfrac{-1}{9}$ Newtons per centimeter

This tells us the magnetic force decreases, on average, by 1/9 Newtons per centimeter over this interval.

Example 6

Find the average rate of change of $g(t) = t^2 + 3t + 1$ on the interval $[0, a]$. Your answer will be an expression involving a.

Using the average rate of change formula

$\dfrac{g(a) - g(0)}{a - 0}$ Evaluating the function

$\dfrac{(a^2 + 3a + 1) - (0^2 + 3(0) + 1)}{a - 0}$ Simplifying

$\dfrac{a^2 + 3a + 1 - 1}{a}$ Simplifying further, and factoring

$$\frac{a(a+3)}{a}$$

Cancelling the common factor a

$$a+3$$

This result tells us the average rate of change between $t = 0$ and any other point $t = a$. For example, on the interval [0, 5], the average rate of change would be 5+3 = 8.

Try it Now

3. Find the average rate of change of $f(x) = x^3 + 2$ on the interval $[a, a+h]$.

Graphical Behavior of Functions

As part of exploring how functions change, it is interesting to explore the graphical behavior of functions.

Increasing/Decreasing

A function is **increasing** on an interval if the function values increase as the inputs increase. More formally, a function is increasing if $f(b) > f(a)$ for any two input values a and b in the interval with $b > a$. The average rate of change of an increasing function is **positive.**

A function is **decreasing** on an interval if the function values decrease as the inputs increase. More formally, a function is decreasing if $f(b) < f(a)$ for any two input values a and b in the interval with $b > a$. The average rate of change of a decreasing function is **negative.**

Example 7

Given the function p(t) graphed here, on what intervals does the function appear to be increasing?

The function appears to be increasing from $t = 1$ to $t = 3$, and from $t = 4$ on.

In interval notation, we would say the function appears to be increasing on the interval $(1, 3)$ and the interval $(4, \infty)$.

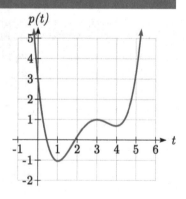

Notice in the last example that we used open intervals (intervals that don't include the endpoints) since the function is neither increasing nor decreasing at $t = 1$, 3, or 4.

Local Extrema
A point where a function changes from increasing to decreasing is called a **local maximum**.
A point where a function changes from decreasing to increasing is called a **local minimum**.
Together, local maxima and minima are called the **local extrema**, or local extreme values, of the function.

Example 8

Using the cost of gasoline function from the beginning of the section, find an interval on which the function appears to be decreasing. Estimate any local extrema using the table.

t	2	3	4	5	6	7	8	9
C(t)	1.47	1.69	1.94	2.30	2.51	2.64	3.01	2.14

It appears that the cost of gas increased from $t = 2$ to $t = 8$. It appears the cost of gas decreased from $t = 8$ to $t = 9$, so the function appears to be decreasing on the interval $(8, 9)$.

Since the function appears to change from increasing to decreasing at $t = 8$, there is local maximum at $t = 8$.

Example 9

Use a graph to estimate the local extrema of the function $f(x) = \dfrac{2}{x} + \dfrac{x}{3}$. Use these to determine the intervals on which the function is increasing.

Using technology to graph the function, it appears there is a local minimum somewhere between $x = 2$ and $x = 3$, and a symmetric local maximum somewhere between $x = -3$ and $x = -2$.

Most graphing calculators and graphing utilities can estimate the location of maxima and minima. Below are screen images from two different technologies, showing the estimate for the local maximum and minimum.

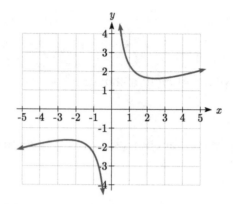

Based on these estimates, the function is increasing on the intervals $(-\infty, -2.449)$ and $(2.449, \infty)$. Notice that while we expect the extrema to be symmetric, the two different technologies agree only up to 4 decimals due to the differing approximation algorithms used by each.

Try it Now

4. Use a graph of the function $f(x) = x^3 - 6x^2 - 15x + 20$ to estimate the local extrema of the function. Use these to determine the intervals on which the function is increasing and decreasing.

Concavity

The total sales, in thousands of dollars, for two companies over 4 weeks are shown.

Company A

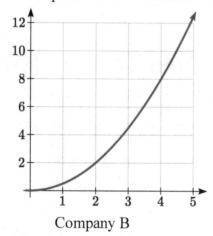

Company B

As you can see, the sales for each company are increasing, but they are increasing in very different ways. To describe the difference in behavior, we can investigate how the average rate of change varies over different intervals. Using tables of values,

Company A

Week	Sales	Rate of Change
0	0	
		5
1	5	
		2.1
2	7.1	
		1.6
3	8.7	
		1.3
4	10	

Company B

Week	Sales	Rate of Change
0	0	
		0.5
1	0.5	
		1.5
2	2	
		2.5
3	4.5	
		3.5
4	8	

From the tables, we can see that the rate of change for company A is *decreasing*, while the rate of change for company B is *increasing*.

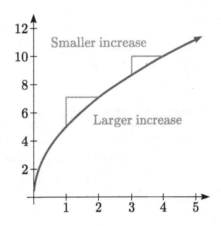

When the rate of change is getting smaller, as with Company A, we say the function is **concave down**. When the rate of change is getting larger, as with Company B, we say the function is **concave up**.

Concavity

A function is **concave up** if the rate of change is increasing.

A function is **concave down** if the rate of change is decreasing.

A point where a function changes from concave up to concave down or vice versa is called an **inflection point**.

Example 10

An object is thrown from the top of a building. The object's height in feet above ground after t seconds is given by the function $h(t) = 144 - 16t^2$ for $0 \le t \le 3$. Describe the concavity of the graph.

Sketching a graph of the function, we can see that the function is decreasing. We can calculate some rates of change to explore the behavior.

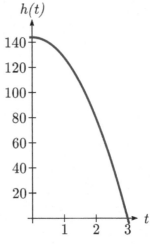

t	$h(t)$	Rate of Change
0	144	
		-16
1	128	
		-48
2	80	
		-80
3	0	

Notice that the rates of change are becoming more negative, so the rates of change are *decreasing*. This means the function is concave down.

Example 11

The value, V, of a car after t years is given in the table below. Is the value increasing or decreasing? Is the function concave up or concave down?

t	0	2	4	6	8
$V(t)$	28000	24342	21162	18397	15994

can compute rates of change to determine concavity.

t	0	2	4	6	8
$V(t)$	28000	24342	21162	18397	15994
Rate of change		-1829	-1590	-1382.5	-1201.5

Since these values are becoming less negative, the rates of change are *increasing*, so this function is concave up.

Try it Now

5. Is the function described in the table below concave up or concave down?

x	0	5	10	15	20
$g(x)$	10000	9000	7000	4000	0

Graphically, concave down functions bend downwards like a frown, and concave up function bend upwards like a smile.

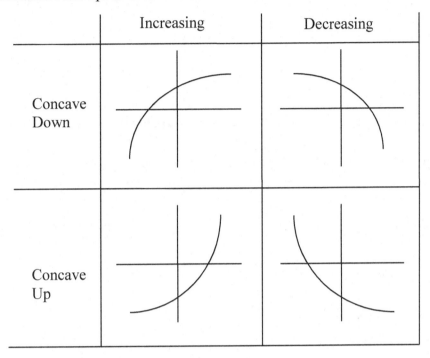

	Increasing	Decreasing
Concave Down		
Concave Up		

Estimate from the graph shown the intervals on which the function is concave down and concave up.

On the far left, the graph is decreasing but concave up, since it is bending upwards. It begins increasing at $x = -2$, but it continues to bend upwards until about $x = -1$.

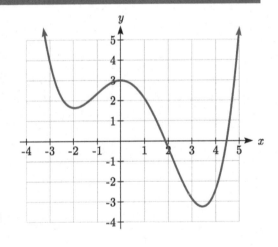

From $x = -1$ the graph starts to bend downward, and continues to do so until about $x = 2$. The graph then begins curving upwards for the remainder of the graph shown.

From this, we can estimate that the graph is concave up on the intervals $(-\infty, -1)$ and $(2, \infty)$, and is concave down on the interval $(-1, 2)$. The graph has inflection points at $x = -1$ and $x = 2$.

Try it Now

6. Using the graph from Try it Now 4, $f(x) = x^3 - 6x^2 - 15x + 20$, estimate the intervals on which the function is concave up and concave down.

Behaviors of the Toolkit Functions

We will now return to our toolkit functions and discuss their graphical behavior.

Function	Increasing/Decreasing	Concavity
Constant Function $f(x) = c$	Neither increasing nor decreasing	Neither concave up nor down
Identity Function $f(x) = x$	Increasing	Neither concave up nor down
Quadratic Function $f(x) = x^2$	Increasing on $(0, \infty)$ Decreasing on $(-\infty, 0)$ Minimum at $x = 0$	Concave up $(-\infty, \infty)$
Cubic Function $f(x) = x^3$	Increasing	Concave down on $(-\infty, 0)$ Concave up on $(0, \infty)$ Inflection point at $(0,0)$
Reciprocal $f(x) = \dfrac{1}{x}$	Decreasing $(-\infty, 0) \cup (0, \infty)$	Concave down on $(-\infty, 0)$ Concave up on $(0, \infty)$
Function	Increasing/Decreasing	Concavity
Reciprocal squared $f(x) = \dfrac{1}{x^2}$	Increasing on $(-\infty, 0)$ Decreasing on $(0, \infty)$	Concave up on $(-\infty, 0) \cup (0, \infty)$
Cube Root $f(x) = \sqrt[3]{x}$	Increasing	Concave down on $(0, \infty)$ Concave up on $(-\infty, 0)$ Inflection point at $(0,0)$
Square Root $f(x) = \sqrt{x}$	Increasing on $(0, \infty)$	Concave down on $(0, \infty)$
Absolute Value $f(x) = |x|$	Increasing on $(0, \infty)$ Decreasing on $(-\infty, 0)$	Neither concave up or down

Important Topics of This Section
Rate of Change
Average Rate of Change
Calculating Average Rate of Change using Function Notation
Increasing/Decreasing
Local Maxima and Minima (Extrema)
Inflection points
Concavity

Try it Now Answers

1. $\dfrac{\$3.01 - \$1.69}{5\,years} = \dfrac{\$1.32}{5\,years} = 0.264$ dollars per year.

2. Average rate of change $= \dfrac{f(9) - f(1)}{9 - 1} = \dfrac{\left(9 - 2\sqrt{9}\right) - \left(1 - 2\sqrt{1}\right)}{9 - 1} = \dfrac{(3) - (-1)}{9 - 1} = \dfrac{4}{8} = \dfrac{1}{2}$

3. $\dfrac{f(a+h) - f(a)}{(a+h) - a} = \dfrac{\left((a+h)^3 + 2\right) - \left(a^3 + 2\right)}{h} = \dfrac{a^3 + 3a^2 h + 3ah^2 + h^3 + 2 - a^3 - 2}{h} =$

   $\dfrac{3a^2 h + 3ah^2 + h^3}{h} = \dfrac{h\left(3a^2 + 3ah + h^2\right)}{h} = 3a^2 + 3ah + h^2$

4. Based on the graph, the local maximum appears to occur at (-1, 28), and the local minimum occurs at (5,-80). The function is increasing on $(-\infty, -1) \cup (5, \infty)$ and decreasing on $(-1, 5)$.

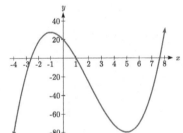

5. Calculating the rates of change, we see the rates of change become *more* negative, so the rates of change are *decreasing*. This function is concave down.

x	0	5	10	15	20
g(x)	10000	9000	7000	4000	0
Rate of change		-1000	-2000	-3000	-4000	

6. Looking at the graph, it appears the function is concave down on $(-\infty, 2)$ and concave up on $(2, \infty)$.

Section 1.3 Exercises

1. The table below gives the annual sales (in millions of dollars) of a product. What was the average rate of change of annual sales...
    a) Between 2001 and 2002?   b) Between 2001 and 2004?

year	1998	1999	2000	2001	2002	2003	2004	2005	2006
sales	201	219	233	243	249	251	249	243	233

2. The table below gives the population of a town, in thousands. What was the average rate of change of population...
    a) Between 2002 and 2004?   b) Between 2002 and 2006?

year	2000	2001	2002	2003	2004	2005	2006	2007	2008
population	87	84	83	80	77	76	75	78	81

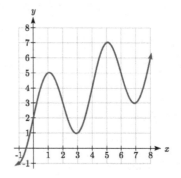

3. Based on the graph shown, estimate the average rate of change from $x = 1$ to $x = 4$.

4. Based on the graph shown, estimate the average rate of change from $x = 2$ to $x = 5$.

Find the average rate of change of each function on the interval specified.

5. $f(x) = x^2$ on $[1, 5]$

6. $q(x) = x^3$ on $[-4, 2]$

7. $g(x) = 3x^3 - 1$ on $[-3, 3]$

8. $h(x) = 5 - 2x^2$ on $[-2, 4]$

9. $k(t) = 6t^2 + \dfrac{4}{t^3}$ on $[-1, 3]$

10. $p(t) = \dfrac{t^2 - 4t + 1}{t^2 + 3}$ on $[-3, 1]$

Find the average rate of change of each function on the interval specified. Your answers will be expressions involving a parameter (b or h).

11. $f(x) = 4x^2 - 7$ on $[1, b]$

12. $g(x) = 2x^2 - 9$ on $[4, b]$

13. $h(x) = 3x + 4$ on $[2, 2+h]$

14. $k(x) = 4x - 2$ on $[3, 3+h]$

15. $a(t) = \dfrac{1}{t+4}$ on $[9, 9+h]$

16. $b(x) = \dfrac{1}{x+3}$ on $[1, 1+h]$

17. $j(x) = 3x^3$ on $[1, 1+h]$

18. $r(t) = 4t^3$ on $[2, 2+h]$

19. $f(x) = 2x^2 + 1$ on $[x, x+h]$

20. $g(x) = 3x^2 - 2$ on $[x, x+h]$

For each function graphed, estimate the intervals on which the function is increasing and decreasing.

21.   22.

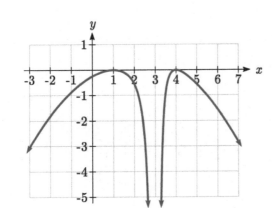

23.   24.

For each table below, select whether the table represents a function that is increasing or decreasing, and whether the function is concave up or concave down.

25.

x	f(x)
1	2
2	4
3	8
4	16
5	32

26.

x	g(x)
1	90
2	80
3	75
4	72
5	70

27.

x	h(x)
1	300
2	290
3	270
4	240
5	200

28.

x	k(x)
1	0
2	15
3	25
4	32
5	35

29.

x	f(x)
1	-10
2	-25
3	-37
4	-47
5	-54

30.

x	g(x)
1	-200
2	-190
3	-160
4	-100
5	0

31.

x	h(x)
1	-100
2	-50
3	-25
4	-10
5	0

32.

x	k(x)
1	-50
2	-100
3	-200
4	-400
5	-900

For each function graphed, estimate the intervals on which the function is concave up and concave down, and the location of any inflection points.

33.   34.

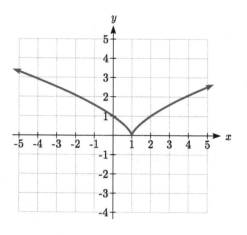

35.   36.

Use a graph to estimate the local extrema and inflection points of each function, and to estimate the intervals on which the function is increasing, decreasing, concave up, and concave down.

37. $f(x) = x^4 - 4x^3 + 5$ 38. $h(x) = x^5 + 5x^4 + 10x^3 + 10x^2 - 1$

39. $g(t) = t\sqrt{t+3}$ 40. $k(t) = 3t^{2/3} - t$

41. $m(x) = x^4 + 2x^3 - 12x^2 - 10x + 4$ 42. $n(x) = x^4 - 8x^3 + 18x^2 - 6x + 2$

Section 1.4 Composition of Functions

Suppose we wanted to calculate how much it costs to heat a house on a particular day of the year. The cost to heat a house will depend on the average daily temperature, and the average daily temperature depends on the particular day of the year. Notice how we have just defined two relationships: The temperature depends on the day, and the cost depends on the temperature. Using descriptive variables, we can notate these two functions.

The first function, $C(T)$, gives the cost C of heating a house when the average daily temperature is T degrees Celsius, and the second, $T(d)$, gives the average daily temperature on day d of the year in some city. If we wanted to determine the cost of heating the house on the 5th day of the year, we could do this by linking our two functions together, an idea called composition of functions. Using the function $T(d)$, we could evaluate $T(5)$ to determine the average daily temperature on the 5th day of the year. We could then use that temperature as the input to the $C(T)$ function to find the cost to heat the house on the 5th day of the year: $C(T(5))$.

Composition of Functions

When the output of one function is used as the input of another, we call the entire operation a **composition of functions**. We write $f(g(x))$, and read this as "f of g of x" or "f composed with g at x".

An alternate notation for composition uses the composition operator: $\circ$

$(f \circ g)(x)$ is read "f of g of x" or "f composed with g at x", just like $f(g(x))$.

Example 1

Suppose $c(s)$ gives the number of calories burned doing s sit-ups, and $s(t)$ gives the number of sit-ups a person can do in t minutes. Interpret $c(s(3))$.

When we are asked to interpret, we are being asked to explain the meaning of the expression in words. The inside expression in the composition is $s(3)$. Since the input to the s function is time, the 3 is representing 3 minutes, and $s(3)$ is the number of sit-ups that can be done in 3 minutes. Taking this output and using it as the input to the $c(s)$ function will gives us the calories that can be burned by the number of sit-ups that can be done in 3 minutes.

Note that it is not important that the same variable be used for the output of the inside function and the input to the outside function. However, it *is* essential that the units on the output of the inside function match the units on the input to the outside function, if the units are specified.

Example 2

Suppose $f(x)$ gives miles that can be driven in x hours, and $g(y)$ gives the gallons of gas used in driving y miles. Which of these expressions is meaningful: $f(g(y))$ or $g(f(x))$?

The expression $g(y)$ takes miles as the input and outputs a number of gallons. The function $f(x)$ is expecting a number of hours as the input; trying to give it a number of gallons as input does not make sense. Remember the units must match, and number of gallons does not match number of hours, so the expression $f(g(y))$ is meaningless.

The expression $f(x)$ takes hours as input and outputs a number of miles driven. The function $g(y)$ is expecting a number of miles as the input, so giving the output of the $f(x)$ function (miles driven) as an input value for g(y), where gallons of gas depends on miles driven, does make sense. The expression $g(f(x))$ makes sense, and will give the number of gallons of gas used, g, driving a certain number of miles, $f(x)$, in x hours.

Try it Now

1. In a department store you see a sign that says 50% off clearance merchandise, so final cost C depends on the clearance price, p, according to the function $C(p)$. Clearance price, p, depends on the original discount, d, given to the clearance item, $p(d)$. Interpret $C(p(d))$.

Composition of Functions using Tables and Graphs

When working with functions given as tables and graphs, we can look up values for the functions using a provided table or graph, as discussed in section 1.1. We start evaluation from the provided input, and first evaluate the inside function. We can then use the output of the inside function as the input to the outside function. To remember this, always work from the inside out.

Example 3

Using the tables below, evaluate $f(g(3))$ and $g(f(4))$

x	$f(x)$
1	6
2	8
3	3
4	1

x	$g(x)$
1	3
2	5
3	2
4	7

To evaluate $f(g(3))$, we start from the inside with the value 3. We then evaluate the inside expression $g(3)$ using the table that defines the function g: $g(3) = 2$.

We can then use that result as the input to the f function, so $g(3)$ is replaced by the equivalent value 2 and we can evaluate $f(2)$. Then using the table that defines the function f, we find that $f(2) = 8$.

$f(g(3)) = f(2) = 8$.

To evaluate $g(f(4))$, we first evaluate the inside expression $f(4)$ using the first table: $f(4) = 1$. Then using the table for g we can evaluate:

$g(f(4)) = g(1) = 3$.

Try it Now

2. Using the tables from the example above, evaluate $f(g(1))$ and $g(f(3))$.

Example 4

Using the graphs below, evaluate $f(g(1))$.

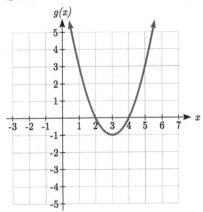

To evaluate $f(g(1))$, we again start with the inside evaluation. We evaluate $g(1)$ using the graph of the $g(x)$ function, finding the input of 1 on the horizontal axis and finding the output value of the graph at that input. Here, $g(1) - 3$.

Using this value as the input to the f function, $f(g(1)) = f(3)$. We can then evaluate this by looking to the graph of the $f(x)$ function, finding the input of 3 on the horizontal axis, and reading the output value of the graph at this input.

$f(3) = 6$, so $f(g(1)) = 6$.

Try it Now

3. Using the graphs from the previous example, evaluate $g(f(2))$.

Composition using Formulas

When evaluating a composition of functions where we have either created or been given formulas, the concept of working from the inside out remains the same. First, we evaluate the inside function using the input value provided, then use the resulting output as the input to the outside function.

Example 5

Given $f(t) = t^2 - t$ and $h(x) = 3x + 2$, evaluate $f(h(1))$.

Since the inside evaluation is $h(1)$ we start by evaluating the $h(x)$ function at 1:
$h(1) = 3(1) + 2 = 5$

Then $f(h(1)) = f(5)$, so we evaluate the $f(t)$ function at an input of 5:
$f(h(1)) = f(5) = 5^2 - 5 = 20$

Try it Now

4. Using the functions from the example above, evaluate $h(f(-2))$.

While we can compose the functions as above for each individual input value, sometimes it would be really helpful to find a single formula which will calculate the result of a composition $f(g(x))$. To do this, we will extend our idea of function evaluation. Recall that when we evaluate a function like $f(t) = t^2 - t$, we put whatever value is inside the parentheses after the function name into the formula wherever we see the input variable.

Example 6

Given $f(t) = t^2 - t$, evaluate $f(3)$ and $f(-2)$.

$f(3) = 3^2 - 3$
$f(-2) = (-2)^2 - (-2)$

We could simplify the results above if we wanted to
$f(3) = 3^2 - 3 = 9 - 3 = 6$
$f(-2) = (-2)^2 - (-2) = 4 + 2 = 6$

We are not limited, however, to using a numerical value as the input to the function. We can put anything into the function: a value, a different variable, or even an algebraic expression, provided we use the input expression everywhere we see the input variable.

Example 7

Using the function from the previous example, evaluate $f(a)$.

This means that the input value for t is some unknown quantity a. As before, we evaluate by replacing the input variable t with the input quantity, in this case a.

$f(a) = a^2 - a$

The same idea can then be applied to expressions more complicated than a single letter.

Example 8

Using the same $f(t)$ function from above, evaluate $f(x+2)$.

Everywhere in the formula for f where there was a t, we would replace it with the input $(x+2)$. Since in the original formula the input t was squared in the first term, the entire input $x+2$ needs to be squared when we substitute, so we need to use grouping parentheses. To avoid problems, it is advisable to always use parentheses around inputs.

$f(x+2) = (x+2)^2 - (x+2)$

We could simplify this expression further to $f(x+2) = x^2 + 3x + 2$ if we wanted to:

$f(x+2) = (x+2)(x+2) - (x+2)$	Use the "FOIL" technique (first, outside, inside, last)
$f(x+2) = x^2 + 2x + 2x + 4 - (x+2)$	distribute the negative sign
$f(x+2) = x^2 + 2x + 2x + 4 - x - 2$	combine like terms
$f(x+2) = x^2 + 3x + 2$	

Example 9

Using the same function, evaluate $f(t^3)$.

Note that in this example, the same variable is used in the input expression and as the input variable of the function. This doesn't matter – we still replace the original input t in the formula with the new input expression, t^3.

$f(t^3) = (t^3)^2 - (t^3) = t^6 - t^3$

Try it Now

5. Given $g(x) = 3x - \sqrt{x}$, evaluate $g(t - 2)$.

This now allows us to find an expression for a composition of functions. If we want to find a formula for $f(g(x))$, we can start by writing out the formula for $g(x)$. We can then evaluate the function $f(x)$ at that expression, as in the examples above.

Example 10

Let $f(x) = x^2$ and $g(x) = \dfrac{1}{x} - 2x$, find $f(g(x))$ and $g(f(x))$.

To find $f(g(x))$, we start by evaluating the inside, writing out the formula for $g(x)$.

$$g(x) = \frac{1}{x} - 2x$$

We then use the expression $\left(\dfrac{1}{x} - 2x\right)$ as input for the function f.

$$f(g(x)) = f\left(\frac{1}{x} - 2x\right)$$

We then evaluate the function $f(x)$ using the formula for $g(x)$ as the input.

Since $f(x) = x^2$, $f\left(\dfrac{1}{x} - 2x\right) = \left(\dfrac{1}{x} - 2x\right)^2$

This gives us the formula for the composition: $f(g(x)) = \left(\dfrac{1}{x} - 2x\right)^2$.

Likewise, to find $g(f(x))$, we evaluate the inside, writing out the formula for $f(x)$

$$g(f(x)) = g\left(x^2\right)$$

Now we evaluate the function $g(x)$ using x^2 as the input.

$$g(f(x)) = \frac{1}{x^2} - 2x^2$$

Try it Now

6. Let $f(x) = x^3 + 3x$ and $g(x) = \sqrt{x}$, find $f(g(x))$ and $g(f(x))$.

Example 11

A city manager determines that the tax revenue, R, in millions of dollars collected on a population of p thousand people is given by the formula $R(p) = 0.03p + \sqrt{p}$, and that the city's population, in thousands, is predicted to follow the formula $p(t) = 60 + 2t + 0.3t^2$, where t is measured in years after 2010. Find a formula for the tax revenue as a function of the year.

Since we want tax revenue as a function of the year, we want year to be our initial input, and revenue to be our final output. To find revenue, we will first have to predict the city population, and then use that result as the input to the tax function. So we need to find $R(p(t))$. Evaluating this,

$$R(p(t)) = R\left(60 + 2t + 0.3t^2\right) = 0.03\left(60 + 2t + 0.3t^2\right) + \sqrt{60 + 2t + 0.3t^2}$$

This composition gives us a single formula which can be used to predict the tax revenue during a given year, without needing to find the intermediary population value.

For example, to predict the tax revenue in 2017, when $t = 7$ (because t is measured in years after 2010),

$$R(p(7)) = 0.03\left(60 + 2(7) + 0.3(7)^2\right) + \sqrt{60 + 2(7) + 0.3(7)^2} \approx 12.079 \text{ million dollars}$$

Domain of Compositions

When we think about the domain of a composition $h(x) = f(g(x))$, we must consider both the domain of the inner function and the domain of the composition itself. While it is tempting to only look at the resulting composite function, if the inner function were undefined at a value of x, the composition would not be possible.

Example 12

Let $f(x) = \dfrac{1}{x^2 - 1}$ and $g(x) = \sqrt{x - 2}$. Find the domain of $f\left(g(x)\right)$.

Since we want to avoid the square root of negative numbers, the domain of $g(x)$ is the set of values where $x - 2 \geq 0$. The domain is $x \geq 2$.

The composition is $f\left(g(x)\right) = \dfrac{1}{\left(\sqrt{x-2}\right)^2 - 1} = \dfrac{1}{(x-2) - 1} = \dfrac{1}{x - 3}$.

The composition is undefined when $x = 3$, so that value must also be excluded from the domain. Notice that the composition doesn't involve a square root, but we still have to consider the domain limitation from the inside function.

Combining the two restrictions, the domain is all values of x greater than or equal to 2, except $x = 3$.

In inequalities, the domain is: $2 \le x < 3$ or $x > 3$.
In interval notation, the domain is: $[2,3) \cup (3,\infty)$.

Try it Now

7. Let $f(x) = \dfrac{1}{x-2}$ and $g(x) = \dfrac{1}{x}$. Find the domain of $f\big(g(x)\big)$.

Decomposing Functions

In some cases, it is desirable to decompose a function – to write it as a composition of two simpler functions.

Example 13

Write $f(x) = 3 + \sqrt{5 - x^2}$ as the composition of two functions.

We are looking for two functions, g and h, so $f(x) = g(h(x))$. To do this, we look for a function inside a function in the formula for f(x). As one possibility, we might notice that $5 - x^2$ is the inside of the square root. We could then decompose the function as:
$$h(x) = 5 - x^2$$
$$g(x) = 3 + \sqrt{x}$$

We can check our answer by recomposing the functions:
$$g(h(x)) = g\big(5 - x^2\big) = 3 + \sqrt{5 - x^2}$$

Note that this is not the only solution to the problem. Another non-trivial decomposition would be $h(x) = x^2$ and $g(x) = 3 + \sqrt{5 - x}$

Important Topics of this Section
Definition of Composition of Functions
Compositions using:
Words
Tables
Graphs
Equations
Domain of Compositions
Decomposition of Functions

Try it Now Answers

1. The final cost, C, depends on the clearance price, p, which is based on the original discount, d. (Or the original discount d, determines the clearance price and the final cost is half of the clearance price.)

2. $f(g(1)) = f(3) = 3$ and $g(f(3)) = g(3) = 2$

3. $g(f(2)) = g(5) = 3$

4. $h(f(-2)) = h(6) = 20$ *did you remember to insert your input values using parentheses?*

5. $g(t-2) = 3(t-2) - \sqrt{(t-2)}$

6. $f(g(x)) = f\left(\sqrt{x}\right) = \left(\sqrt{x}\right)^3 + 3\left(\sqrt{x}\right)$

   $g(f(x)) = g\left(x^3 + 3x\right) = \sqrt{\left(x^3 + 3x\right)}$

7. $g(x) = \dfrac{1}{x}$ is undefined at $x = 0$.

   The composition, $f\left(g(x)\right) = f\left(\dfrac{1}{x}\right) = \dfrac{1}{\dfrac{1}{x} - 2} = \dfrac{1}{\dfrac{1}{x} - \dfrac{2x}{x}} = \dfrac{1}{\dfrac{1-2x}{x}} = \dfrac{x}{1-2x}$ is undefined

   when $1 - 2x = 0$, when $x = \dfrac{1}{2}$.

   Restricting these two values, the domain is $(-\infty, 0) \cup \left(0, \dfrac{1}{2}\right) \cup \left(\dfrac{1}{2}, \infty\right)$.

Section 1.4 Exercises

Given each pair of functions, calculate $f(g(0))$ and $g(f(0))$.

1. $f(x) = 4x + 8$, $g(x) = 7 - x^2$

2. $f(x) = 5x + 7$, $g(x) = 4 - 2x^2$

3. $f(x) = \sqrt{x+4}$, $g(x) = 12 - x^3$

4. $f(x) = \dfrac{1}{x+2}$, $g(x) = 4x + 3$

Use the table of values to evaluate each expression

5. $f(g(8))$

6. $f(g(5))$

7. $g(f(5))$

8. $g(f(3))$

9. $f(f(4))$

10. $f(f(1))$

11. $g(g(2))$

12. $g(g(6))$

x	$f(x)$	$g(x)$
0	7	9
1	6	5
2	5	6
3	8	2
4	4	1
5	0	8
6	2	7
7	1	3
8	9	4
9	3	0

Use the graphs to evaluate the expressions below.

13. $f(g(3))$

14. $f(g(1))$

15. $g(f(1))$

16. $g(f(0))$

17. $f(f(5))$

18. $f(f(4))$

19. $g(g(2))$

20. $g(g(0))$

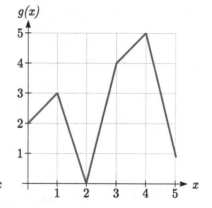

For each pair of functions, find $f(g(x))$ and $g(f(x))$. Simplify your answers.

21. $f(x) = \dfrac{1}{x-6}$, $g(x) = \dfrac{7}{x} + 6$

22. $f(x) = \dfrac{1}{x-4}$, $g(x) = \dfrac{2}{x} + 4$

23. $f(x) = x^2 + 1$, $g(x) = \sqrt{x+2}$

24. $f(x) = \sqrt{x} + 2$, $g(x) = x^2 + 3$

25. $f(x) = |x|$, $g(x) = 5x + 1$

26. $f(x) = \sqrt[3]{x}$, $g(x) = \dfrac{x+1}{x^3}$

27. If $f(x)=x^4+6$, $g(x)=x-6$ and $h(x)=\sqrt{x}$, find $f(g(h(x)))$

28. If $f(x)=x^2+1$, $g(x)=\dfrac{1}{x}$ and $h(x)=x+3$, find $f(g(h(x)))$

29. The function $D(p)$ gives the number of items that will be demanded when the price is p. The production cost, $C(x)$ is the cost of producing x items. To determine the cost of production when the price is \$6, you would do which of the following:
    a. Evaluate $D(C(6))$ b. Evaluate $C(D(6))$
    c. Solve $D(C(x))=6$ d. Solve $C(D(p))=6$

30. The function $A(d)$ gives the pain level on a scale of 0-10 experienced by a patient with d milligrams of a pain reduction drug in their system. The milligrams of drug in the patient's system after t minutes is modeled by $m(t)$. To determine when the patient will be at a pain level of 4, you would need to:
    a. Evaluate $A\big(m(4)\big)$ b. Evaluate $m\big(A(4)\big)$
    c. Solve $A\big(m(t)\big)=4$ d. Solve $m\big(A(d)\big)=4$

31. The radius r, in inches, of a spherical balloon is related to the volume, V, by
    $r(V)=\sqrt[3]{\dfrac{3V}{4\pi}}$. Air is pumped into the balloon, so the volume after t seconds is given by $V(t)=10+20t$.
    a. Find the composite function $r\big(V(t)\big)$
    b. Find the radius after 20 seconds

32. The number of bacteria in a refrigerated food product is given by
    $N(T)=23T^2-56T+1$, $3<T<33$, where T is the temperature of the food. When the food is removed from the refrigerator, the temperature is given by $T(t)=5t+1.5$, where t is the time in hours.
    a. Find the composite function $N\big(T(t)\big)$
    b. Find the bacteria count after 4 hours

33. Given $p(x)=\dfrac{1}{\sqrt{x}}$ and $m(x)=x^2-4$, find the domain of $m(p(x))$.

34. Given $p(x)=\dfrac{1}{\sqrt{x}}$ and $m(x)=9-x^2$, find the domain of $m(p(x))$.

35. Given $f(x)=\dfrac{1}{x+3}$ and $g(x)=\dfrac{2}{x-1}$, find the domain of $f\big(g(x)\big)$.

36. Given $f(x) = \dfrac{x}{x+1}$ and $g(x) = \dfrac{4}{x}$, find the domain of $f(g(x))$.

37. Given $f(x) = \sqrt{x-2}$ and $g(x) = \dfrac{2}{x^2 - 3}$, find the domain of $g(f(x))$.

38. Given $f(x) = \sqrt{4-x}$ and $g(x) = \dfrac{1}{x^2 - 2}$, find the domain of $g(f(x))$.

Find functions $f(x)$ and $g(x)$ so the given function can be expressed as $h(x) = f(g(x))$.

39. $h(x) = (x+2)^2$

40. $h(x) = (x-5)^3$

41. $h(x) = \dfrac{3}{x-5}$

42. $h(x) = \dfrac{4}{(x+2)^2}$

43. $h(x) = 3 + \sqrt{x-2}$

44. $h(x) = 4 + \sqrt[3]{x}$

45. Let $f(x)$ be a linear function, with form $f(x) = ax + b$ for constants a and b. [UW]

   a. Show that $f(f(x))$ is a linear function

   b. Find a function $g(x)$ such that $g(g(x)) = 6x - 8$

46. Let $f(x) = \dfrac{1}{2}x + 3$ [UW]

   a. Sketch the graphs of $f(x), f(f(x)), f(f(f(x)))$ on the interval $-2 \le x \le 10$.

   b. Your graphs should all intersect at the point (6, 6). The value x = 6 is called a fixed point of the function $f(x)$ since $f(6) = 6$; that is, 6 is fixed - it doesn't move when f is applied to it. Give an explanation for why 6 is a fixed point for any function $f(f(f(...f(x)...)))$.

   c. Linear functions (with the exception of $f(x) = x$) can have at most one fixed point. Quadratic functions can have at most two. Find the fixed points of the function $g(x) = x^2 - 2$.

   d. Give a quadratic function whose fixed points are $x = -2$ and $x = 3$.

47. A car leaves Seattle heading east. The speed of the car in mph after m minutes is given by the function $C(m) = \dfrac{70m^2}{10 + m^2}$. [UW]

a. Find a function $m = f(s)$ that converts seconds s into minutes m. Write out the formula for the new function $C(f(s))$; what does this function calculate?

b. Find a function $m = g(h)$ that converts hours h into minutes m. Write out the formula for the new function $C(g(h))$; what does this function calculate?

c. Find a function $z = v(s)$ that converts mph s into ft/sec z. Write out the formula for the new function $v(C(m))$; what does this function calculate?

Section 1.5 Transformation of Functions

Often when given a problem, we try to model the scenario using mathematics in the form of words, tables, graphs and equations in order to explain or solve it. When building models, it is often helpful to build off of existing formulas or models. Knowing the basic graphs of your tool-kit functions can help you solve problems by being able to model new behavior by adapting something you already know. Unfortunately, the models and existing formulas we know are not always exactly the same as the ones presented in the problems we face.

Fortunately, there are systematic ways to shift, stretch, compress, flip and combine functions to help them become better models for the problems we are trying to solve. We can transform what we already know into what we need, hence the name, "Transformation of functions." When we have a story problem, formula, graph, or table, we can then transform that function in a variety of ways to form new functions.

Shifts

Example 1

To regulate temperature in a green building, air flow vents near the roof open and close throughout the day to allow warm air to escape. The graph below shows the open vents V (in square feet) throughout the day, t in hours after midnight. During the summer, the facilities staff decides to try to better regulate temperature by increasing the amount of open vents by 20 square feet throughout the day. Sketch a graph of this new function.

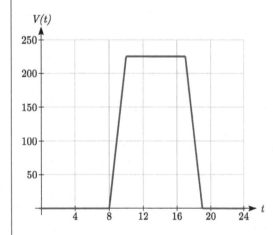

We can sketch a graph of this new function by adding 20 to each of the output values of the original function. This will have the effect of shifting the graph up.

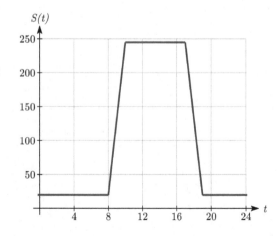

Notice that in the second graph, for each input value, the output value has increased by twenty, so if we call the new function $S(t)$, we could write $S(t) = V(t) + 20$.

Note that this notation tells us that for any value of t, $S(t)$ can be found by evaluating the V function at the same input, then adding twenty to the result.
This defines S as a transformation of the function V, in this case a vertical shift up 20 units.

Notice that with a vertical shift the input values stay the same and only the output values change.

Vertical Shift

Given a function $f(x)$, if we define a new function $g(x)$ as

$g(x) = f(x) + k$, where k is a constant

then $g(x)$ is a **vertical shift** of the function $f(x)$, where all the output values have been increased by k.

If k is positive, then the graph will shift up

If k is negative, then the graph will shift down

Example 2

A function $f(x)$ is given as a table below. Create a table for the function $g(x) = f(x) - 3$

x	2	4	6	8
$f(x)$	1	3	7	11

The formula $g(x) = f(x) - 3$ tells us that we can find the output values of the g function by subtracting 3 from the output values of the f function. For example,

$f(2) = 1$ is found from the given table

$g(x) - f(x) - 3$ is our given transformation

$g(2) = f(2) - 3 = 1 - 3 = -2$

Subtracting 3 from each $f(x)$ value, we can complete a table of values for $g(x)$

x	2	4	6	8
$g(x)$	-2	0	4	8

As with the earlier vertical shift, notice the input values stay the same and only the output values change.

1. The function $h(t) = -4.9t^2 + 30t$ gives the height h of a ball (in meters) thrown upwards from the ground after t seconds. Suppose the ball was instead thrown from the top of a 10 meter building. Relate this new height function $b(t)$ to $h(t)$, then find a formula for $b(t)$.

The vertical shift is a change to the output, or outside, of the function. We will now look at how changes to input, on the inside of the function, change its graph and meaning.

Example 3

Returning to our building air flow example from the beginning of the section, suppose that in Fall, the facilities staff decides that the original venting plan starts too late, and they want to move the entire venting program to start two hours earlier. Sketch a graph of the new function.

$V(t)$ = the original venting plan

$F(t)$ = starting 2 hours sooner

In the new graph, which we can call $F(t)$, at each time, the air flow is the same as the original function $V(t)$ was two hours later. For example, in the original function V, the air flow starts to change at 8am, while for the function $F(t)$ the air flow starts to change at 6am. The comparable function values are $V(8) = F(6)$.

Notice also that the vents first opened to 220 sq. ft. at 10 a.m. under the original plan, while under the new plan the vents reach 220 sq. ft. at 8 a.m., so $V(10) = F(8)$.

In both cases we see that since $F(t)$ starts 2 hours sooner, the same output values are reached when, $F(t) = V(t+2)$

Note that $V(t+2)$ had the effect of shifting the graph to the *left*.

Horizontal changes or "inside changes" affect the domain of a function (the input) instead of the range and often seem counterintuitive. The new function $F(t)$ uses the same outputs as $V(t)$, but matches those outputs to inputs two hours earlier than those of $V(t)$. Said another way, we must add 2 hours to the input of V to find the corresponding output for F: $F(t) = V(t+2)$.

Horizontal Shift

Given a function $f(x)$, if we define a new function $g(x)$ as

$g(x) = f(x+k)$, where k is a constant

then $g(x)$ is a **horizontal shift** of the function $f(x)$

If k is positive, then the graph will shift left

If k is negative, then the graph will shift right

Example 4

A function $f(x)$ is given as a table below. Create a table for the function $g(x) = f(x-3)$

x	2	4	6	8
$f(x)$	1	3	7	11

The formula $g(x) = f(x-3)$ tells us that the output values of g are the same as the output value of f with an input value three smaller. For example, we know that $f(2) = 1$. To get the same output from the g function, we will need an input value that is 3 *larger*: We input a value that is three larger for $g(x)$ because the function takes three away before evaluating the function f.

$g(5) = f(5-3) = f(2) = 1$

x	5	7	9	11
$g(x)$	1	3	7	11

The result is that the function $g(x)$ has been shifted to the right by 3. Notice the output values for $g(x)$ remain the same as the output values for $f(x)$ in the chart, but the corresponding input values, x, have shifted to the right by 3: 2 shifted to 5, 4 shifted to 7, 6 shifted to 9 and 8 shifted to 11.

Example 5

The graph shown is a transformation of the toolkit function $f(x) = x^2$. Relate this new function $g(x)$ to $f(x)$, and then find a formula for $g(x)$.

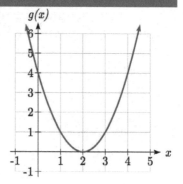

Notice that the graph looks almost identical in shape to the $f(x) = x^2$ function, but the x values are shifted to the right two units. The vertex used to be at $(0, 0)$ but now the vertex is at $(2, 0)$. The graph is the basic quadratic function shifted two to the right, so
$g(x) = f(x-2)$

Notice how we must input the value $x = 2$, to get the output value $y = 0$; the x values must be two units larger, because of the shift to the right by 2 units.

We can then use the definition of the *f(x)* function to write a formula for *g(x)* by evaluating $f(x-2)$:
Since $f(x) = x^2$ and $g(x) = f(x-2)$
$g(x) = f(x-2) = (x-2)^2$

If you find yourself having trouble determining whether the shift is +2 or -2, it might help to consider a single point on the graph. For a quadratic, looking at the bottom-most point is convenient. In the original function, $f(0) = 0$. In our shifted function, $g(2) = 0$. To obtain the output value of 0 from the *f* function, we need to decide whether a +2 or -2 will work to satisfy $g(2) = f(2?2) = f(0) = 0$. For this to work, we will need to subtract 2 from our input values.

When thinking about horizontal and vertical shifts, it is good to keep in mind that vertical shifts are affecting the output values of the function, while horizontal shifts are affecting the input values of the function.

Example 6

The function *G(m)* gives the number of gallons of gas required to drive *m* miles. Interpret $G(m)+10$ and $G(m+10)$.

$G(m)+10$ is adding 10 to the output, gallons. This is 10 gallons of gas more than is required to drive *m* miles. So, this is the gas required to drive *m* miles, plus another 10 gallons of gas.

$G(m+10)$ is adding 10 to the input, miles. This is the number of gallons of gas required to drive 10 miles more than *m* miles.

Try it Now

2. Given the function $f(x) = \sqrt{x}$ graph the original function $f(x)$ and the transformation $g(x) = f(x+2)$.
   a. Is this a horizontal or a vertical change?
   b. Which way is the graph shifted and by how many units?
   c. Graph $f(x)$ and $g(x)$ on the same axes.

Now that we have two transformations, we can combine them together.

Remember:
Vertical Shifts are outside changes that affect the output (vertical) axis values shifting the transformed function up or down.

Horizontal Shifts are inside changes that affect the input (horizontal) axis values shifting the transformed function left or right.

Example 7

Given $f(x) = |x|$, sketch a graph of $h(x) = f(x+1) - 3$.

The function f is our toolkit absolute value function. We know that this graph has a V shape, with the point at the origin. The graph of h has transformed f in two ways: $f(x+1)$ is a change on the inside of the function, giving a horizontal shift left by 1, then the subtraction by 3 in $f(x+1) - 3$ is a change to the outside of the function, giving a vertical shift down by 3. Transforming the graph gives

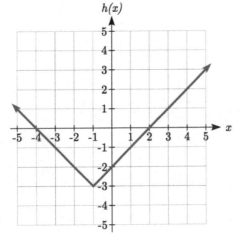

We could also find a formula for this transformation by evaluating the expression for $h(x)$:
$$h(x) = f(x+1) - 3$$
$$h(x) = |x+1| - 3$$

Example 8

Write a formula for the graph shown, a transformation of the toolkit square root function.

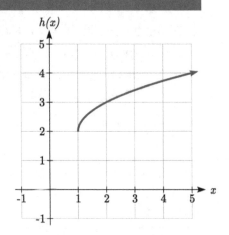

The graph of the toolkit function starts at the origin, so this graph has been shifted 1 to the right, and up 2. In function notation, we could write that as
$h(x) = f(x-1)+2$. Using the formula for the square root function we can write
$$h(x) = \sqrt{x-1}+2$$

Note that this transformation has changed the domain and range of the function. This new graph has domain $[1,\infty)$ and range $[2,\infty)$.

Reflections

Another transformation that can be applied to a function is a reflection over the horizontal or vertical axis.

Example 9

Reflect the graph of $s(t) = \sqrt{t}$ both vertically and horizontally.

Reflecting the graph vertically, each output value will be reflected over the horizontal t axis:

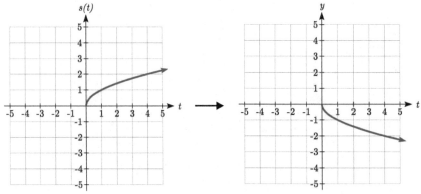

Since each output value is the opposite of the original output value, we can write
$$V(t) = -s(t)$$
$$V(t) = -\sqrt{t}$$

Notice this is an outside change or vertical change that affects the output *s(t)* values so the negative sign belongs outside of the function.

Reflecting horizontally, each input value will be reflected over the vertical axis.

Since each input value is the opposite of the original input value, we can write

$$H(t) = s(-t)$$

$$H(t) = \sqrt{-t}$$

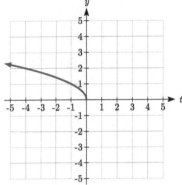

Notice this is an inside change or horizontal change that affects the input values so the negative sign is on the inside of the function.

Note that these transformations can affect the domain and range of the functions. While the original square root function has domain $[0, \infty)$ and range $[0, \infty)$, the vertical reflection gives the $V(t)$ function the range $(-\infty, 0]$, and the horizontal reflection gives the $H(t)$ function the domain $(-\infty, 0]$.

Reflections

Given a function $f(x)$, if we define a new function $g(x)$ as

$$g(x) = -f(x),$$

then $g(x)$ is a **vertical reflection** of the function $f(x)$, sometimes called a reflection about the x-axis

If we define a new function $g(x)$ as

$$g(x) = f(-x),$$

then $g(x)$ is a **horizontal reflection** of the function $f(x)$, sometimes called a reflection about the y-axis

---

Example 10

A function $f(x)$ is given as a table below. Create a table for the function $g(x) = -f(x)$ and $h(x) = f(-x)$

x	2	4	6	8
$f(x)$	1	3	7	11

For $g(x)$, this is a vertical reflection, so the x values stay the same and each output value will be the opposite of the original output value

For $h(x)$, this is a horizontal reflection, and each input value will be the opposite of the original input value and the $h(x)$ values stay the same as the $f(x)$ values:

x	-2	-4	-6	-8
$h(x)$	1	3	7	11

Example 11

A common model for learning has an equation similar to $k(t) = -2^{-t} + 1$, where k is the percentage of mastery that can be achieved after t practice sessions. This is a transformation of the function $f(t) = 2^t$ shown here. Sketch a graph of $k(t)$.

This equation combines three transformations into one equation.

A horizontal reflection: $f(-t) = 2^{-t}$ combined with

A vertical reflection: $-f(-t) = -2^{-t}$ combined with

A vertical shift up 1: $-f(-t) + 1 = -2^{-t} + 1$

We can sketch a graph by applying these transformations one at a time to the original function:

The original graph	Horizontally reflected	Then vertically reflected
		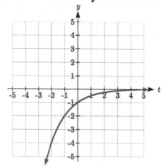

Then, after shifting up 1, we get the final graph.

$k(t) = -f(-t) + 1 = -2^{-t} + 1$.

Note: As a model for learning, this function would be limited to a domain of $t \geq 0$, with corresponding range $[0,1)$.

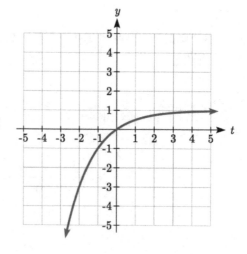

Try it Now

3. Given the toolkit function $f(x) = x^2$, graph $g(x) = -f(x)$ and $h(x) = f(-x)$.
   Do you notice anything surprising?

Some functions exhibit symmetry, in which reflections result in the original graph. For example, reflecting the toolkit functions $f(x) = x^2$ or $f(x) = |x|$ about the y-axis will result in the original graph. We call these types of graphs symmetric about the y-axis.

Likewise, if the graphs of $f(x) = x^3$ or $f(x) = \dfrac{1}{x}$ were reflected over both axes, the result would be the original graph:

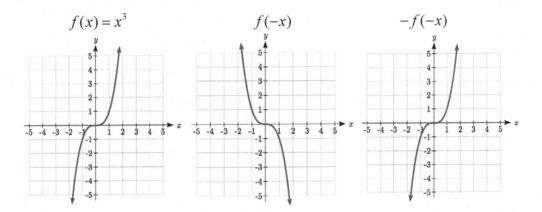

We call these graphs symmetric about the origin.

Even and Odd Functions
A function is called an **even function** if $f(x) = f(-x)$ The graph of an even function is symmetric about the vertical axis A function is called an **odd function** if $f(x) = -f(-x)$ The graph of an odd function is symmetric about the origin

Note: A function can be neither even nor odd if it does not exhibit either symmetry. For example, the $f(x) = 2^x$ function is neither even nor odd.

Example 12

Is the function $f(x) = x^3 + 2x$ even, odd, or neither?

Without looking at a graph, we can determine this by finding formulas for the reflections, and seeing if they return us to the original function:

$$f(-x) = (-x)^3 + 2(-x) = -x^3 - 2x$$
This does not return us to the original function, so this function is not even.

We can now try also applying a horizontal reflection:
$$-f(-x) = -\left(-x^3 - 2x\right) = x^3 + 2x$$

Since $-f(-x) = f(x)$, this is an odd function.

Stretches and Compressions

With shifts, we saw the effect of adding or subtracting to the inputs or outputs of a function. We now explore the effects of multiplying the inputs or outputs.

Remember, we can transform the inside (input values) of a function or we can transform the outside (output values) of a function. Each change has a specific effect that can be seen graphically.

Example 13

A function $P(t)$ models the growth of a population of fruit flies. The growth is shown in the graph. A scientist is comparing this to another population, Q, that grows the same way, but starts twice as large. Sketch a graph of this population.

Since the population is always twice as large, the new population's output values are always twice the original function output values. Graphically, this would look like the second graph shown.

Symbolically, $Q(t) = 2P(t)$

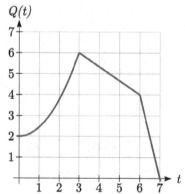

This means that for any input t, the value of the Q function is twice the value of the P function. Notice the effect on the graph is a vertical stretching of the graph, where every point doubles its distance from the horizontal axis. The input values, t, stay the same while the output values are twice as large as before.

Vertical Stretch/Compression

Given a function *f(x)*, if we define a new function *g(x)* as

$g(x) = kf(x)$, where k is a constant

then *g(x)* is a **vertical stretch or compression** of the function *f(x)*.

If $k > 1$, then the graph will be stretched

If $0 < k < 1$, then the graph will be compressed

If $k < 0$, then there will be combination of a vertical stretch or compression with a vertical reflection

Example 14

A function *f(x)* is given as a table below. Create a table for the function $g(x) = \dfrac{1}{2} f(x)$

x	2	4	6	8
f(x)	1	3	7	11

The formula $g(x) = \dfrac{1}{2} f(x)$ tells us that the output values of *g* are half of the output values of *f* with the same inputs. For example, we know that $f(4) = 3$. Then

$$g(4) = \frac{1}{2} f(4) = \frac{1}{2}(3) = \frac{3}{2}$$

x	2	4	6	8
g(x)	1/2	3/2	7/2	11/2

The result is that the function *g(x)* has been compressed vertically by ½. Each output value has been cut in half, so the graph would now be half the original height.

Example 15

The graph shown is a transformation of the toolkit function $f(x) = x^3$. Relate this new function *g(x)* to *f(x)*, then find a formula for *g(x)*.

When trying to determine a vertical stretch or shift, it is helpful to look for a point on the graph that is relatively clear. In this graph, it appears that $g(2) = 2$. With the basic cubic function at the same input, $f(2) = 2^3 = 8$.

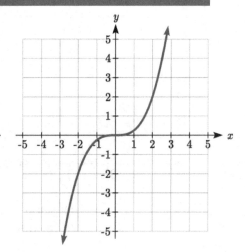

Based on that, it appears that the outputs of g are ¼ the outputs of the function f, since $g(2) = \dfrac{1}{4} f(2)$.

From this we can fairly safely conclude that:

$$g(x) = \dfrac{1}{4} f(x)$$

We can write a formula for g by using the definition of the function f

$$g(x) = \dfrac{1}{4} f(x) = \dfrac{1}{4} x^3$$

Now we consider changes to the inside of a function.

Example 16

Returning to the fruit fly population we looked at earlier, suppose the scientist is now comparing it to a population that progresses through its lifespan twice as fast as the original population. In other words, this new population, R, will progress in 1 hour the same amount the original population did in 2 hours, and in 2 hours, will progress as much as the original population did in 4 hours. Sketch a graph of this population.

Symbolically, we could write
$R(1) = P(2)$
$R(2) = P(4)$, and in general,
$R(t) = P(2t)$

Graphing this,

Original population, $P(t)$

Transformed, $R(t)$

Note the effect on the graph is a horizontal compression, where all input values are half their original distance from the vertical axis.

Horizontal Stretch/Compression

Given a function $f(x)$, if we define a new function $g(x)$ as

$g(x) = f(kx)$, where k is a constant

then $g(x)$ is a **horizontal stretch or compression** of the function $f(x)$.

If $k > 1$, then the graph will be compressed by $\frac{1}{k}$

If $0 < k < 1$, then the graph will be stretched by $\frac{1}{k}$

If $k < 0$, then there will be combination of a horizontal stretch or compression with a horizontal reflection.

Example 17

A function $f(x)$ is given as a table below. Create a table for the function $g(x) = f\left(\frac{1}{2}x\right)$

x	2	4	6	8
$f(x)$	1	3	7	11

The formula $g(x) = f\left(\frac{1}{2}x\right)$ tells us that the output values for g are the same as the output values for the function f at an input half the size. Notice that we don't have enough information to determine $g(2)$ since $g(2) = f\left(\frac{1}{2}\cdot 2\right) = f(1)$, and we do not have a value for $f(1)$ in our table. Our input values to g will need to be twice as large to get inputs for f that we can evaluate. For example, we can determine $g(4)$ since

$g(4) = f\left(\frac{1}{2}\cdot 4\right) = f(2) = 1$.

x	4	8	12	16
$g(x)$	1	3	7	11

Since each input value has been doubled, the result is that the function $g(x)$ has been stretched horizontally by 2.

Example 18

Two graphs are shown below. Relate the function *g(x)* to *f(x)*.

The graph of *g(x)* looks like the graph of *f(x)* horizontally compressed. Since *f(x)* ends at (6,4) and *g(x)* ends at (2,4) we can see that the *x* values have been compressed by 1/3, because $6(1/3) = 2$. We might also notice that $g(2) = f(6)$, and $g(1) = f(3)$. Either way, we can describe this relationship as $g(x) = f(3x)$. This is a horizontal compression by 1/3.

Notice that the coefficient needed for a horizontal stretch or compression is the *reciprocal* of the stretch or compression. To stretch the graph horizontally by 4, we need a coefficient of 1/4 in our function: $f\left(\dfrac{1}{4}x\right)$. This means the input values must be four times larger to produce the same result, requiring the input to be larger, causing the horizontal stretching.

Try it Now

4. Write a formula for the toolkit square root function horizontally stretched by three.

It is useful to note that for most toolkit functions, a horizontal stretch or vertical stretch can be represented in other ways. For example, a horizontal compression of the function $f(x) = x^2$ by ½ would result in a new function $g(x) = (2x)^2$, but this can also be written as $g(x) = 4x^2$, a vertical stretch of *f(x)* by 4. When writing a formula for a transformed toolkit, we only need to find one transformation that would produce the graph.

Combining Transformations

When combining transformations, it is very important to consider the order of the transformations. For example, vertically shifting by 3 and then vertically stretching by 2 does not create the same graph as vertically stretching by 2 then vertically shifting by 3.

When we see an expression like $2f(x) + 3$, which transformation should we start with? The answer here follows nicely from order of operations, for outside transformations. Given the output value of $f(x)$, we first multiply by 2, causing the vertical stretch, then add 3, causing the vertical shift. (Multiplication before Addition)

Combining Vertical Transformations

When combining vertical transformations written in the form $af(x) + k$,

first vertically stretch by a, then vertically shift by k.

Horizontal transformations are a little trickier to think about. When we write $g(x) = f(2x + 3)$ for example, we have to think about how the inputs to the g function relate to the inputs to the f function. Suppose we know $f(7) = 12$. What input to g would produce that output? In other words, what value of x will allow $g(x) = f(2x + 3) = f(12)$? We would need $2x + 3 = 12$. To solve for x, we would first subtract 3, resulting in horizontal shift, then divide by 2, causing a horizontal compression.

Combining Horizontal Transformations

When **combining horizontal transformations** written in the form $f(bx + p)$,

first horizontally shift by p, then horizontally stretch by $1/b$.

This format ends up being very difficult to work with, since it is usually much easier to horizontally stretch a graph before shifting. We can work around this by factoring inside the function.

$$f(bx + p) = f\left(b\left(x + \frac{p}{b}\right)\right)$$

Factoring in this way allows us to horizontally stretch first, then shift horizontally.

Combining Horizontal Transformations (Factored Form)

When **combining horizontal transformations** written in the form $f(b(x + h))$,

first horizontally stretch by $1/b$, then horizontally shift by h.

> **Independence of Horizontal and Vertical Transformations**
>
> **Horizontal and vertical transformations are independent.** It does not matter whether horizontal or vertical transformations are done first.

Example 19

Given the table of values for the function *f(x)* below, create a table of values for the function $g(x) = 2f(3x) + 1$

x	6	12	18	24
f(x)	10	14	15	17

There are 3 steps to this transformation and we will work from the inside out. Starting with the horizontal transformations, $f(3x)$ is a horizontal compression by 1/3, which means we multiply each *x* value by 1/3.

x	2	4	6	8
f(3x)	10	14	15	17

Looking now to the vertical transformations, we start with the vertical stretch, which will multiply the output values by 2. We apply this to the previous transformation.

x	2	4	6	8
2f(3x)	20	28	30	34

Finally, we can apply the vertical shift, which will add 1 to all the output values.

x	2	4	6	8
g(x) = 2f(3x) + 1	21	29	31	35

Example 20

Using the graph of *f(x)* below, sketch a graph of $k(x) = f\left(\frac{1}{2}x + 1\right) - 3$

To make things simpler, we'll start by factoring out the inside of the function

$$f\left(\frac{1}{2}x + 1\right) - 3 = f\left(\frac{1}{2}(x+2)\right) - 3$$

By factoring the inside, we can first horizontally stretch by 2, as indicated by the ½ on the inside of the function. Remember twice the size of 0 is still 0, so the point (0,2) remains at (0,2) while the point (2,0) will stretch to (4,0).

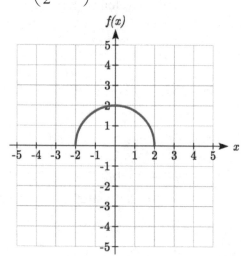

Next, we horizontally shift left by 2 units, as indicated by the $x+2$.

Last, we vertically shift down by 3 to complete our sketch, as indicated by the -3 on the outside of the function.

Horizontal stretch by 2   Horizontal shift left by 2   Vertical shift down 3

      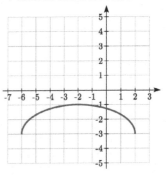

Example 21

Write an equation for the transformed graph of the quadratic function shown.

Since this is a quadratic function, first consider what the basic quadratic tool kit function looks like and how this has changed. Observing the graph, we notice several transformations:

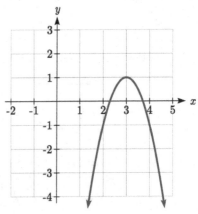

The original tool kit function has been flipped over the x axis, some kind of stretch or compression has occurred, and we can see a shift to the right 3 units and a shift up 1 unit.

In total there are four operations:
Vertical reflection, requiring a negative sign outside the function
Vertical Stretch *or* Horizontal Compression*
Horizontal Shift Right 3 units, which tells us to put $x-3$ on the inside of the function
Vertical Shift up 1 unit, telling us to add 1 on the outside of the function

*It is unclear from the graph whether it is showing a vertical stretch or a horizontal compression. For the quadratic, it turns out we could represent it either way, so we'll use a vertical stretch. You may be able to determine the vertical stretch by observation.

By observation, the basic tool kit function has a vertex at (0, 0) and symmetrical points at (1, 1) and (-1, 1). These points are one unit up and one unit over from the vertex. The new points on the transformed graph are one unit away horizontally but 2 units away vertically. They have been stretched vertically by two.

Not everyone can see this by simply looking at the graph. If you can, great, but if not, we can solve for it. First, we will write the equation for this graph, with an unknown vertical stretch.

$f(x) = x^2$ The original function

$-f(x) = -x^2$ Vertically reflected

$-af(x) = -ax^2$ Vertically stretched

$-af(x-3) = -a(x-3)^2$ Shifted right 3

$-af(x-3)+1 = -a(x-3)^2 +1$ Shifted up 1

We now know our graph is going to have an equation of the form $g(x) = -a(x-3)^2 + 1$. To find the vertical stretch, we can identify any point on the graph (other than the highest point), such as the point (2,-1), which tells us $g(2) = -1$. Using our general formula, and substituting 2 for x, and -1 for $g(x)$

$-1 = -a(2-3)^2 + 1$

$-1 = -a + 1$

$-2 = -a$

$2 = a$

This tells us that to produce the graph we need a vertical stretch by two.
The function that produces this graph is therefore $g(x) = -2(x-3)^2 + 1$.

Try it Now

5. Consider the linear function $g(x) = -2x + 1$. Describe its transformation in words using the identity tool kit function $f(x) = x$ as a reference.

Example 22

On what interval(s) is the function $g(x) = \dfrac{-2}{(x-1)^2} + 3$ increasing and decreasing?

This is a transformation of the toolkit reciprocal squared function, $f(x) = \dfrac{1}{x^2}$:

$-2f(x) = \dfrac{-2}{x^2}$ A vertical flip and vertical stretch by 2

$-2f(x-1) = \dfrac{-2}{(x-1)^2}$ A shift right by 1

$-2f(x-1)+3 = \dfrac{-2}{(x-1)^2} + 3$ A shift up by 3

The basic reciprocal squared function is increasing on $(-\infty,0)$ and decreasing on $(0,\infty)$. Because of the vertical flip, the *g(x)* function will be decreasing on the left and increasing on the right. The horizontal shift right by 1 will also shift these intervals to the right one. From this, we can determine *g(x)* will be increasing on $(1,\infty)$ and decreasing on $(-\infty,1)$. We also could graph the transformation to help us determine these intervals.

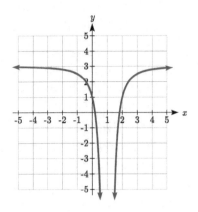

Try it Now

6. On what interval(s) is the function $h(t) = (t-3)^3 + 2$ concave up and down?

Important Topics of This Section
Transformations
Vertical Shift (up & down)
Horizontal Shifts (left & right)
Reflections over the vertical & horizontal axis
Even & Odd functions
Vertical Stretches & Compressions
Horizontal Stretches & Compressions
Combinations of Transformation

Try it Now Answers

1. $b(t) = h(t) + 10 = -4.9t^2 + 30t + 10$

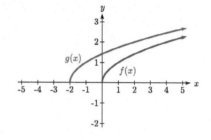

2. a. Horizontal shift
   b. The function is shifted to the LEFT by 2 units.
   c. Shown to the right

3. Shown to the right
   Notice: $g(x) = f(-x)$ looks the same as $f(x)$

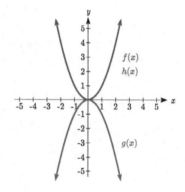

4. $g(x) = f\left(\dfrac{1}{3}x\right)$ so using the square root function we get

$$g(x) = \sqrt{\dfrac{1}{3}x}$$

5. The identity tool kit function $f(x) = x$ has been transformed in 3 steps
   a. Vertically stretched by 2.
   b. Vertically reflected over the x axis.
   c. Vertically shifted up by 1 unit.

6. $h(t)$ is concave down on $(-\infty, 3)$ and concave up on $(3, \infty)$

Section 1.5 Exercises

Describe how each function is a transformation of the original function $f(x)$

1. $f(x-49)$
2. $f(x+43)$
3. $f(x+3)$
4. $f(x-4)$
5. $f(x)+5$
6. $f(x)+8$
7. $f(x)-2$
8. $f(x)-7$
9. $f(x-2)+3$
10. $f(x+4)-1$

11. Write a formula for $f(x)=\sqrt{x}$ shifted up 1 unit and left 2 units.

12. Write a formula for $f(x)=|x|$ shifted down 3 units and right 1 unit.

13. Write a formula for $f(x)=\dfrac{1}{x}$ shifted down 4 units and right 3 units.

14. Write a formula for $f(x)=\dfrac{1}{x^2}$ shifted up 2 units and left 4 units.

15. Tables of values for $f(x)$, $g(x)$, and $h(x)$ are given below. Write $g(x)$ and $h(x)$ as transformations of $f(x)$.

x	$f(x)$
-2	-2
-1	-1
0	-3
1	1
2	2

x	$g(x)$
-1	-2
0	-1
1	-3
2	1
3	2

x	$h(x)$
-2	-1
-1	0
0	-2
1	2
2	3

16. Tables of values for $f(x)$, $g(x)$, and $h(x)$ are given below. Write $g(x)$ and $h(x)$ as transformations of $f(x)$.

x	$f(x)$
-2	-1
-1	-3
0	4
1	2
2	1

x	$g(x)$
-3	-1
-2	-3
-1	4
0	2
1	1

x	$h(x)$
-2	-2
-1	-4
0	3
1	1
2	0

The graph of $f(x) = 2^x$ is shown. Sketch a graph of each transformation of $f(x)$.

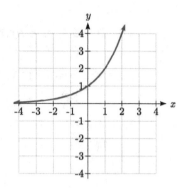

17. $g(x) = 2^x + 1$

18. $h(x) = 2^x - 3$

19. $w(x) = 2^{x-1}$

20. $q(x) = 2^{x+3}$

Sketch a graph of each function as a transformation of a toolkit function.

21. $f(t) = (t+1)^2 - 3$

22. $h(x) = |x-1| + 4$

23. $k(x) = (x-2)^3 - 1$

24. $m(t) = 3 + \sqrt{t+2}$

Write an equation for each function graphed below.

25.

26.

27.

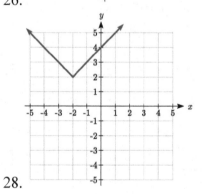

28.

Find a formula for each of the transformations of the square root whose graphs are given below.

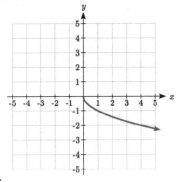

29.

30.

The graph of $f(x) = 2^x$ is shown. Sketch a graph of each transformation of $f(x)$

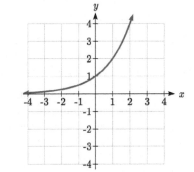

31. $g(x) = -2^x + 1$

32. $h(x) = 2^{-x}$

33. Starting with the graph of $f(x) = 6^x$ write the equation of

the graph that results from
a. reflecting $f(x)$ about the x-axis and the y-axis
b. reflecting $f(x)$ about the x-axis, shifting left 2 units, and down 3 units

34. Starting with the graph of $f(x) = 4^x$ write the equation of the graph that results from

a. reflecting $f(x)$ about the x-axis
b. reflecting $f(x)$ about the y-axis, shifting right 4 units, and up 2 units

Write an equation for each function graphed below.

35.

36.

37.

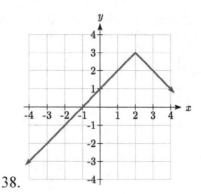

38.

39. For each equation below, determine if the function is Odd, Even, or Neither.

    a. $f(x) = 3x^4$

    b. $g(x) = \sqrt{x}$

    c. $h(x) = \dfrac{1}{x} + 3x$

40. For each equation below, determine if the function is Odd, Even, or Neither.

    a. $f(x) = (x-2)^2$

    b. $g(x) = 2x^4$

    c. $h(x) = 2x - x^3$

Describe how each function is a transformation of the original function $f(x)$.

41. $-f(x)$ 42. $f(-x)$

43. $4f(x)$ 44. $6f(x)$

45. $f(5x)$ 46. $f(2x)$

47. $f\left(\dfrac{1}{3}x\right)$ 48. $f\left(\dfrac{1}{5}x\right)$

49. $3f(-x)$ 50. $-f(3x)$

Write a formula for the function that results when the given toolkit function is transformed as described.

51. $f(x) = |x|$ reflected over the y axis and horizontally compressed by a factor of $\dfrac{1}{4}$.

52. $f(x) = \sqrt{x}$ reflected over the x axis and horizontally stretched by a factor of 2.

53. $f(x) = \dfrac{1}{x^2}$ vertically compressed by a factor of $\dfrac{1}{3}$, then shifted to the left 2 units and down 3 units.

54. $f(x) = \dfrac{1}{x}$ vertically stretched by a factor of 8, then shifted to the right 4 units and up 2 units.

55. $f(x) = x^2$ horizontally compressed by a factor of $\dfrac{1}{2}$, then shifted to the right 5 units and up 1 unit.

56. $f(x) = x^2$ horizontally stretched by a factor of 3, then shifted to the left 4 units and down 3 units.

Describe how each formula is a transformation of a toolkit function. Then sketch a graph of the transformation.

57. $f(x) = 4(x+1)^2 - 5$ 58. $g(x) = 5(x+3)^2 - 2$

59. $h(x) = -2|x-4| + 3$ 60. $k(x) = -3\sqrt{x} - 1$

61. $m(x) = \dfrac{1}{2}x^3$ 62. $n(x) = \dfrac{1}{3}|x-2|$

63. $p(x) = \left(\dfrac{1}{3}x\right)^2 - 3$ 64. $q(x) = \left(\dfrac{1}{4}x\right)^3 + 1$

65. $a(x) = \sqrt{-x+4}$ 66. $b(x) = \sqrt[3]{-x-6}$

Determine the interval(s) on which the function is increasing and decreasing.

67. $f(x) = 4(x+1)^2 - 5$ 68. $g(x) = 5(x+3)^2 - 2$

69. $a(x) = \sqrt{-x+4}$ 70. $k(x) = -3\sqrt{x} - 1$

Determine the interval(s) on which the function is concave up and concave down.

71. $m(x) = -2(x+3)^3 + 1$ 72. $b(x) = \sqrt[3]{-x-6}$

73. $p(x) = \left(\dfrac{1}{3}x\right)^2 - 3$ 74. $k(x) = -3\sqrt{x} - 1$

The function $f(x)$ is graphed here. Write an equation for each graph below as a transformation of $f(x)$.

75.

76.

77.

78.

79.

80.

81.

82.

83.

84.

85.

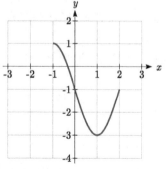

86.

Write an equation for each transformed toolkit function graphed below.

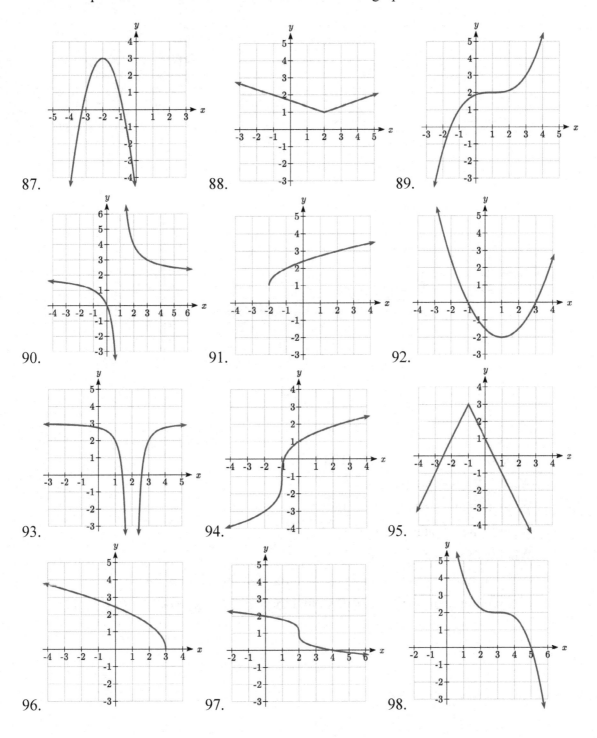

87.

88.

89.

90.

91.

92.

93.

94.

95.

96.

97.

98.

Write a formula for the piecewise function graphed below.

99.

100.

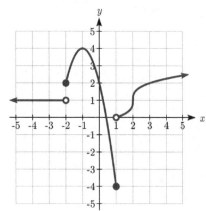

101.

102.

103. Suppose you have a function $y = f(x)$ such that the domain of $f(x)$ is $1 \le x \le 6$ and the range of $f(x)$ is $-3 \le y \le 5$. [UW]
  a. What is the domain of $f(2(x-3))$?
  b. What is the range of $f(2(x-3))$?
  c. What is the domain of $2f(x) - 3$?
  d. What is the range of $2f(x) - 3$?
  e. Can you find constants B and C so that the domain of $f(B(x-C))$ is $8 \le x \le 9$?
  f. Can you find constants A and D so that the range of $Af(x) + D$ is $0 \le y \le 1$?

Section 1.6 Inverse Functions

A fashion designer is travelling to Milan for a fashion show. He asks his assistant, Betty, what 75 degrees Fahrenheit is in Celsius, and after a quick search on Google, she finds the formula $C = \frac{5}{9}(F - 32)$. Using this formula, she calculates $\frac{5}{9}(75 - 32) \approx 24$ degrees Celsius. The next day, the designer sends his assistant the week's weather forecast for Milan, and asks her to convert the temperatures to Fahrenheit.

Mon	Tue	Wed	Thu
26°C | 19°C	29°C | 19°C	30°C | 20°C	26°C | 18°C

At first, Betty might consider using the formula she has already found to do the conversions. After all, she knows her algebra, and can easily solve the equation for F after substituting a value for C. For example, to convert 26 degrees Celsius, she could write:

$$26 = \frac{5}{9}(F - 32)$$

$$26 \cdot \frac{9}{5} = F - 32$$

$$F = 26 \cdot \frac{9}{5} + 32 \approx 79$$

After considering this option for a moment, she realizes that solving the equation for each of the temperatures would get awfully tedious, and realizes that since evaluation is easier than solving, it would be much more convenient to have a different formula, one which takes the Celsius temperature and outputs the Fahrenheit temperature. This is the idea of an inverse function, where the input becomes the output and the output becomes the input.

Inverse Function

If $f(a) = b$, then a function *g(x)* is an **inverse** of *f* if $g(b) = a$.

The inverse of *f(x)* is typically notated $f^{-1}(x)$, which is read "*f* inverse of *x*", so equivalently, if $f(a) = b$ then $f^{-1}(b) = a$.

Important: The raised -1 used in the notation for inverse functions is simply a notation, and does not designate an exponent or power of -1.

Example 1

If for a particular function, $f(2) = 4$, what do we know about the inverse?

The inverse function reverses which quantity is input and which quantity is output, so if $f(2) = 4$, then $f^{-1}(4) = 2$.

Alternatively, if you want to re-name the inverse function $g(x)$, then $g(4) = 2$

Try it Now

1. Given that $h^{-1}(6) = 2$, what do we know about the original function $h(x)$?

Notice that original function and the inverse function *undo* each other. If $f(a) = b$, then $f^{-1}(b) = a$, returning us to the original input. More simply put, if you compose these functions together you get the original input as your answer.

$$f^{-1}\left(f(a)\right) = a \quad \text{and} \quad f\left(f^{-1}(b)\right) = b$$

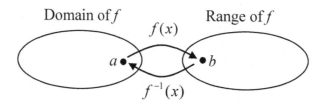

Since the outputs of the function f are the inputs to f^{-1}, the range of f is also the domain of f^{-1}. Likewise, since the inputs to f are the outputs of f^{-1}, the domain of f is the range of f^{-1}.

Basically, like how the input and output values switch, the domain & ranges switch as well. But be careful, because sometimes a function doesn't even have an inverse function, or only has an inverse on a limited domain. For example, the inverse of $f(x) = \sqrt{x}$ is $f^{-1}(x) = x^2$, since a square "undoes" a square root, but it is only the inverse of $f(x)$ on the domain $[0,\infty)$, since that is the range of $f(x) = \sqrt{x}$.

Example 2

The function $f(x) = 2^x$ has domain $(-\infty, \infty)$ and range $(0, \infty)$, what would we expect the domain and range of f^{-1} to be?

We would expect f^{-1} to swap the domain and range of f, so f^{-1} would have domain $(0,\infty)$ and range $(-\infty, \infty)$.

Example 3

A function $f(t)$ is given as a table below, showing distance in miles that a car has traveled in t minutes. Find and interpret $f^{-1}(70)$

t (minutes)	30	50	70	90
$f(t)$ (miles)	20	40	60	70

The inverse function takes an output of f and returns an input for f. So in the expression $f^{-1}(70)$, the 70 is an output value of the original function, representing 70 miles. The inverse will return the corresponding input of the original function f, 90 minutes, so $f^{-1}(70) = 90$. Interpreting this, it means that to drive 70 miles, it took 90 minutes.

Alternatively, recall the definition of the inverse was that if $f(a) = b$ then $f^{-1}(b) = a$. By this definition, if you are given $f^{-1}(70) = a$ then you are looking for a value a so that $f(a) = 70$. In this case, we are looking for a t so that $f(t) = 70$, which is when $t = 90$.

Try it Now

2. Using the table below

t (minutes)	30	50	60	70	90
$f(t)$ (miles)	20	40	50	60	70

Find and interpret the following
   a. $f(60)$
   b. $f^{-1}(60)$

Example 4

A function $g(x)$ is given as a graph below. Find $g(3)$ and $g^{-1}(3)$

To evaluate $g(3)$, we find 3 on the horizontal axis and find the corresponding output value on the vertical axis. The point $(3, 1)$ tells us that $g(3) = 1$

To evaluate $g^{-1}(3)$, recall that by definition $g^{-1}(3)$ means $g(x) = 3$. By looking for the output value 3 on the vertical axis we find the point $(5, 3)$ on the graph, which means $g(5) = 3$, so by definition $g^{-1}(3) = 5$.

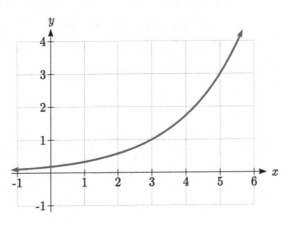

3. Using the graph in Example 4 above

    a. find $g^{-1}(1)$

    b. estimate $g^{-1}(4)$

Example 5

Returning to our designer's assistant, find a formula for the inverse function that gives Fahrenheit temperature given a Celsius temperature.

A quick Google search would find the inverse function, but alternatively, Betty might look back at how she solved for the Fahrenheit temperature for a specific Celsius value, and repeat the process in general

$$C = \frac{5}{9}(F - 32)$$

$$C \cdot \frac{9}{5} = F - 32$$

$$F = \frac{9}{5}C + 32$$

By solving in general, we have uncovered the inverse function. If

$$C = h(F) = \frac{5}{9}(F - 32)$$

Then

$$F = h^{-1}(C) = \frac{9}{5}C + 32$$

In this case, we introduced a function h to represent the conversion since the input and output variables are descriptive, and writing C^{-1} could get confusing.

It is important to note that not all functions will have an inverse function. Since the inverse $f^{-1}(x)$ takes an output of f and returns an input of f, in order for f^{-1} to itself be a function, then each output of f (input to f^{-1}) must correspond to exactly one input of f (output of f^{-1}) in order for f^{-1} to be a function. You might recall that this is the definition of a one-to-one function.

Properties of Inverses

In order **for a function to have an inverse**, it must be a one-to-one function.

In some cases, it is desirable to have an inverse for a function even though the function is not one-to-one. In those cases, we can often limit the domain of the original function to an interval on which the function *is* one-to-one, then find an inverse only on that interval.

If you have not already done so, go back to the toolkit functions that were not one-to-one and limit or restrict the domain of the original function so that it is one-to-one. If you are not sure how to do this, proceed to Example 6.

Example 6

The quadratic function $h(x) = x^2$ is not one-to-one. Find a domain on which this function is one-to-one, and find the inverse on that domain.

We can limit the domain to $[0, \infty)$ to restrict the graph to a portion that is one-to-one, and find an inverse on this limited domain.

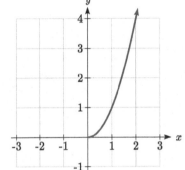

You may have already guessed that since we undo a square with a square root, the inverse of $h(x) = x^2$ on this domain is $h^{-1}(x) = \sqrt{x}$.

You can also solve for the inverse function algebraically. If $h(x) = x^2$, we can introduce the variable y to represent the output values, allowing us to write $y = x^2$. To find the inverse we solve for the input variable

To solve for x we take the square root of each side. $\sqrt{y} = \sqrt{x^2}$ and get $\sqrt{y} = |x|$, so $x = \pm\sqrt{y}$. We have restricted x to being non-negative, so we'll use the positive square root, $x = \sqrt{y}$ or $h^{-1}(y) = \sqrt{y}$. In cases like this where the variables are not descriptive, it is common to see the inverse function rewritten with the variable x: $h^{-1}(x) = \sqrt{x}$. Rewriting the inverse using the variable x is often required for graphing inverse functions using calculators or computers.

Note that the domain and range of the square root function do correspond with the range and domain of the quadratic function on the limited domain. In fact, if we graph $h(x)$ on the restricted domain and $h^{-1}(x)$ on the same axes, we can notice symmetry: the graph of $h^{-1}(x)$ is the graph of $h(x)$ reflected over the line $y = x$.

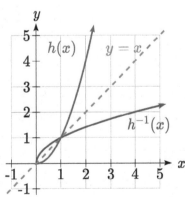

Example 7

Given the graph of *f(x)* shown, sketch a graph of $f^{-1}(x)$.

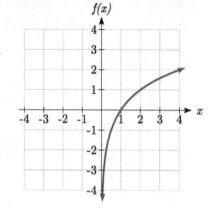

This is a one-to-one function, so we will be able to sketch an inverse. Note that the graph shown has an apparent domain of $(0,\infty)$ and range of $(-\infty,\infty)$, so the inverse will have a domain of $(-\infty,\infty)$ and range of $(0,\infty)$.

Reflecting this graph of the line $y = x$, the point (1, 0) reflects to (0, 1), and the point (4, 2) reflects to (2, 4). Sketching the inverse on the same axes as the original graph:

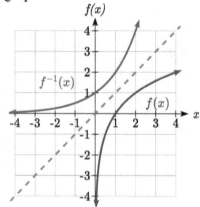

Important Topics of this Section
Definition of an inverse function
Composition of inverse functions yield the original input value
Not every function has an inverse function
To have an inverse a function must be one-to-one
Restricting the domain of functions that are not one-to-one.

Try it Now Answers

1. $g(2) = 6$

2.a. $f(60) = 50$. In 60 minutes, 50 miles are traveled.

  b. $f^{-1}(60) = 70$. To travel 60 miles, it will take 70 minutes.

3. a. $g^{-1}(1) = 3$

  b. $g^{-1}(4) = 5.5$ (this is an approximation – answers may vary slightly)

Section 1.6 Exercises

Assume that the function *f* is a one-to-one function.

1. If $f(6) = 7$, find $f^{-1}(7)$

2. If $f(3) = 2$, find $f^{-1}(2)$

3. If $f^{-1}(-4) = -8$, find $f(-8)$

4. If $f^{-1}(-2) = -1$, find $f(-1)$

5. If $f(5) = 2$, find $\left(f(5) \right)^{-1}$

6. If $f(1) = 4$, find $\left(f(1) \right)^{-1}$

7. Using the graph of $f(x)$ shown

    a. Find $f(0)$

    b. Solve $f(x) = 0$

    c. Find $f^{-1}(0)$

    d. Solve $f^{-1}(x) = 0$

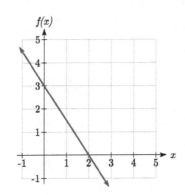

8. Using the graph shown

    a. Find $g(1)$

    b. Solve $g(x) = 1$

    c. Find $g^{-1}(1)$

    d. Solve $g^{-1}(x) = 1$

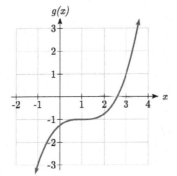

9. Use the table below to find the indicated quantities.

x	0	1	2	3	4	5	6	7	8	9
f(x)	8	0	7	4	2	6	5	3	9	1

    a. Find $f(1)$

    b. Solve $f(x) = 3$

    c. Find $f^{-1}(0)$

    d. Solve $f^{-1}(x) = 7$

10. Use the table below to fill in the missing values.

t	0	1	2	3	4	5	6	7	8
h(t)	6	0	1	7	2	3	5	4	9

    a. Find $h(6)$

    b. Solve $h(t) = 0$

    c. Find $h^{-1}(5)$

    d. Solve $h^{-1}(t) = 1$

For each table below, create a table for $f^{-1}(x)$.

11.

x	3	6	9	13	14
f(x)	1	4	7	12	16

12.

x	3	5	7	13	15
f(x)	2	6	9	11	16

For each function below, find $f^{-1}(x)$

13. $f(x) = x + 3$

14. $f(x) = x + 5$

15. $f(x) = 2 - x$

16. $f(x) = 3 - x$

17. $f(x) = 11x + 7$

18. $f(x) = 9 + 10x$

For each function, find a domain on which f is one-to-one and non-decreasing, then find the inverse of f restricted to that domain.

19. $f(x) = (x + 7)^2$

20. $f(x) = (x - 6)^2$

21. $f(x) = x^2 - 5$

22. $f(x) = x^2 + 1$

23. If $f(x) = x^3 - 5$ and $g(x) = \sqrt[3]{x + 5}$, find

    a. $f(g(x))$

    b. $g(f(x))$

    c. What does this tell us about the relationship between $f(x)$ and $g(x)$?

24. If $f(x) = \dfrac{x}{2 + x}$ and $g(x) = \dfrac{2x}{1 - x}$, find

    a. $f(g(x))$

    b. $g(f(x))$

    c. What does this tell us about the relationship between $f(x)$ and $g(x)$?

Chapter 2:
Linear Functions

Chapter one was a window that gave us a peek into the entire course. Our goal was to understand the basic structure of functions and function notation, the toolkit functions, domain and range, how to recognize and understand composition and transformations of functions and how to understand and utilize inverse functions. With these basic components in hand we will further research the specific details and intricacies of each type of function in our toolkit and use them to model the world around us.

Mathematical Modeling
As we approach day to day life we often need to quantify the things around us, giving structure and numeric value to various situations. This ability to add structure enables us to make choices based on patterns we see that are weighted and systematic. With this structure in place we can model and even predict behavior to make decisions. Adding a numerical structure to a real world situation is called **Mathematical Modeling.**

When modeling real world scenarios, there are some common growth patterns that are regularly observed. We will devote this chapter and the rest of the book to the study of the functions used to model these growth patterns.

Section 2.1 Linear Functions

As you hop into a taxicab in Las Vegas, the meter will immediately read $3.50; this is the "drop" charge made when the taximeter is activated. After that initial fee, the taximeter will add $2.76 for each mile the taxi drives[1]. In this scenario, the total taxi fare depends upon the number of miles ridden in the taxi, and we can ask whether it is possible to model this type of scenario with a function. Using descriptive variables, we choose m for miles and C for Cost in dollars as a function of miles: $C(m)$.

---

[1] Nevada Taxicab Authority, retrieved May 8, 2017. There is also a waiting fee assessed when the taxi is waiting at red lights, but we'll ignore that in this discussion.

We know for certain that $C(0) = 3.50$, since the $3.50 drop charge is assessed regardless of how many miles are driven. Since $2.67 is added for each mile driven, then $C(1) = 3.50 + 2.67 = 6.17$.

If we then drove a second mile, another $2.67 would be added to the cost:
$C(2) = 3.50 + 2.67 + 2.67 = 3.50 + 2.67(2) = 8.84$

If we drove a third mile, another $2.67 would be added to the cost:
$C(3) = 3.50 + 2.67 + 2.67 + 2.67 = 3.50 + 2.67(3) = 11.51$

From this we might observe the pattern, and conclude that if m miles are driven, $C(m) = 3.50 + 2.67m$ because we start with a $3.50 drop fee and then for each mile increase we add $2.67.

It is good to verify that the units make sense in this equation. The $3.50 drop charge is measured in dollars; the $2.67 charge is measured in dollars per mile.

$$C(m) = 3.50\,dollars + \left(2.67\frac{dollars}{mile}\right)(m\ miles)$$

When dollars per mile are multiplied by a number of miles, the result is a number of dollars, matching the units on the 3.50, and matching the desired units for the C function.

Notice this equation $C(m) = 3.50 + 2.67m$ consisted of two quantities. The first is the fixed $3.50 charge which does not change based on the value of the input. The second is the $2.67 dollars per mile value, which is a **rate of change**. In the equation, this rate of change is multiplied by the input value.

Looking at this same problem in table format we can also see the cost changes by $2.67 for every 1 mile increase.

m	0	1	2	3
$C(m)$	3.50	6.17	8.84	11.51

It is important here to note that in this equation, the **rate of change is constant**; over any interval, the rate of change is the same.

Graphing this equation, $C(m) = 3.50 + 2.67m$ we see the shape is a line, which is how these functions get their name: **linear functions**.

When the number of miles is zero the cost is $3.50, giving the point (0, 3.50) on the graph. This is the vertical or $C(m)$ intercept. The graph is increasing in a straight line from left to right because for each mile the cost goes up by $2.67; this rate remains consistent.

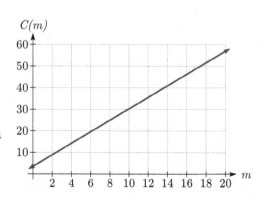

In this example, you have seen the taxicab cost modeled in words, an equation, a table and in graphical form. Whenever possible, ensure that you can link these four representations together to continually build your skills. It is important to note that you will not always be able to find all 4 representations for a problem and so being able to work with all 4 forms is very important.

Linear Function

A **linear function** is a function whose graph produces a line. Linear functions can always be written in the form

$$f(x) = b + mx \qquad \text{or} \qquad f(x) = mx + b; \text{ they're equivalent}$$

where

b is the initial or starting value of the function (when input, $x = 0$), and

m is the constant rate of change of the function

Many people like to write linear functions in the form $f(x) = b + mx$ because it corresponds to the way we tend to speak: "The output starts at b and increases at a rate of m."

For this reason alone we will use the $f(x) = b + mx$ form for many of the examples, but remember they are equivalent and can be written correctly both ways.

Slope and Increasing/Decreasing

m is the constant rate of change of the function (also called **slope**). The slope determines if the function is an increasing function or a decreasing function.

$f(x) = b + mx$ is an **increasing** function if $m > 0$

$f(x) = b + mx$ is a **decreasing** function if $m < 0$

If $m = 0$, the rate of change zero, and the function $f(x) = b + 0x = b$ is just a horizontal line passing through the point $(0, b)$, neither increasing nor decreasing.

Example 1

Marcus currently owns 200 songs in his iTunes collection. Every month, he adds 15 new songs. Write a formula for the number of songs, N, in his iTunes collection as a function of the number of months, m. How many songs will he own in a year?

The initial value for this function is 200, since he currently owns 200 songs, so $N(0) = 200$. The number of songs increases by 15 songs per month, so the rate of change is 15 songs per month. With this information, we can write the formula:
$N(m) = 200 + 15m$.

N(m) is an increasing linear function. With this formula we can predict how many songs he will have in 1 year (12 months):

$N(12) = 200 + 15(12) = 200 + 180 = 380$. Marcus will have 380 songs in 12 months.

Try it Now

1. If you earn $30,000 per year and you spend $29,000 per year write an equation for the amount of money you save after *y* years, if you start with nothing.
   "The most important thing, spend less than you earn![2]"

Calculating Rate of Change

Given two values for the input, x_1 and x_2, and two corresponding values for the output, y_1 and y_2, or a set of points, $(x_1, \ y_1)$ and $(x_2, \ y_2)$, if we wish to find a linear function that contains both points we can calculate the rate of change, *m*:

$$m = \frac{\text{change in output}}{\text{change in input}} = \frac{\Delta y}{\Delta x} = \frac{y_2 - y_1}{x_2 - x_1}$$

Rate of change of a linear function is also called the **slope** of the line.

Note in function notation, $y_1 = f(x_1)$ and $y_2 = f(x_2)$, so we could equivalently write

$$m = \frac{f(x_2) - f(x_1)}{x_2 - x_1}$$

Example 2

The population of a city increased from 23,400 to 27,800 between 2002 and 2006. Find the rate of change of the population during this time span.

The rate of change will relate the change in population to the change in time. The population increased by $27800 - 23400 = 4400$ people over the 4 year time interval. To find the rate of change, the number of people per year the population changed by:

$$\frac{4400 \, people}{4 \, years} = 1100 \frac{people}{year} = 1100 \text{ people per year}$$

Notice that we knew the population was increasing, so we would expect our value for *m* to be positive. This is a quick way to check to see if your value is reasonable.

---

[2] http://www.thesimpledollar.com/2009/06/19/rule-1-spend-less-than-you-earn/

Example 3

The pressure, P, in pounds per square inch (PSI) on a diver depends upon their depth below the water surface, d, in feet, following the equation $P(d) = 14.696 + 0.434d$. Interpret the components of this function.

The rate of change, or slope, 0.434 would have units $\dfrac{\text{output}}{\text{input}} = \dfrac{\text{pressure}}{\text{depth}} = \dfrac{\text{PSI}}{\text{ft}}$. This tells us the pressure on the diver increases by 0.434 PSI for each foot their depth increases.

The initial value, 14.696, will have the same units as the output, so this tells us that at a depth of 0 feet, the pressure on the diver will be 14.696 PSI.

Example 4

If $f(x)$ is a linear function, $f(3) = -2$, and $f(8) = 1$, find the rate of change.

$f(3) = -2$ tells us that the input 3 corresponds with the output -2, and $f(8) = 1$ tells us that the input 8 corresponds with the output 1. To find the rate of change, we divide the change in output by the change in input:

$m = \dfrac{\text{change in output}}{\text{change in input}} = \dfrac{1 - (-2)}{8 - 3} = \dfrac{3}{5}$. If desired we could also write this as $m = 0.6$

Note that it is not important which pair of values comes first in the subtractions so long as the first output value used corresponds with the first input value used.

Try it Now

2. Given the two points (2, 3) and (0, 4), find the rate of change. Is this function increasing or decreasing?

We can now find the rate of change given two input-output pairs, and can write an equation for a linear function once we have the rate of change and initial value. If we have two input-output pairs and they do not include the initial value of the function, then we will have to solve for it.

Example 5

Write an equation for the linear function graphed to the right.

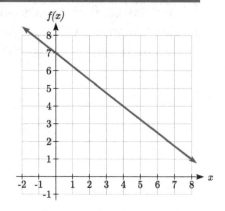

Looking at the graph, we might notice that it passes through the points (0, 7) and (4, 4). From the first value, we know the initial value of the function is $b = 7$, so in this case we will only need to calculate the rate of change:

$$m = \frac{4-7}{4-0} = \frac{-3}{4}$$

This allows us to write the equation:

$$f(x) = 7 - \frac{3}{4}x$$

Example 6

If $f(x)$ is a linear function, $f(3) = -2$, and $f(8) = 1$, find an equation for the function.

In example 3, we computed the rate of change to be $m = \frac{3}{5}$. In this case, we do not know the initial value $f(0)$, so we will have to solve for it. Using the rate of change, we know the equation will have the form $f(x) = b + \frac{3}{5}x$. Since we know the value of the function when $x = 3$, we can evaluate the function at 3.

$f(3) = b + \frac{3}{5}(3)$ Since we know that $f(3) = -2$, we can substitute on the left side

$-2 = b + \frac{3}{5}(3)$ This leaves us with an equation we can solve for the initial value

$b = -2 - \frac{9}{5} = \frac{-19}{5}$

Combining this with the value for the rate of change, we can now write a formula for this function:

$$f(x) = \frac{-19}{5} + \frac{3}{5}x$$

Example 7

Working as an insurance salesperson, Ilya earns a base salary and a commission on each new policy, so Ilya's weekly income, I, depends on the number of new policies, n, he sells during the week. Last week he sold 3 new policies, and earned $760 for the week. The week before, he sold 5 new policies, and earned $920. Find an equation for $I(n)$, and interpret the meaning of the components of the equation.

The given information gives us two input-output pairs: $(3,760)$ and $(5,920)$. We start by finding the rate of change.
$$m = \frac{920 - 760}{5 - 3} = \frac{160}{2} = 80$$

Keeping track of units can help us interpret this quantity. Income increased by $160 when the number of policies increased by 2, so the rate of change is $80 per policy; Ilya earns a commission of $80 for each policy sold during the week.

We can then solve for the initial value
$I(n) = b + 80n$ then when $n = 3$, $I(3) = 760$, giving
$760 = b + 80(3)$ this allows us to solve for b
$b = 760 - 80(3) = 520$

This value is the starting value for the function. This is Ilya's income when $n = 0$, which means no new policies are sold. We can interpret this as Ilya's base salary for the week, which does not depend upon the number of policies sold.

Writing the final equation:
$I(n) = 520 + 80n$
Our final interpretation is: Ilya's base salary is $520 per week and he earns an additional $80 commission for each policy sold each week.

Flashback

Looking at Example 7:
    Determine the independent and dependent variables.
    What is a reasonable domain and range?
    Is this function one-to-one?

Try it Now

3. The balance in your college payment account, C, is a function of the number of quarters, q, you attend. Interpret the function $C(a) = 20000 - 4000q$ in words. How many quarters of college can you pay for until this account is empty?

Example 8

Given the table below write a linear equation that represents the table values

w, number of weeks	0	2	4	6
$P(w)$, number of rats	1000	1080	1160	1240

We can see from the table that the initial value of rats is 1000 so in the linear format $P(w) = b + mw$, $b = 1000$.

Rather than solving for m, we can notice from the table that the population goes up by 80 for every 2 weeks that pass. This rate is consistent from week 0, to week 2, 4, and 6. The rate of change is 80 rats per 2 weeks. This can be simplified to 40 rats per week and we can write
$P(w) = b + mw$ as $P(w) = 1000 + 40w$

If you didn't notice this from the table you could still solve for the slope using any two points from the table. For example, using (2, 1080) and (6, 1240),
$$m = \frac{1240 - 1080}{6 - 2} = \frac{160}{4} = 40 \text{ rats per week}$$

Important Topics of this Section
Definition of Modeling
Definition of a linear function
Structure of a linear function
Increasing & Decreasing functions
Finding the vertical intercept (0, b)
Finding the slope/rate of change, m
Interpreting linear functions

Try it Now Answers

1. $S(y) = 30,000y - 29,000y = 1000y$ $1000 is saved each year.

2. $m = \frac{4-3}{0-2} = \frac{1}{-2} = -\frac{1}{2}$; Decreasing because $m < 0$

3. Your College account starts with $20,000 in it and you withdraw $4,000 each quarter (or your account contains $20,000 and decreases by $4000 each quarter.)
   Solving $C(a) = 0$ gives $a = 5$. You can pay for 5 quarters before the money in this account is gone.

Flashback Answers
n (number of policies sold) is the independent variable
I(n) (weekly income as a function of policies sold) is the dependent variable.

A reasonable domain is (0, 15)[*]
A reasonable range is ($540, $1740)[*]
[*]answers may vary given reasoning is stated; 15 is an arbitrary upper limit based on selling 3 policies per day in a 5 day work week and $1740 corresponds with the domain.

Yes this function is one-to-one

Section 2.1 Exercises

1. A town's population has been growing linearly. In 2003, the population was 45,000, and the population has been growing by 1700 people each year. Write an equation, $P(t)$, for the population t years after 2003.

2. A town's population has been growing linearly. In 2005, the population was 69,000, and the population has been growing by 2500 people each year. Write an equation, $P(t)$, for the population t years after 2005.

3. Sonya is currently 10 miles from home, and is walking further away at 2 miles per hour. Write an equation for her distance from home t hours from now.

4. A boat is 100 miles away from the marina, sailing directly towards it at 10 miles per hour. Write an equation for the distance of the boat from the marina after t hours.

5. Timmy goes to the fair with $40. Each ride costs $2. How much money will he have left after riding n rides?

6. At noon, a barista notices she has $20 in her tip jar. If she makes an average of $0.50 from each customer, how much will she have in her tip jar if she serves n more customers during her shift?

Determine if each function is increasing or decreasing

7. $f(x) = 4x + 3$ 8. $g(x) = 5x + 6$

9. $a(x) = 5 - 2x$ 10. $b(x) = 8 - 3x$

11. $h(x) = -2x + 4$ 12. $k(x) = -4x + 1$

13. $j(x) = \dfrac{1}{2}x - 3$ 14. $p(x) = \dfrac{1}{4}x - 5$

15. $n(x) = -\dfrac{1}{3}x - 2$ 16. $m(x) = -\dfrac{3}{8}x + 3$

Find the slope of the line that passes through the two given points
17. (2, 4) and (4, 10)   18. (1, 5) and (4, 11)
19. (-1,4) and (5, 2)   20. (-2, 8) and (4, 6)
21. (6,11) and (-4,3)   22. (9,10) and (-6,-12)

Find the slope of the lines graphed

23.

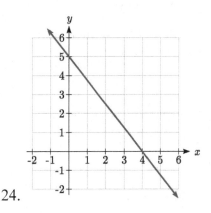

24.

25. Sonya is walking home from a friend's house. After 2 minutes she is 1.4 miles from home. Twelve minutes after leaving, she is 0.9 miles from home. What is her rate?

26. A gym membership with two personal training sessions costs $125, while gym membership with 5 personal training sessions costs $260. What is the rate for personal training sessions?

27. A city's population in the year 1960 was 287,500. In 1989 the population was 275,900. Compute the slope of the population growth (or decline) and make a statement about the population rate of change in people per year.

28. A city's population in the year 1958 was 2,113,000. In 1991 the population was 2,099,800. Compute the slope of the population growth (or decline) and make a statement about the population rate of change in people per year.

29. A phone company charges for service according to the formula: $C(n) = 24 + 0.1n$, where n is the number of minutes talked, and $C(n)$ is the monthly charge, in dollars. Find and interpret the rate of change and initial value.

30. A phone company charges for service according to the formula: $C(n) = 26 + 0.04n$, where n is the number of minutes talked, and $C(n)$ is the monthly charge, in dollars. Find and interpret the rate of change and initial value.

31. Terry is skiing down a steep hill. Terry's elevation, $E(t)$, in feet after t seconds is given by $E(t) = 3000 - 70t$. Write a complete sentence describing Terry's starting elevation and how it is changing over time.

32. Maria is climbing a mountain. Maria's elevation, $E(t)$, in feet after t minutes is given by $E(t)=1200+40t$. Write a complete sentence describing Maria's starting elevation and how it is changing over time.

Given each set of information, find a linear equation satisfying the conditions, if possible

33. $f(-5)=-4$, and $f(5)=2$ 34. $f(-1)=4$, and $f(5)=1$

35. Passes through (2, 4) and (4, 10)   36. Passes through (1, 5) and (4, 11)

37. Passes through (-1,4) and (5, 2)   38. Passes through (-2, 8) and (4, 6)

39. x intercept at (-2, 0) and y intercept at (0, -3)

40. x intercept at (-5, 0) and y intercept at (0, 4)

Find an equation for the function graphed

41.

42.

43.

44.

45. A clothing business finds there is a linear relationship between the number of shirts, n, it can sell and the price, p, it can charge per shirt. In particular, historical data shows that 1000 shirts can be sold at a price of $30, while 3000 shirts can be sold at a price of $22. Find a linear equation in the form $p = mn + b$ that gives the price p they can charge for n shirts.

46. A farmer finds there is a linear relationship between the number of bean stalks, n, she plants and the yield, y, each plant produces. When she plants 30 stalks, each plant yields 30 oz of beans. When she plants 34 stalks, each plant produces 28 oz of beans. Find a linear relationships in the form $y = mn + b$ that gives the yield when n stalks are planted.

47. Which of the following tables could represent a linear function? For each that could be linear, find a linear equation models the data.

x	$g(x)$
0	5
5	-10
10	-25
15	-40

x	$h(x)$
0	5
5	30
10	105
15	230

x	$f(x)$
0	-5
5	20
10	45
15	70

x	$k(x)$
5	13
10	28
20	58
25	73

48. Which of the following tables could represent a linear function? For each that could be linear, find a linear equation models the data.

x	$g(x)$
0	6
2	-19
4	-44
6	-69

x	$h(x)$
2	13
4	23
8	43
10	53

x	$f(x)$
2	-4
4	16
6	36
8	56

x	$k(x)$
0	6
2	31
6	106
8	231

49. While speaking on the phone to a friend in Oslo, Norway, you learned that the current temperature there was -23 Celsius (-23°C). After the phone conversation, you wanted to convert this temperature to Fahrenheit degrees, °F, but you could not find a reference with the correct formulas. You then remembered that the relationship between °F and °C is linear. [UW]

   a. Using this and the knowledge that 32°F = 0 °C and 212 °F = 100 °C, find an equation that computes Celsius temperature in terms of Fahrenheit; i.e. an equation of the form C = "an expression involving only the variable F."

   b. Likewise, find an equation that computes Fahrenheit temperature in terms of Celsius temperature; i.e. an equation of the form F = "an expression involving only the variable C."

   c. How cold was it in Oslo in °F?

Section 2.2 Graphs of Linear Functions

When we are working with a new function, it is useful to know as much as we can about the function: its graph, where the function is zero, and any other special behaviors of the function. We will begin this exploration of linear functions with a look at graphs.

When graphing a linear function, there are three basic ways to graph it:
1) By plotting points (at least 2) and drawing a line through the points
2) Using the initial value (output when $x = 0$) and rate of change (slope)
3) Using transformations of the identity function $f(x) = x$

Example 1

Graph $f(x) = 5 - \dfrac{2}{3}x$ by plotting points

In general, we evaluate the function at two or more inputs to find at least two points on the graph. Usually it is best to pick input values that will "work nicely" in the equation.

In this equation, multiples of 3 will work nicely due to the $\dfrac{2}{3}$ in the equation, and of course using $x = 0$ to get the vertical intercept. Evaluating $f(x)$ at $x = 0, 3$ and 6:

$$f(0) = 5 - \frac{2}{3}(0) = 5$$

$$f(3) = 5 - \frac{2}{3}(3) = 3$$

$$f(6) = 5 - \frac{2}{3}(6) = 1$$

These evaluations tell us that the points (0,5), (3,3), and (6,1) lie on the graph of the line. Plotting these points and drawing a line through them gives us the graph.

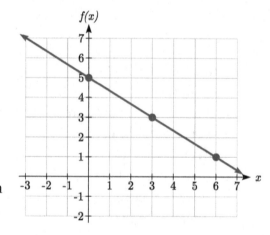

When using the initial value and rate of change to graph, we need to consider the graphical interpretation of these values. Remember the initial value of the function is the output when the input is zero, so in the equation $f(x) = b + mx$, the graph includes the point $(0, b)$. On the graph, this is the vertical intercept – the point where the graph crosses the vertical axis.

For the rate of change, it is helpful to recall that we calculated this value as

$$m = \frac{\text{change of output}}{\text{change of input}}$$

From a graph of a line, this tells us that if we divide the vertical difference, or rise, of the function outputs by the horizontal difference, or run, of the inputs, we will obtain the rate of change, also called slope of the line.

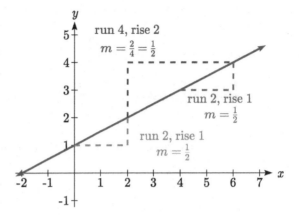

$$m = \frac{\text{change of output}}{\text{change of input}} = \frac{rise}{run}$$

Notice that this ratio is the same regardless of which two points we use.

Graphical Interpretation of a Linear Equation

Graphically, in the equation $f(x) = b + mx$,

b is the **vertical intercept** of the graph and tells us we can start our graph at $(0, b)$

m is the **slope of the line** and tells us how far to rise & run to get to the next point

Once we have at least 2 points, we can extend the graph of the line to the left and right.

Example 2

Graph $f(x) = 5 - \dfrac{2}{3}x$ using the vertical intercept and slope.

The vertical intercept of the function is $(0, 5)$, giving us a point on the graph of the line.

The slope is $-\dfrac{2}{3}$. This tells us that for every 3 units the graph "runs" in the horizontal, the vertical "rise" decreases by 2 units.

In graphing, we can use this by first plotting our vertical intercept on the graph, then using the slope to find a second point. From the initial value $(0, 5)$ the slope tells us that if we move to the right 3, we will move down 2, moving us to the point $(3, 3)$. We can continue this again to find a third point at $(6, 1)$. Finally, extend the line to the left and right, containing these points.

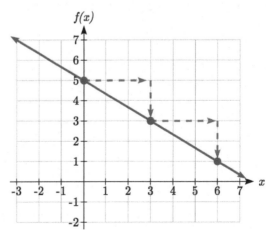

1. Consider that the slope $-\dfrac{2}{3}$ could also be written as $\dfrac{2}{-3}$. Using $\dfrac{2}{-3}$, find another point on the graph that has a negative x value.

Another option for graphing is to use transformations of the identity function $f(x) = x$.

In the equation $f(x) = mx$, the m is acting as the vertical stretch of the identity function. When m is negative, there is also a vertical reflection of the graph. Looking at some examples:

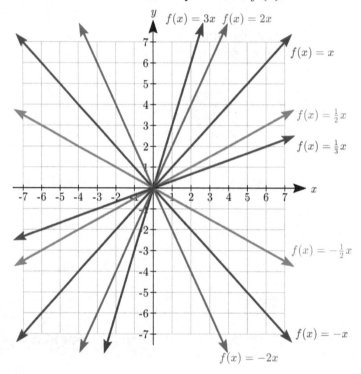

In $f(x) = mx + b$, the b acts as the vertical shift, moving the graph up and down without affecting the slope of the line. Some examples:

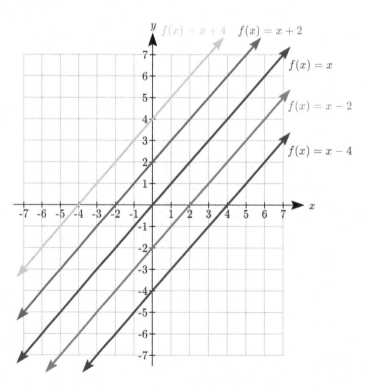

Using Vertical Stretches or Compressions along with Vertical Shifts is another way to look at identifying different types of linear functions. Although this may not be the easiest way for you to graph this type of function, make sure you practice each method.

Example 3

Graph $f(x) = -3 + \frac{1}{2}x$ using transformations.

The equation is the graph of the identity function vertically compressed by ½ and vertically shifted down 3.

Vertical compression   combined with Vertical shift

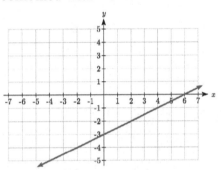

Notice how this nicely compares to the other method where the vertical intercept is found at (0, -3) and to get to another point we rise (go up vertically) by 1 unit and run (go horizontally) by 2 units to get to the next point (2, -2), and the next one (4, -1). In these three points (0, -3), (2, -2), and (4, -1), the output values change by +1, and the x values change by +2, corresponding with the slope $m = ½$.

Example 4

Match each equation with one of the lines in the graph below

$f(x) = 2x + 3$

$g(x) = 2x - 3$

$h(x) = -2x + 3$

$j(x) = \frac{1}{2}x + 3$

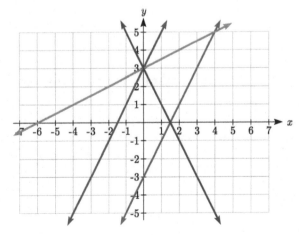

Only one graph has a vertical intercept of -3, so we can immediately match that graph with $g(x)$.

For the three graphs with a vertical intercept at 3, only one has a negative slope, so we can match that line with *h(x)*. Of the other two, the steeper line would have a larger slope, so we can match that graph with equation *f(x)*, and the flatter line with the equation *j(x)*.

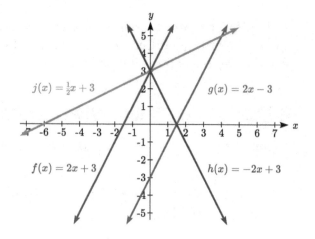

In addition to understanding the basic behavior of a linear function (increasing or decreasing, recognizing the slope and vertical intercept), it is often helpful to know the horizontal intercept of the function – where it crosses the horizontal axis.

Finding Horizontal Intercepts

The **horizontal intercept** of the function is where the graph crosses the horizontal axis. If a function has a horizontal intercept, you can always find it by solving $f(x) = 0$.

Example 5

Find the horizontal intercept of $f(x) = -3 + \dfrac{1}{2}x$

Setting the function equal to zero to find what input will put us on the horizontal axis,

$$0 = -3 + \frac{1}{2}x$$

$$3 = \frac{1}{2}x$$

$$x = 6$$

The graph crosses the horizontal axis at (6,0)

There are two special cases of lines: a horizontal line and a vertical line. In a horizontal line like the one graphed to the right, notice that between any two points, the change in the outputs is 0. In the slope equation, the numerator will be 0, resulting in a slope of 0. Using a slope of 0 in the $f(x) = b + mx$, the equation simplifies to $f(x) = b$.

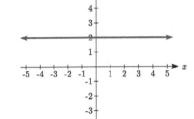

Notice a horizontal line has a vertical intercept, but no horizontal intercept (unless it's the line $f(x) = 0$).

In the case of a vertical line, notice that between any two points, the change in the inputs is zero. In the slope equation, the denominator will be zero, and you may recall that we cannot divide by the zero; the slope of a vertical line is undefined. You might also notice that a vertical line is not a function. To write the equation of vertical line, we simply write input=value, like $x = b$.

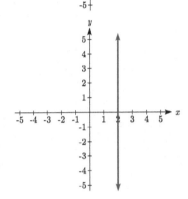

Notice a vertical line has a horizontal intercept, but no vertical intercept (unless it's the line $x = 0$).

Horizontal and Vertical Lines
Horizontal lines have equations of the form $f(x) = b$
Vertical lines have equations of the form $x = a$

Example 6

Write an equation for the horizontal line graphed above.

This line would have equation $f(x) = 2$

Example 7

Write an equation for the vertical line graphed above.

This line would have equation $x = 2$

Try it Now
2. Describe the function $f(x) = 6 - 3x$ in terms of transformations of the identity function and find its horizontal intercept.

Parallel and Perpendicular Lines

When two lines are graphed together, the lines will be **parallel** if they are increasing at the same rate – if the rates of change are the same. In this case, the graphs will never cross (unless they're the same line).

Parallel Lines

Two lines are **parallel** if the slopes are equal (or, if both lines are vertical).

In other words, given two linear equations $f(x) = b + m_1 x$ and $g(x) = b + m_2 x$, the lines will be parallel if $m_1 = m_2$.

Example 8

Find a line parallel to $f(x) = 6 + 3x$ that passes through the point (3, 0)

We know the line we're looking for will have the same slope as the given line, $m = 3$. Using this and the given point, we can solve for the new line's vertical intercept:

$g(x) = b + 3x$ then at (3, 0),

$0 = b + 3(3)$

$b = -9$

The line we're looking for is $g(x) = -9 + 3x$

If two lines are not parallel, one other interesting possibility is that the lines are perpendicular, which means the lines form a right angle (90 degree angle – a square corner) where they meet. In this case, the slopes when multiplied together will equal -1. Solving for one slope leads us to the definition:

Perpendicular Lines

Given two linear equations $f(x) = b + m_1 x$ and $g(x) = b + m_2 x$

The lines will be **perpendicular** if $m_1 m_2 = -1$, and so $m_2 = \dfrac{-1}{m_1}$

We often say the slope of a perpendicular line is the "negative reciprocal" of the other line's slope.

Example 9

Find the slope of a line perpendicular to a line with:

a) a slope of 2.   b) a slope of -4.   c) a slope of $\dfrac{2}{3}$.

If the original line had slope 2, the perpendicular line's slope would be $m_2 = \dfrac{-1}{2}$

If the original line had slope -4, the perpendicular line's slope would be $m_2 = \dfrac{-1}{-4} = \dfrac{1}{4}$

If the original line had slope $\dfrac{2}{3}$, the perpendicular line's slope would be $m_2 = \dfrac{-1}{{}^{2}\!/_{3}} = \dfrac{-3}{2}$

Example 10

Find the equation of a line perpendicular to $f(x) = 6 + 3x$ and passing through the point $(3, 0)$

The original line has slope $m = 3$. The perpendicular line will have slope $m = \dfrac{-1}{3}$.

Using this and the given point, we can find the equation for the line.

$g(x) = b - \dfrac{1}{3}x$ then at $(3, 0)$,

$0 = b - \dfrac{1}{3}(3)$

$b = 1$

The line we're looking for is $g(x) = 1 - \dfrac{1}{3}x$.

Try it Now

3. Given the line $h(t) = -4 + 2t$, find an equation for the line passing through $(0, 0)$ that
   is: a) parallel to $h(t)$.   b) perpendicular to $h(t)$.

Example 12

A line passes through the points $(-2, 6)$ and $(4, 5)$. Find the equation of a perpendicular line that passes through the point $(4, 5)$.

From the two given points on the reference line, we can calculate the slope of that line:

$m_1 = \dfrac{5 - 6}{4 - (-2)} = \dfrac{-1}{6}$

The perpendicular line will have slope

$$m_2 = \frac{-1}{-\frac{1}{6}} = 6$$

We can then solve for the vertical intercept that makes the line pass through the desired point:

$g(x) = b + 6x$ then at (4, 5),

$5 = b + 6(4)$

$b = -19$

Giving the line $g(x) = -19 + 6x$

Intersections of Lines

The graphs of two lines will intersect if they are not parallel. They will intersect at the point that satisfies both equations. To find this point when the equations are given as functions, we can solve for an input value so that $f(x) = g(x)$. In other words, we can set the formulas for the lines equal, and solve for the input that satisfies the equation.

Example 13

Find the intersection of the lines $h(t) = 3t - 4$ and $j(t) = 5 - t$

Setting $h(t) = j(t)$,

$3t - 4 = 5 - t$

$4t = 9$

$t = \frac{9}{4}$

This tells us the lines intersect when the input is $\frac{9}{4}$.

We can then find the output value of the intersection point by evaluating either function at this input

$$j\left(\frac{9}{4}\right) = 5 - \frac{9}{4} = \frac{11}{4}$$

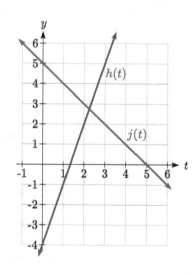

These lines intersect at the point $\left(\frac{9}{4}, \frac{11}{4}\right)$. Looking at the graph, this result seems reasonable.

Two parallel lines can also intersect if they happen to be the same line. In that case, they intersect at every point on the lines.

---

4. Look at the graph in example 13 above and answer the following for the function $h(t)$:
   a. Vertical intercept coordinates
   b. Horizontal intercepts coordinates
   c. Slope
   d. Is $j(t)$ parallel or perpendicular to $h(t)$ (or neither)
   e. Is $h(t)$ an Increasing or Decreasing function (or neither)
   f. Write a transformation description from the identity toolkit function $f(x) = x$

Finding the intersection allows us to answer other questions as well, such as discovering when one function is larger than another.

Example 14

Using the functions from the previous example, for what values of t is $h(t) > j(t)$

To answer this question, it is helpful first to know where the functions are equal, since that is the point where $h(t)$ could switch from being greater to smaller than $j(t)$ or vice-versa. From the previous example, we know the functions are equal at $t = \dfrac{9}{4}$.

By examining the graph, we can see that $h(t)$, the function with positive slope, is going to be larger than the other function to the right of the intersection. So $h(t) > j(t)$ when

$$t > \frac{9}{4}$$

Important Topics of this Section
Methods for graphing linear functions
Another name for slope = rise/run
Horizontal intercepts (a,0)
Horizontal lines
Vertical lines
Parallel lines
Perpendicular lines
Intersecting lines

---

1. (-3,7) found by starting at the vertical intercept, going up 2 units and 3 in the negative
   horizontal direction. You could have also answered, (-6, 9) or (-9, 11) etc…

2. Vertically stretched by a factor of 3, Vertically flipped (flipped over the x axis),
   Vertically shifted up by 6 units.
   Horizontal intercept: $6-3x=0$ when $x=2$

3. Parallel $f(t) = 2t$; Perpendicular $g(t) = -\dfrac{1}{2}t$

4. Given $h(t) = 3t - 4$
   a. (0,-4)
   b. $\left(\dfrac{4}{3}, 0\right)$
   c. Slope 3
   d. Neither parallel nor perpendicular
   e. Increasing function
   f. Given the identity function, vertically stretch by 3 and shift down 4 units.

Section 2.2 Exercises

Match each linear equation with its graph

1. $f(x) = -x - 1$
2. $f(x) = -2x - 1$
3. $f(x) = -\dfrac{1}{2}x - 1$
4. $f(x) = 2$
5. $f(x) = 2 + x$
6. $f(x) = 3x + 2$

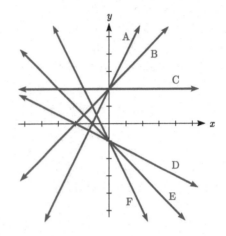

Sketch a line with the given features

7. An x-intercept of (-4, 0) and y-intercept of (0, -2)

8. An x-intercept of (-2, 0) and y-intercept of (0, 4)

9. A vertical intercept of (0, 7) and slope $-\dfrac{3}{2}$

10. A vertical intercept of (0, 3) and slope $\dfrac{2}{5}$

11. Passing through the points (-6,-2) and (6,-6)

12. Passing through the points (-3,-4) and (3,0)

Sketch the graph of each equation

13. $f(x) = -2x - 1$ 14. $g(x) = -3x + 2$

15. $h(x) = \dfrac{1}{3}x + 2$ 16. $k(x) = \dfrac{2}{3}x - 3$

17. $k(t) = 3 + 2t$ 18. $p(t) = -2 + 3t$

19. $x = 3$ 20. $x = -2$

21. $r(x) = 4$ 22. $q(x) = 3$

23. If $g(x)$ is the transformation of $f(x) = x$ after a vertical compression by $3/4$, a shift left by 2, and a shift down by 4
   a. Write an equation for $g(x)$
   b. What is the slope of this line?
   c. Find the vertical intercept of this line.

24. If $g(x)$ is the transformation of $f(x) = x$ after a vertical compression by $1/3$, a shift right by 1, and a shift up by 3
   a. Write an equation for $g(x)$
   b. What is the slope of this line?
   c. Find the vertical intercept of this line.

Write the equation of the line shown

25.

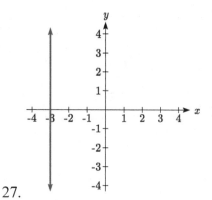

27.

26.

28.

Find the horizontal and vertical intercepts of each equation

29. $f(x) = -x + 2$

30. $g(x) = 2x + 4$

31. $h(x) = 3x - 5$

32. $k(x) = -5x + 1$

33. $-2x + 5y = 20$

34. $7x + 2y = 56$

Given below are descriptions of two lines. Find the slopes of Line 1 and Line 2. Is each pair of lines parallel, perpendicular or neither?

35. Line 1: Passes through $(0,6)$ and $(3,-24)$
    Line 2: Passes through $(-1,19)$ and $(8,-71)$

36. Line 1: Passes through $(-8,-55)$ and $(10,89)$
    Line 2: Passes through $(9,-44)$ and $(4,-14)$

37. Line 1: Passes through $(2,3)$ and $(4,-1)$
    Line 2: Passes through $(6,3)$ and $(8,5)$

38. Line 1: Passes through $(1,7)$ and $(5,5)$
    Line 2: Passes through $(-1,-3)$ and $(1,1)$

39. Line 1: Passes through $(0,5)$ and $(3,3)$
    Line 2: Passes through $(1,-5)$ and $(3,-2)$

40. Line 1: Passes through $(2,5)$ and $(5,-1)$
    Line 2: Passes through $(-3,7)$ and $(3,-5)$

41. Write an equation for a line parallel to $f(x)=-5x-3$ and passing through the point $(2,-12)$

42. Write an equation for a line parallel to $g(x)=3x-1$ and passing through the point $(4,9)$

43. Write an equation for a line perpendicular to $h(t)=-2t+4$ and passing through the point $(-4,-1)$

44. Write an equation for a line perpendicular to $p(t)=3t+4$ and passing through the point $(3,1)$

45. Find the point at which the line $f(x)=-2x-1$ intersects the line $g(x)=-x$

46. Find the point at which the line $f(x)=2x+5$ intersects the line $g(x)=-3x-5$

47. Use algebra to find the point at which the line $f(x) = -\dfrac{4}{5}x + \dfrac{274}{25}$ intersects the line

$$h(x) = \dfrac{9}{4}x + \dfrac{73}{10}$$

48. Use algebra to find the point at which the line $f(x) = \dfrac{7}{4}x + \dfrac{457}{60}$ intersects the line

$$g(x) = \dfrac{4}{3}x + \dfrac{31}{5}$$

49. A car rental company offers two plans for renting a car.
    Plan A: 30 dollars per day and 18 cents per mile
    Plan B: 50 dollars per day with free unlimited mileage
    How many miles would you need to drive for plan B to save you money?

50. You're comparing two cell phone companies.
    Company A: $20/month for unlimited talk and text, and $10/GB for data.
    Company B: $65/month for unlimited talk, text, and data.
    Under what circumstances will company A save you money?

Find a formula for each piecewise defined function.

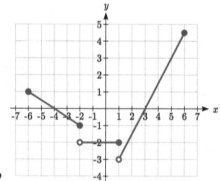

51.

52.

53. Sketch an accurate picture of the line having equation $f(x) = 2 - \dfrac{1}{2}x$. Let c be an unknown constant. [UW]

    a. Find the point of intersection between the line you have graphed and the line $g(x) = 1 + cx$; your answer will be a point in the xy plane whose coordinates involve the unknown c.
    b. Find c so that the intersection point in (a) has x-coordinate 10.
    c. Find c so that the intersection point in (a) lies on the x-axis.

Section 2.3 Modeling with Linear Functions

When modeling scenarios with a linear function and solving problems involving quantities changing linearly, we typically follow the same problem solving strategies that we would use for any type of function:

Problem solving strategy
1) Identify changing quantities, and then carefully and clearly define descriptive variables to represent those quantities. When appropriate, sketch a picture or define a coordinate system.
2) Carefully read the problem to identify important information. Look for information giving values for the variables, or values for parts of the functional model, like slope and initial value.
3) Carefully read the problem to identify what we are trying to find, identify, solve, or interpret.
4) Identify a solution pathway from the provided information to what we are trying to find. Often this will involve checking and tracking units, building a table or even finding a formula for the function being used to model the problem.
5) When needed, find a formula for the function.
6) Solve or evaluate using the formula you found for the desired quantities.
7) Reflect on whether your answer is reasonable for the given situation and whether it makes sense mathematically.
8) Clearly convey your result using appropriate units, and answer in full sentences when appropriate.

Example 1

Emily saved up $3500 for her summer visit to Seattle. She anticipates spending $400 each week on rent, food, and fun. Find and interpret the horizontal intercept and determine a reasonable domain and range for this function.

In the problem, there are two changing quantities: time and money. The amount of money she has remaining while on vacation depends on how long she stays. We can define our variables, including units.
Output: M, money remaining, in dollars
Input: t, time, in weeks

Reading the problem, we identify two important values. The first, $3500, is the initial value for M. The other value appears to be a rate of change – the units of dollars per week match the units of our output variable divided by our input variable. She is spending money each week, so you should recognize that the amount of money remaining is decreasing each week and the slope is negative.

To answer the first question, looking for the horizontal intercept, it would be helpful to have an equation modeling this scenario. Using the intercept and slope provided in the problem, we can write the equation: $M(t) = 3500 - 400t$.

To find the horizontal intercept, we set the output to zero, and solve for the input:
$$0 = 3500 - 400t$$
$$t = \frac{3500}{400} = 8.75$$

The horizontal intercept is 8.75 weeks. Since this represents the input value where the output will be zero, interpreting this, we could say: Emily will have no money left after 8.75 weeks.

When modeling any real life scenario with functions, there is typically a limited domain over which that model will be valid – almost no trend continues indefinitely. In this case, it certainly doesn't make sense to talk about input values less than zero. It is also likely that this model is not valid after the horizontal intercept (unless Emily's going to start using a credit card and go into debt).

The domain represents the set of input values and so the reasonable domain for this function is $0 \le t \le 8.75$.

However, in a real world scenario, the rental might be weekly or nightly. She may not be able to stay a partial week and so all options should be considered. Emily could stay in Seattle for 0 to 8 full weeks (and a couple of days), but would have to go into debt to stay 9 full weeks, so restricted to whole weeks, a reasonable domain without going in to debt would be $0 \le t \le 8$, or $0 \le t \le 9$ if she went into debt to finish out the last week.

The range represents the set of output values and she starts with $3500 and ends with $0 after 8.75 weeks so the corresponding range is $0 \le M(t) \le 3500$. If we limit the rental to whole weeks, however, the range would change. If she left after 8 weeks because she didn't have enough to stay for a full 9 weeks, she would have $M(8) = 3500-400(8) =$ $300 dollars left after 8 weeks, giving a range of $300 \le M(t) \le 3500$. If she wanted to stay the full 9 weeks she would be $100 in debt giving a range of $-100 \le M(t) \le 3500$.

Most importantly remember that domain and range are tied together, and what ever you decide is most appropriate for the domain (the independent variable) will dictate the requirements for the range (the dependent variable).

Try it Now

1. A database manager is loading a large table from backups. Getting impatient, she notices 1.2 million rows had been loaded. Ten minutes later, 2.5 million rows had been loaded. How much longer will she have to wait for all 80 million rows to load?

Example 2

Jamal is choosing between two moving companies. The first, U-Haul, charges an up-front fee of $20, then 59 cents a mile. The second, Budget, charges an up-front fee of $16, then 63 cents a mile[3]. When will U-Haul be the better choice for Jamal?

The two important quantities in this problem are the cost, and the number of miles that are driven. Since we have two companies to consider, we will define two functions:

Input: m, miles driven
Outputs:
$Y(m)$: cost, in dollars, for renting from U-Haul
$B(m)$: cost, in dollars, for renting from Budget

Reading the problem carefully, it appears that we were given an initial cost and a rate of change for each company. Since our outputs are measured in dollars but the costs per mile given in the problem are in cents, we will need to convert these quantities to match our desired units: $0.59 a mile for U-Haul, and $0.63 a mile for Budget.

Looking to what we're trying to find, we want to know when U-Haul will be the better choice. Since all we have to make that decision from is the costs, we are looking for when U-Haul will cost less, or when $Y(m) < B(m)$. The solution pathway will lead us to find the equations for the two functions, find the intersection, then look to see where the $Y(m)$ function is smaller. Using the rates of change and initial charges, we can write the equations:

$Y(m) = 20 + 0.59m$

$B(m) = 16 + 0.63m$

These graphs are sketched to the right, with $Y(m)$ drawn dashed.

To find the intersection, we set the equations equal and solve:
$Y(m) = B(m)$
$20 + 0.59m = 16 + 0.63m$

$4 = 0.04m$

$m = 100$

This tells us that the cost from the two companies will be the same if 100 miles are driven. Either by looking at the graph, or noting that $Y(m)$ is growing at a slower rate, we can conclude that U-Haul will be the cheaper price when more than 100 miles are driven.

---

[3] Rates retrieved Aug 2, 2010 from http://www.budgettruck.com and http://www.uhaul.com/

Example 3

A town's population has been growing linearly. In 2004 the population was 6,200. By 2009 the population had grown to 8,100. If this trend continues,
a. Predict the population in 2013
b. When will the population reach 15000?

The two changing quantities are the population and time. While we could use the actual year value as the input quantity, doing so tends to lead to very ugly equations, since the vertical intercept would correspond to the year 0, more than 2000 years ago!
To make things a little nicer, and to make our lives easier too, we will define our input as years since 2004:
Input: t, years since 2004
Output: $P(t)$, the town's population

The problem gives us two input-output pairs. Converting them to match our defined variables, the year 2004 would correspond to $t = 0$, giving the point (0, 6200). Notice that through our clever choice of variable definition, we have "given" ourselves the vertical intercept of the function. The year 2009 would correspond to $t = 5$, giving the point (5, 8100).

To predict the population in 2013 ($t = 9$), we would need an equation for the population. Likewise, to find when the population would reach 15000, we would need to solve for the input that would provide an output of 15000. Either way, we need an equation. To find it, we start by calculating the rate of change:

$$m = \frac{8100 - 6200}{5 - 0} = \frac{1900}{5} = 380 \text{ people per year}$$

Since we already know the vertical intercept of the line, we can immediately write the equation:
$P(t) = 6200 + 380t$

To predict the population in 2013, we evaluate our function at $t = 9$
$P(9) = 6200 + 380(9) = 9620$
If the trend continues, our model predicts a population of 9,620 in 2013.

To find when the population will reach 15,000, we can set $P(t) = 15000$ and solve for t.
$15000 = 6200 + 380t$

$8800 = 380t$

$t \approx 23.158$

Our model predicts the population will reach 15,000 in a little more than 23 years after 2004, or somewhere around the year 2027.

Example 4

Anna and Emanuel start at the same intersection. Anna walks east at 4 miles per hour while Emanuel walks south at 3 miles per hour. They are communicating with a two-way radio with a range of 2 miles. How long after they start walking will they fall out of radio contact?

In essence, we can partially answer this question by saying they will fall out of radio contact when they are 2 miles apart, which leads us to ask a new question: how long will it take them to be 2 miles apart?

In this problem, our changing quantities are time and the two peoples' positions, but ultimately we need to know how long will it take for them to be 2 miles apart. We can see that time will be our input variable, so we'll define
Input: t, time in hours.

Since it is not obvious how to define our output variables, we'll start by drawing a picture.

Because of the complexity of this question, it may be helpful to introduce some intermediary variables. These are quantities that we aren't directly interested in, but seem important to the problem. For this problem, Anna's and Emanuel's distances from the starting point seem important. To notate these, we are going to define a coordinate system, putting the "starting point" at the intersection where they both started, then we're going to introduce a variable, A, to represent Anna's position, and define it to be a measurement from the starting point, in the eastward direction. Likewise, we'll introduce a variable, E, to represent Emanuel's position, measured from the starting point in the southward direction. Note that in defining the coordinate system we specified both the origin, or starting point, of the measurement, as well as the direction of measure.

While we're at it, we'll define a third variable, D, to be the measurement of the distance between Anna and Emanuel. Showing the variables on the picture is often helpful: Looking at the variables on the picture, we remember we need to know how long it takes for D, the distance between them, to equal 2 miles.

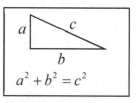

Seeing this picture we remember that in order to find the distance between the two, we can use the Pythagorean Theorem, a property of right triangles.

From here, we can now look back at the problem for relevant information. Anna is walking 4 miles per hour, and Emanuel is walking 3 miles per hour, which are rates of change. Using those, we can write formulas for the distance each has walked.

They both start at the same intersection and so when $t = 0$, the distance travelled by each person should also be 0, so given the rate for each, and the initial value for each, we get:

$A(t) = 4t$

$E(t) = 3t$

Using the Pythagorean theorem we get:

$D(t)^2 = A(t)^2 + E(t)^2$ substitute in the function formulas

$D(t)^2 = (4t)^2 + (3t)^2 = 16t^2 + 9t^2 = 25t^2$ solve for $D(t)$ using the square root

$D(t) = \pm\sqrt{25t^2} = \pm 5|t|$

Since in this scenario we are only considering positive values of t and our distance $D(t)$ will always be positive, we can simplify this answer to $D(t) = 5t$

Interestingly, the distance between them is also a linear function. Using it, we can now answer the question of when the distance between them will reach 2 miles:

$D(t) = 2$

$5t = 2$

$t = \dfrac{2}{5} = 0.4$

They will fall out of radio contact in 0.4 hours, or 24 minutes.

Example 5

There is currently a straight road leading from the town of Westborough to a town 30 miles east and 10 miles north. Partway down this road, it junctions with a second road, perpendicular to the first, leading to the town of Eastborough. If the town of Eastborough is located 20 miles directly east of the town of Westborough, how far is the road junction from Westborough?

It might help here to draw a picture of the situation. It would then be helpful to introduce a coordinate system. While we could place the origin anywhere, placing it at Westborough seems convenient. This puts the other town at coordinates (30, 10), and Eastborough at (20, 0).

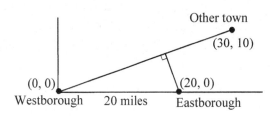

Using this point along with the origin, we can find the slope of the line from Westborough to the other town: $m = \dfrac{10-0}{30-0} = \dfrac{1}{3}$. This gives the equation of the road from Westborough to the other town to be $W(x) = \dfrac{1}{3}x$.

From this, we can determine the perpendicular road to Eastborough will have slope $m = -3$. Since the town of Eastborough is at the point (20, 0), we can find the equation:

$E(x) = -3x + b$ plug in the point (20, 0)

$0 = -3(20) + b$

$b = 60$

$E(x) = -3x + 60$

We can now find the coordinates of the junction of the roads by finding the intersection of these lines. Setting them equal,

$\dfrac{1}{3}x = -3x + 60$

$\dfrac{10}{3}x - 60$

$10x = 180$

$x = 18$ Substituting this back into $W(x)$

$y = W(18) = \dfrac{1}{3}(18) = 6$

The roads intersect at the point (18, 6). Using the distance formula, we can now find the distance from Westborough to the junction:

$dist = \sqrt{(18-0)^2 + (6-0)^2} \approx 18.934$ miles.

Important Topics of this Section

The problem solving process

1) Identify changing quantities, and then carefully and clearly define descriptive variables to represent those quantities. When appropriate, sketch a picture or define a coordinate system.

2) Carefully read the problem to identify important information. Look for information giving values for the variables, or values for parts of the functional model, like slope and initial value.

3) Carefully read the problem to identify what we are trying to find, identify, solve, or interpret.

4) Identify a solution pathway from the provided information to what we are trying to find. Often this will involve checking and tracking units, building a table or even finding a formula for the function being used to model the problem.

5) When needed, find a formula for the function.

6) Solve or evaluate using the formula you found for the desired quantities.

7) Reflect on whether your answer is reasonable for the given situation and whether it makes sense mathematically.

8) Clearly convey your result using appropriate units, and answer in full sentences when appropriate.

Try it Now

1. Letting t be the number of minutes since she got impatient, and N be the number rows loaded, in millions, we have two points: (0, 1.2) and (10, 2.5).

The slope is $m = \dfrac{2.5 - 1.2}{10 - 0} = \dfrac{1.3}{10} = 0.13$ million rows per minute.

We know the N intercept, so we can write the equation:
$N = 0.13t + 1.2$

To determine how long she will have to wait, we need to solve for when $N = 80$.
$N = 0.13t + 1.2 = 80$
$0.13t = 78.8$
$t = \dfrac{78.8}{0.13} \approx 606$. She'll have to wait another 606 minutes, about 10 hours.

Section 2.3 Exercises

1. In 2004, a school population was 1001. By 2008 the population had grown to 1697. Assume the population is changing linearly.
   a. How much did the population grow between the year 2004 and 2008?
   b. How long did it take the population to grow from 1001 students to 1697 students?
   c. What is the average population growth per year?
   d. What was the population in the year 2000?
   e. Find an equation for the population, P, of the school t years after 2000.
   f. Using your equation, predict the population of the school in 2011.

2. In 2003, a town's population was 1431. By 2007 the population had grown to 2134. Assume the population is changing linearly.
   a. How much did the population grow between the year 2003 and 2007?
   b. How long did it take the population to grow from 1431 people to 2134?
   c. What is the average population growth per year?
   d. What was the population in the year 2000?
   e. Find an equation for the population, P, of the town t years after 2000.
   f. Using your equation, predict the population of the town in 2014.

3. A phone company has a monthly cellular plan where a customer pays a flat monthly fee and then a certain amount of money per minute used on the phone. If a customer uses 410 minutes, the monthly cost will be $71.50. If the customer uses 720 minutes, the monthly cost will be $118.
   a. Find a linear equation for the monthly cost of the cell plan as a function of x, the number of monthly minutes used.
   b. Interpret the slope and vertical intercept of the equation.
   c. Use your equation to find the total monthly cost if 687 minutes are used.

4. A phone company has a monthly cellular data plan where a customer pays a flat monthly fee and then a certain amount of money per megabyte (MB) of data used on the phone. If a customer uses 20 MB, the monthly cost will be $11.20. If the customer uses 130 MB, the monthly cost will be $17.80.
   a. Find a linear equation for the monthly cost of the data plan as a function of x, the number of MB used.
   b. Interpret the slope and vertical intercept of the equation.
   c. Use your equation to find the total monthly cost if 250 MB are used.

5. In 1991, the moose population in a park was measured to be 4360. By 1999, the population was measured again to be 5880. If the population continues to change linearly,
    a. Find a formula for the moose population, P.
    b. What does your model predict the moose population to be in 2003?

6. In 2003, the owl population in a park was measured to be 340. By 2007, the population was measured again to be 285. If the population continues to change linearly,
    a. Find a formula for the owl population, P.
    b. What does your model predict the owl population to be in 2012?

7. The Federal Helium Reserve held about 16 billion cubic feet of helium in 2010, and is being depleted by about 2.1 billion cubic feet each year.
    a. Give a linear equation for the remaining federal helium reserves, R, in terms of t, the number of years since 2010.
    b. In 2015, what will the helium reserves be?
    c. If the rate of depletion doesn't change, when will the Federal Helium Reserve be depleted?

8. Suppose the world's current oil reserves are 1820 billion barrels. If, on average, the total reserves is decreasing by 25 billion barrels of oil each year:
    a. Give a linear equation for the remaining oil reserves, R, in terms of t, the number of years since now.
    b. Seven years from now, what will the oil reserves be?
    c. If the rate of depletion isn't change, when will the world's oil reserves be depleted?

9. You are choosing between two different prepaid cell phone plans. The first plan charges a rate of 26 cents per minute. The second plan charges a monthly fee of $19.95 *plus* 11 cents per minute. How many minutes would you have to use in a month in order for the second plan to be preferable?

10. You are choosing between two different window washing companies. The first charges $5 per window. The second charges a base fee of $40 plus $3 per window. How many windows would you need to have for the second company to be preferable?

11. When hired at a new job selling jewelry, you are given two pay options:
    Option A: Base salary of $17,000 a year, with a commission of 12% of your sales
    Option B: Base salary of $20,000 a year, with a commission of 5% of your sales
    How much jewelry would you need to sell for option A to produce a larger income?

12. When hired at a new job selling electronics, you are given two pay options:
    Option A: Base salary of $14,000 a year, with a commission of 10% of your sales
    Option B: Base salary of $19,000 a year, with a commission of 4% of your sales
    How much electronics would you need to sell for option A to produce a larger income?

13. Find the area of a triangle bounded by the y axis, the line $f(x) = 9 - \dfrac{6}{7}x$, and the line perpendicular to $f(x)$ that passes through the origin.

14. Find the area of a triangle bounded by the x axis, the line $f(x) = 12 - \dfrac{1}{3}x$, and the line perpendicular to $f(x)$ that passes through the origin.

15. Find the area of a parallelogram bounded by the y axis, the line $x = 3$, the line $f(x) = 1 + 2x$, and the line parallel to $f(x)$ passing through (2, 7)

16. Find the area of a parallelogram bounded by the x axis, the line $g(x) = 2$, the line $f(x) = 3x$, and the line parallel to $f(x)$ passing through (6, 1)

17. If $b > 0$ and $m < 0$, then the line $f(x) = b + mx$ cuts off a triangle from the first quadrant. Express the area of that triangle in terms of m and b. [UW]

18. Find the value of m so the lines $f(x) = mx + 5$ and $g(x) = x$ and the y-axis form a triangle with an area of 10. [UW]

19. The median home values in Mississippi and Hawaii (adjusted for inflation) are shown below. If we assume that the house values are changing linearly,

Year	Mississippi	Hawaii
1950	25200	74400
2000	71400	272700

   a. In which state have home values increased at a higher rate?
   b. If these trends were to continue, what would be the median home value in Mississippi in 2010?
   c. If we assume the linear trend existed before 1950 and continues after 2000, the two states' median house values will be (or were) equal in what year? (The answer might be absurd)

20. The median home value ins Indiana and Alabama (adjusted for inflation) are shown below. If we assume that the house values are changing linearly,

Year	Indiana	Alabama
1950	37700	27100
2000	94300	85100

  a. In which state have home values increased at a higher rate?
  b. If these trends were to continue, what would be the median home value in Indiana in 2010?
  c. If we assume the linear trend existed before 1950 and continues after 2000, the two states' median house values will be (or were) equal in what year? (The answer might be absurd)

21. Pam is taking a train from the town of Rome to the town of Florence. Rome is located 30 miles due West of the town of Paris. Florence is 25 miles East, and 45 miles North of Rome. On her trip, how close does Pam get to Paris? [UW]

22. You're flying from Joint Base Lewis-McChord (JBLM) to an undisclosed location 226 km south and 230 km east. Mt. Rainier is located approximately 56 km east and 40 km south of JBLM. If you are flying at a constant speed of 800 km/hr, how long after you depart JBLM will you be the closest to Mt. Rainier?

Section 2.4 Fitting Linear Models to Data

In the real world, rarely do things follow trends perfectly. When we expect the trend to behave linearly, or when inspection suggests the trend is behaving linearly, it is often desirable to find an equation to approximate the data. Finding an equation to approximate the data helps us understand the behavior of the data and allows us to use the linear model to make predictions about the data, inside and outside of the data range.

Example 1

The table below shows the number of cricket chirps in 15 seconds, and the air temperature, in degrees Fahrenheit[4]. Plot this data, and determine whether the data appears to be linearly related.

chirps	44	35	20.4	33	31	35	18.5	37	26
Temp	80.5	70.5	57	66	68	72	52	73.5	53

Plotting this data, it appears there may be a trend, and that the trend appears roughly linear, though certainly not perfectly so.

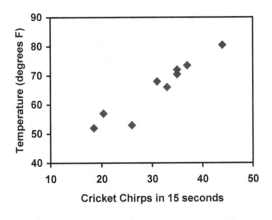

The simplest way to find an equation to approximate this data is to try to "eyeball" a line that seems to fit the data pretty well, then find an equation for that line based on the slope and intercept.

You can see from the trend in the data that the number of chirps increases as the temperature increases. As you consider a function for this data you should know that you are looking at an increasing function or a function with a positive slope.

---

[4] Selected data from http://classic.globe.gov/fsl/scientistsblog/2007/10/. Retrieved Aug 3, 2010

1. a. What descriptive variables would you choose to represent Temperature & Chirps?
   b. Which variable is the independent variable and which is the dependent variable?
   c. Based on this data and the graph, what is a reasonable domain & range?
   d. Based on the data alone, is this function one-to-one, explain?

Example 2

Using the table of values from the previous example, find a linear function that fits the data by "eyeballing" a line that seems to fit.

On a graph, we could try sketching in a line. Note the scale on the axes have been adjusted to start at zero to include the vertical axis and vertical intercept in the graph.

Using the starting and ending points of our "hand drawn" line, points (0, 30) and (50, 90), this graph has a slope of $m = \dfrac{60}{50} = 1.2$ and a vertical intercept at 30, giving an equation of

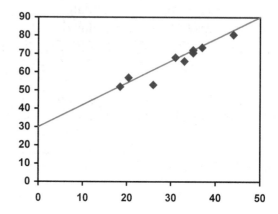

$T(c) = 30 + 1.2c$

where c is the number of chirps in 15 seconds, and $T(c)$ is the temperature in degrees Fahrenheit.

This linear equation can then be used to approximate the solution to various questions we might ask about the trend. While the data does not perfectly fall on the linear equation, the equation is our best guess as to how the relationship will behave outside of the values we have data for. There is a difference, though, between making predictions inside the domain and range of values we have data for, and outside that domain and range.

Interpolation and Extrapolation
Interpolation: When we predict a value inside the domain and range of the data
Extrapolation: When we predict a value outside the domain and range of the data

For the Temperature as a function of chirps in our hand drawn model above,
- Interpolation would occur if we used our model to predict temperature when the values for chirps are between 18.5 and 44.
- Extrapolation would occur if we used our model to predict temperature when the values for chirps are less than 18.5 or greater than 44.

Example 3

a) Would predicting the temperature when crickets are chirping 30 times in 15 seconds be interpolation or extrapolation? Make the prediction, and discuss if it is reasonable.

b) Would predicting the number of chirps crickets will make at 40 degrees be interpolation or extrapolation? Make the prediction, and discuss if it is reasonable.

With our cricket data, our number of chirps in the data provided varied from 18.5 to 44. A prediction at 30 chirps per 15 seconds is inside the domain of our data, so would be interpolation. Using our model:
$T(30) = 30 + 1.2(30) = 66$ degrees.

Based on the data we have, this value seems reasonable.

The temperature values varied from 52 to 80.5. Predicting the number of chirps at 40 degrees is extrapolation since 40 is outside the range of our data. Using our model:
$40 = 30 + 1.2c$

$10 = 1.2c$

$c \approx 8.33$

Our model predicts the crickets would chirp 8.33 times in 15 seconds. While this might be possible, we have no reason to believe our model is valid outside the domain and range. In fact, generally crickets stop chirping altogether below around 50 degrees.

When our model no longer applies after some point, it is sometimes called **model breakdown**.

Try it Now

1. What temperature would you predict if you counted 20 chirps in 15 seconds?

Fitting Lines with Technology

While eyeballing a line works reasonably well, there are statistical techniques for fitting a line to data that minimize the differences between the line and data values[5]. This technique is called **least-square regression**, and can be computed by many graphing calculators, spreadsheet software like Excel or Google Docs, statistical software, and many web-based calculators[6].

---

[5] Technically, the method minimizes the sum of the squared differences in the vertical direction between the line and the data values.
[6] For example, http://www.shodor.org/unchem/math/lls/leastsq.html

Example 4

Find the least-squares regression line using the cricket chirp data from above.

Using the cricket chirp data from earlier, with technology we obtain the equation:
$T(c) = 30.281 + 1.143c$

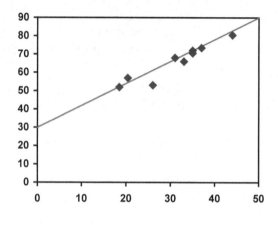

Notice that this line is quite similar to the equation we "eyeballed", but should fit the data better. Notice also that using this equation would change our prediction for the temperature when hearing 30 chirps in 15 seconds from 66 degrees to:
$T(30) = 30.281 + 1.143(30) = 64.571 \approx 64.6$
degrees.

Most calculators and computer software will also provide you with the **correlation coefficient**, a measure of how closely the line fits the data.

Correlation Coefficient

The **correlation coefficient** is a value, r, between -1 and 1.

$r > 0$ suggests a positive (increasing) relationship

$r < 0$ suggests a negative (decreasing) relationship

The closer the value is to 0, the more scattered the data

The closer the value is to 1 or -1, the less scattered the data is

The correlation coefficient provides an easy way to get some idea of how close to a line the data falls.

We should only compute the correlation coefficient for data that follows a linear pattern; if the data exhibits a non-linear pattern, the correlation coefficient is meaningless. To get a sense for the relationship between the value of r and the graph of the data, here are some large data sets with their correlation coefficients:

Examples of Correlation Coefficient Values

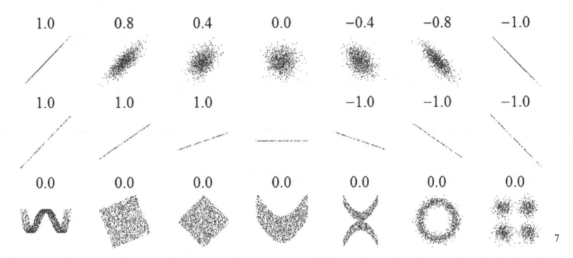

7

Example 5

Calculate the correlation coefficient for our cricket data.

Because the data appears to follow a linear pattern, we can use technology to calculate $r = 0.9509$. Since this value is very close to 1, it suggests a strong increasing linear relationship.

Example 6

Gasoline consumption in the US has been increasing steadily. Consumption data from 1994 to 2004 is shown below.[8] Determine if the trend is linear, and if so, find a model for the data. Use the model to predict the consumption in 2008.

Year	'94	'95	'96	'97	'98	'99	'00	'01	'02	'03	'04
Consumption (billions of gallons)	113	116	118	119	123	125	126	128	131	133	136

To make things simpler, a new input variable is introduced, t, representing years since 1994.

Using technology, the correlation coefficient was calculated to be 0.9965, suggesting a very strong increasing linear trend.

7 http://en.wikipedia.org/wiki/File:Correlation_examples.png
8 http://www.bts.gov/publications/national_transportation_statistics/2005/html/table_04_10.html

The least-squares regression equation is:
$C(t) = 113.318 + 2.209t$.

Using this to predict consumption in 2008 ($t = 14$),
$C(14) = 113.318 + 2.209(14) = 144.244$ billions of gallons

The model predicts 144.244 billion gallons of gasoline will be consumed in 2008.

Try it Now
2. Use the model created by technology in example 6 to predict the gas consumption in 2011. Is this an interpolation or an extrapolation?

Important Topics of this Section
Fitting linear models to data by hand
Fitting linear models to data using technology
Interpolation
Extrapolation
Correlation coefficient

Flashback Answers
1. a. T = Temperature, C = Chirps (answers may vary)
   b. Independent (Chirps) , Dependent (Temperature)
   c. Reasonable Domain (18.5, 44) , Reasonable Range (52, 80.5) (answers may vary)
   d. NO, it is not one-to-one, there are two different output values for 35 chirps.

Try it Now Answers
1. 54 degrees Fahrenheit
2. 150.871 billion gallons; extrapolation

Section 2.4 Exercises

1. The following is data for the first and second quiz scores for 8 students in a class. Plot the points, then sketch a line that fits the data.

First Quiz	11	20	24	25	33	42	46	49
Second Quiz	10	16	23	28	30	39	40	49

2. Eight students were asked to estimate their score on a 10 point quiz. Their estimated and actual scores are given. Plot the points, then sketch a line that fits the data.

Predicted	5	7	6	8	10	9	10	7
Actual	6	6	7	8	9	9	10	6

Based on each set of data given, calculate the regression line using your calculator or other technology tool, and determine the correlation coefficient.

3.

x	y
5	4
7	12
10	17
12	22
15	24

4.

x	y
8	23
15	41
26	53
31	72
56	103

5.

x	y
3	21.9
4	22.22
5	22.74
6	22.26
7	20.78
8	17.6
9	16.52
10	18.54
11	15.76
12	13.68
13	14.1
14	14.02
15	11.94
16	12.76
17	11.28
18	9.1

6.

x	y
4	44.8
5	43.1
6	38.8
7	39
8	38
9	32.7
10	30.1
11	29.3
12	27
13	25.8
14	24.7
15	22
16	20.1
17	19.8
18	16.8

7. A regression was run to determine if there is a relationship between hours of TV watched per day (x) and number of situps a person can do (y). The results of the regression are given below. Use this to predict the number of situps a person who watches 11 hours of TV can do.

```
y=ax+b
a=-1.341
b=32.234
r²=0.803
r=-0.896
```

8. A regression was run to determine if there is a relationship between the diameter of a tree (x, in inches) and the tree's age (y, in years). The results of the regression are given below. Use this to predict the age of a tree with diameter 10 inches.

```
y=ax+b
a=6.301
b=-1.044
r²=0.940
r=0.970
```

Match each scatterplot shown below with one of the four specified correlations.
9. $r = 0.95$ 10. $r = -0.89$ 11. $r = 0.26$ 12. $r = -0.39$

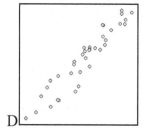

A   B   C   D

13. The US census tracks the percentage of persons 25 years or older who are college graduates. That data for several years is given below. Determine if the trend appears linear. If so and the trend continues, in what year will the percentage exceed 35%?

Year	1990	1992	1994	1996	1998	2000	2002	2004	2006	2008
Percent Graduates	21.3	21.4	22.2	23.6	24.4	25.6	26.7	27.7	28	29.4

14. The US import of wine (in hectoliters) for several years is given below. Determine if the trend appears linear. If so and the trend continues, in what year will imports exceed 12,000 hectoliters?

Year	1992	1994	1996	1998	2000	2002	2004	2006	2008	2009
Imports	2665	2688	3565	4129	4584	5655	6549	7950	8487	9462

Section 2.5 Absolute Value Functions

So far in this chapter we have been studying the behavior of linear functions. The Absolute Value Function is a piecewise-defined function made up of two linear functions. The name, Absolute Value Function, should be familiar to you from Section 1.2. In its basic form $f(x) = |x|$ it is one of our toolkit functions.

Absolute Value Function

The absolute value function can be defined as

$$f(x) = |x| = \begin{cases} x & if \quad x \geq 0 \\ -x & if \quad x < 0 \end{cases}$$

The absolute value function is commonly used to determine the distance between two numbers on the number line. Given two values a and b, then $|a - b|$ will give the distance, a positive quantity, between these values, regardless of which value is larger.

Example 1

Describe all values, x, within a distance of 4 from the number 5.

We want the distance between x and 5 to be less than or equal to 4. The distance can be represented using the absolute value, giving the expression

$|x - 5| \leq 4$

Example 2

A 2010 poll reported 78% of Americans believe that people who are gay should be able to serve in the US military, with a reported margin of error of 3%[9]. The margin of error tells us how far off the actual value could be from the survey value[10]. Express the set of possible values using absolute values.

Since we want the size of the difference between the actual percentage, p, and the reported percentage to be less than 3%,

$|p - 78| \leq 3$

---

[9] http://www.pollingreport.com/civil.htm, retrieved August 4, 2010
[10] Technically, margin of error usually means that the surveyors are 95% confident that actual value falls within this range.

1. Students who score within 20 points of 80 will pass the test. Write this as a distance from 80 using the absolute value notation.

Important Features

The most significant feature of the absolute value graph is the corner point where the graph changes direction. When finding the equation for a transformed absolute value function, this point is very helpful for determining the horizontal and vertical shifts.

Example 3

Write an equation for the function graphed.

The basic absolute value function changes direction at the origin, so this graph has been shifted to the right 3 and down 2 from the basic toolkit function.

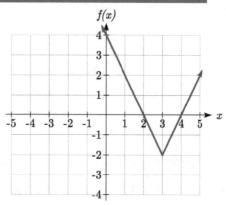

We might also notice that the graph appears stretched, since the linear portions have slopes of 2 and -2. From this information we could write the write the equation in two ways:

$f(x) = 2|x - 3| - 2$, treating the stretch as a vertical stretch

$f(x) = |2(x - 3)| - 2$, treating the stretch as a horizontal compression

Note that these equations are algebraically equivalent – the stretch for an absolute value function can be written interchangeably as a vertical or horizontal stretch/compression.

If you had not been able to determine the stretch based on the slopes of the lines, you can solve for the stretch factor by putting in a known pair of values for x and f(x)

$f(x) = a|x - 3| - 2$ Now substituting in the point (1, 2)

$2 = a|1 - 3| - 2$

$4 = 2a$

$a = 2$

2. Given the description of the transformed absolute value function write the equation. The absolute value function is horizontally shifted left 2 units, is vertically flipped, and vertically shifted up 3 units.

The graph of an absolute value function will have a vertical intercept when the input is zero. The graph may or may not have horizontal intercepts, depending on how the graph has been shifted and reflected. It is possible for the absolute value function to have zero, one, or two horizontal intercepts.

Zero horizontal intercepts   One   Two

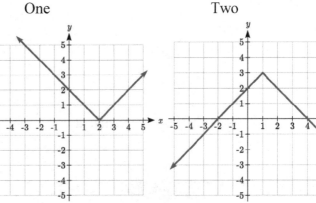

To find the horizontal intercepts, we will need to solve an equation involving an absolute value.

Notice that the absolute value function is not one-to-one, so typically inverses of absolute value functions are not discussed.

Solving Absolute Value Equations

To solve an equation like $8 = |2x - 6|$, we can notice that the absolute value will be equal to eight if the quantity *inside* the absolute value were 8 or -8. This leads to two different equations we can solve independently:

$$2x - 6 = 8 \quad \text{or} \quad 2x - 6 = -8$$
$$2x = 14 \qquad\qquad 2x = -2$$
$$x = 7 \qquad\qquad\quad x = -1$$

Solutions to Absolute Value Equations
An equation of the form $|A| = B$, with $B \geq 0$, will have solutions when $A = B$ or $A = -B$

Example 4
Find the horizontal intercepts of the graph of $f(x) = |4x + 1| - 7$ The horizontal intercepts will occur when $f(x) = 0$. Solving,

$0 = |4x+1| - 7$ Isolate the absolute value on one side of the equation

$7 = |4x+1|$ Now we can break this into two separate equations:

$7 = 4x+1$ $-7 = 4x+1$

$6 = 4x$ or $-8 = 4x$

$x = \dfrac{6}{4} = \dfrac{3}{2}$ $x = \dfrac{-8}{4} = -2$

The graph has two horizontal intercepts, at $x = \dfrac{3}{2}$ and $x = -2$

Example 5

Solve $1 = 4|x-2| + 2$

Isolating the absolute value on one side the equation,

$1 = 4|x-2| + 2$

$-1 = 4|x-2|$

$-\dfrac{1}{4} = |x-2|$

At this point, we notice that this equation has no solutions – the absolute value always returns a positive value, so it is impossible for the absolute value to equal a negative value.

Try it Now

3. Find the horizontal & vertical intercepts for the function $f(x) = -|x+2| + 3$

Solving Absolute Value Inequalities

When absolute value inequalities are written to describe a set of values, like the inequality $|x-5| \le 4$ we wrote earlier, it is sometimes desirable to express this set of values without the absolute value, either using inequalities, or using interval notation.

We will explore two approaches to solving absolute value inequalities:
1) Using the graph
2) Using test values

Example 6

Solve $|x-5| \leq 4$

With both approaches, we will need to know first where the corresponding *equality* is true. In this case, we first will find where $|x-5| = 4$. We do this because the absolute value is a nice friendly function with no breaks, so the only way the function values can switch from being less than 4 to being greater than 4 is by passing through where the values equal 4. Solve $|x-5| = 4$,

$$x - 5 = 4 \qquad\qquad x - 5 = -4$$
$$\text{or}$$
$$x = 9 \qquad\qquad\qquad x = 1$$

To use a graph, we can sketch the function $f(x) = |x-5|$. To help us see where the outputs are 4, the line $g(x) = 4$ could also be sketched.

On the graph, we can see that indeed the output values of the absolute value are equal to 4 at $x = 1$ and $x = 9$. Based on the shape of the graph, we can determine the absolute value is less than or equal to 4 between these two points, when $1 \leq x \leq 9$. In interval notation, this would be the interval [1,9].

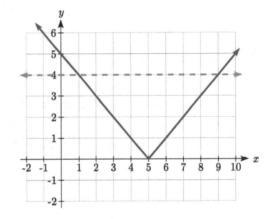

As an alternative to graphing, after determining that the absolute value is equal to 4 at $x = 1$ and $x = 9$, we know the graph can only change from being less than 4 to greater than 4 at these values. This divides the number line up into three intervals: $x<1$, $1<x<9$, and $x>9$. To determine when the function is less than 4, we could pick a value in each interval and see if the output is less than or greater than 4.

Interval	Test x	$f(x)$	<4 or >4?
$x<1$	0	$|0-5| = 5$	greater
$1<x<9$	6	$|6-5| = 1$	less
$x>9$	11	$|11-5| = 6$	greater

Since $1 \leq x \leq 9$ is the only interval in which the output at the test value is less than 4, we can conclude the solution to $|x-5| \leq 4$ is $1 \leq x \leq 9$.

Example 7

Given the function $f(x) = -\dfrac{1}{2}\left|4x - 5\right| + 3$, determine for what x values the function values are negative.

We are trying to determine where $f(x) < 0$, which is when $-\dfrac{1}{2}\left|4x - 5\right| + 3 < 0$. We begin by isolating the absolute value:

$-\dfrac{1}{2}\left|4x - 5\right| < -3$ when we multiply both sides by -2, it reverses the inequality

$\left|4x - 5\right| > 6$

Next we solve for the equality $\left|4x - 5\right| = 6$

$$4x - 5 = 6 \qquad\qquad 4x - 5 = -6$$
$$4x = 11 \qquad \text{or} \qquad 4x = -1$$
$$x = \frac{11}{4} \qquad\qquad x = \frac{-1}{4}$$

We can now either pick test values or sketch a graph of the function to determine on which intervals the original function value are negative. Notice that it is not even really important exactly what the graph looks like, as long as we know that it crosses the horizontal axis at $x = \dfrac{-1}{4}$ and $x = \dfrac{11}{4}$, and that the graph has been reflected vertically.

From the graph of the function, we can see the function values are negative to the left of the first horizontal intercept at $x = \dfrac{-1}{4}$, and negative to the right of the second intercept at $x = \dfrac{11}{4}$. This gives us the solution to the inequality:

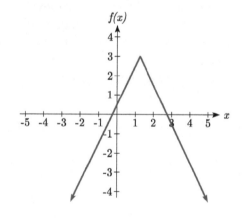

$$x < \frac{-1}{4} \quad \text{or} \quad x > \frac{11}{4}$$

In interval notation, this would be $\left(-\infty, \dfrac{-1}{4}\right) \cup \left(\dfrac{11}{4}, \infty\right)$

Try it Now

4. Solve $-2|k-4| \le -6$

Important Topics of this Section
The properties of the absolute value function
Solving absolute value equations
Finding intercepts
Solving absolute value inequalities

Try it Now Answers

1. Using the variable p, for passing, $|p-80| \le 20$

2. $f(x) = -|x+2| + 3$

3. $f(0) = 1$, so the vertical intercept is at (0,1).
   $f(x) = 0$ when
   $-|x+2| + 3 = 0$
   $|x+2| = 3$
   $x+2 = 3$ or $x+2 = -3$
   $x = 1$ or $x = -5$ so the horizontal intercepts are at (-5,0) & (1,0)

4. $-2|k-4| \le -6$
   $|k-4| \ge 3$
   Solving the equality $|k-4| = 3$, $k-4 = 3$ or $k-4 = -3$, so
   $k = 1$ or $k = 7$.
   Using a graph or test values, we can determine the
   intervals that satisfy the inequality are $k \le 1$ or $k \ge 7$; in
   interval notation this would be $(-\infty, 1] \cup [7, \infty)$

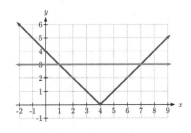

Section 2.5 Exercises

Write an equation for each transformation of $f(x) = |x|$

1.

2.

3.

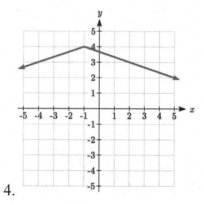

4.

Sketch a graph of each function

5. $f(x) = -|x-1| - 1$

6. $f(x) = -|x+3| + 4$

7. $f(x) = 2|x+3| + 1$

8. $f(x) = 3|x-2| - 3$

9. $f(x) = |2x-4| - 3$

10. $f(x) = |3x+9| + 2$

Solve each the equation

11. $|5x-2| = 11$

12. $|4x+2| = 15$

13. $2|4-x| = 7$

14. $3|5-x| = 5$

15. $3|x+1| - 4 = -2$

16. $5|x-4| - 7 = 2$

Find the horizontal and vertical intercepts of each function

17. $f(x) = 2|x+1| - 10$

18. $f(x) = 4|x-3| + 4$

19. $f(x) = -3|x-2| - 1$

20. $f(x) = -2|x+1| + 6$

Solve each inequality

21. $|x+5| < 6$

22. $|x-3| < 7$

23. $|x-2| \geq 3$

24. $|x+4| \geq 2$

25. $|3x+9| < 4$

26. $|2x-9| \leq 8$

Chapter 3: Polynomial and Rational Functions

Section 3.1 Power Functions & Polynomial Functions

A square is cut out of cardboard, with each side having length L. If we wanted to write a function for the area of the square, with L as the input and the area as output, you may recall that the area of a rectangle can be found by multiplying the length times the width. Since our shape is a square, the length & the width are the same, giving the formula:

$A(L) = L \cdot L = L^2$

Likewise, if we wanted a function for the volume of a cube with each side having some length L, you may recall volume of a rectangular box can be found by multiplying length by width by height, which are all equal for a cube, giving the formula:

$V(L) = L \cdot L \cdot L = L^3$

These two functions are examples of **power functions**, functions that are some power of the variable.

Power Function
A **power function** is a function that can be represented in the form
$f(x) = x^p$
Where the base is a variable and the exponent, p, is a number.

Example 1

Which of our toolkit functions are power functions?

The constant and identity functions are power functions, since they can be written as $f(x) = x^0$ and $f(x) = x^1$ respectively.

The quadratic and cubic functions are both power functions with whole number powers: $f(x) = x^2$ and $f(x) = x^3$.

The reciprocal and reciprocal squared functions are both power functions with negative whole number powers since they can be written as $f(x) = x^{-1}$ and $f(x) = x^{-2}$.

The square and cube root functions are both power functions with fractional powers since they can be written as $f(x) = x^{1/2}$ or $f(x) = x^{1/3}$.

Try it Now
1. What point(s) do the toolkit power functions have in common?

Characteristics of Power Functions

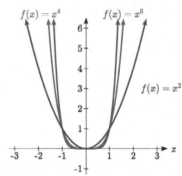

Shown to the right are the graphs of
$f(x) = x^2$, $f(x) = x^4$, and $f(x) = x^6$, all even whole number powers. Notice that all these graphs have a fairly similar shape, very similar to the quadratic toolkit, but as the power increases the graphs flatten somewhat near the origin, and become steeper away from the origin.

To describe the behavior as numbers become larger and larger, we use the idea of infinity. The symbol for positive infinity is ∞, and $-\infty$ for negative infinity. When we say that "x approaches infinity", which can be symbolically written as $x \to \infty$, we are describing a behavior – we are saying that x is getting large in the positive direction.

With the even power functions, as the x becomes large in either the positive or negative direction, the output values become very large positive numbers. Equivalently, we could describe this by saying that as x approaches positive or negative infinity, the *f(x)* values approach positive infinity. In symbolic form, we could write: as $x \to \pm\infty$, $f(x) \to \infty$.

Shown here are the graphs of
$f(x) = x^3$, $f(x) = x^5$, and $f(x) = x^7$, all odd whole number powers. Notice all these graphs look similar to the cubic toolkit, but again as the power increases the graphs flatten near the origin and become steeper away from the origin.

For these odd power functions, as x approaches negative infinity, *f(x)* approaches negative infinity. As x approaches positive infinity, *f(x)* approaches positive infinity. In symbolic form we write: as $x \to -\infty$, $f(x) \to -\infty$ and as $x \to \infty$, $f(x) \to \infty$.

Long Run Behavior
The behavior of the graph of a function as the input takes on large negative values, $x \to -\infty$, and large positive values, $x \to \infty$, is referred to as the **long run behavior** of the function.

Example 2

Describe the long run behavior of the graph of $f(x) = x^8$.

Since $f(x) = x^8$ has a whole, even power, we would expect this function to behave somewhat like the quadratic function. As the input gets large positive or negative, we would expect the output to grow without bound in the positive direction. In symbolic form, as $x \to \pm\infty$, $f(x) \to \infty$.

Example 3

Describe the long run behavior of the graph of $f(x) = -x^9$

Since this function has a whole odd power, we would expect it to behave somewhat like the cubic function. The negative in front of the x^9 will cause a vertical reflection, so as the inputs grow large positive, the outputs will grow large in the negative direction, and as the inputs grow large negative, the outputs will grow large in the positive direction. In symbolic form, for the long run behavior we would write: as $x \to \infty$, $f(x) \to -\infty$ and as $x \to -\infty$, $f(x) \to \infty$.

You may use words or symbols to describe the long run behavior of these functions.

Try it Now

2. Describe in words and symbols the long run behavior of $f(x) = -x^4$

Treatment of the rational and radical forms of power functions will be saved for later.

Polynomials

An oil pipeline bursts in the Gulf of Mexico, causing an oil slick in a roughly circular shape. The slick is currently 24 miles in radius, but that radius is increasing by 8 miles each week. If we wanted to write a formula for the area covered by the oil slick, we could do so by composing two functions together. The first is a formula for the radius, r, of the spill, which depends on the number of weeks, w, that have passed.

Hopefully you recognized that this relationship is linear:
$r(w) = 24 + 8w$

We can combine this with the formula for the area, A, of a circle:
$A(r) = \pi r^2$

Composing these functions gives a formula for the area in terms of weeks:
$A(w) = A(r(w)) = A(24 + 8w) = \pi(24 + 8w)^2$

Multiplying this out gives the formula
$A(w) = 576\pi + 384\pi w + 64\pi w^2$

This formula is an example of a **polynomial**. A polynomial is simply the sum of terms each consisting of a vertically stretched or compressed power function with non-negative whole number power.

Terminology of Polynomial Functions

A **polynomial** is function that can be written as $f(x) = a_0 + a_1 x + a_2 x^2 + \cdots + a_n x^n$

Each of the a_i constants are called **coefficients** and can be positive, negative, or zero, and be whole numbers, decimals, or fractions.

A **term** of the polynomial is any one piece of the sum, that is any $a_i x^i$. Each individual term is a transformed power function.

The **degree** of the polynomial is the highest power of the variable that occurs in the polynomial.

The **leading term** is the term containing the highest power of the variable: the term with the highest degree.

The **leading coefficient** is the coefficient of the leading term.

Because of the definition of the "leading" term we often rearrange polynomials so that the powers are descending.

$f(x) = a_n x^n + \ldots + a_2 x^2 + a_1 x + a_0$

Example 4

Identify the degree, leading term, and leading coefficient of these polynomials:

a) $f(x) = 3 + 2x^2 - 4x^3$ b) $g(t) = 5t^5 - 2t^3 + 7t$ c) $h(p) = 6p - p^3 - 2$

a) For the function *f(x)*, the degree is 3, the highest power on *x*. The leading term is the term containing that power, $-4x^3$. The leading coefficient is the coefficient of that term, -4.

b) For *g(t)*, the degree is 5, the leading term is $5t^5$, and the leading coefficient is 5.

c) For *h(p)*, the degree is 3, the leading term is $-p^3$, so the leading coefficient is -1.

Long Run Behavior of Polynomials

For any polynomial, the **long run behavior** of the polynomial will match the long run behavior of the leading term.

Example 5

What can we determine about the long run behavior and degree of the equation for the polynomial graphed here?

Since the output grows large and positive as the inputs grow large and positive, we describe the long run behavior symbolically by writing: as $x \to \infty$, $f(x) \to \infty$. Similarly, as $x \to -\infty$, $f(x) \to -\infty$.

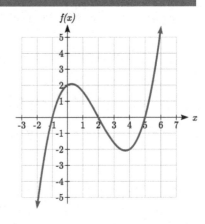

In words, we could say that as *x* values approach infinity, the function values approach infinity, and as *x* values approach negative infinity the function values approach negative infinity.

We can tell this graph has the shape of an odd degree power function which has not been reflected, so the degree of the polynomial creating this graph must be odd, and the leading coefficient would be positive.

Try it Now

3. Given the function $f(x) = 0.2(x - 2)(x + 1)(x - 5)$ use your algebra skills to write the function in standard polynomial form (as a sum of terms) and determine the leading term, degree, and long run behavior of the function.

Short Run Behavior

Characteristics of the graph such as vertical and horizontal intercepts and the places the graph changes direction are part of the short run behavior of the polynomial.

Like with all functions, the vertical intercept is where the graph crosses the vertical axis, and occurs when the input value is zero. Since a polynomial is a function, there can only be one vertical intercept, which occurs at the point $(0, a_0)$. The horizontal intercepts occur at the input values that correspond with an output value of zero. It is possible to have more than one horizontal intercept.

Horizontal intercepts are also called **zeros**, or **roots** of the function.

Example 6

Given the polynomial function $f(x) = (x-2)(x+1)(x-4)$, written in factored form for your convenience, determine the vertical and horizontal intercepts.

The vertical intercept occurs when the input is zero.
$f(0) = (0-2)(0+1)(0-4) = 8$.

The graph crosses the vertical axis at the point (0, 8).

The horizontal intercepts occur when the output is zero.
$0 = (x-2)(x+1)(x-4)$ when $x = 2$, -1, or 4.
f(x) has zeros, or roots, at $x = 2$, -1, and 4.

The graph crosses the horizontal axis at the points (2, 0), (-1, 0), and (4, 0)

Notice that the polynomial in the previous example, which would be degree three if multiplied out, had three horizontal intercepts and two turning points – places where the graph changes direction. We will now make a general statement without justifying it – the reasons will become clear later in this chapter.

Intercepts and Turning Points of Polynomials
A polynomial of degree *n* will have:
At most *n* horizontal intercepts. An odd degree polynomial will always have at least one.
At most $n-1$ turning points

Example 7

What can we conclude about the graph of the polynomial shown here?

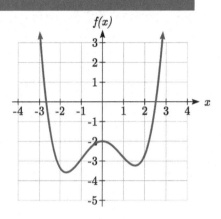

Based on the long run behavior, with the graph becoming large positive on both ends of the graph, we can determine that this is the graph of an even degree polynomial. The graph has 2 horizontal intercepts, suggesting a degree of 2 or greater, and 3 turning points, suggesting a degree of 4 or greater. Based on this, it would be reasonable to conclude that the degree is even and at least 4, so it is probably a fourth degree polynomial.

Try it Now

4. Given the function $f(x) = 0.2(x-2)(x+1)(x-5)$, determine the short run behavior.

Important Topics of this Section

Power Functions

Polynomials

Coefficients

Leading coefficient

Term

Leading Term

Degree of a polynomial

Long run behavior

Short run behavior

Try it Now Answers

1. (0, 0) and (1, 1) are common to all power functions.

2. As x approaches positive and negative infinity, $f(x)$ approaches negative infinity: as $x \to \pm\infty$, $f(x) \to -\infty$ because of the vertical flip.

3. The leading term is $0.2x^3$, so it is a degree 3 polynomial.
   As x approaches infinity (or gets very large in the positive direction) $f(x)$ approaches infinity; as x approaches negative infinity (or gets very large in the negative direction) $f(x)$ approaches negative infinity. (Basically the long run behavior is the same as the cubic function).

4. Horizontal intercepts are (2, 0) (-1, 0) and (5, 0), the vertical intercept is (0, 2) and there are 2 turns in the graph.

Section 3.1 Exercises

Find the long run behavior of each function as $x \to \infty$ and $x \to -\infty$

1. $f(x) = x^4$ 2. $f(x) = x^6$ 3. $f(x) = x^3$ 4. $f(x) = x^5$

5. $f(x) = -x^2$ 6. $f(x) = -x^4$ 7. $f(x) = -x^7$ 8. $f(x) = -x^9$

Find the degree and leading coefficient of each polynomial

9. $4x^7$ 10. $5x^6$

11. $5 - x^2$ 12. $6 + 3x - 4x^3$

13. $-2x^4 - 3x^2 + x - 1$ 14. $6x^5 - 2x^4 + x^2 + 3$

15. $(2x + 3)(x - 4)(3x + 1)$ 16. $(3x + 1)(x + 1)(4x + 3)$

Find the long run behavior of each function as $x \to \infty$ and $x \to -\infty$

17. $-2x^4 - 3x^2 + x - 1$ 18. $6x^5 - 2x^4 + x^2 + 3$

19. $3x^2 + x - 2$ 20. $-2x^3 + x^2 - x + 3$

21. What is the maximum number of x-intercepts and turning points for a polynomial of degree 5?

22. What is the maximum number of x-intercepts and turning points for a polynomial of degree 8?

What is the least possible degree of the polynomial function shown in each graph?

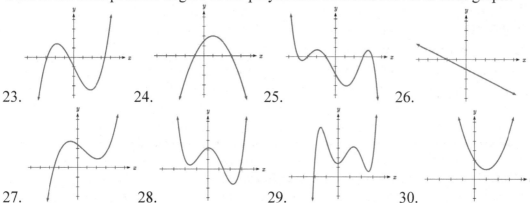

23.   24.   25.   26.

27.   28.   29.   30.

Find the vertical and horizontal intercepts of each function.

31. $f(t) = 2(t - 1)(t + 2)(t - 3)$ 32. $f(x) = 3(x + 1)(x - 4)(x + 5)$

33. $g(n) = -2(3n - 1)(2n + 1)$ 34. $k(u) = -3(4 - n)(4n + 3)$

Section 3.2 Quadratic Functions

In this section, we will explore the family of 2nd degree polynomials, the quadratic functions. While they share many characteristics of polynomials in general, the calculations involved in working with quadratics is typically a little simpler, which makes them a good place to start our exploration of short run behavior. In addition, quadratics commonly arise from problems involving area and projectile motion, providing some interesting applications.

Example 1

A backyard farmer wants to enclose a rectangular space for a new garden. She has purchased 80 feet of wire fencing to enclose 3 sides, and will put the 4th side against the backyard fence. Find a formula for the area enclosed by the fence if the sides of fencing perpendicular to the existing fence have length L.

In a scenario like this involving geometry, it is often helpful to draw a picture. It might also be helpful to introduce a temporary variable, W, to represent the side of fencing parallel to the 4th side or backyard fence.

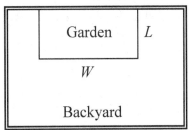

Since we know we only have 80 feet of fence available, we know that $L + W + L = 80$, or more simply, $2L + W = 80$. This allows us to represent the width, W, in terms of L: $W = 80 - 2L$

Now we are ready to write an equation for the area the fence encloses. We know the area of a rectangle is length multiplied by width, so
$A = LW = L(80 - 2L)$
$A(L) = 80L - 2L^2$
This formula represents the area of the fence in terms of the variable length L.

Short run Behavior: Vertex

We now explore the interesting features of the graphs of quadratics. In addition to intercepts, quadratics have an interesting feature where they change direction, called the **vertex**. You probably noticed that all quadratics are related to transformations of the basic quadratic function $f(x) = x^2$.

Example 2

Write an equation for the quadratic graphed below as a transformation of $f(x) = x^2$, then expand the formula and simplify terms to write the equation in standard polynomial form.

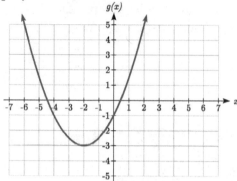

We can see the graph is the basic quadratic shifted to the left 2 and down 3, giving a formula in the form $g(x) = a(x+2)^2 - 3$. By plugging in a point that falls on the grid, such as (0,-1), we can solve for the stretch factor:

$$-1 = a(0+2)^2 - 3$$

$$2 = 4a$$

$$a = \frac{1}{2}$$

Written as a transformation, the equation for this formula is $g(x) = \frac{1}{2}(x+2)^2 - 3$. To write this in standard polynomial form, we can expand the formula and simplify terms:

$$g(x) = \frac{1}{2}(x+2)^2 - 3$$

$$g(x) = \frac{1}{2}(x+2)(x+2) - 3$$

$$g(x) = \frac{1}{2}(x^2 + 4x + 4) - 3$$

$$g(x) = \frac{1}{2}x^2 + 2x + 2 - 3$$

$$g(x) = \frac{1}{2}x^2 + 2x - 1$$

Notice that the horizontal and vertical shifts of the basic quadratic determine the location of the vertex of the parabola; the vertex is unaffected by stretches and compressions.

Try it Now

1. A coordinate grid has been superimposed over the quadratic path of a basketball[1]. Find an equation for the path of the ball. Does he make the basket?

Forms of Quadratic Functions

The **standard form** of a quadratic function is $f(x) = ax^2 + bx + c$

The **transformation form** of a quadratic function is $f(x) = a(x-h)^2 + k$

The **vertex** of the quadratic function is located at (h, k), where h and k are the numbers in the transformation form of the function. Because the vertex appears in the transformation form, it is often called the **vertex form**.

In the previous example, we saw that it is possible to rewrite a quadratic function given in transformation form and rewrite it in standard form by expanding the formula. It would be useful to reverse this process, since the transformation form reveals the vertex.

Expanding out the general transformation form of a quadratic gives:

$$f(x) = a(x-h)^2 + k = a(x-h)(x-h) + k$$
$$f(x) = a(x^2 - 2xh + h^2) + k = ax^2 - 2ahx + ah^2 + k$$

This should be equal to the standard form of the quadratic:

$$ax^2 - 2ahx + ah^2 + k = ax^2 + bx + c$$

The second degree terms are already equal. For the linear terms to be equal, the coefficients must be equal:

$$-2ah = b \text{, so } h = -\frac{b}{2a}$$

This provides us a method to determine the horizontal shift of the quadratic from the standard form. We could likewise set the constant terms equal to find:

$$ah^2 + k = c \text{, so } k = c - ah^2 = c - a\left(-\frac{b}{2a}\right)^2 = c - a\frac{b^2}{4a^2} = c - \frac{b^2}{4a}$$

---

[1] From http://blog.mrmeyer.com/?p=4778,

In practice, though, it is usually easier to remember that k is the output value of the function when the input is h, so $k = f(h)$.

Finding the Vertex of a Quadratic

For a quadratic given in standard form, the vertex (h, k) is located at:

$$h = -\frac{b}{2a}, \quad k = f(h) = f\left(\frac{-b}{2a}\right)$$

Example 3

Find the vertex of the quadratic $f(x) = 2x^2 - 6x + 7$. Rewrite the quadratic into transformation form (vertex form).

The horizontal coordinate of the vertex will be at $h = -\dfrac{b}{2a} = -\dfrac{-6}{2(2)} = \dfrac{6}{4} = \dfrac{3}{2}$

The vertical coordinate of the vertex will be at $f\left(\dfrac{3}{2}\right) = 2\left(\dfrac{3}{2}\right)^2 - 6\left(\dfrac{3}{2}\right) + 7 = \dfrac{5}{2}$

Rewriting into transformation form, the stretch factor will be the same as the a in the original quadratic. Using the vertex to determine the shifts,

$$f(x) = 2\left(x - \frac{3}{2}\right)^2 + \frac{5}{2}$$

Try it Now

2. Given the equation $g(x) = 13 + x^2 - 6x$ write the equation in standard form and then in transformation/vertex form.

As an alternative to using a formula for finding the vertex, the equation can also be written into vertex form by **completing the square**. This process is most easily explained through example. In most cases, using the formula for finding the vertex will be quicker and easier than completing the square, but completing the square is a useful technique when faced with some other algebraic problems.

Example 4

Rewrite $f(x) = 2x^2 - 12x + 14$ into vertex form by completing the square.

We start by factoring the leading coefficient from the quadratic and linear terms.
$$2\left(x^2 - 6x\right) + 14$$

Next, we are going to add something inside the parentheses so that the quadratic inside the parentheses becomes a perfect square. In other words, we are looking for values p and q so that $\left(x^2 - 6x + p\right) = (x - q)^2$.

Notice that if multiplied out on the right, the middle term would be $-2q$, so q must be half of the middle term on the left; $q = -3$. In that case, p must be $(-3)^2 = 9$.
$$\left(x^2 - 6x + 9\right) = (x - 3)^2$$

Now, we can't just add 9 into the expression – that would change the value of the expression. In fact, adding 9 inside the parentheses actually adds 18 to the expression, since the 2 outside the parentheses will distribute. To keep the expression balanced, we can subtract 18.
$$2\left(x^2 - 6x + 9\right) + 14 - 18$$

Simplifying, we are left with vertex form.
$$2(x - 3)^2 - 4$$

In addition to enabling us to more easily graph a quadratic written in standard form, finding the vertex serves another important purpose – it allows us to determine the maximum or minimum value of the function, depending on which way the graph opens.

Example 5

Returning to our backyard farmer from the beginning of the section, what dimensions should she make her garden to maximize the enclosed area?

Earlier we determined the area she could enclose with 80 feet of fencing on three sides was given by the equation $A(L) = 80L - 2L^2$. Notice that quadratic has been vertically reflected, since the coefficient on the squared term is negative, so the graph will open downwards, and the vertex will be a maximum value for the area.

In finding the vertex, we take care since the equation is not written in standard polynomial form with decreasing powers. But we know that a is the coefficient on the squared term, so $a = -2$, $b = 80$, and $c = 0$.
Finding the vertex:
$$h = -\frac{80}{2(-2)} = 20, \quad k = A(20) = 80(20) - 2(20)^2 = 800$$

The maximum value of the function is an area of 800 square feet, which occurs when L = 20 feet. When the shorter sides are 20 feet, that leaves 40 feet of fencing for the longer side. To maximize the area, she should enclose the garden so the two shorter sides have length 20 feet, and the longer side parallel to the existing fence has length 40 feet.

Example 6

A local newspaper currently has 84,000 subscribers, at a quarterly charge of $30. Market research has suggested that if they raised the price to $32, they would lose 5,000 subscribers. Assuming that subscriptions are linearly related to the price, what price should the newspaper charge for a quarterly subscription to maximize their revenue?

Revenue is the amount of money a company brings in. In this case, the revenue can be found by multiplying the charge per subscription times the number of subscribers. We can introduce variables, C for charge per subscription and S for the number subscribers, giving us the equation
Revenue $= CS$

Since the number of subscribers changes with the price, we need to find a relationship between the variables. We know that currently $S = 84,000$ and $C = 30$, and that if they raise the price to $32 they would lose 5,000 subscribers, giving a second pair of values, $C = 32$ and $S = 79,000$. From this we can find a linear equation relating the two quantities. Treating C as the input and S as the output, the equation will have form $S = mC + b$. The slope will be

$$m = \frac{79,000 - 84,000}{32 - 30} = \frac{-5,000}{2} = -2,500$$

This tells us the paper will lose 2,500 subscribers for each dollar they raise the price. We can then solve for the vertical intercept

$S = -2500C + b$ Plug in the point $S = 84,000$ and $C = 30$
$84,000 = -2500(30) + b$ Solve for b
$b = 159,000$

This gives us the linear equation $S = -2,500C + 159,000$ relating cost and subscribers. We now return to our revenue equation.

Revenue $= CS$ Substituting the equation for S from above
Revenue $= C(-2,500C + 159,000)$ Expanding
Revenue $= -2,500C^2 + 159,000C$

We now have a quadratic equation for revenue as a function of the subscription charge. To find the price that will maximize revenue for the newspaper, we can find the vertex:

$$h = -\frac{159,000}{2(-2,500)} = 31.8$$

The model tells us that the maximum revenue will occur if the newspaper charges $31.80 for a subscription. To find what the maximum revenue is, we can evaluate the revenue equation:
Maximum Revenue $= -2,500(31.8)^2 + 159,000(31.8) = \$2,528,100$

Short run Behavior: Intercepts

As with any function, we can find the vertical intercepts of a quadratic by evaluating the function at an input of zero, and we can find the horizontal intercepts by solving for when the output will be zero. Notice that depending upon the location of the graph, we might have zero, one, or two horizontal intercepts.

zero horizontal intercepts

one horizontal intercept
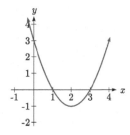
two horizontal intercepts

Example 7

Find the vertical and horizontal intercepts of the quadratic $f(x) = 3x^2 + 5x - 2$

We can find the vertical intercept by evaluating the function at an input of zero:
$f(0) = 3(0)^2 + 5(0) - 2 = -2$ Vertical intercept at (0,-2)

For the horizontal intercepts, we solve for when the output will be zero
$0 = 3x^2 + 5x - 2$

In this case, the quadratic can be factored easily, providing the simplest method for solution
$0 = (3x - 1)(x + 2)$
$0 = 3x - 1$
             $0 = x + 2$
$x = \dfrac{1}{3}$ or $x = -2$ Horizontal intercepts at $\left(\dfrac{1}{3}, 0\right)$ and (-2,0)

Notice that in the standard form of a quadratic, the constant term c reveals the vertical intercept of the graph.

Example 8

Find the horizontal intercepts of the quadratic $f(x) = 2x^2 + 4x - 4$

Again we will solve for when the output will be zero
$0 = 2x^2 + 4x - 4$

Since the quadratic is not easily factorable in this case, we solve for the intercepts by first rewriting the quadratic into transformation form.

$$h = -\frac{b}{2a} = -\frac{4}{2(2)} = -1 \qquad k = f(-1) = 2(-1)^2 + 4(-1) - 4 = -6$$

$$f(x) = 2(x+1)^2 - 6$$

Now we can solve for when the output will be zero

$$0 = 2(x+1)^2 - 6$$

$$6 = 2(x+1)^2$$

$$3 = (x+1)^2$$

$$x + 1 = \pm\sqrt{3}$$

$$x = -1 \pm \sqrt{3}$$

The graph has horizontal intercepts at $(-1-\sqrt{3}, 0)$ and $(-1+\sqrt{3}, 0)$

Try it Now

3. In Try it Now problem 2 we found the standard & transformation form for the function $g(x) = 13 + x^2 - 6x$. Now find the Vertical & Horizontal intercepts (if any).

The process in the last example is done commonly enough that sometimes people find it easier to solve the problem once in general and remember the formula for the result, rather than repeating the process each time. Based on our previous work we showed that any quadratic in standard form can be written into transformation form as:

$$f(x) = a\left(x + \frac{b}{2a}\right)^2 + c - \frac{b^2}{4a}$$

Solving for the horizontal intercepts using this general equation gives:

$$0 = a\left(x + \frac{b}{2a}\right)^2 + c - \frac{b^2}{4a} \qquad \text{start to solve for } x \text{ by moving the constants to the other side}$$

$$\frac{b^2}{4a} - c = a\left(x + \frac{b}{2a}\right)^2 \qquad \text{divide both sides by } a$$

$$\frac{b^2}{4a^2} - \frac{c}{a} = \left(x + \frac{b}{2a}\right)^2 \qquad \text{find a common denominator to combine fractions}$$

$$\frac{b^2}{4a^2} - \frac{4ac}{4a^2} = \left(x + \frac{b}{2a}\right)^2 \qquad \text{combine the fractions on the left side of the equation}$$

$$\frac{b^2 - 4ac}{4a^2} = \left(x + \frac{b}{2a}\right)^2 \qquad \text{take the square root of both sides}$$

$$\pm\sqrt{\frac{b^2 - 4ac}{4a^2}} = x + \frac{b}{2a} \qquad \text{subtract } b/2a \text{ from both sides}$$

$$-\frac{b}{2a} \pm \frac{\sqrt{b^2 - 4ac}}{2a} = x \qquad \text{combining the fractions}$$

$$x = \frac{-b \pm \sqrt{b^2 - 4ac}}{2a} \qquad \text{Notice that this can yield two different answers for } x$$

Quadratic Formula

For a quadratic function given in standard form $f(x) = ax^2 + bx + c$, the **quadratic formula** gives the horizontal intercepts of the graph of this function.

$$x = \frac{-b \pm \sqrt{b^2 - 4ac}}{2a}$$

Example 9

A ball is thrown upwards from the top of a 40-foot-tall building at a speed of 80 feet per second. The ball's height above ground can be modeled by the equation
$H(t) = -16t^2 + 80t + 40$.
What is the maximum height of the ball?
When does the ball hit the ground?

To find the maximum height of the ball, we would need to know the vertex of the quadratic.

$$h = -\frac{80}{2(-16)} = \frac{80}{32} = \frac{5}{2}, \quad k = H\left(\frac{5}{2}\right) = -16\left(\frac{5}{2}\right)^2 + 80\left(\frac{5}{2}\right) + 40 = 140$$

The ball reaches a maximum height of 140 feet after 2.5 seconds.
To find when the ball hits the ground, we need to determine when the height is zero – when $H(t) = 0$. While we could do this using the transformation form of the quadratic, we can also use the quadratic formula:

$$t = \frac{-80 \pm \sqrt{80^2 - 4(-16)(40)}}{2(-16)} = \frac{-80 \pm \sqrt{8960}}{-32}$$

Since the square root does not simplify nicely, we can use a calculator to approximate the values of the solutions:

$$t = \frac{-80 - \sqrt{8960}}{-32} \approx 5.458 \quad \text{or} \quad t = \frac{-80 + \sqrt{8960}}{-32} \approx -0.458$$

The second answer is outside the reasonable domain of our model, so we conclude the ball will hit the ground after about 5.458 seconds.

Try it Now

4. For these two equations determine if the vertex will be a maximum value or a minimum value.

   a. $g(x) = -8x + x^2 + 7$

   b. $g(x) = -3(3 - x)^2 + 2$

Important Topics of this Section

Quadratic functions
    Standard form
    Transformation form/Vertex form
    Vertex as a maximum / Vertex as a minimum
Short run behavior
    Vertex / Horizontal & Vertical intercepts
Quadratic formula

Try it Now Answers

1. The path passes through the origin with vertex at (-4, 7).

   $h(x) = -\frac{7}{16}(x + 4)^2 + 7$. To make the shot, $h(-7.5)$ would

   need to be about 4. $h(-7.5) \approx 1.64$; he doesn't make it.

2. $g(x) = x^2 - 6x + 13$ in Standard form;

   Finding the vertex, $h = \frac{-(-6)}{2(1)} = 3$. $k = g(3) = 3^2 - 6(3) + 13 = 4$.

   $g(x) = (x - 3)^2 + 4$ in Transformation form

3. Vertical intercept at (0, 13), No horizontal intercepts since the vertex is above the x-axis and the graph opens upwards.

4. a. Vertex is a minimum value, since $a > 0$ and the graph opens upwards
   b. Vertex is a maximum value, since $a < 0$ and the graph opens downwards

Section 3.2 Exercises

Write an equation for the quadratic function graphed.

1.

2.

3.

4.

5.

6.

For each of the follow quadratic functions, find a) the vertex, b) the vertical intercept, and c) the horizontal intercepts.

7. $y(x) = 2x^2 + 10x + 12$

8. $z(p) = 3x^2 + 6x - 9$

9. $f(x) = 2x^2 - 10x + 4$

10. $g(x) = -2x^2 - 14x + 12$

11. $h(t) = -4t^2 + 6t - 1$

12. $k(t) = 2x^2 + 4x - 15$

Rewrite the quadratic function into vertex form.

13. $f(x) = x^2 - 12x + 32$

14. $g(x) = x^2 + 2x - 3$

15. $h(x) = 2x^2 + 8x - 10$

16. $k(x) = 3x^2 - 6x - 9$

17. Find the values of b and c so $f(x) = -8x^2 + bx + c$ has vertex $(2, -7)$

18. Find the values of b and c so $f(x) = 6x^2 + bx + c$ has vertex $(7, -9)$

Write an equation for a quadratic with the given features

19. x-intercepts (-3, 0) and (1, 0), and y intercept (0, 2)
20. x-intercepts (2, 0) and (-5, 0), and y intercept (0, 3)
21. x-intercepts (2, 0) and (5, 0), and y intercept (0, 6)
22. x-intercepts (1, 0) and (3, 0), and y intercept (0, 4)
23. Vertex at (4, 0), and y intercept (0, -4)
24. Vertex at (5, 6), and y intercept (0, -1)
25. Vertex at (-3, 2), and passing through (3, -2)
26. Vertex at (1, -3), and passing through (-2, 3)

27. A rocket is launched in the air. Its height, in meters above sea level, as a function of time, in seconds, is given by $h(t) = -4.9t^2 + 229t + 234$.

    a. From what height was the rocket launched?
    b. How high above sea level does the rocket reach its peak?
    c. Assuming the rocket will splash down in the ocean, at what time does splashdown occur?

28. A ball is thrown in the air from the top of a building. Its height, in meters above ground, as a function of time, in seconds, is given by $h(t) = -4.9t^2 + 24t + 8$.

    a. From what height was the ball thrown?
    b. How high above ground does the ball reach its peak?
    c. When does the ball hit the ground?

29. The height of a ball thrown in the air is given by $h(x) = -\dfrac{1}{12}x^2 + 6x + 3$, where x is the horizontal distance in feet from the point at which the ball is thrown.

    a. How high is the ball when it was thrown?
    b. What is the maximum height of the ball?
    c. How far from the thrower does the ball strike the ground?

30. A javelin is thrown in the air. Its height is given by $h(x) = -\dfrac{1}{20}x^2 + 8x + 6$, where x is the horizontal distance in feet from the point at which the javelin is thrown.

    a. How high is the javelin when it was thrown?
    b. What is the maximum height of the javelin?
    c. How far from the thrower does the javelin strike the ground?

31. A box with a square base and no top is to be made from a square piece of cardboard by cutting 6 in. squares out of each corner and folding up the sides. The box needs to hold 1000 in^3. How big a piece of cardboard is needed?

32. A box with a square base and no top is to be made from a square piece of cardboard by cutting 4 in. squares out of each corner and folding up the sides. The box needs to hold 2700 in^3. How big a piece of cardboard is needed?

33. A farmer wishes to enclose two pens with fencing, as shown. If the farmer has 500 feet of fencing to work with, what dimensions will maximize the area enclosed?

34. A farmer wishes to enclose three pens with fencing, as shown. If the farmer has 700 feet of fencing to work with, what dimensions will maximize the area enclosed?

35. You have a wire that is 56 cm long. You wish to cut it into two pieces. One piece will be bent into the shape of a square. The other piece will be bent into the shape of a circle. Let A represent the total area enclosed by the square and the circle. What is the circumference of the circle when A is a minimum?

36. You have a wire that is 71 cm long. You wish to cut it into two pieces. One piece will be bent into the shape of a right triangle with legs of equal length. The other piece will be bent into the shape of a circle. Let A represent the total area enclosed by the triangle and the circle. What is the circumference of the circle when A is a minimum?

37. A soccer stadium holds 62,000 spectators. With a ticket price of $11, the average attendance has been 26,000. When the price dropped to $9, the average attendance rose to 31,000. Assuming that attendance is linearly related to ticket price, what ticket price would maximize revenue?

38. A farmer finds that if she plants 75 trees per acre, each tree will yield 20 bushels of fruit. She estimates that for each additional tree planted per acre, the yield of each tree will decrease by 3 bushels. How many trees should she plant per acre to maximize her harvest?

39. A hot air balloon takes off from the edge of a mountain lake. Impose a coordinate system as pictured and assume that the path of the balloon follows the graph of

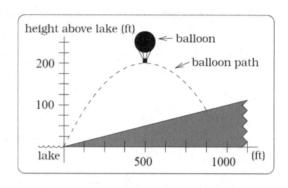

$f(x) = -\dfrac{2}{2500}x^2 + \dfrac{4}{5}x$. The land rises at a constant incline from the lake at the rate of 2 vertical feet for each 20 horizontal feet. [UW]

   a.   What is the maximum height of the balloon above water level?
   b.   What is the maximum height of the balloon above ground level?
   c.   Where does the balloon land on the ground?
   d.   Where is the balloon 50 feet above the ground?

40. A hot air balloon takes off from the edge of a plateau. Impose a coordinate system as pictured below and assume that the path the balloon follows is the graph of the quadratic function

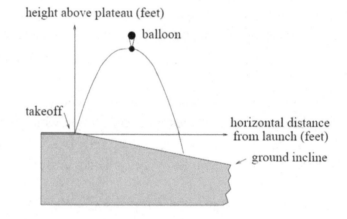

$f(x) = -\dfrac{4}{2500}x^2 + \dfrac{4}{5}x$. The land drops at a constant incline from the plateau at the rate of 1 vertical foot for each 5 horizontal feet. [UW]

   a.   What is the maximum height of the balloon above plateau level?
   b.   What is the maximum height of the balloon above ground level?
   c.   Where does the balloon land on the ground?
   d.   Where is the balloon 50 feet above the ground?

Section 3.3 Graphs of Polynomial Functions

In the previous section, we explored the short run behavior of quadratics, a special case of polynomials. In this section, we will explore the short run behavior of polynomials in general.

Short run Behavior: Intercepts

As with any function, the vertical intercept can be found by evaluating the function at an input of zero. Since this is evaluation, it is relatively easy to do it for a polynomial of any degree.

To find horizontal intercepts, we need to solve for when the output will be zero. For general polynomials, this can be a challenging prospect. While quadratics can be solved using the relatively simple quadratic formula, the corresponding formulas for cubic and 4th degree polynomials are not simple enough to remember, and formulas do not exist for general higher-degree polynomials. Consequently, we will limit ourselves to three cases:

1) The polynomial can be factored using known methods: greatest common factor and trinomial factoring.
2) The polynomial is given in factored form.
3) Technology is used to determine the intercepts.

Other techniques for finding the intercepts of general polynomials will be explored in the next section.

Example 1

Find the horizontal intercepts of $f(x) = x^6 - 3x^4 + 2x^2$.

We can attempt to factor this polynomial to find solutions for $f(x) = 0$.

$x^6 - 3x^4 + 2x^2 = 0$ Factoring out the greatest common factor

$x^2(x^4 - 3x^2 + 2) = 0$ Factoring the inside as a quadratic in x^2

$x^2(x^2 - 1)(x^2 - 2) = 0$ Then break apart to find solutions

$$(x^2 - 1) = 0 \qquad\qquad (x^2 - 2) = 0$$

$x^2 = 0$

        or $x^2 = 1$ or $x^2 = 2$

$x = 0$

             $x = \pm 1$ $x = \pm\sqrt{2}$

This gives us 5 horizontal intercepts.

Example 2

Find the vertical and horizontal intercepts of $g(t) = (t-2)^2(2t+3)$

The vertical intercept can be found by evaluating $g(0)$.
$g(0) = (0-2)^2(2(0)+3) = 12$

The horizontal intercepts can be found by solving $g(t) = 0$
$(t-2)^2(2t+3) = 0$ Since this is already factored, we can break it apart:
$(t-2)^2 = 0$ $(2t+3) = 0$
$t-2 = 0$ or $t = \dfrac{-3}{2}$
$t = 2$

We can always check our answers are reasonable by graphing the polynomial.

Example 3

Find the horizontal intercepts of $h(t) = t^3 + 4t^2 + t - 6$

Since this polynomial is not in factored form, has no common factors, and does not appear to be factorable using techniques we know, we can turn to technology to find the intercepts.

Graphing this function, it appears there are horizontal intercepts at $t = -3, -2$, and 1.

We could check these are correct by plugging in these values for t and verifying that $h(-3) = h(-2) = h(1) = 0$.

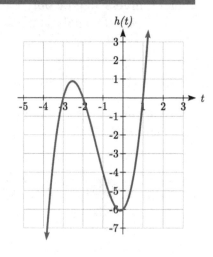

Try it Now

1. Find the vertical and horizontal intercepts of the function $f(t) = t^4 - 4t^2$.

Graphical Behavior at Intercepts

If we graph the function $f(x) = (x+3)(x-2)^2(x+1)^3$, notice that the behavior at each of the horizontal intercepts is different.

At the horizontal intercept $x = -3$, coming from the $(x+3)$ factor of the polynomial, the graph passes directly through the horizontal intercept.

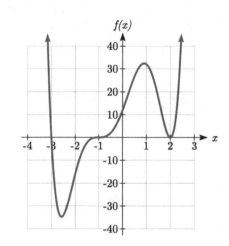

The factor $(x + 3)$ is linear (has a power of 1), so the behavior near the intercept is like that of a line - it passes directly through the intercept. We call this a single zero, since the zero corresponds to a single factor of the function.

At the horizontal intercept $x = 2$, coming from the $(x - 2)^2$ factor of the polynomial, the graph touches the axis at the intercept and changes direction. The factor is quadratic (degree 2), so the behavior near the intercept is like that of a quadratic – it bounces off the horizontal axis at the intercept. Since $(x - 2)^2 = (x - 2)(x - 2)$, the factor is repeated twice, so we call this a double zero. We could also say the zero has **multiplicity** 2.

At the horizontal intercept $x = -1$, coming from the $(x + 1)^3$ factor of the polynomial, the graph passes through the axis at the intercept, but flattens out a bit first. This factor is cubic (degree 3), so the behavior near the intercept is like that of a cubic, with the same "S" type shape near the intercept that the toolkit x^3 has. We call this a triple zero. We could also say the zero has multiplicity 3.

By utilizing these behaviors, we can sketch a reasonable graph of a factored polynomial function without needing technology.

Graphical Behavior of Polynomials at Horizontal Intercepts

If a polynomial contains a factor of the form $(x - h)^p$, the behavior near the horizontal intercept h is determined by the power on the factor.

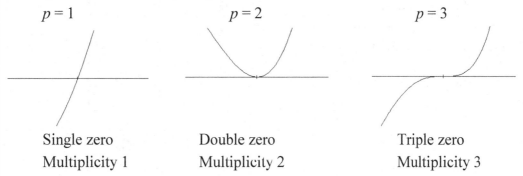

$p = 1$	$p = 2$	$p = 3$
Single zero	Double zero	Triple zero
Multiplicity 1	Multiplicity 2	Multiplicity 3

For higher even powers 4,6,8 etc…. the graph will still bounce off the horizontal axis but the graph will appear flatter with each increasing even power as it approaches and leaves the axis.

For higher odd powers, 5,7,9 etc… the graph will still pass through the horizontal axis but the graph will appear flatter with each increasing odd power as it approaches and leaves the axis.

Example 4

Sketch a graph of $f(x) = -2(x+3)^2(x-5)$.

This graph has two horizontal intercepts. At $x = -3$, the factor is squared, indicating the graph will bounce at this horizontal intercept. At $x = 5$, the factor is not squared, indicating the graph will pass through the axis at this intercept.

Additionally, we can see the leading term, if this polynomial were multiplied out, would be $-2x^3$, so the long-run behavior is that of a vertically reflected cubic, with the outputs decreasing as the inputs get large positive, and the inputs increasing as the inputs get large negative.

To sketch this we consider the following:
As $x \to -\infty$ the function $f(x) \to \infty$ so we know the graph starts in the 2nd quadrant and is decreasing toward the horizontal axis.

At (-3, 0) the graph bounces off the horizontal axis and so the function must start increasing.

At (0, 90) the graph crosses the vertical axis at the vertical intercept.

Somewhere after this point, the graph must turn back down or start decreasing toward the horizontal axis since the graph passes through the next intercept at (5,0).

As $x \to \infty$ the function $f(x) \to -\infty$ so we know the graph continues to decrease and we can stop drawing the graph in the 4th quadrant.

Using technology we can verify the shape of the graph.

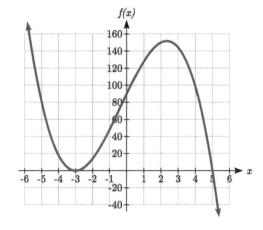

Try it Now

2. Given the function $g(x) = x^3 - x^2 - 6x$ use the methods that we have learned so far to find the vertical & horizontal intercepts, determine where the function is negative and positive, describe the long run behavior and sketch the graph without technology.

Solving Polynomial Inequalities

One application of our ability to find intercepts and sketch a graph of polynomials is the ability to solve polynomial inequalities. It is a very common question to ask when a function will be positive and negative. We can solve polynomial inequalities by either utilizing the graph, or by using test values.

Example 5

Solve $(x+3)(x+1)^2(x-4) > 0$

As with all inequalities, we start by solving the equality $(x+3)(x+1)^2(x-4) = 0$, which has solutions at $x = $ -3, -1, and 4. We know the function can only change from positive to negative at these values, so these divide the inputs into 4 intervals.

We could choose a test value in each interval and evaluate the function $f(x) = (x+3)(x+1)^2(x-4)$ at each test value to determine if the function is positive or negative in that interval

Interval	Test x in interval	$f($ test value$)$	>0 or <0?
$x < -3$	-4	72	> 0
$-3 < x < -1$	-2	-6	< 0
$-1 < x < 4$	0	-12	< 0
$x > 4$	5	288	> 0

On a number line this would look like:

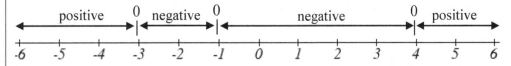

From our test values, we can determine this function is positive when $x < $ -3 or $x > 4$, or in interval notation, $(-\infty, -3) \cup (4, \infty)$

We could have also determined on which intervals the function was positive by sketching a graph of the function. We illustrate that technique in the next example

Example 6

Find the domain of the function $v(t) = \sqrt{6 - 5t - t^2}$.

A square root is only defined when the quantity we are taking the square root of, the quantity inside the square root, is zero or greater. Thus, the domain of this function will be when $6 - 5t - t^2 \geq 0$.

We start by solving the equality $6 - 5t - t^2 = 0$. While we could use the quadratic formula, this equation factors nicely to $(6 + t)(1 - t) = 0$, giving horizontal intercepts $t = 1$ and $t = -6$.

Sketching a graph of this quadratic will allow us to determine when it is positive.

From the graph we can see this function is positive for inputs between the intercepts. So $6 - 5t - t^2 \geq 0$ for $-6 \leq t \leq 1$, and this will be the domain of the $v(t)$ function.

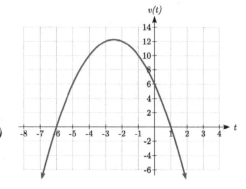

Writing Equations using Intercepts

Since a polynomial function written in factored form will have a horizontal intercept where each factor is equal to zero, we can form a function that will pass through a set of horizontal intercepts by introducing a corresponding set of factors.

Factored Form of Polynomials

If a polynomial has horizontal intercepts at $x = x_1, x_2, \ldots, x_n$, then the polynomial can be written in the factored form

$$f(x) = a(x - x_1)^{p_1}(x - x_2)^{p_2} \cdots (x - x_n)^{p_n}$$

where the powers p_i on each factor can be determined by the behavior of the graph at the corresponding intercept, and the stretch factor a can be determined given a value of the function other than the horizontal intercept.

Example 7

Write a formula for the polynomial function graphed here.

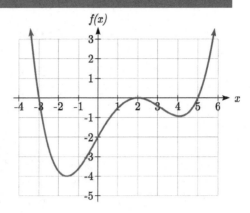

This graph has three horizontal intercepts: $x = -3$, 2, and 5. At $x = -3$ and 5 the graph passes through the axis, suggesting the corresponding factors of the polynomial will be linear. At $x = 2$ the graph bounces at the intercept, suggesting the corresponding factor of the polynomial will be 2nd degree (quadratic).

Together, this gives us:

$$f(x) = a(x+3)(x-2)^2(x-5)$$

To determine the stretch factor, we can utilize another point on the graph. Here, the vertical intercept appears to be (0,-2), so we can plug in those values to solve for a:

$$-2 = a(0+3)(0-2)^2(0-5)$$
$$-2 = -60a$$
$$a = \frac{1}{30}$$

The graphed polynomial appears to represent the function

$$f(x) = \frac{1}{30}(x+3)(x-2)^2(x-5).$$

Try it Now

3. Given the graph, write a formula for the function shown.

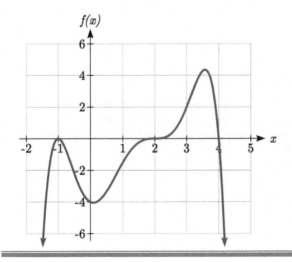

Estimating Extrema

With quadratics, we were able to algebraically find the maximum or minimum value of the function by finding the vertex. For general polynomials, finding these turning points is not possible without more advanced techniques from calculus. Even then, finding where extrema occur can still be algebraically challenging. For now, we will estimate the locations of turning points using technology to generate a graph.

Example 8

An open-top box is to be constructed by cutting out squares from each corner of a 14cm by 20cm sheet of plastic then folding up the sides. Find the size of squares that should be cut out to maximize the volume enclosed by the box.

We will start this problem by drawing a picture, labeling the width of the cut-out squares with a variable, w.

Notice that after a square is cut out from each end, it leaves a $(14 - 2w)$ cm by $(120 - 2w)$ cm rectangle for the base of the box, and the box will be w cm tall. This gives the volume:
$$V(w) = (14 - 2w)(20 - 2w)w = 280w - 68w^2 + 4w^3$$

Using technology to sketch a graph allows us to estimate the maximum value for the volume, restricted to reasonable values for w: values from 0 to 7.

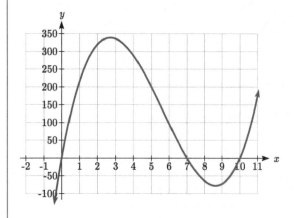

From this graph, we can estimate the maximum value is around 340, and occurs when the squares are about 2.75cm square. To improve this estimate, we could use advanced features of our technology, if available, or simply change our window to zoom in on our graph.

From this zoomed-in view, we can refine our estimate for the max volume to about 339, when the squares are 2.7cm square.

Try it Now

4. Use technology to find the maximum and minimum values on the interval [-1, 4] of the function $f(x) = -0.2(x-2)^3 (x+1)^2 (x-4)$.

Important Topics of this Section
Short Run Behavior
Intercepts (Horizontal & Vertical)
Methods to find Horizontal intercepts
Factoring Methods
Factored Forms
Technology
Graphical Behavior at intercepts
Single, Double and Triple zeros (or multiplicity 1, 2, and 3 behaviors)
Solving polynomial inequalities using test values & graphing techniques
Writing equations using intercepts
Estimating extrema

Try it Now Answers

1. Vertical intercept (0, 0). $0 = t^4 - 4t^2$ factors as $0 = t^2(t^2 - 4) = t^2(t-2)(t+2)$

   Horizontal intercepts (0, 0), (-2, 0), (2, 0)

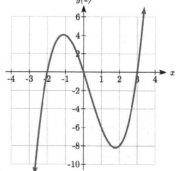

2. Vertical intercept (0, 0),

   Horizontal intercepts (-2, 0), (0, 0), (3, 0)

   The function is negative on $(-\infty, -2)$ and (0, 3)

   The function is positive on (-2, 0) and $(3, \infty)$

   The leading term is x^3 so as $x \to -\infty$, $g(x) \to -\infty$ and as

   $x \to \infty$, $g(x) \to \infty$

3. Double zero at $x=-1$, triple zero at $x=2$. Single zero at $x=4$.

   $f(x) = a(x-2)^3(x+1)^2(x-4)$. Substituting (0,-4) and solving for a,

   $f(x) = -\dfrac{1}{8}(x-2)^3(x+1)^2(x-4)$

4. The minimum occurs at approximately the point (0, -6.5), and the maximum occurs at approximately the point (3.5, 7).

Section 3.3 Exercises

Find the C and t intercepts of each function.

1. $C(t) = 2(t-4)(t+1)(t-6)$

2. $C(t) = 3(t+2)(t-3)(t+5)$

3. $C(t) = 4t(t-2)^2(t+1)$

4. $C(t) = 2t(t-3)(t+1)^2$

5. $C(t) = 2t^4 - 8t^3 + 6t^2$

6. $C(t) = 4t^4 + 12t^3 - 40t^2$

Use your calculator or other graphing technology to solve graphically for the zeros of the function.

7. $f(x) = x^3 - 7x^2 + 4x + 30$

8. $g(x) = x^3 - 6x^2 + x + 28$

Find the long run behavior of each function as $t \to \infty$ and $t \to -\infty$

9. $h(t) = 3(t-5)^3(t-3)^3(t-2)$

10. $k(t) = 2(t-3)^2(t+1)^3(t+2)$

11. $p(t) = -2t(t-1)(3-t)^2$

12. $q(t) = -4t(2-t)(t+1)^3$

Sketch a graph of each equation.

13. $f(x) = (x+3)^2(x-2)$

14. $g(x) = (x+4)(x-1)^2$

15. $h(x) = (x-1)^3(x+3)^2$

16. $k(x) = (x-3)^3(x-2)^2$

17. $m(x) = -2x(x-1)(x+3)$

18. $n(x) = -3x(x+2)(x-4)$

Solve each inequality.

19. $(x-3)(x-2)^2 > 0$

20. $(x-5)(x+1)^2 > 0$

21. $(x-1)(x+2)(x-3) < 0$

22. $(x-4)(x+3)(x+6) < 0$

Find the domain of each function.

23. $f(x) = \sqrt{-42 + 19x - 2x^2}$

24. $g(x) = \sqrt{28 - 17x - 3x^2}$

25. $h(x) = \sqrt{4 - 5x + x^2}$

26. $k(x) = \sqrt{2 + 7x + 3x^2}$

27. $n(x) = \sqrt{(x-3)(x+2)^2}$

28. $m(x) = \sqrt{(x-1)^2(x+3)}$

29. $p(t) = \dfrac{1}{t^2 + 2t - 8}$

30. $q(t) = \dfrac{4}{x^2 - 4x - 5}$

Write an equation for a polynomial the given features.

31. Degree 3. Zeros at $x = -2$, $x = 1$, and $x = 3$. Vertical intercept at $(0, -4)$

32. Degree 3. Zeros at $x = -5$, $x = -2$, and $x = 1$. Vertical intercept at $(0, 6)$

33. Degree 5. Roots of multiplicity 2 at $x = 3$ and $x = 1$, and a root of multiplicity 1 at $x = -3$. Vertical intercept at $(0, 9)$

34. Degree 4. Root of multiplicity 2 at $x = 4$, and a roots of multiplicity 1 at $x = 1$ and $x = -2$. Vertical intercept at $(0, -3)$

35. Degree 5. Double zero at $x = 1$, and triple zero at $x = 3$. Passes through the point $(2, 15)$

36. Degree 5. Single zero at $x = -2$ and $x = 3$, and triple zero at $x = 1$. Passes through the point $(2, 4)$

Write a formula for each polynomial function graphed.

37.   38.   39.

40.   41.   42.

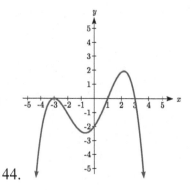

43.   44.

Write a formula for each polynomial function graphed.

45.

46.

47.

48.

49.

50.

51. A rectangle is inscribed with its base on the x axis and its upper corners on the parabola $y = 5 - x^2$. What are the dimensions of such a rectangle that has the greatest possible area?

52. A rectangle is inscribed with its base on the x axis and its upper corners on the curve $y = 16 - x^4$. What are the dimensions of such a rectangle that has the greatest possible area?

Section 3.4 Factor Theorem and Remainder Theorem

In the last section, we limited ourselves to finding the intercepts, or zeros, of polynomials that factored simply, or we turned to technology. In this section, we will look at algebraic techniques for finding the zeros of polynomials like $h(t) = t^3 + 4t^2 + t - 6$.

Long Division

In the last section we saw that we could write a polynomial as a product of factors, each corresponding to a horizontal intercept. If we knew that $x = 2$ was an intercept of the polynomial $x^3 + 4x^2 - 5x - 14$, we might guess that the polynomial could be factored as $x^3 + 4x^2 - 5x - 14 = (x - 2)(\text{something})$.   To find that "something," we can use polynomial division.

Example 1

Divide $x^3 + 4x^2 - 5x - 14$ by $x - 2$

Start by writing the problem out in long division form

$$x - 2 \overline{)x^3 + 4x^2 - 5x - 14}$$

Now we divide the leading terms: $x^3 \div x = x^2$. It is best to align it above the same-powered term in the dividend. Now, multiply that x^2 by $x - 2$ and write the result below the dividend.

$$
\begin{array}{r}
x^2 \\
x - 2 \overline{)x^3 + 4x^2 - 5x - 14} \\
x^3 - 2x^2
\end{array}
$$
Now subtract that expression from the dividend.

$$
\begin{array}{r}
x^2 \\
x - 2 \overline{) x^3 + 4x^2 - 5x - 14} \\
-\left(x^3 - 2x^2\right) \\
\hline
6x^2 - 5x - 14
\end{array}
$$

Again, divide the leading term of the remainder by the leading term of the divisor. $6x^2 \div x = 6x$. We add this to the result, multiply $6x$ by $x - 2$, and subtract.

$$
\begin{array}{r}
x^2 + 6x \\
x - 2 \overline{)\ x^3 + 4x^2 - 5x - 14} \\
\underline{-\left(x^3 - 2x^2\right)} \\
6x^2 - 5x - 14 \\
\underline{-\left(6x^2 - 12x\right)} \\
7x - 14
\end{array}
$$

Repeat the process one last time.

$$
\begin{array}{r}
x^2 + 6x + 7 \\
x - 2 \overline{)\ x^3 + 4x^2 - 5x - 14} \\
\underline{-\left(x^3 - 2x^2\right)} \\
6x^2 - 5x - 14 \\
\underline{-\left(6x^2 - 12x\right)} \\
7x - 14 \\
\underline{-\left(7x - 14\right)} \\
0
\end{array}
$$

This tells us $x^3 + 4x^2 - 5x - 14$ divided by $x - 2$ is $x^2 + 6x + 7$, with a remainder of zero. This also means that we can factor $x^3 + 4x^2 - 5x - 14$ as $(x - 2)(x^2 + 6x + 7)$.

This gives us a way to find the intercepts of this polynomial.

Example 2

Find the horizontal intercepts of $h(x) = x^3 + 4x^2 - 5x - 14$.

To find the horizontal intercepts, we need to solve h(x) = 0. From the previous example, we know the function can be factored as $h(x) = (x - 2)(x^2 + 6x + 7)$.

$h(x) = (x - 2)(x^2 + 6x + 7) = 0$ when $x = 2$ or when $x^2 + 6x + 7 = 0$. This doesn't factor nicely, but we could use the quadratic formula to find the remaining two zeros.

$$x = \frac{-6 \pm \sqrt{6^2 - 4(1)(7)}}{2(1)} = -3 \pm \sqrt{2}.$$

The horizontal intercepts will be at $(2,0)$, $\left(-3 - \sqrt{2}, 0\right)$, and $\left(-3 + \sqrt{2}, 0\right)$.

1. Divide $2x^3 - 7x + 3$ by $x + 3$ using long division.

The Factor and Remainder Theorems

When we divide a polynomial, $p(x)$ by some divisor polynomial $d(x)$, we will get a quotient polynomial $q(x)$ and possibly a remainder $r(x)$. In other words,
$p(x) = d(x)q(x) + r(x)$.

Because of the division, the remainder will either be zero, or a polynomial of lower degree than $d(x)$. Because of this, if we divide a polynomial by a term of the form $x - c$, then the remainder will be zero or a constant.

If $p(x) = (x - c)q(x) + r$, then $p(c) = (c - c)q(c) + r = 0 + r = r$, which establishes the Remainder Theorem.

The Remainder Theorem

If $p(x)$ is a polynomial of degree 1 or greater and c is a real number, then when $p(x)$ is divided by $x - c$, the remainder is $p(c)$.

If $x - c$ is a factor of the polynomial p, then $p(x) = (x - c)q(x)$ for some polynomial q. Then $p(c) = (c - c)q(c) = 0$, showing c is a zero of the polynomial. This shouldn't surprise us - we already knew that if the polynomial factors it reveals the roots.

If $p(c) = 0$, then the remainder theorem tells us that if p is divided by $x - c$, then the remainder will be zero, which means $x - c$ is a factor of p.

The Factor Theorem

If $p(x)$ is a nonzero polynomial, then the real number c is a zero of $p(x)$ if and only if $x - c$ is a factor of $p(x)$.

Synthetic Division

Since dividing by $x - c$ is a way to check if a number is a zero of the polynomial, it would be nice to have a faster way to divide by $x - c$ than having to use long division every time. Happily, quicker ways have been discovered.

Let's look back at the long division we did in Example 1 and try to streamline it. First, let's change all the subtractions into additions by distributing through the negatives.

$$
\begin{array}{r}
x^2 + 6x + 7 \\
x-2{\overline{\smash{\big)}\,x^3 + 4x^2 - 5x - 14}} \\
\underline{-x^3 + 2x^2 } \\
6x^2 - 5x - 14 \\
\underline{-6x^2 + 12x } \\
7x - 14 \\
\underline{-7x + 14} \\
0
\end{array}
$$

Next, observe that the terms $-x^3$, $-6x^2$, and $-7x$ are the exact opposite of the terms above them. The algorithm we use ensures this is always the case, so we can omit them without losing any information. Also note that the terms we 'bring down' (namely the $-5x$ and -14) aren't really necessary to recopy, so we omit them, too.

$$
\begin{array}{r}
x^2 + 6x + 7 \\
x-2{\overline{\smash{\big)}\,x^3 + 4x^2 - 5x - 14}} \\
\underline{2x^2 } \\
6x^2 \\
\underline{12x } \\
7x \\
\underline{14} \\
0
\end{array}
$$

Now, let's move things up a bit and, for reasons which will become clear in a moment, copy the x^3 into the last row.

$$
\begin{array}{r}
x^2 + 6x + 7 \\
x-2{\overline{\smash{\big)}\,x^3 + 4x^2 - 5x - 14}} \\
\underline{2x^2 \; 12x \; 14} \\
x^3 \; 6x^2 \; 7x \; 0
\end{array}
$$

Note that by arranging things in this manner, each term in the last row is obtained by adding the two terms above it. Notice also that the quotient polynomial can be obtained by dividing each of the first three terms in the last row by x and adding the results. If you take the time to work back through the original division problem, you will find that this is exactly the way we determined the quotient polynomial.

This means that we no longer need to write the quotient polynomial down, nor the x in the divisor, to determine our answer.

$$x - 2 \overline{)\ x^3 + 4x^2 - 5x - 14}$$

$$\underline{\hspace{1.2cm} 2x^2 \quad 12x \quad 14}$$

$$x^3 \quad 6x^2 \quad 7x \quad 0$$

We've streamlined things quite a bit so far, but we can still do more. Let's take a moment to remind ourselves where the $2x^2$, $12x$ and 14 came from in the second row. Each of these terms was obtained by multiplying the terms in the quotient, x^2, $6x$ and 7, respectively, by the -2 in $x - 2$, then by -1 when we changed the subtraction to addition. Multiplying by -2 then by -1 is the same as multiplying by 2, so we replace the -2 in the divisor by 2. Furthermore, the coefficients of the quotient polynomial match the coefficients of the first three terms in the last row, so we now take the plunge and write only the coefficients of the terms to get

$$
\begin{array}{r|rrrr}
2 & 1 & 4 & -5 & -14 \\
  &   & 2 & 12 & 14 \\
\hline
  & 1 & 6 & 7 & 0
\end{array}
$$

We have constructed a **synthetic division** tableau for this polynomial division problem. Let's re-work our division problem using this tableau to see how it greatly streamlines the division process. To divide $x^3 + 4x^2 - 5x - 14$ by $x - 2$, we write 2 in the place of the divisor and the coefficients of $x^3 + 4x^2 - 5x - 14$ in for the dividend. Then "bring down" the first coefficient of the dividend.

$$
\begin{array}{r|rrrr}
2 & 1 & 4 & -5 & -14 \\
\hline
\\
\end{array}
\qquad
\begin{array}{r|rrrr}
2 & 1 & 4 & -5 & -14 \\
  & \downarrow & & & \\
\hline
  & 1 & & &
\end{array}
$$

Next, take the 2 from the divisor and multiply by the 1 that was "brought down" to get 2. Write this underneath the 4, then add to get 6.

$$
\begin{array}{r|rrrr}
2 & 1 & 4 & -5 & -14 \\
  & \downarrow & 2 & & \\
\hline
  & 1 & & &
\end{array}
\qquad
\begin{array}{r|rrrr}
2 & 1 & 4 & -5 & -14 \\
  & \downarrow & 2 & & \\
\hline
  & 1 & 6 & &
\end{array}
$$

Now take the 2 from the divisor times the 6 to get 12, and add it to the -5 to get 7.

$$
\begin{array}{r|rrrr}
2 & 1 & 4 & -5 & -14 \\
  & \downarrow & 2 & 12 & \\
\hline
  & 1 & 6 & &
\end{array}
\qquad
\begin{array}{r|rrrr}
2 & 1 & 4 & -5 & -14 \\
  & \downarrow & 2 & 12 & \\
\hline
  & 1 & 6 & 7 &
\end{array}
$$

Finally, take the 2 in the divisor times the 7 to get 14, and add it to the -14 to get 0.

$$
\begin{array}{r|rrrr}
2 & 1 & 4 & -5 & -14 \\
  & \downarrow & 2 & 12 & 14 \\
\hline
  & 1 & 6 & 7 \\
\end{array}
\qquad
\begin{array}{r|rrrr}
2 & 1 & 4 & -5 & -14 \\
  & \downarrow & 2 & 12 & 14 \\
\hline
  & 1 & 6 & 7 & \boxed{0} \\
\end{array}
$$

The first three numbers in the last row of our tableau are the coefficients of the quotient polynomial. Remember, we started with a third degree polynomial and divided by a first degree polynomial, so the quotient is a second degree polynomial. Hence the quotient is $x^2 + 6x + 7$. The number in the box is the remainder. Synthetic division is our tool of choice for dividing polynomials by divisors of the form $x - c$. It is important to note that it works only for these kinds of divisors. Also take note that when a polynomial (of degree at least 1) is divided by $x - c$, the result will be a polynomial of exactly one less degree. Finally, it is worth the time to trace each step in synthetic division back to its corresponding step in long division.

Example 3

Use synthetic division to divide $5x^3 - 2x^2 + 1$ by $x - 3$.

When setting up the synthetic division tableau, we need to enter 0 for the coefficient of x in the dividend. Doing so gives

$$
\begin{array}{r|rrrr}
3 & 5 & -2 & 0 & 1 \\
  & \downarrow & 15 & 39 & 117 \\
\hline
  & 5 & 13 & 39 & \boxed{118} \\
\end{array}
$$

Since the dividend was a third degree polynomial, the quotient is a quadratic polynomial with coefficients 5, 13 and 39. Our quotient is $q(x) = 5x^2 + 13x + 39$ and the remainder is $r(x) = 118$. This means
$5x^3 - 2x^2 + 1 = (x - 3)(5x^2 + 13x + 39) + 118$.

It also means that $x - 3$ is *not* a factor of $5x^3 - 2x^2 + 1$.

Example 4

Divide $x^3 + 8$ by $x + 2$

For this division, we rewrite $x + 2$ as $x - (-2)$ and proceed as before.

$$
\begin{array}{r|rrrr}
-2 & 1 & 0 & 0 & 8 \\
   & \downarrow & -2 & 4 & -8 \\
\hline
   & 1 & -2 & 4 & \boxed{0} \\
\end{array}
$$

The quotient is $x^2 - 2x + 4$ and the remainder is zero. Since the remainder is zero, $x + 2$ is a factor of $x^3 + 8$.

$$x^3 + 8 = (x + 2)(x^2 - 2x + 4)$$

Try it Now

2. Divide $4x^4 - 8x^2 - 5x$ by $x - 3$ using synthetic division.

Using this process allows us to find the real zeros of polynomials, presuming we can figure out at least one root. We'll explore how to do that in the next section.

Example 5

The polynomial $p(x) = 4x^4 - 4x^3 - 11x^2 + 12x - 3$ has a horizontal intercept at $x = \dfrac{1}{2}$ with multiplicity 2. Find the other intercepts of $p(x)$.

Since $x = \dfrac{1}{2}$ is an intercept with multiplicity 2, then $x - \dfrac{1}{2}$ is a factor twice. Use synthetic division to divide by $x - \dfrac{1}{2}$ twice.

```
1/2 |  4   -4   -11   12   -3
    |↓      2    -1   -6    3
       ────────────────────────
       4   -2    -1   -6 |  0

1/2 |  4   -2    -1   -6
    |↓      2     0   -6
       ──────────────────
       4    0   -12 |  0
```

From the first division, we get $4x^4 - 4x^3 - 11x^2 + 12x - 3 = \left(x - \dfrac{1}{2}\right)\left(4x^3 - 2x^2 - x - 6\right)$

The second division tells us

$$4x^4 - 4x^3 - 11x^2 + 12x - 3 = \left(x - \dfrac{1}{2}\right)\left(x - \dfrac{1}{2}\right)\left(4x^2 - 12\right).$$

To find the remaining intercepts, we set $4x^2 - 12 = 0$ and get $x = \pm\sqrt{3}$.

Note this also means $4x^4 - 4x^3 - 11x^2 + 12x - 3 = 4\left(x - \dfrac{1}{2}\right)\left(x - \dfrac{1}{2}\right)\left(x - \sqrt{3}\right)\left(x + \sqrt{3}\right)$.

Important Topics of this Section
Long division of polynomials
Remainder Theorem
Factor Theorem
Synthetic division of polynomials

Try it Now Answers

1.

$$
\begin{array}{r}
2x^2 - 6x + 11 \\
x+3 \overline{) \ 2x^3 + 0x^2 - 7x + 3} \\
-\left(2x^3 + 6x^2\right) \\
\hline
-6x^2 - 7x + 3 \\
-\left(-6x^2 - 18x\right) \\
\hline
11x + 3 \\
-\left(11x + 33\right) \\
\hline
-30
\end{array}
$$

The quotient is $2x^2 - 6x + 11$ with remainder -30.

2.

$$
\begin{array}{r|rrrr}
3 & 4 & 0 & -8 & -5 & 0 \\
 & \downarrow & 12 & 36 & 84 & 237 \\
\hline
 & 4 & 12 & 28 & 79 & \boxed{237}
\end{array}
$$

$4x^4 - 8x^2 - 5x$ divided by $x - 3$ is $4x^3 + 12x^2 + 28x + 79$ with remainder 237

Section 3.4 Exercises

Use polynomial long division to perform the indicated division.

1. $\left(4x^2 + 3x - 1\right) \div (x - 3)$
2. $\left(2x^3 - x + 1\right) \div \left(x^2 + x + 1\right)$
3. $\left(5x^4 - 3x^3 + 2x^2 - 1\right) \div \left(x^2 + 4\right)$
4. $\left(-x^5 + 7x^3 - x\right) \div \left(x^3 - x^2 + 1\right)$
5. $\left(9x^3 + 5\right) \div (2x - 3)$
6. $\left(4x^2 - x - 23\right) \div \left(x^2 - 1\right)$

Use synthetic division to perform the indicated division.

7. $\left(3x^2 - 2x + 1\right) \div (x - 1)$
8. $\left(x^2 - 5\right) \div (x - 5)$
9. $\left(3 - 4x - 2x^2\right) \div (x + 1)$
10. $\left(4x^2 - 5x + 3\right) \div (x + 3)$
11. $\left(x^3 + 8\right) \div (x + 2)$
12. $\left(4x^3 + 2x - 3\right) \div (x - 3)$
13. $\left(18x^2 - 15x - 25\right) \div \left(x - \dfrac{5}{3}\right)$
14. $\left(4x^2 - 1\right) \div \left(x - \dfrac{1}{2}\right)$
15. $\left(2x^3 + x^2 + 2x + 1\right) \div \left(x + \dfrac{1}{2}\right)$
16. $\left(3x^3 - x + 4\right) \div \left(x - \dfrac{2}{3}\right)$
17. $\left(2x^3 - 3x + 1\right) \div \left(x - \dfrac{1}{2}\right)$
18. $\left(4x^4 - 12x^3 + 13x^2 - 12x + 9\right) \div \left(x - \dfrac{3}{2}\right)$
19. $\left(x^4 - 6x^2 + 9\right) \div \left(x - \sqrt{3}\right)$
20. $\left(x^6 - 6x^4 + 12x^2 - 8\right) \div \left(x + \sqrt{2}\right)$

Below you are given a polynomial and one of its zeros. Use the techniques in this section to find the rest of the real zeros and factor the polynomial.

21. $x^3 - 6x^2 + 11x - 6$, $c = 1$
22. $x^3 - 24x^2 + 192x - 512$, $c = 8$
23. $3x^3 + 4x^2 - x - 2$, $c = \dfrac{2}{3}$
24. $2x^3 - 3x^2 - 11x + 6$, $c = \dfrac{1}{2}$
25. $x^3 + 2x^2 - 3x - 6$, $c = -2$
26. $2x^3 - x^2 - 10x + 5$, $c = \dfrac{1}{2}$

27. $4x^4 - 28x^3 + 61x^2 - 42x + 9$, $c = \dfrac{1}{2}$ is a zero of multiplicity 2

28. $x^5 + 2x^4 - 12x^3 - 38x^2 - 37x - 12$, $c = -1$ is a zero of multiplicity 3

Section 3.5 Real Zeros of Polynomials

In the last section, we saw how to determine if a real number was a zero of a polynomial. In this section, we will learn how to find good candidates to test using synthetic division. In the days before graphing technology was commonplace, mathematicians discovered a lot of clever tricks for determining the likely locations of zeros. Technology has provided a much simpler approach to narrow down potential candidates, but it is not always sufficient by itself. For example, the function shown to the right does not have any clear intercepts.

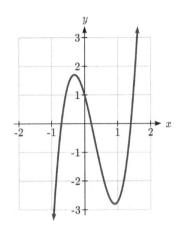

There are two results that can help us identify where the zeros of a polynomial are. The first gives us an interval on which all the real zeros of a polynomial can be found.

Cauchy's Bound

Given a polynomial $f(x) = a_n x^n + a_{n-1} x^{n-1} + \cdots + a_1 x + a_0$, let M be the largest of the coefficients in absolute value. Then all the real zeros of $f(x)$ lie in the interval

$$\left[-\frac{M}{|a_n|} - 1, \quad \frac{M}{|a_n|} + 1 \right]$$

Example 1

Let $f(x) = 2x^4 + 4x^3 - x^2 - 6x - 3$. Determine an interval which contains all the real zeros of f.

To find the M from Cauchy's Bound, we take the absolute value of the coefficients and pick the largest, in this case $|-6| = 6$. Divide this by the absolute value of the leading coefficient, 2, to get 3. All the real zeros of f lie in the interval

$$\left[-\frac{6}{|2|} - 1, \quad \frac{6}{|2|} + 1 \right] = \left[-3 - 1, \quad 3 + 1 \right] = \left[-4, 4 \right].$$

Knowing this bound can be very helpful when using a graphing calculator, since we can use it to set the display bounds. This helps avoid missing a zero because it is graphed outside of the viewing window.

Now that we know *where* we can find the real zeros, we still need a list of *possible* real zeros. The Rational Roots Theorem provides us a list of potential integer and rational zeros.

Rational Roots Theorem

Given a polynomial $f(x) = a_n x^n + a_{n-1} x^{n-1} + \cdots + a_1 x + a_0$ with integer coefficients, if r is a rational zero of f, then r is of the form $r = \pm \dfrac{p}{q}$, where p is a factor of the constant term a_0, and q is a factor of the leading coefficient, a_n.

This gives us a list of numbers to try in our synthetic division, which is a nicer place to start than simply guessing. If none of the numbers in the list are zeros, then either the polynomial has no real zeros at all, or all the real zeros are irrational numbers.

Example 2

Let $f(x) = 2x^4 + 4x^3 - x^2 - 6x - 3$. Use the Rational Roots Theorem to list all the possible rational zeros of *f(x)*.

To generate a complete list of rational zeros, we need to take each of the factors of the constant term, $a_0 = -3$, and divide them by each of the factors of the leading coefficient $a_4 = 2$. The factors of -3 are ± 1 and ± 3. Since the Rational Roots Theorem tacks on a $\pm$ anyway, for the moment, we consider only the positive factors 1 and 3. The factors of 2 are 1 and 2, so the Rational Roots Theorem gives the list

$$\left\{ \pm \frac{1}{1}, \pm \frac{1}{2}, \pm \frac{3}{1}, \pm \frac{3}{2} \right\}, \text{ or } \left\{ \pm 1, \pm \frac{1}{2}, \pm 3, \pm \frac{3}{2} \right\}$$

Now we can use synthetic division to test these possible zeros. To narrow the list first, we could use graphing technology to help us identify some good possibilities.

Example 3

Find the horizontal intercepts of $f(x) = 2x^4 + 4x^3 - x^2 - 6x - 3$.

From Example 1, we know that the real zeros lie in the interval [-4, 4]. Using a graphing calculator, we could set the window accordingly and get the graph below.

In Example 2, we learned that any rational zero must be on the list $\left\{ \pm 1, \pm \frac{1}{2}, \pm 3, \pm \frac{3}{2} \right\}$.

From the graph, it looks like -1 is a good possibility, so we try that using synthetic division.

$$
\begin{array}{r|rrrrr}
-1 & 2 & 4 & -1 & -6 & -3 \\
   &   & -2 & -2 & 3 & 3 \\
\hline
   & 2 & 2 & -3 & -3 & \boxed{0}
\end{array}
$$

Success! Remembering that f was a fourth degree polynomial, we know that our quotient is a third degree polynomial. If we can do one more successful division, we will have knocked the quotient down to a quadratic, and, if all else fails, we can use the quadratic formula to find the last two zeros. Since there seems to be no other rational zeros to try, we continue with -1. Also, the shape of the crossing at x = -1 leads us to wonder if the zero $x = -1$ has multiplicity 3.

$$
\begin{array}{r|rrrr}
-1 & 2 & 2 & -3 & -3 \\
   &   & -2 & 0 & 3 \\
\hline
   & 2 & 0 & -3 & \boxed{0}
\end{array}
$$

Success again! Our quotient polynomial is now $2x^2 - 3$. Setting this to zero gives $2x^2 - 3 = 0$, giving $x = \pm\sqrt{\frac{3}{2}} = \pm\frac{\sqrt{6}}{2}$. Since a fourth degree polynomial can have at most four zeros, including multiplicities, then the intercept $x = -1$ must only have multiplicity 2, which we had found through division, and not 3 as we had guessed.

It is interesting to note that we could greatly improve on the graph of $y = f(x)$ in the previous example given to us by the calculator. For instance, from our determination of the zeros of f and their multiplicities, we know the graph crosses at $x = -\dfrac{\sqrt{6}}{2} \approx -1.22$ then turns back upwards to touch the x–axis at $x = -1$. This tells us that, despite what the calculator showed us the first time, there is a relative maximum occurring at $x = -1$ and not a "flattened crossing" as we originally believed.

After resizing the window, we see not only the relative maximum but also a relative minimum just to the left of $x = -1$.

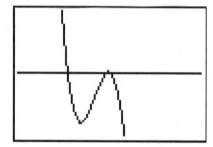

In this case, mathematics helped reveal something that was hidden in the initial graph.

Example 4

Find the real zeros of $f(x) = 4x^3 - 10x^2 - 2x + 2$.

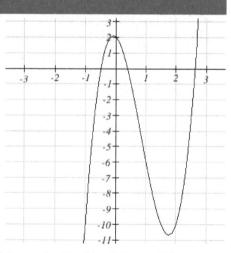

Cauchy's Bound tells us that the real zeros lie in the interval $\left[-\dfrac{10}{|4|} - 1, \ \dfrac{10}{|4|} + 1 \right] = [-3.5, 3.5]$.

Graphing on this interval reveals no clear integer zeros. Turning to the rational roots theorem, we need to take each of the factors of the constant term, $a_0 = 2$, and divide them by each of the factors of the leading coefficient $a_3 = 4$. The factors of 2 are 1 and 2. The factors of 4 are 1, 2, and 4, so the Rational Roots Theorem gives the list

$$\left\{ \pm\frac{1}{1}, \pm\frac{1}{2}, \pm\frac{1}{4}, \pm\frac{2}{1}, \pm\frac{2}{2}, \pm\frac{2}{4} \right\}, \text{ or } \left\{ \pm 1, \pm\frac{1}{2}, \pm\frac{1}{4}, \pm 2 \right\}$$

The two most likely candidates are $\pm\dfrac{1}{2}$.

Trying $\dfrac{1}{2}$,

$$\begin{array}{r|rrrr}
1/2 & 4 & -10 & -2 & 2 \\
    & \downarrow & 2 & -4 & -3 \\
\hline
    & 4 & -8 & -6 & \boxed{-1}
\end{array}$$

The remainder is not zero, so this is not a zero. Trying $-\dfrac{1}{2}$,

$$\begin{array}{r|rrrr}
-1/2 & 4 & -10 & -2 & 2 \\
     & \downarrow & -2 & 6 & -2 \\
\hline
     & 4 & -12 & 4 & \boxed{0}
\end{array}$$

Success! This tells us $4x^3 - 10x^2 - 2x + 2 = \left(x + \dfrac{1}{2} \right)\left(4x^2 - 12x + 4\right)$, and that the graph

has a horizontal intercept at $x = -\dfrac{1}{2}$.

To find the remaining two intercepts, we can use the quadratic equation, setting $4x^2 - 12x + 4 = 0$. First, we might pull out the common factor, $4\left(x^2 - 3x + 1\right) = 0$.

$$x = \frac{3 \pm \sqrt{(-3)^2 - 4(1)(1)}}{2(1)} = \frac{3 \pm \sqrt{5}}{2} \approx 2.618,\ 0.382$$

Try it Now

2. Find the real zeros of $f(x) = 3x^3 - x^2 - 6x + 2$

Important Topics of this Section
Cauchy's Bound for all real zeros of a polynomial
Rational Roots Theorem
Finding real zeros of a polynomial

Try it Now Answers

1. The maximum coefficient in absolute value is 12. Cauchy's Bound for all real zeros is

$$\left[-\frac{12}{|3|}-1, \ \frac{12}{|3|}+1\right]=[-5,5]$$

2. Cauchy's Bound tells us the zeros lie in the interval $\left[-\frac{6}{|3|}-1, \ \frac{6}{|3|}+1\right]=[-3,3]$.

The rational roots theorem tells us the possible rational zeros of the polynomial are on the list

$$\left\{\pm\frac{1}{1},\pm\frac{1}{3},\pm\frac{2}{1},\pm\frac{2}{3}\right\}=\left\{\pm1,\pm\frac{1}{3},\pm2,\pm\frac{2}{3}\right\}.$$

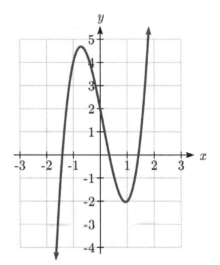

Looking at a graph, the only likely candidate is $\frac{1}{3}$

Using synthetic division,

$$
\begin{array}{r|rrrr}
1/3 & 3 & -1 & -6 & 2 \\
    & \downarrow & 1 & 0 & -2 \\
\hline
    & 3 & 0 & -6 & \boxed{0} \\
\end{array}
$$

$$3x^3-x^2-6x+2=\left(x-\frac{1}{3}\right)(3x^2-6)=3\left(x-\frac{1}{3}\right)(x^2-2).$$

Solving $x^2-2=0$ gives zeros $x=\pm\sqrt{2}$.

The real zeros of the polynomial are $x=\sqrt{2},-\sqrt{2},\frac{1}{3}$.

Section 3.5 Exercises

For each of the following polynomials, use Cauchy's Bound to find an interval containing all the real zeros, then use Rational Roots Theorem to make a list of possible rational zeros.

1. $f(x) = x^3 - 2x^2 - 5x + 6$

2. $f(x) = x^4 + 2x^3 - 12x^2 - 40x - 32$

3. $f(x) = x^4 - 9x^2 - 4x + 12$

4. $f(x) = x^3 + 4x^2 - 11x + 6$

5. $f(x) = x^3 - 7x^2 + x - 7$

6. $f(x) = -2x^3 + 19x^2 - 49x + 20$

7. $f(x) = -17x^3 + 5x^2 + 34x - 10$

8. $f(x) = 36x^4 - 12x^3 - 11x^2 + 2x + 1$

9. $f(x) = 3x^3 + 3x^2 - 11x - 10$

10. $f(x) = 2x^4 + x^3 - 7x^2 - 3x + 3$

Find the real zeros of each polynomial.

11. $f(x) = x^3 - 2x^2 - 5x + 6$

12. $f(x) = x^4 + 2x^3 - 12x^2 - 40x - 32$

13. $f(x) = x^4 - 9x^2 - 4x + 12$

14. $f(x) = x^3 + 4x^2 - 11x + 6$

15. $f(x) = x^3 - 7x^2 + x - 7$

16. $f(x) = -2x^3 + 19x^2 - 49x + 20$

17. $f(x) = -17x^3 + 5x^2 + 34x - 10$

18. $f(x) = 36x^4 - 12x^3 - 11x^2 + 2x + 1$

19. $f(x) = 3x^3 + 3x^2 - 11x - 10$

20. $f(x) = 2x^4 + x^3 - 7x^2 - 3x + 3$

21. $f(x) = 9x^3 - 5x^2 - x$

22. $f(x) = 6x^4 - 5x^3 - 9x^2$

23. $f(x) = x^4 + 2x^2 - 15$

24. $f(x) = x^4 - 9x^2 + 14$

25. $f(x) = 3x^4 - 14x^2 - 5$

26. $f(x) = 2x^4 - 7x^2 + 6$

27. $f(x) = x^6 - 3x^3 - 10$

28. $f(x) = 2x^6 - 9x^3 + 10$

29. $f(x) = x^5 - 2x^4 - 4x + 8$

30. $f(x) = 2x^5 + 3x^4 - 18x - 27$

31. $f(x) = x^5 - 60x^3 - 80x^2 + 960x + 2304$

32. $f(x) = 25x^5 - 105x^4 + 174x^3 - 142x^2 + 57x - 9$

Section 3.6 Complex Zeros

When finding the zeros of polynomials, at some point you're faced with the problem $x^2 = -1$. While there are clearly no real numbers that are solutions to this equation, leaving things there has a certain feel of incompleteness. To address that, we will need utilize the imaginary unit, i.

Imaginary Number i

The most basic complex number is i, defined to be $i = \sqrt{-1}$, commonly called an **imaginary number**. Any real multiple of i is also an imaginary number.

Example 1

Simplify $\sqrt{-9}$.

We can separate $\sqrt{-9}$ as $\sqrt{9}\sqrt{-1}$. We can take the square root of 9, and write the square root of -1 as i.
$$\sqrt{-9} = \sqrt{9}\sqrt{-1} = 3i$$

A complex number is the sum of a real number and an imaginary number.

Complex Number

A **complex number** is a number $z = a + bi$, where a and b are real numbers
a is the real part of the complex number
b is the imaginary part of the complex number
$i = \sqrt{-1}$

Arithmetic on Complex Numbers

Before we dive into the more complicated uses of complex numbers, let's make sure we remember the basic arithmetic involved. To add or subtract complex numbers, we simply add the like terms, combining the real parts and combining the imaginary parts.

Example 3

Add $3 - 4i$ and $2 + 5i$.

Adding $(3 - 4i) + (2 + 5i)$, we add the real parts and the imaginary parts
$3 + 2 - 4i + 5i$
$5 + i$

Try it Now
1. Subtract $2 + 5i$ from $3 - 4i$.

We can also multiply and divide complex numbers.

Example 4

Multiply: $4(2 + 5i)$.

To multiply the complex number by a real number, we simply distribute as we would when multiplying polynomials.

$4(2 + 5i)$
$= 4 \cdot 2 + 4 \cdot 5i$
$= 8 + 20i$

Example 5

Divide $\dfrac{(2 + 5i)}{(4 - i)}$.

To divide two complex numbers, we have to devise a way to write this as a complex number with a real part and an imaginary part.

We start this process by eliminating the complex number in the denominator. To do this, we multiply the numerator and denominator by a special complex number so that the result in the denominator is a real number. The number we need to multiply by is called the **complex conjugate**, in which the sign of the imaginary part is changed.

Here, $4+i$ is the complex conjugate of $4-i$. Of course, obeying our algebraic rules, we must multiply by $4+i$ on both the top and bottom.
$\dfrac{(2 + 5i)}{(4 - i)} \cdot \dfrac{(4 + i)}{(4 + i)}$

To multiply two complex numbers, we expand the product as we would with polynomials (the process commonly called FOIL – "first outer inner last"). In the numerator:

$(2+5i)(4+i)$ $\qquad$ Expand

$= 8 + 20i + 2i + 5i^2$ $\qquad$ Since $i = \sqrt{-1}$, $i^2 = -1$
$= 8 + 20i + 2i + 5(-1)$ $\qquad$ Simplify
$= 3 + 22i$

Following the same process to multiply the denominator

$(4-i)(4+i)$ $\qquad$ Expand

$= (16 - 4i + 4i - i^2)$ $\qquad$ Since $i = \sqrt{-1}$, $i^2 = -1$
$= (16 - (-1))$
$= 17$

Combining this we get $\dfrac{3+22i}{17} = \dfrac{3}{17} + \dfrac{22i}{17}$

Try it Now

2. Multiply $3 - 4i$ and $2 + 3i$.

In the last example, we used the conjugate of a complex number

Complex Conjugate
The **conjugate** of a complex number $a + bi$ is the number $a - bi$.
The notation commonly used for conjugation is a bar: $\overline{a + bi} = a - bi$

Complex Zeros of Polynomials

Complex numbers allow us a way to write solutions to quadratic equations that do not have real solutions.

Example 6

Find the zeros of $f(x) = x^2 - 2x + 5$.

Using the quadratic formula,

$$x = \frac{2 \pm \sqrt{(-2)^2 - 4(1)(5)}}{2(1)} = \frac{2 \pm \sqrt{-16}}{2} = \frac{2 \pm 4i}{2} = 1 \pm 2i.$$

Try it Now

3. Find the zeros of $f(x) = 2x^2 + 3x + 4$.

Two things are important to note. First, the zeros $1 + 2i$ and $1 - 2i$ are complex conjugates. This will always be the case when we find non-real zeros to a quadratic function with real coefficients.

Second, we could write $f(x) = x^2 - 2x + 5 = (x - (1 + 2i))(x - (1 - 2i))$ if we really wanted to, so the Factor and Remainder Theorems hold.

How do we know if a general polynomial has any complex zeros? We have seen examples of polynomials with no real zeros; can there be polynomials with no zeros at all? The answer to that last question, which comes from the Fundamental Theorem of Algebra, is "No."

Fundamental Theorem of Algebra

A non-constant polynomial f with real or complex coefficients will have at least one real or complex zero.

This theorem is an example of an "existence" theorem in mathematics. It guarantees the existence of at least one zero, but provides no algorithm to use for finding it.

Now suppose we have a polynomial $f(x)$ of degree n. The Fundamental Theorem of Algebra guarantees at least one zero z_1, then the Factor Theorem guarantees that f can be factored as $f(x) = (x - z_1)q_1(x)$, where the quotient $q_1(x)$ will be of degree $n-1$.

If this function is non-constant, than the Fundamental Theorem of Algebra applies to it, and we can find another zero. This can be repeated n times.

Complex Factorization Theorem

If f is a polynomial f with real or complex coefficients with degree $n \geq 1$, then f has exactly n real or complex zeros, counting multiplicities.

If $z_1, z_2, \ldots, z_k$ are the distinct zero of f with multiplicities $m_1, m_2, \ldots, m_k$ respectively, then $f(x) = a(x - z_1)^{m_1} (x - z_2)^{m_2} \cdots (x - z_k)^{m_k}$

Example 7

Find all the real and complex zeros of $f(x) = 12x^5 - 20x^4 + 19x^3 - 6x^2 - 2x + 1$.

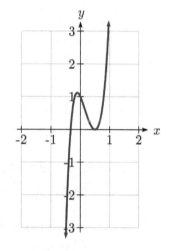

Using the Rational Roots Theorem, the possible real rational roots are

$$\left\{ \pm \frac{1}{1}, \pm \frac{1}{2}, \pm \frac{1}{3}, \pm \frac{1}{4}, \pm \frac{1}{6}, \pm \frac{1}{12} \right\}$$

Testing $\frac{1}{2}$,

```
1/2 |  12   -20    19    -6    -2     1
     ↓         6    -7     6     0    -1
        12   -14    12     0    -2 |  0
```

Success! Because the graph bounces at this intercept, it is likely that this zero has multiplicity 2. We can try synthetic division again to test that.

```
1/2 |  12   -14    12     0    -2
     ↓         6    -4     4     2
        12    -8     8    -4 |  0
```

The other real root appears to be $-\frac{1}{3}$ or $-\frac{1}{4}$. Testing $-\frac{1}{3}$,

```
-1/3 |  12    -8     8    -4
      ↓        -4     4    -4
         12   -12    12 |  0
```

Excellent! So far, we have factored the polynomial to

$$f(x) = \left(x - \frac{1}{2}\right)^2 \left(x + \frac{1}{3}\right)\left(12x^2 - 12x + 12\right) = 12\left(x - \frac{1}{2}\right)^2 \left(x + \frac{1}{3}\right)\left(x^2 - x + 1\right)$$

We can use the quadratic formula to find the two remaining zeros by setting $x^2 - x + 1 = 0$, which are likely complex zeros.

$$x = \frac{1 \pm \sqrt{(-1)^2 - 4(1)(1)}}{2(1)} = \frac{1 \pm \sqrt{-3}}{2} = \frac{1 \pm i\sqrt{3}}{2}.$$

The zeros of the function are $x = \dfrac{1}{2}, -\dfrac{1}{3}, \dfrac{1 + i\sqrt{3}}{2}, \dfrac{1 - i\sqrt{3}}{2}$. We could write the function

fully factored as $f(x) = 12\left(x - \dfrac{1}{2}\right)^2 \left(x + \dfrac{1}{3}\right)\left(x - \dfrac{1 + i\sqrt{3}}{2}\right)\left(x - \dfrac{1 - i\sqrt{3}}{2}\right)$.

When factoring a polynomial like we did at the end of the last example, we say that it is **factored completely over the complex numbers**, meaning it is impossible to factor the polynomial any further using complex numbers. If we wanted to factor the function over

the **real numbers**, we would have stopped at $f(x) = 12\left(x - \dfrac{1}{2}\right)^2 \left(x + \dfrac{1}{3}\right)\left(x^2 - x + 1\right)$. Since

the zeros of $x^2 - x + 1$ are nonreal, we call $x^2 - x + 1$ an **irreducible quadratic** meaning it is impossible to break it down any further using real numbers.

It turns out that a polynomial with real number coefficients can be factored into a product of linear factors corresponding to the real zeros of the function and irreducible quadratic factors which give the nonreal zeros of the function. Consequently, any nonreal zeros will come in conjugate pairs, so if z is a zero of the polynomial, so is $\bar{z}$.

Try it Now

4. Find the real and complex zeros of $f(x) = x^3 - 4x^2 + 9x - 10$.

Important Topics of This Section
Complex and Imaginary numbers
Finding Complex zeros of polynomials

Try it Now Answers

1. $(3-4i)-(2+5i)=1-9i$

2. $(3-4i)(2+3i)=18+i$

3. $x = \dfrac{-3 \pm \sqrt{(3)^2 - 4(2)(4)}}{2(2)} = \dfrac{-3 \pm \sqrt{-23}}{4} = \dfrac{-3 \pm i\sqrt{23}}{4} = \dfrac{-3}{4} \pm \dfrac{\sqrt{23}}{4}i$

4. Cauchy's Bound limits us to the interval [-11, 11]. The rational roots theorem gives a list of potential zeros: $\{\pm 1, \pm 2, \pm 5, \pm 10\}$. A quick graph shows that the likely rational root is $x = 2$.

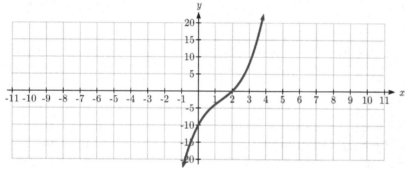

Verifying this,

$$
\begin{array}{r|rrrr}
2 & 1 & -4 & 9 & -10 \\
  & \downarrow & 2 & -4 & 10 \\
\hline
  & 1 & -2 & 5 & \boxed{0}
\end{array}
$$

So $f(x) = (x-2)(x^2 - 2x + 5)$

Using quadratic formula, we can find the complex roots from the irreducible quadratic.

$$x = \dfrac{-(-2) \pm \sqrt{(-2)^2 - 4(1)(5)}}{2(1)} = \dfrac{2 \pm \sqrt{-16}}{2} = \dfrac{-2 \pm 4i}{2} = -1 \pm 2i \;.$$

The zeros of this polynomial are $x = 2, -1 + 2i, -1 - 2i$

Section 3.6 Exercises

Simplify each expression to a single complex number.

1. $\sqrt{-9}$

2. $\sqrt{-16}$

3. $\sqrt{-6}\sqrt{-24}$

4. $\sqrt{-3}\sqrt{-75}$

5. $\dfrac{2+\sqrt{-12}}{2}$

6. $\dfrac{4+\sqrt{-20}}{2}$

Simplify each expression to a single complex number.

7. $(3+2i)+(5-3i)$

8. $(-2-4i)+(1+6i)$

9. $(-5+3i)-(6-i)$

10. $(2-3i)-(3+2i)$

11. $(2+3i)(4i)$

12. $(5-2i)(3i)$

13. $(6-2i)(5)$

14. $(-2+4i)(8)$

15. $(2+3i)(4-i)$

16. $(-1+2i)(-2+3i)$

17. $(4-2i)(4+2i)$

18. $(3+4i)(3-4i)$

19. $\dfrac{3+4i}{2}$

20. $\dfrac{6-2i}{3}$

21. $\dfrac{-5+3i}{2i}$

22. $\dfrac{6+4i}{i}$

23. $\dfrac{2-3i}{4+3i}$

24. $\dfrac{3+4i}{2-i}$

Find all of the zeros of the polynomial then completely factor it over the real numbers and completely factor it over the complex numbers.

25. $f(x)=x^2-4x+13$

26. $f(x)=x^2-2x+5$

27. $f(x)=3x^2+2x+10$

28. $f(x)=x^3-2x^2+9x-18$

29. $f(x)=x^3+6x^2+6x+5$

30. $f(x)=3x^3-13x^2+43x-13$

31. $f(x)=x^3+3x^2+4x+12$

32. $f(x)=4x^3-6x^2-8x+15$

33. $f(x)=x^3+7x^2+9x-2$

34. $f(x)=9x^3+2x+1$

35. $f(x)=4x^4-4x^3+13x^2-12x+3$

36. $f(x)=2x^4-7x^3+14x^2-15x+6$

37. $f(x)=x^4+x^3+7x^2+9x-18$

38. $f(x)=6x^4+17x^3-55x^2+16x+12$

39. $f(x)=-3x^4-8x^3-12x^2-12x-5$

40. $f(x)=8x^4+50x^3+43x^2+2x-4$

41. $f(x)=x^4+9x^2+20$

42. $f(x)=x^4+5x^2-24$

Section 3.7 Rational Functions

In the previous sections, we have built polynomials based on the positive whole number power functions. In this section, we explore functions based on power functions with negative integer powers, called rational functions.

Example 1

You plan to drive 100 miles. Find a formula for the time the trip will take as a function of the speed you drive.

You may recall that multiplying speed by time will give you distance. If we let t represent the drive time in hours, and v represent the velocity (speed or rate) at which we drive, then $vt = $ distance. Since our distance is fixed at 100 miles, $vt = 100$. Solving this relationship for the time gives us the function we desired:

$$t(v) = \frac{100}{v} = 100v^{-1}$$

While this type of relationship can be written using the negative exponent, it is more common to see it written as a fraction.

This particular example is one of an **inversely proportional** relationship – where one quantity is a constant divided by the other quantity, like $y = \frac{5}{x}$.

Notice that this is a transformation of the reciprocal toolkit function, $f(x) = \frac{1}{x}$.

Several natural phenomena, such as gravitational force and volume of sound, behave in a manner **inversely proportional to the square** of another quantity. For example, the volume, V, of a sound heard at a distance d from the source would be related by $V = \frac{k}{d^2}$ for some constant value k.

These functions are transformations of the reciprocal squared toolkit function $f(x) = \frac{1}{x^2}$.

We have seen the graphs of the basic reciprocal function and the squared reciprocal function from our study of toolkit functions. These graphs have several important features.

$f(x) = \dfrac{1}{x}$

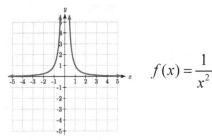 $f(x) = \dfrac{1}{x^2}$

Let's begin by looking at the reciprocal function, $f(x) = \dfrac{1}{x}$. As you well know, dividing by zero is not allowed and therefore zero is not in the domain, and so the function is undefined at an input of zero.

Short run behavior:
As the input values approach zero from the left side (taking on very small, negative values), the function values become very large in the negative direction (in other words, they approach negative infinity).
We write: as $x \to 0^-$, $f(x) \to -\infty$.

As we approach zero from the right side (small, positive input values), the function values become very large in the positive direction (approaching infinity).
We write: as $x \to 0^+$, $f(x) \to \infty$.

This behavior creates a **vertical asymptote**. An asymptote is a line that the graph approaches. In this case the graph is approaching the vertical line $x = 0$ as the input becomes close to zero.

Long run behavior:
As the values of x approach infinity, the function values approach 0.
As the values of x approach negative infinity, the function values approach 0.
Symbolically: as $x \to \pm\infty$, $f(x) \to 0$

Based on this long run behavior and the graph we can see that the function approaches 0 but never actually reaches 0, it just "levels off" as the inputs become large. This behavior creates a **horizontal asymptote**. In this case the graph is approaching the horizontal line $f(x) = 0$ as the input becomes very large in the negative and positive directions.

Vertical and Horizontal Asymptotes

A **vertical asymptote** of a graph is a vertical line $x = a$ where the graph tends towards positive or negative infinity as the inputs approach a. As $x \to a$, $f(x) \to \pm\infty$.

A **horizontal asymptote** of a graph is a horizontal line $y = b$ where the graph approaches the line as the inputs get large. As $x \to \pm\infty$, $f(x) \to b$.

Try it Now:
1. Use symbolic notation to describe the long run behavior and
   short run behavior for the reciprocal squared function.

Example 2

Sketch a graph of the reciprocal function shifted two units to the left and up three units.
Identify the horizontal and vertical asymptotes of the graph, if any.

Transforming the graph left 2 and up 3 would result in the function

$f(x) = \dfrac{1}{x+2} + 3$, or equivalently, by giving the terms a common denominator,

$f(x) = \dfrac{3x+7}{x+2}$.

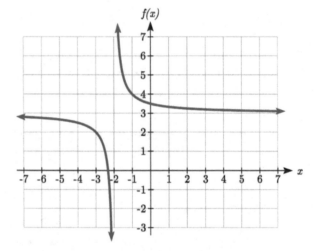

Shifting the toolkit function would give us
this graph. Notice that this equation is
undefined at $x = -2$, and the graph also is
showing a vertical asymptote at $x = -2$.

As $x \to -2^-$, $f(x) \to -\infty$, and

as $x \to -2^+$, $f(x) \to \infty$

As the inputs grow large, the graph appears
to be leveling off at output values of 3,
indicating a horizontal asymptote at $y = 3$.

As $x \to \pm\infty$, $f(x) \to 3$.

Notice that horizontal and vertical asymptotes get shifted left 2 and up 3 along with the
function.

Try it Now
2. Sketch the graph and find the horizontal and vertical asymptotes of the reciprocal
   squared function that has been shifted right 3 units and down 4 units.

In the previous example, we shifted a toolkit function in a way that resulted in a function

of the form $f(x) = \dfrac{3x+7}{x+2}$. This is an example of a more general rational function.

> **Rational Function**
>
> A **rational function** is a function that can be written as the ratio of two polynomials, $P(x)$ and $Q(x)$.
>
> $$f(x) = \frac{P(x)}{Q(x)} = \frac{a_0 + a_1 x + a_2 x^2 + \cdots + a_p x^p}{b_0 + b_1 x + b_2 x^2 + \cdots + b_q x^q}$$

Example 3

A large mixing tank currently contains 100 gallons of water, into which 5 pounds of sugar have been mixed. A tap will open pouring 10 gallons per minute of water into the tank at the same time sugar is poured into the tank at a rate of 1 pound per minute. Find the concentration (pounds per gallon) of sugar in the tank after t minutes.

Notice that the amount of water in the tank is changing linearly, as is the amount of sugar in the tank. We can write an equation independently for each:
$water = 100 + 10t$
$sugar = 5 + 1t$

The concentration, C, will be the ratio of pounds of sugar to gallons of water

$$C(t) = \frac{5 + t}{100 + 10t}$$

Finding Asymptotes and Intercepts

Given a rational function, as part of investigating the short run behavior we are interested in finding any vertical and horizontal asymptotes, as well as finding any vertical or horizontal intercepts, as we have done in the past.

To find vertical asymptotes, we notice that the vertical asymptotes in our examples occur when the denominator of the function is undefined. With one exception, a vertical asymptote will occur whenever the denominator is undefined.

Example 4

Find the vertical asymptotes of the function $k(x) = \dfrac{5 + 2x^2}{2 - x - x^2}$

To find the vertical asymptotes, we determine where this function will be undefined by setting the denominator equal to zero:
$2 - x - x^2 = 0$
$(2 + x)(1 - x) = 0$
$x = -2, 1$

This indicates two vertical asymptotes, which a look at a graph confirms.

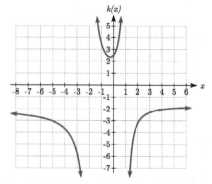

The exception to this rule can occur when both the numerator and denominator of a rational function are zero at the same input.

Example 5

Find the vertical asymptotes of the function $k(x) = \dfrac{x-2}{x^2-4}$.

To find the vertical asymptotes, we determine where this function will be undefined by setting the denominator equal to zero:

$x^2 - 4 = 0$

$x^2 = 4$

$x = -2, 2$

However, the numerator of this function is also equal to zero when $x = 2$. Because of this, the function will still be undefined at 2, since $\dfrac{0}{0}$ is undefined, but the graph will not have a vertical asymptote at $x = 2$.

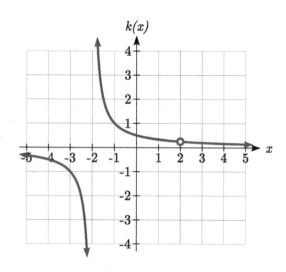

The graph of this function will have the vertical asymptote at $x = -2$, but at $x = 2$ the graph will have a hole: a single point where the graph is not defined, indicated by an open circle.

Vertical Asymptotes and Holes of Rational Functions

The **vertical asymptotes** of a rational function will occur where the denominator of the function is equal to zero and the numerator is not zero.

A **hole** occurs in the graph of a rational function if an input causes both numerator and denominator to be zero. In this case, factor the numerator and denominator and simplify; if the simplified expression still has a zero in the denominator at the original input the original function has a vertical asymptote at the input, otherwise it has a hole.

To find horizontal asymptotes, we are interested in the behavior of the function as the input grows large, so we consider long run behavior of the numerator and denominator separately. Recall that a polynomial's long run behavior will mirror that of the leading term. Likewise, a rational function's long run behavior will mirror that of the ratio of the leading terms of the numerator and denominator functions.

There are three distinct outcomes when this analysis is done:

Case 1: The degree of the denominator > degree of the numerator

Example: $f(x) = \dfrac{3x+2}{x^2+4x-5}$

In this case, the long run behavior is $f(x) \approx \dfrac{3x}{x^2} = \dfrac{3}{x}$. This tells us that as the inputs grow large, this function will behave similarly to the function $g(x) = \dfrac{3}{x}$. As the inputs grow large, the outputs will approach zero, resulting in a horizontal asymptote at $y = 0$.

As $x \to \pm\infty$, $f(x) \to 0$

Case 2: The degree of the denominator < degree of the numerator

Example: $f(x) = \dfrac{3x^2+2}{x-5}$

In this case, the long run behavior is $f(x) \approx \dfrac{3x^2}{x} = 3x$. This tells us that as the inputs grow large, this function will behave similarly to the function $g(x) = 3x$. As the inputs grow large, the outputs will grow and not level off, so this graph has no horizontal asymptote.

As $x \to \pm\infty$, $f(x) \to \pm\infty$, respectively.

Case 3: The degree of the denominator = degree of the numerator

Example: $f(x) = \dfrac{3x^2+2}{x^2+4x-5}$

In this case, the long run behavior is $f(x) \approx \dfrac{3x^2}{x^2} = 3$. This tells us that as the inputs grow large, this function will behave like the function $g(x) = 3$, which is a horizontal line. As $x \to \pm\infty$, $f(x) \to 3$, resulting in a horizontal asymptote at $y = 3$.

> ### Horizontal Asymptote of Rational Functions
>
> The **horizontal asymptote** of a rational function can be determined by looking at the degrees of the numerator and denominator.
>
> Degree of denominator > degree of numerator: Horizontal asymptote at $y = 0$
>
> Degree of denominator < degree of numerator: No horizontal asymptote
>
> Degree of denominator = degree of numerator: Horizontal asymptote at ratio of leading coefficients.

Example 6

In the sugar concentration problem from earlier, we created the equation

$$C(t) = \frac{5+t}{100+10t}.$$

Find the horizontal asymptote and interpret it in context of the scenario.

Both the numerator and denominator are linear (degree 1), so since the degrees are equal, there will be a horizontal asymptote at the ratio of the leading coefficients. In the numerator, the leading term is t, with coefficient 1. In the denominator, the leading term is $10t$, with coefficient 10. The horizontal asymptote will be at the ratio of these values: As $t \to \infty$, $C(t) \to \frac{1}{10}$. This function will have a horizontal asymptote at

$$y = \frac{1}{10}.$$

This tells us that as the input gets large, the output values will approach 1/10. In context, this means that as more time goes by, the concentration of sugar in the tank will approach one tenth of a pound of sugar per gallon of water or 1/10 pounds per gallon.

Example 7

Find the horizontal and vertical asymptotes of the function

$$f(x) = \frac{(x-2)(x+3)}{(x-1)(x+2)(x-5)}$$

First, note this function has no inputs that make both the numerator and denominator zero, so there are no potential holes. The function will have vertical asymptotes when the denominator is zero, causing the function to be undefined. The denominator will be zero at $x = 1$, -2, and 5, indicating vertical asymptotes at these values.

The numerator has degree 2, while the denominator has degree 3. Since the degree of the denominator is greater than that of the numerator, the denominator will grow faster than the numerator, causing the outputs to tend towards zero as the inputs get large, and so as $x \to \pm\infty$, $f(x) \to 0$. This function will have a horizontal asymptote at $y = 0$.

Try it Now

3. Find the vertical and horizontal asymptotes of the function $f(x) = \dfrac{(2x-1)(2x+1)}{(x-2)(x+3)}$

Intercepts

As with all functions, a rational function will have a vertical intercept when the input is zero, if the function is defined at zero. It is possible for a rational function to not have a vertical intercept if the function is undefined at zero.

Likewise, a rational function will have horizontal intercepts at the inputs that cause the output to be zero (unless that input corresponds to a hole). It is possible there are no horizontal intercepts. Since a fraction is only equal to zero when the numerator is zero, horizontal intercepts will occur when the numerator of the rational function is equal to zero.

Example 8

Find the intercepts of $f(x) = \dfrac{(x-2)(x+3)}{(x-1)(x+2)(x-5)}$

We can find the vertical intercept by evaluating the function at zero

$$f(0) = \frac{(0-2)(0+3)}{(0-1)(0+2)(0-5)} = \frac{-6}{10} = -\frac{3}{5}$$

The horizontal intercepts will occur when the function is equal to zero:

$0 = \dfrac{(x-2)(x+3)}{(x-1)(x+2)(x-5)}$ This is zero when the numerator is zero

$0 = (x-2)(x+3)$

$x = 2, -3$

Try it Now

4. Given the reciprocal squared function that is shifted right 3 units and down 4 units, write this as a rational function and find the horizontal and vertical intercepts and the horizontal and vertical asymptotes.

From the previous example, you probably noticed that the numerator of a rational function reveals the horizontal intercepts of the graph, while the denominator reveals the vertical asymptotes of the graph. As with polynomials, factors of the numerator may have integer powers greater than one. Happily, the effect on the shape of the graph at those intercepts is the same as we saw with polynomials.

When factors of the denominator have integer powers greater than one, the behavior at the corresponding vertical asymptote will mirror one of the two toolkit reciprocal functions.

We get this behavior when the degree of the factor in the denominator is odd. The distinguishing characteristic is that on one side of the vertical asymptote the graph heads towards positive infinity, and on the other side the graph heads towards negative infinity.

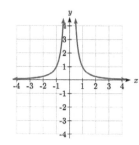

We get this behavior when the degree of the factor in the denominator is even. The distinguishing characteristic is that the graph either heads toward positive infinity on both sides of the vertical asymptote, or heads toward negative infinity on both sides.

For example, the graph of

$$f(x) = \frac{(x+1)^2(x-3)}{(x+3)^2(x-2)}$$ is shown here.

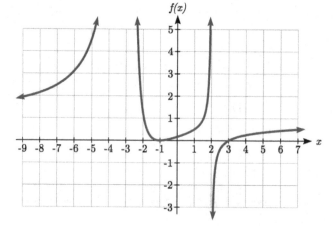

At the horizontal intercept $x = -1$ corresponding to the $(x+1)^2$ factor of the numerator, the graph bounces at the intercept, consistent with the quadratic nature of the factor.

At the horizontal intercept $x = 3$ corresponding to the $(x-3)$ factor of the numerator, the graph passes through the axis as we'd expect from a linear factor.

At the vertical asymptote $x = -3$ corresponding to the $(x+3)^2$ factor of the denominator, the graph heads towards positive infinity on both sides of the asymptote, consistent with the behavior of the $\frac{1}{x^2}$ toolkit.

At the vertical asymptote $x = 2$ corresponding to the $(x - 2)$ factor of the denominator, the graph heads towards positive infinity on the left side of the asymptote and towards negative infinity on the right side, consistent with the behavior of the $\dfrac{1}{x}$ toolkit.

Example 9

Sketch a graph of $f(x) = \dfrac{(x+2)(x-3)}{(x+1)^2(x-2)}$.

We can start our sketch by finding intercepts and asymptotes. Evaluating the function at zero gives the vertical intercept:
$$f(0) = \frac{(0+2)(0-3)}{(0+1)^2(0-2)} = 3$$

Looking at when the numerator of the function is zero, we can determine the graph will have horizontal intercepts at $x = -2$ and $x = 3$. At each, the behavior will be linear, with the graph passing through the intercept.

Looking at when the denominator of the function is zero, we can determine the graph will have vertical asymptotes at $x = -1$ and $x = 2$.

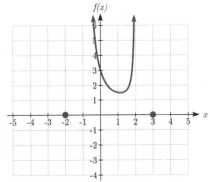

Finally, the degree of denominator is larger than the degree of the numerator, telling us this graph has a horizontal asymptote at $y = 0$.
To sketch the graph, we might start by plotting the three intercepts. Since the graph has no horizontal intercepts between the vertical asymptotes, and the vertical intercept is positive, we know the function must remain positive between the asymptotes, letting us fill in the middle portion of the graph.

Since the factor associated with the vertical asymptote at $x = -1$ was squared, we know the graph will have the same behavior on both sides of the asymptote. Since the graph heads towards positive infinity as the inputs approach the asymptote on the right, the graph will head towards positive infinity on the left as well. For the vertical asymptote at $x = 2$, the factor was not squared, so the graph will have opposite behavior on either side of the asymptote.

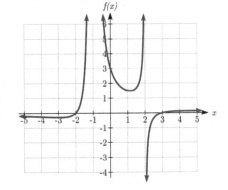

After passing through the horizontal intercepts, the graph will then level off towards an output of zero, as indicated by the horizontal asymptote.

5. Given the function $f(x) = \dfrac{(x+2)^2(x-2)}{2(x-1)^2(x-3)}$, use the characteristics of polynomials and rational functions to describe its behavior and sketch the function.

Since a rational function written in factored form will have a horizontal intercept where each factor of the numerator is equal to zero, we can form a numerator that will pass through a set of horizontal intercepts by introducing a corresponding set of factors. Likewise, since the function will have a vertical asymptote where each factor of the denominator is equal to zero, we can form a denominator that will produce the vertical asymptotes by introducing a corresponding set of factors.

Writing Rational Functions from Intercepts and Asymptotes

If a rational function has horizontal intercepts at $x = x_1, x_2, \ldots, x_n$, and vertical asymptotes at $x = v_1, v_2, \ldots, v_m$ then the function can be written in the form

$$f(x) = a\frac{(x-x_1)^{p_1}(x-x_2)^{p_2}\cdots(x-x_n)^{p_n}}{(x-v_1)^{q_1}(x-v_2)^{q_2}\cdots(x-v_m)^{q_n}}$$

where the powers p_i or q_i on each factor can be determined by the behavior of the graph at the corresponding intercept or asymptote, and the stretch factor a can be determined given a value of the function other than the horizontal intercept, or by the horizontal asymptote if it is nonzero.

Example 10

Write an equation for the rational function graphed here.

The graph appears to have horizontal intercepts at $x = -2$ and $x = 3$. At both, the graph passes through the intercept, suggesting linear factors.

The graph has two vertical asymptotes. The one at $x = -1$ seems to exhibit the basic behavior similar to $\dfrac{1}{x}$, with the graph heading toward positive infinity on one side and heading toward negative infinity on the other.

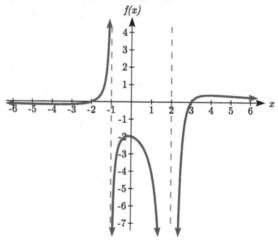

The asymptote at $x = 2$ is exhibiting a behavior similar to $\dfrac{1}{x^2}$, with the graph heading toward negative infinity on both sides of the asymptote.

Utilizing this information indicates an function of the form

$$f(x) = a\frac{(x+2)(x-3)}{(x+1)(x-2)^2}$$

To find the stretch factor, we can use another clear point on the graph, such as the vertical intercept (0,-2):

$$-2 = a\frac{(0+2)(0-3)}{(0+1)(0-2)^2}$$

$$-2 = a\frac{-6}{4}$$

$$a = \frac{-8}{-6} = \frac{4}{3}$$

This gives us a final function of $f(x) = \dfrac{4(x+2)(x-3)}{3(x+1)(x-2)^2}$

Oblique Asymptotes

Earlier we saw graphs of rational functions that had no horizontal asymptote, which occurs when the degree of the numerator is larger than the degree of the denominator. We can, however, describe in more detail the long-run behavior of a rational function.

Example 11

Describe the long-run behavior of $f(x) = \dfrac{3x^2 + 2}{x - 5}$

Earlier we explored this function when discussing horizontal asymptotes. We found the long-run behavior is $f(x) \approx \dfrac{3x^2}{x} = 3x$, meaning that $x \to \pm\infty$, $f(x) \to \pm\infty$, respectively, and there is no horizontal asymptote.

If we were to do polynomial long division, we could get a better understanding of the behavior as $x \to \pm\infty$.

$$
\begin{array}{r}
3x+15 \\
x-5{\overline{\smash{\big)}\,3x^2+\ 0x+2}} \\
\underline{-\left(3x^2-15x\right)} \\
15x+2 \\
\underline{-\left(15x-75\right)} \\
77
\end{array}
$$

This means $f(x)=\dfrac{3x^2+2}{x-5}$ can be rewritten as

$f(x)=3x+15+\dfrac{77}{x-5}$.

As $x\to\pm\infty$, the term $\dfrac{77}{x-5}$ will become very

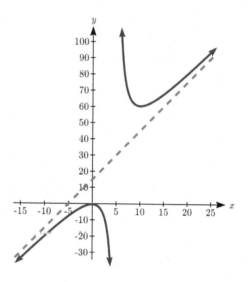

small and approach zero, becoming insignificant. The remaining $3x+15$ then describes the long-run behavior of the function: as $x\to\pm\infty$, $f(x)\to3x+15$.

We call this equation $y=3x+15$ the **oblique asymptote** of the function.

In the graph, you can see how the function is approaching the line on the far left and far right.

Oblique Asymptotes

To explore the long-run behavior of a rational function,
1) Perform polynomial long division (or synthetic division)
2) The quotient will describe the asymptotic behavior of the function

When this result is a line, we call it an **oblique asymptote**, or **slant asymptote**.

Example 12

Find the oblique asymptote of $f(x)=\dfrac{-x^2+2x+1}{x+1}$

Performing polynomial long division:

$$\begin{array}{r}
-x+3 \\
x+1{\overline{\smash{\big)}\,-x^2+2x+1}} \\
\underline{-\left(-x^2-x\right)} \\
3x+1 \\
\underline{-\left(3x+3\right)} \\
-2
\end{array}$$

This allows us to rewrite the function as

$$f(x) = -x+3-\frac{2}{x+1}.$$

The quotient, $y=-x+3$, is the oblique asymptote of $f(x)$. Just like functions we saw earlier approached their horizontal asymptote in the long run, this function will approach this oblique asymptote in the long run.

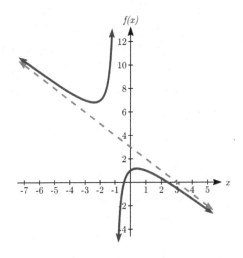

Try it Now

6. Find the oblique asymptote of $f(x) = \dfrac{1+7x-2x^2}{x-2}$

While we primarily concern ourselves with oblique asymptotes, this same approach can describe other asymptotic behavior.

Example 13

Describe the long-run shape of $f(x) = \dfrac{-x^3-x^2+4x+2}{x+1}$

We could rewrite this using long division as

$$f(x) = -x^2+4+\frac{2}{x+1}.$$

Just looking at the quotient gives us the asymptote, $y=-x^2+4$.

This suggests that in the long run, the function will behave like a downwards opening parabola. The function will also have a vertical asymptote at $x=-1$.

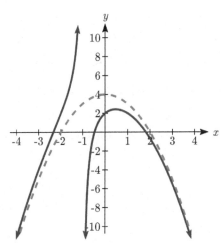

Important Topics of this Section
Inversely proportional; Reciprocal toolkit function
Inversely proportional to the square; Reciprocal squared toolkit function
Horizontal Asymptotes
Vertical Asymptotes
Rational Functions
Finding intercepts, asymptotes, and holes.
Given equation sketch the graph
Identifying a function from its graph
Oblique Asymptotes

Try it Now Answers

1. Long run behavior, as $x \to \pm\infty$, $f(x) \to 0$

   Short run behavior, as $x \to 0$, $f(x) \to \infty$ (there are no horizontal or vertical intercepts)

2. The function and the asymptotes are shifted 3 units right and 4 units down.

   As $x \to 3$, $f(x) \to \infty$ and as $x \to \pm\infty$, $f(x) \to -4$

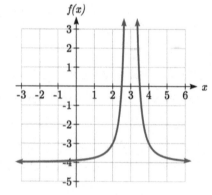

3. Vertical asymptotes at $x = 2$ and $x = -3$; horizontal asymptote at $y = 4$

4. For the transformed reciprocal squared function, we find the rational form.

$$f(x) = \frac{1}{(x-3)^2} - 4 = \frac{1 - 4(x-3)^2}{(x-3)^2} = \frac{1 - 4(x^2 - 6x + 9)}{(x-3)(x-3)} = \frac{-4x^2 + 24x - 35}{x^2 - 6x + 9}$$

Since the numerator is the same degree as the denominator we know that as $x \to \pm\infty$, $f(x) \to -4$. $y = -4$ is the horizontal asymptote. Next, we set the denominator equal to zero to find the vertical asymptote at $x = 3$, because as $x \to 3$, $f(x) \to \infty$. We set the numerator equal to 0 and find the horizontal intercepts are at $(2.5, 0)$ and $(3.5, 0)$, then we evaluate at 0 and the vertical intercept is at $\left(0, \dfrac{-35}{9}\right)$

Try it Now Answers, Continued

5. Horizontal asymptote at $y = 1/2$.
    Vertical asymptotes are at $x = 1$, and $x = 3$.
    Vertical intercept at $(0, 4/3)$,
    Horizontal intercepts $(2, 0)$ and $(-2, 0)$
    $(-2, 0)$ is a double zero and the graph bounces off the
    axis at this point.
    $(2, 0)$ is a single zero and crosses the axis at this point.

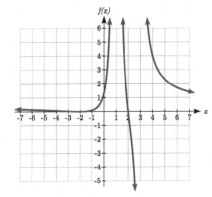

6. Using long division:

$$
\begin{array}{r}
-2x+3 \\
x-2\overline{)\,-2x^2+7x+1} \\
-\left(-2x^2+4x\right) \\
\hline
3x+1 \\
-\left(3x-6\right) \\
\hline
7
\end{array}
$$

$$f(x) = \frac{1+7x-2x^2}{x-2} = -2x+3+\frac{7}{x-2}$$

The oblique asymptote is $y = -2x+3$

Section 3.7 Exercises

Match each equation form with one of the graphs.

1. $f(x) = \dfrac{x-A}{x-B}$ 2. $g(x) = \dfrac{(x-A)^2}{x-B}$ 3. $h(x) = \dfrac{x-A}{(x-B)^2}$ 4. $k(x) = \dfrac{(x-A)^2}{(x-B)^2}$

A   B   C   D

For each function, find the horizontal intercepts, the vertical intercept, the vertical asymptotes, and the horizontal asymptote. Use that information to sketch a graph.

5. $p(x) = \dfrac{2x-3}{x+4}$

6. $q(x) = \dfrac{x-5}{3x-1}$

7. $s(x) = \dfrac{4}{(x-2)^2}$

8. $r(x) = \dfrac{5}{(x+1)^2}$

9. $f(x) = \dfrac{3x^2 - 14x - 5}{3x^2 + 8x - 16}$

10. $g(x) = \dfrac{2x^2 + 7x - 15}{3x^2 - 14x + 15}$

11. $a(x) = \dfrac{x^2 + 2x - 3}{x^2 - 1}$

12. $b(x) = \dfrac{x^2 - x - 6}{x^2 - 4}$

13. $h(x) = \dfrac{2x^2 + x - 1}{x - 4}$

14. $k(x) = \dfrac{2x^2 - 3x - 20}{x - 5}$

15. $n(x) = \dfrac{3x^2 + 4x - 4}{x^3 - 4x^2}$

16. $m(x) = \dfrac{5 - x}{2x^2 + 7x + 3}$

17. $w(x) = \dfrac{(x-1)(x+3)(x-5)}{(x+2)^2 (x-4)}$

18. $z(x) = \dfrac{(x+2)^2 (x-5)}{(x-3)(x+1)(x+4)}$

Write an equation for a rational function with the given characteristics.

19. Vertical asymptotes at $x = 5$ and $x = -5$
    x intercepts at $(2,0)$ and $(-1,0)$ y intercept at $(0,4)$

20. Vertical asymptotes at $x = -4$ and $x = -1$
    x intercepts at $(1,0)$ and $(5,0)$ y intercept at $(0,7)$

21. Vertical asymptotes at $x = -4$ and $x = -5$
    x intercepts at $(4,0)$ and $(-6,0)$ Horizontal asymptote at $y = 7$

22. Vertical asymptotes at $x = -3$ and $x = 6$
    x intercepts at $(-2,0)$ and $(1,0)$ Horizontal asymptote at $y = -2$

23. Vertical asymptote at $x = -1$
    Double zero at $x = 2$ y intercept at $(0,2)$

24. Vertical asymptote at $x = 3$
    Double zero at $x = 1$ y intercept at $(0,4)$

Write an equation for the function graphed.

25.

26.

27.

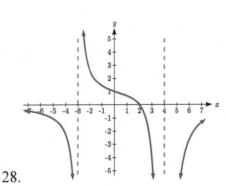

28.

Write an equation for the function graphed.

29.

30.

31.

32.

33.

34.

35.

36.

Write an equation for the function graphed.

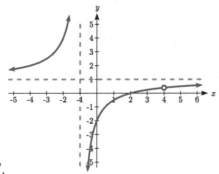

37.   38.

Find the oblique asymptote of each function.

39. $f(x) = \dfrac{3x^2 + 4x}{x+2}$

40. $g(x) = \dfrac{2x^2 + 3x - 8}{x-1}$

41. $h(x) = \dfrac{x^2 - x - 3}{2x - 6}$

42. $k(x) = \dfrac{5 + x - 2x^2}{2x + 1}$

43. $m(x) = \dfrac{-2x^3 + x^2 - 6x + 7}{x^2 + 3}$

44. $n(x) = \dfrac{2x^3 + x^2 + x}{x^2 + x + 1}$

45. A scientist has a beaker containing 20 mL of a solution containing 20% acid. To dilute this, she adds pure water.
    a. Write an equation for the concentration in the beaker after adding n mL of water.
    b. Find the concentration if 10 mL of water has been added.
    c. How many mL of water must be added to obtain a 4% solution?
    d. What is the behavior as $n \to \infty$, and what is the physical significance of this?

46. A scientist has a beaker containing 30 mL of a solution containing 3 grams of potassium hydroxide. To this, she mixes a solution containing 8 milligrams per mL of potassium hydroxide.
    a. Write an equation for the concentration in the tank after adding n mL of the second solution.
    b. Find the concentration if 10 mL of the second solution has been added.
    c. How many mL of water must be added to obtain a 50 mg/mL solution?
    d. What is the behavior as $n \to \infty$, and what is the physical significance of this?

47. Oscar is hunting magnetic fields with his gauss meter, a device for measuring the strength and polarity of magnetic fields. The reading on the meter will increase as Oscar gets closer to a magnet. Oscar is in a long hallway at the end of which is a room containing an extremely strong magnet. When he is far down the hallway from the room, the meter reads a level of 0.2. He then walks down the hallway and enters the room. When he has gone 6 feet into the room, the meter reads 2.3. Eight feet into the room, the meter reads 4.4. [UW]

    a. Give a rational model of form $m(x) = \dfrac{ax+b}{cx+d}$ relating the meter reading $m(x)$ to how many feet x Oscar has gone into the room.

    b. How far must he go for the meter to reach 10? 100?

    c. Considering your function from part (a) and the results of part (b), how far into the room do you think the magnet is?

48. The more you study for a certain exam, the better your performance on it. If you study for 10 hours, your score will be 65%. If you study for 20 hours, your score will be 95%. You can get as close as you want to a perfect score just by studying long enough. Assume your percentage score, $p(n)$, is a function of the number of hours, n, that you study in the form $p(n) = \dfrac{an+b}{cn+d}$. If you want a score of 80%, how long do you need to study? [UW]

49. A street light is 10 feet north of a straight bike path that runs east-west. Olav is bicycling down the path at a rate of 15 miles per hour. At noon, Olav is 33 feet west of the point on the bike path closest to the street light. (See the picture). The relationship between the intensity C of light (in candlepower) and the

(Top View–looking down)

10ft

olav —▸

path

33ft

distance d (in feet) from the light source is given by $C = \dfrac{k}{d^2}$, where k is a constant depending on the light source. [UW]

    a. From 20 feet away, the street light has an intensity of 1 candle. What is k?

    b. Find a function which gives the intensity of the light shining on Olav as a function of time, in seconds.

    c. When will the light on Olav have maximum intensity?

    d. When will the intensity of the light be 2 candles?

Section 3.8 Inverses and Radical Functions

In this section, we will explore the inverses of polynomial and rational functions, and in particular the radical functions that arise in the process.

Example 1

A water runoff collector is built in the shape of a parabolic trough as shown below. Find the surface area of the water in the trough as a function of the depth of the water.

Since it will be helpful to have an equation for the parabolic cross-sectional shape, we will impose a coordinate system at the cross section, with x measured horizontally and y measured vertically, with the origin at the vertex of the parabola.

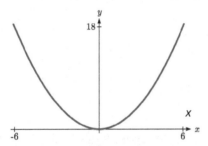

From this we find an equation for the parabolic shape. Since we placed the origin at the vertex of the parabola, we know the equation will have form $y(x) = ax^2$. Our equation will need to pass through the point (6,18), from which we can solve for the stretch factor a:

$$18 = a6^2$$

$$a = \frac{18}{36} = \frac{1}{2}$$

Our parabolic cross section has equation $y(x) = \frac{1}{2}x^2$

Since we are interested in the surface area of the water, we are interested in determining the width at the top of the water as a function of the water depth. For any depth y the width will be given by $2x$, so we need to solve the equation above for x. However notice that the original function is not one-to-one, and indeed given any output there are two inputs that produce the same output, one positive and one negative.

To find an inverse, we can restrict our original function to a limited domain on which it *is* one-to-one. In this case, it makes sense to restrict ourselves to positive x values. On this domain, we can find an inverse by solving for the input variable:

$$y = \frac{1}{2}x^2$$

$$2y = x^2$$

$$x = \pm\sqrt{2y}$$

This is not a function as written. Since we are limiting ourselves to positive x values, we eliminate the negative solution, giving us the inverse function we're looking for

$$x(y) = \sqrt{2y}$$

Since x measures from the center out, the entire width of the water at the top will be $2x$. Since the trough is 3 feet (36 inches) long, the surface area will then be $36(2x)$, or in terms of y:

$$Area = 72x = 72\sqrt{2y}$$

The previous example illustrated two important things:
1) When finding the inverse of a quadratic, we have to limit ourselves to a domain on which the function is one-to-one.
2) The Inverse of a quadratic function is a square root function. Both are toolkit functions and different types of power functions.

Functions involving roots are often called **radical functions**.

Example 2

Find the inverse of $f(x) = (x-2)^2 - 3 = x^2 - 4x + 1$

From the transformation form of the function, we can see this is a transformed quadratic with vertex at (2,-3) that opens upwards. Since the graph will be decreasing on one side of the vertex, and increasing on the other side, we can restrict this function to a domain on which it will be one-to-one by limiting the domain to $x \geq 2$.

To find the inverse, we will use the vertex form of the quadratic. We start by replacing the *f(x)* with a simple variable y, then solve for x.

$$y = (x-2)^2 - 3 \qquad \text{Add 3 to both sides}$$

$$y + 3 = (x-2)^2 \qquad \text{Take the square root}$$

$$\pm\sqrt{y+3} = x - 2 \qquad \text{Add 2 to both sides}$$

$$2 \pm \sqrt{y+3} = x$$

Of course, as written this is not a function. Since we restricted our original function to a domain of $x \geq 2$, the outputs of the inverse should be the same, telling us to utilize the positive case:

$$x = f^{-1}(y) = 2 + \sqrt{y+3}$$

If the quadratic had not been given in vertex form, rewriting it into vertex form is probably the best approach. Alternatively, we could have taken the standard equation and rewritten it equal to zero:

$$0 = x^2 - 4x + 1 - y$$

We would then be able to use the quadratic formula with $a = 1$, $b = -4$, and $c = (1-y)$, resulting in the same solutions we found above:

$$x = \frac{-(-4) \pm \sqrt{(-4)^2 - 4(1)(1-y)}}{2} = 2 \pm \frac{\sqrt{12 + 4y}}{2} = 2 \pm \sqrt{3+y}$$

Try it Now

1. Find the inverse of the function $f(x) = x^2 + 1$, on the domain $x \geq 0$.

While it is not possible to find an inverse of most polynomial functions, some other basic polynomials are invertible.

Example 3

Find the inverse of the function $f(x) = 5x^3 + 1$.

This is a transformation of the basic cubic toolkit function, and based on our knowledge of that function, we know it is one-to-one. Solving for the inverse by solving for x

$$y = 5x^3 + 1$$

$$y - 1 = 5x^3$$

$$\frac{y-1}{5} = x^3$$

$$x = f^{-1}(y) = \sqrt[3]{\frac{y-1}{5}}$$

Notice that this inverse is also a transformation of a power function with a fractional power, $x^{1/3}$.

Try it Now
2. Which toolkit functions have inverse functions without restricting their domain?

Besides being important as an inverse function, radical functions are common in important physical models.

Example 4

The velocity, v in feet per second, of a car that slammed on its brakes can be determined based on the length of skid marks that the tires left on the ground. This relationship is given by

$$v(d) = \sqrt{2gfd}$$

In this formula, g represents acceleration due to gravity (32 ft/sec²), d is the length of the skid marks in feet, and f is a constant representing the friction of the surface. A car lost control on wet asphalt, with a friction coefficient of 0.5, leaving 200 foot skid marks. How fast was the car travelling when it lost control?

Using the given values of $f = 0.5$ and $d = 200$, we can evaluate the given formula:
$v(200) = \sqrt{2(32)(0.5)(200)} = 80$ ft/sec, which is about 54.5 miles per hour.

When radical functions are composed with other functions, determining domain can become more complicated.

Example 5

Find the domain of the function $f(x) = \sqrt{\dfrac{(x+2)(x-3)}{(x-1)}}$.

Since a square root is only defined when the quantity under the radical is non-negative, we need to determine where $\dfrac{(x+2)(x-3)}{(x-1)} \geq 0$. A rational function can change signs (change from positive to negative or vice versa) at horizontal intercepts and at vertical asymptotes. For this equation, the graph could change signs at $x = $ -2, 1, and 3.

To determine on which intervals the rational expression is positive, we could evaluate the expression at test values, or sketch a graph. While both approaches work equally well, for this example we will use a graph.

This function has two horizontal intercepts, both of which exhibit linear behavior, where the graph will pass through the intercept. There is one vertical asymptote, corresponding to a linear factor, leading to a behavior similar to the basic reciprocal toolkit function. There is a vertical intercept at (0, 6). This graph does not have a horizontal asymptote, since the degree of the numerator is larger than the degree of the denominator.

From the vertical intercept and horizontal intercept at $x = -2$, we can sketch the left side of the graph. From the behavior at the asymptote, we can sketch the right side of the graph.

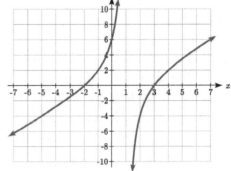

From the graph, we can now tell on which intervals this expression will be non-negative, so the original function $f(x)$ will be defined.
$f(x)$ has domain $-2 \le x < 1 \ or \ x \ge 3$, or in interval notation, $[-2,1) \cup [3,\infty)$.

Like with finding inverses of quadratic functions, it is sometimes desirable to find the inverse of a rational function, particularly of rational functions that are the ratio of linear functions, such as our concentration examples.

The function $C(n) = \dfrac{20 + 0.4n}{100 + n}$ was used in the previous section to represent the concentration of an acid solution after n mL of 40% solution has been added to 100 mL of a 20% solution. We might want to be able to determine instead how much 40% solution has been added based on the current concentration of the mixture.

To do this, we would want the inverse of this function:

$C = \dfrac{20 + 0.4n}{100 + n}$ multiply both sides by the denominator

$C(100 + n) = 20 + 0.4n$ distribute

$100C + Cn = 20 + 0.4n$ group everything with n on one side

$100C - 20 = 0.4n - Cn$ factor out n

$100C - 20 = (0.4 - C)n$ divide to find the inverse

$n(C) = \dfrac{100C - 20}{0.4 - C}$

If, for example, we wanted to know how many mL of 40% solution need to be added to obtain a concentration of 35%, we can simply evaluate the inverse rather than solving an equation involving the original function:

$n(0.35) = \dfrac{100(0.35) - 20}{0.4 - 0.35} = \dfrac{15}{0.05} = 300\,\text{mL}$ of 40% solution would need to be added.

Try it Now

3. Find the inverse of the function $f(x) = \dfrac{x+3}{x-2}$.

Important Topics of this Section
Imposing a coordinate system
Finding an inverse function
Restricting the domain
Invertible toolkit functions
Radical Functions
Inverses of rational functions

Try it Now Answers

1. $y = x^2 + 1$

$y - 1 = x^2$

$x = f^{-1}(y) = \sqrt{y-1}$

2. identity, cubic, square root, cube root

3. $y = \dfrac{x+3}{x-2}$

$y(x-2) = x+3$

$yx - 2y = x+3$

$yx - x = 2y+3$

$x(y-1) = 2y+3$

$f^{-1}(y) = \dfrac{2y+3}{y-1}$

Section 3.8 Exercises

For each function, find a domain on which the function is one-to-one and non-decreasing, then find an inverse of the function on this domain.

1. $f(x) = (x-4)^2$ 2. $f(x) = (x+2)^2$

3. $f(x) = 12 - x^2$ 4. $f(x) = 9 - x^2$

5. $f(x) = 3x^3 + 1$ 6. $f(x) = 4 - 2x^3$

Find the inverse of each function.

7. $f(x) = 9 + \sqrt{4x - 4}$ 8. $f(x) = \sqrt{6x - 8} + 5$

9. $f(x) = 9 + 2\sqrt[3]{x}$ 10. $f(x) = 3 - \sqrt[3]{x}$

11. $f(x) = \dfrac{2}{x+8}$ 12. $f(x) = \dfrac{3}{x-4}$

13. $f(x) = \dfrac{x+3}{x+7}$ 14. $f(x) = \dfrac{x-2}{x+7}$

15. $f(x) = \dfrac{3x+4}{5-4x}$ 16. $f(x) = \dfrac{5x+1}{2-5x}$

Police use the formula $v = \sqrt{20L}$ to estimate the speed of a car, v, in miles per hour, based on the length, L, in feet, of its skid marks when suddenly braking on a dry, asphalt road.

17. At the scene of an accident, a police officer measures a car's skid marks to be 215 feet long. Approximately how fast was the car traveling?

18. At the scene of an accident, a police officer measures a car's skid marks to be 135 feet long. Approximately how fast was the car traveling?

The formula $v = \sqrt{2.7r}$ models the maximum safe speed, v, in miles per hour, at which a car can travel on a curved road with radius of curvature r, in feet.

19. A highway crew measures the radius of curvature at an exit ramp on a highway as 430 feet. What is the maximum safe speed?

20. A highway crew measures the radius of curvature at a tight corner on a highway as 900 feet. What is the maximum safe speed?

21. A drainage canal has a cross-section in the shape of a parabola. Suppose that the canal is 10 feet deep and 20 feet wide at the top. If the water depth in the ditch is 5 feet, how wide is the surface of the water in the ditch? [UW]

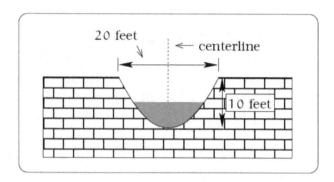

22. Brooke is located 5 miles out from the nearest point A along a straight shoreline in her sea kayak. Hunger strikes and she wants to make it to Kono's for lunch; see picture. Brooke can paddle 2 mph and walk 4 mph. [UW]

   a. If she paddles along a straight line course to the shore, find an expression that computes the total time to reach lunch in terms of the location where Brooke beaches her kayak.
   b. Determine the total time to reach Kono's if she paddles directly to the point A.
   c. Determine the total time to reach Kono's if she paddles directly to Kono's.
   d. Do you think your answer to b or c is the minimum time required for Brooke to reach lunch?
   e. Determine the total time to reach Kono's if she paddles directly to a point on the shore half way between point A and Kono's. How does this time compare to the times in parts b or c? Do you need to modify your answer to part d?

23. Clovis is standing at the edge of a dropoff, which slopes 4 feet downward from him for every 1 horizontal foot. He launches a small model rocket from where he is standing. With the origin of the coordinate system located where he is standing, and the x-axis extending horizontally, the path of the rocket is described by the formula $y = -2x^2 + 120x$. [UW]

   a. Give a function $h = f(x)$ relating the height h of the rocket above the sloping ground to its x-coordinate.
   b. Find the maximum height of the rocket above the sloping ground. What is its x-coordinate when it is at its maximum height?
   c. Clovis measures the height h of the rocket above the sloping ground while it is going up. Give a function $x = g(h)$ relating the x-coordinate of the rocket to h.
   d. Does the function from (c) still work when the rocket is going down? Explain.

24. A trough has a semicircular cross section with a radius of 5 feet. Water starts flowing into the trough in such a way that the depth of the water is increasing at a rate of 2 inches per hour. [UW]

water

5 ft

cross-section of trough

a. Give a function $w = f(t)$ relating the width w of the surface of the water to the time t, in hours. Make sure to specify the domain and compute the range too.

b. After how many hours will the surface of the water have width of 6 feet?

c. Give a function $t = f^{-1}(w)$ relating the time to the width of the surface of the water. Make sure to specify the domain and compute the range too.

Chapter 4:
Exponential and Logarithmic Functions

Section 4.1 Exponential Functions

India is the second most populous country in the world, with a population in 2008 of about 1.14 billion people. The population is growing by about 1.34% each year[1]. We might ask if we can find a formula to model the population, P, as a function of time, t, in years after 2008, if the population continues to grow at this rate.

In linear growth, we had a constant rate of change – a constant *number* that the output increased for each increase in input. For example, in the equation $f(x) = 3x + 4$, the slope tells us the output increases by three each time the input increases by one. This population scenario is different – we have a *percent* rate of change rather than a constant number of people as our rate of change.

To see the significance of this difference consider these two companies:

Company A has 100 stores, and expands by opening 50 new stores a year

Company B has 100 stores, and expands by increasing the number of stores by 50% of their total each year.

Looking at a few years of growth for these companies:

---

[1] World Bank, World Development Indicators, as reported on http://www.google.com/publicdata, retrieved August 20, 2010

Year	Stores, company A		Stores, company B
0	100	Starting with 100 each	100
1	$100 + 50 = 150$	They both grow by 50 stores in the first year.	100 + 50% of 100 $100 + 0.50(100) = 150$
2	$150 + 50 = 200$	Store A grows by 50, Store B grows by 75	150 + 50% of 150 $150 + 0.50(150) = 225$
3	$200 + 50 = 250$	Store A grows by 50, Store B grows by 112.5	225 + 50% of 225 $225 + 0.50(225) = 337.5$

Notice that with the percent growth, each year the company is grows by 50% of the current year's total, so as the company grows larger, the number of stores added in a year grows as well.

To try to simplify the calculations, notice that after 1 year the number of stores for company B was:
$100 + 0.50(100)$ or equivalently by factoring
$100(1 + 0.50) = 150$

We can think of this as "the new number of stores is the original 100% plus another 50%".

After 2 years, the number of stores was:
$150 + 0.50(150)$ or equivalently by factoring
$150(1 + 0.50)$ now recall the 150 came from 100(1+0.50). Substituting that,
$100(1 + 0.50)(1 + 0.50) = 100(1 + 0.50)^2 = 225$

After 3 years, the number of stores was:
$225 + 0.50(225)$ or equivalently by factoring
$225(1 + 0.50)$ now recall the 225 came from $100(1 + 0.50)^2$. Substituting that,
$100(1 + 0.50)^2(1 + 0.50) = 100(1 + 0.50)^3 = 337.5$

From this, we can generalize, noticing that to show a 50% increase, each year we multiply by a factor of (1+0.50), so after n years, our equation would be
$B(n) = 100(1 + 0.50)^n$

In this equation, the 100 represented the initial quantity, and the 0.50 was the percent growth rate. Generalizing further, we arrive at the general form of exponential functions.

Exponential Function

An **exponential growth or decay function** is a function that grows or shrinks at a constant percent growth rate. The equation can be written in the form

$$f(x) = a(1+r)^x \quad \text{or} \quad f(x) = ab^x \quad \text{where } b = 1+r$$

Where

a is the initial or starting value of the function

r is the percent growth or decay rate, written as a decimal

b is the growth factor or growth multiplier. Since powers of negative numbers behave strangely, we limit b to positive values.

To see more clearly the difference between exponential and linear growth, compare the two tables and graphs below, which illustrate the growth of company A and B described above over a longer time frame if the growth patterns were to continue.

years	Company A	Company B
2	200	225
4	300	506
6	400	1139
8	500	2563
10	600	5767

Example 1

Write an exponential function for India's population, and use it to predict the population in 2020.

At the beginning of the chapter we were given India's population of 1.14 billion in the year 2008 and a percent growth rate of 1.34%. Using 2008 as our starting time ($t = 0$), our initial population will be 1.14 billion. Since the percent growth rate was 1.34%, our value for r is 0.0134.

Using the basic formula for exponential growth $f(x) = a(1+r)^x$ we can write the formula, $f(t) = 1.14(1 + 0.0134)^t$

To estimate the population in 2020, we evaluate the function at $t = 12$, since 2020 is 12 years after 2008.

$f(12) = 1.14(1 + 0.0134)^{12} \approx 1.337$ billion people in 2020

1. Given the three statements below, identify which represent exponential functions.

A. The cost of living allowance for state employees increases salaries by 3.1% each year.
B. State employees can expect a $300 raise each year they work for the state.
C. Tuition costs have increased by 2.8% each year for the last 3 years.

Example 2

A certificate of deposit (CD) is a type of savings account offered by banks, typically offering a higher interest rate in return for a fixed length of time you will leave your money invested. If a bank offers a 24 month CD with an annual interest rate of 1.2% compounded monthly, how much will a $1000 investment grow to over those 24 months?

First, we must notice that the interest rate is an annual rate, but is compounded monthly, meaning interest is calculated and added to the account monthly. To find the monthly interest rate, we divide the annual rate of 1.2% by 12 since there are 12 months in a year: 1.2%/12 = 0.1%. Each month we will earn 0.1% interest. From this, we can set up an exponential function, with our initial amount of $1000 and a growth rate of $r = 0.001$, and our input m measured in months.

$$f(m) = 1000\left(1 + \frac{.012}{12}\right)^m$$

$$f(m) = 1000(1 + 0.001)^m$$

After 24 months, the account will have grown to $f(24) = 1000(1 + 0.001)^{24} = \$1024.28$

2. Looking at these two equations that represent the balance in two different savings accounts, which account is growing faster, and which account will have a higher balance after 3 years?

$$A(t) = 1000(1.05)^t \qquad B(t) = 900(1.075)^t$$

In all the preceding examples, we saw exponential growth. Exponential functions can also be used to model quantities that are decreasing at a constant percent rate. An example of this is radioactive decay, a process in which radioactive isotopes of certain atoms transform to an atom of a different type, causing a percentage decrease of the original material over time.

Example 3

Bismuth-210 is an isotope that radioactively decays by about 13% each day, meaning 13% of the remaining Bismuth-210 transforms into another atom (polonium-210 in this case) each day. If you begin with 100 mg of Bismuth-210, how much remains after one week?

With radioactive decay, instead of the quantity increasing at a percent rate, the quantity is decreasing at a percent rate. Our initial quantity is $a = 100$ mg, and our growth rate will be negative 13%, since we are decreasing: $r = -0.13$. This gives the equation:
$$Q(d) = 100(1 - 0.13)^d = 100(0.87)^d$$
This can also be explained by recognizing that if 13% decays, then 87 % remains.

After one week, 7 days, the quantity remaining would be
$$Q(7) = 100(0.87)^7 = 37.73 \text{ mg of Bismuth-210 remains.}$$

Try it Now

3. A population of 1000 is decreasing 3% each year. Find the population in 30 years.

Example 4

T(q) represents the total number of Android smart phone contracts, in thousands, held by a certain Vcrizon store region measured quarterly since January 1, 2016,
Interpret all the parts of the equation $T(2) = 86(1.64)^2 = 231.3056$.

Interpreting this from the basic exponential form, we know that 86 is our initial value. This means that on Jan. 1, 2016 this region had 86,000 Android smart phone contracts. Since $b = 1 + r = 1.64$, we know that every quarter the number of smart phone contracts grows by 64%. $T(2) = 231.3056$ means that in the 2nd quarter (or at the end of the second quarter) there were approximately 231,306 Android smart phone contracts.

Finding Equations of Exponential Functions

In the previous examples, we were able to write equations for exponential functions since we knew the initial quantity and the growth rate. If we do not know the growth rate, but instead know only some input and output pairs of values, we can still construct an exponential function.

Example 5

In 2009, 80 deer were reintroduced into a wildlife refuge area from which the population had previously been hunted to elimination. By 2015, the population had grown to 180 deer. If this population grows exponentially, find a formula for the function.

By defining our input variable to be t, years after 2009, the information listed can be written as two input-output pairs: (0,80) and (6,180). Notice that by choosing our input variable to be measured as years after the first year value provided, we have effectively "given" ourselves the initial value for the function: $a = 80$. This gives us an equation of the form

$$f(t) = 80b^t.$$

Substituting in our second input-output pair allows us to solve for b:

$180 = 80b^6$ Divide by 80

$b^6 = \dfrac{180}{80} = \dfrac{9}{4}$ Take the 6th root of both sides.

$b = \sqrt[6]{\dfrac{9}{4}} = 1.1447$

This gives us our equation for the population:

$$f(t) = 80(1.1447)^t$$

Recall that since $b = 1+r$, we can interpret this to mean that the population growth rate is $r = 0.1447$, and so the population is growing by about 14.47% each year.

In this example, you could also have used $(9/4)^{\wedge}(1/6)$ to evaluate the 6th root if your calculator doesn't have an nth root button.

In the previous example, we chose to use the $f(x) = ab^x$ form of the exponential function rather than the $f(x) = a(1+r)^x$ form. This choice was entirely arbitrary – either form would be fine to use.

When finding equations, the value for b or r will usually have to be rounded to be written easily. To preserve accuracy, it is important to not over-round these values. Typically, you want to be sure to preserve at least 3 significant digits in the growth rate. For example, if your value for b was 1.00317643, you would want to round this no further than to 1.00318.

In the previous example, we were able to "give" ourselves the initial value by clever definition of our input variable. Next, we consider a situation where we can't do this.

Example 6

Find a formula for an exponential function passing through the points (-2,6) and (2,1).

Since we don't have the initial value, we will take a general approach that will work for any function form with unknown parameters: we will substitute in both given input-output pairs in the function form $f(x) = ab^x$ and solve for the unknown values, a and b.

Substituting in (-2, 6) gives $6 = ab^{-2}$

Substituting in (2, 1) gives $1 = ab^2$

We now solve these as a system of equations. To do so, we could try a substitution approach, solving one equation for a variable, then substituting that expression into the second equation.

Solving $6 = ab^{-2}$ for a:

$$a = \frac{6}{b^{-2}} = 6b^2$$

In the second equation, $1 = ab^2$, we substitute the expression above for a:

$$1 = (6b^2)b^2$$

$$1 = 6b^4$$

$$\frac{1}{6} = b^4$$

$$b = \sqrt[4]{\frac{1}{6}} \approx 0.6389$$

Going back to the equation $a = 6b^2$ lets us find a:

$$a = 6b^2 = 6(0.6389)^2 = 2.4492$$

Putting this together gives the equation $f(x) = 2.4492(0.6389)^x$

Try it Now

4. Given the two points (1, 3) and (2, 4.5) find the equation of an exponential function that passes through these two points.

Example 7

Find an equation for the exponential function graphed.

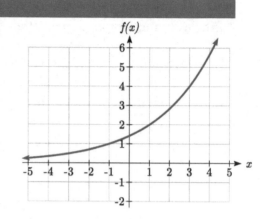

The initial value for the function is not clear in this graph, so we will instead work using two clearer points. There are three clear points: (-1, 1), (1, 2), and (3, 4). As we saw in the last example, two points are sufficient to find the equation for a standard exponential, so we will use the latter two points.

Substituting in (1,2) gives $2 = ab^1$
Substituting in (3,4) gives $4 = ab^3$

Solving the first equation for a gives $a = \dfrac{2}{b}$.

Substituting this expression for a into the second equation:
$4 = ab^3$
$4 = \dfrac{2}{b}b^3 = \dfrac{2b^3}{b}$ Simplify the right-hand side
$4 = 2b^2$
$2 = b^2$
$b = \pm\sqrt{2}$

Since we restrict ourselves to positive values of b, we will use $b = \sqrt{2}$. We can then go back and find a:
$a = \dfrac{2}{b} = \dfrac{2}{\sqrt{2}} = \sqrt{2}$

This gives us a final equation of $f(x) = \sqrt{2}(\sqrt{2})^x$.

Compound Interest

In the bank certificate of deposit (CD) example earlier in the section, we encountered compound interest. Typically bank accounts and other savings instruments in which earnings are reinvested, such as mutual funds and retirement accounts, utilize compound interest. The term *compounding* comes from the behavior that interest is earned not on the original value, but on the accumulated value of the account.

In the example from earlier, the interest was compounded monthly, so we took the annual interest rate, usually called the **nominal rate** or **annual percentage rate (APR)** and divided by 12, the number of compounds in a year, to find the monthly interest. The exponent was then measured in months.

Generalizing this, we can form a general formula for compound interest. If the APR is written in decimal form as r, and there are k compounding periods per year, then the interest per compounding period will be r/k. Likewise, if we are interested in the value after t years, then there will be kt compounding periods in that time.

Compound Interest Formula

Compound Interest can be calculated using the formula

$$A(t) = a\left(1 + \frac{r}{k}\right)^{kt}$$

Where

$A(t)$ is the account value

t is measured in years

a is the starting amount of the account, often called the principal

r is the annual percentage rate (APR), also called the nominal rate

k is the number of compounding periods in one year

Example 8

If you invest \$3,000 in an investment account paying 3% interest compounded quarterly, how much will the account be worth in 10 years?

Since we are starting with \$3000, $a = 3000$
Our interest rate is 3%, so $r = 0.03$
Since we are compounding quarterly, we are compounding 4 times per year, so $k = 4$
We want to know the value of the account in 10 years, so we are looking for $A(10)$, the value when $t = 10$.

$$A(10) = 3000\left(1 + \frac{0.03}{4}\right)^{4(10)} = \$4045.05$$

The account will be worth \$4045.05 in 10 years.

Example 9

A 529 plan is a college savings plan in which a relative can invest money to pay for a child's later college tuition, and the account grows tax free. If Lily wants to set up a 529 account for her new granddaughter, wants the account to grow to $40,000 over 18 years, and she believes the account will earn 6% compounded semi-annually (twice a year), how much will Lily need to invest in the account now?

Since the account is earning 6%, $r = 0.06$
Since interest is compounded twice a year, $k = 2$

In this problem, we don't know how much we are starting with, so we will be solving for a, the initial amount needed. We do know we want the end amount to be $40,000, so we will be looking for the value of a so that $A(18) = 40,000$.

$$40,000 = A(18) = a\left(1 + \frac{0.06}{2}\right)^{2(18)}$$

$$40,000 = a(2.8983)$$

$$a = \frac{40,000}{2.8983} \approx \$13,801$$

Lily will need to invest $13,801 to have $40,000 in 18 years.

Try it now

5. Recalculate example 2 from above with quarterly compounding.

Because of compounding throughout the year, with compound interest the actual increase in a year is *more* than the annual percentage rate. If $1,000 were invested at 10%, the table below shows the value after 1 year at different compounding frequencies:

Frequency	Value after 1 year
Annually	$1100
Semiannually	$1102.50
Quarterly	$1103.81
Monthly	$1104.71
Daily	$1105.16

If we were to compute the actual percentage increase for the daily compounding, there was an increase of $105.16 from an original amount of $1,000, for a percentage increase of $\frac{105.16}{1000} = 0.10516 = 10.516\%$ increase. This quantity is called the **annual percentage yield (APY)**.

Notice that given any starting amount, the amount after 1 year would be

$A(1) = a\left(1 + \dfrac{r}{k}\right)^k$. To find the total change, we would subtract the original amount, then

to find the percentage change we would divide that by the original amount:

$$\frac{a\left(1 + \dfrac{r}{k}\right)^k - a}{a} = \left(1 + \dfrac{r}{k}\right)^k - 1$$

Annual Percentage Yield

The **annual percentage yield** is the actual percent a quantity increases in one year. It can be calculated as

$$APY = \left(1 + \dfrac{r}{k}\right)^k - 1$$

This is equivalent to finding the value of $1 after 1 year, and subtracting the original dollar.

Example 10

Bank A offers an account paying 1.2% compounded quarterly. Bank B offers an account paying 1.1% compounded monthly. Which is offering a better rate?

We can compare these rates using the annual percentage yield – the actual percent increase in a year.

Bank A: $APY = \left(1 + \dfrac{0.012}{4}\right)^4 - 1 = 0.012054 = 1.2054\%$

Bank B: $APY = \left(1 + \dfrac{0.011}{12}\right)^{12} - 1 = 0.011056 = 1.1056\%$

Bank B's monthly compounding is not enough to catch up with Bank A's better APR. Bank A offers a better rate.

A Limit to Compounding

As we saw earlier, the amount we earn increases as we increase the compounding frequency. The table, though, shows that the increase from annual to semi-annual compounding is larger than the increase from monthly to daily compounding. This might lead us to believe that although increasing the frequency of compounding will increase our result, there is an upper limit to this process.

To see this, let us examine the value of $1 invested at 100% interest for 1 year.

Frequency	Value
Annual	$2
Quarterly	$2.441406
Monthly	$2.613035
Daily	$2.714567
Hourly	$2.718127
Once per minute	$2.718279
Once per second	$2.718282

These values do indeed appear to be approaching an upper limit. This value ends up being so important that it gets represented by its own letter, much like how π represents a number.

Euler's Number: e

e is the letter used to represent the value that $\left(1+\dfrac{1}{k}\right)^k$ approaches as k gets big.

$e \approx 2.718282$

Because e is often used as the base of an exponential, most scientific and graphing calculators have a button that can calculate powers of e, usually labeled e^x. Some computer software instead defines a function $exp(x)$, where $exp(x) = e^x$.

Because e arises when the time between compounds becomes very small, e allows us to define **continuous growth** and allows us to define a new toolkit function, $f(x) = e^x$.

Continuous Growth Formula

Continuous Growth can be calculated using the formula

$f(x) = ae^{rx}$

where

a is the starting amount

r is the continuous growth rate

This type of equation is commonly used when describing quantities that change more or less continuously, like chemical reactions, growth of large populations, and radioactive decay.

Example 11

Radon-222 decays at a continuous rate of 17.3% per day. How much will 100mg of Radon-222 decay to in 3 days?

Since we are given a continuous decay rate, we use the continuous growth formula. Since the substance is decaying, we know the growth rate will be negative: $r = -0.173$

$f(3) = 100e^{-0.173(3)} \approx 59.512 \, \text{mg}$ of Radon-222 will remain.

Try it Now

6. Interpret the following: $S(t) = 20e^{0.12t}$ if $S(t)$ represents the growth of a substance in grams, and time is measured in days.

Continuous growth is also often applied to compound interest, allowing us to talk about continuous compounding.

Example 12

If $1000 is invested in an account earning 10% compounded continuously, find the value after 1 year.

Here, the continuous growth rate is 10%, so $r = 0.10$. We start with $1000, so $a = 1000$. To find the value after 1 year,

$f(1) = 1000e^{0.10(1)} \approx \$1105.17$

Notice this is a $105.17 increase for the year. As a percent increase, this is

$\dfrac{105.17}{1000} = 0.10517 = 10.517\%$ increase over the original $1000.

Notice that this value is slightly larger than the amount generated by daily compounding in the table computed earlier.

The continuous growth rate is like the nominal growth rate (or APR) – it reflects the growth rate before compounding takes effect. This is different than the annual growth rate used in the formula $f(x) = a(1+r)^x$, which is like the annual percentage yield – it reflects the *actual* amount the output grows in a year.

While the continuous growth rate in the example above was 10%, the actual annual yield was 10.517%. This means we could write two different looking but equivalent formulas for this account's growth:

$f(t) = 1000e^{0.10t}$ using the 10% continuous growth rate

$f(t) = 1000(1.10517)^t$ using the 10.517% actual annual yield rate.

Important Topics of this Section
Percent growth
Exponential functions
Finding formulas
Interpreting equations
Graphs
Exponential Growth & Decay
Compound interest
Annual Percent Yield
Continuous Growth

Try it Now Answers

1. A & C are exponential functions, they grow by a % not a constant number.

2. B(t) is growing faster ($r = 0.075 > 0.05$), but after 3 years A(t) still has a higher account balance

3. $P(t) = 1000(1 - 0.03)^t = 1000(0.97)^t$
$P(30) = 1000(0.97)^{30} = 401.0071$

4. $3 = ab^1$, so $a = \dfrac{3}{b}$,

$4.5 = ab^2$, so $4.5 = \dfrac{3}{b}b^2$. $4.5 = 3b$

$b = 1.5$. $a = \dfrac{3}{1.5} = 2$

$f(x) = 2(1.5)^x$

5. 24 months = 2 years. $1000\left(1 + \dfrac{.012}{4}\right)^{4(2)} = \$1024.25$

6. An initial substance weighing 20g is growing at a continuous rate of 12% per day.

Section 4.1 Exercises

For each table below, could the table represent a function that is linear, exponential, or neither?

1.
x	1	2	3	4
f(x)	70	40	10	-20

2.
x	1	2	3	4
g(x)	40	32	26	22

3.
x	1	2	3	4
h(x)	70	49	34.3	24.01

4.
x	1	2	3	4
k(x)	90	80	70	60

5.
x	1	2	3	4
m(x)	80	61	42.9	25.61

6.
x	1	2	3	4
n(x)	90	81	72.9	65.61

7. A population numbers 11,000 organisms initially and grows by 8.5% each year. Write an exponential model for the population.

8. A population is currently 6,000 and has been increasing by 1.2% each day. Write an exponential model for the population.

9. The fox population in a certain region has an annual growth rate of 9 percent per year. It is estimated that the population in the year 2010 was 23,900. Estimate the fox population in the year 2018.

10. The amount of area covered by blackberry bushes in a park has been growing by 12% each year. It is estimated that the area covered in 2009 was 4,500 square feet. Estimate the area that will be covered in 2020.

11. A vehicle purchased for $32,500 depreciates at a constant rate of 5% each year. Determine the approximate value of the vehicle 12 years after purchase.

12. A business purchases $125,000 of office furniture which depreciates at a constant rate of 12% each year. Find the residual value of the furniture 6 years after purchase.

Find a formula for an exponential function passing through the two points.

13. $(0,6), (3,750)$ 14. $(0,3), (2,75)$

15. $(0,2000), (2,20)$ 16. $(0,9000), (3,72)$

17. $\left(-1, \dfrac{3}{2}\right), (3,24)$ 18. $\left(-1, \dfrac{2}{5}\right), (1,10)$

19. $(-2,6), (3,1)$ 20. $(-3,4), (3,2)$

21. $(3,1), (5,4)$ 22. $(2,5), (6,9)$

23. A radioactive substance decays exponentially. A scientist begins with 100 milligrams of a radioactive substance. After 35 hours, 50 mg of the substance remains. How many milligrams will remain after 54 hours?

24. A radioactive substance decays exponentially. A scientist begins with 110 milligrams of a radioactive substance. After 31 hours, 55 mg of the substance remains. How many milligrams will remain after 42 hours?

25. A house was valued at $110,000 in the year 1985. The value appreciated to $145,000 by the year 2005. What was the annual growth rate between 1985 and 2005? Assume that the house value continues to grow by the same percentage. What did the value equal in the year 2010?

26. An investment was valued at $11,000 in the year 1995. The value appreciated to $14,000 by the year 2008. What was the annual growth rate between 1995 and 2008? Assume that the value continues to grow by the same percentage. What did the value equal in the year 2012?

27. A car was valued at $38,000 in the year 2003. The value depreciated to $11,000 by the year 2009. Assume that the car value continues to drop by the same percentage. What was the value in the year 2013?

28. A car was valued at $24,000 in the year 2006. The value depreciated to $20,000 by the year 2009. Assume that the car value continues to drop by the same percentage. What was the value in the year 2014?

29. If $4,000 is invested in a bank account at an interest rate of 7 per cent per year, find the amount in the bank after 9 years if interest is compounded annually, quarterly, monthly, and continuously.

30. If $6,000 is invested in a bank account at an interest rate of 9 per cent per year, find the amount in the bank after 5 years if interest is compounded annually, quarterly, monthly, and continuously.

31. Find the annual percentage yield (APY) for a savings account with annual percentage rate of 3% compounded quarterly.

32. Find the annual percentage yield (APY) for a savings account with annual percentage rate of 5% compounded monthly.

33. A population of bacteria is growing according to the equation $P(t)=1600e^{0.21t}$, with t measured in years. Estimate when the population will exceed 7569.

34. A population of bacteria is growing according to the equation $P(t)=1200e^{0.17t}$, with t measured in years. Estimate when the population will exceed 3443.

35. In 1968, the U.S. minimum wage was $1.60 per hour. In 1976, the minimum wage was $2.30 per hour. Assume the minimum wage grows according to an exponential model $w(t)$, where t represents the time in years after 1960. [UW]

    a. Find a formula for $w(t)$.
    b. What does the model predict for the minimum wage in 1960?
    c. If the minimum wage was $5.15 in 1996, is this above, below or equal to what the model predicts?

36. In 1989, research scientists published a model for predicting the cumulative number of AIDS cases (in thousands) reported in the United States: $a(t)=155\left(\dfrac{t-1980}{10}\right)^3$, where t is the year. This paper was considered a "relief", since there was a fear the correct model would be of exponential type. Pick two data points predicted by the research model $a(t)$ to construct a new exponential model $b(t)$ for the number of cumulative AIDS cases. Discuss how the two models differ and explain the use of the word "relief." [UW]

37. You have a chess board as pictured, with squares numbered 1 through 64. You also have a huge change jar with an unlimited number of dimes. On the first square you place one dime. On the second square you stack 2 dimes. Then you continue, always doubling the number from the previous square. [UW]

						63	64
						10	9
1	2	3					8

   a. How many dimes will you have stacked on the 10th square?
   b. How many dimes will you have stacked on the nth square?
   c. How many dimes will you have stacked on the 64th square?
   d. Assuming a dime is 1 mm thick, how high will this last pile be?
   e. The distance from the earth to the sun is approximately 150 million km. Relate the height of the last pile of dimes to this distance.

Section 4.2 Graphs of Exponential Functions

Like with linear functions, the graph of an exponential function is determined by the values for the parameters in the function's formula.

To get a sense for the behavior of exponentials, let us begin by looking more closely at the function $f(x) = 2^x$. Listing a table of values for this function:

x	-3	-2	-1	0	1	2	3
$f(x)$	$\dfrac{1}{8}$	$\dfrac{1}{4}$	$\dfrac{1}{2}$	1	2	4	8

Notice that:
1) This function is positive for all values of x.
2)   As x increases, the function grows faster and faster (the rate of change increases).
3) As x decreases, the function values grow smaller, approaching zero.
4) This is an example of exponential growth.

Looking at the function $g(x) = \left(\dfrac{1}{2}\right)^x$

x	-3	-2	-1	0	1	2	3
$g(x)$	8	4	2	1	$\dfrac{1}{2}$	$\dfrac{1}{4}$	$\dfrac{1}{8}$

Note this function is also positive for all values of x, but in this case grows as x decreases, and decreases towards zero as x increases. This is an example of exponential decay. You may notice from the table that this function appears to be the horizontal reflection of the $f(x) = 2^x$ table. This is in fact the case:

$$f(-x) = 2^{-x} = (2^{-1})^x = \left(\dfrac{1}{2}\right)^x = g(x)$$

Looking at the graphs also confirms this relationship.

Consider a function of the form $f(x) = ab^x$. Since a, which we called the initial value in the last section, is the function value at an input of zero, a will give us the vertical intercept of the graph.

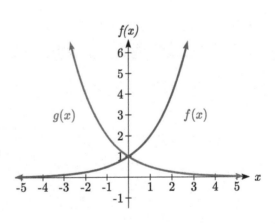

From the graphs above, we can see that an exponential graph will have a horizontal asymptote on one side of the graph, and can either increase or decrease, depending upon the growth factor. This horizontal asymptote will also help us determine the long run behavior and is easy to determine from the graph.

The graph will grow when the growth rate is positive, which will make the growth factor b larger than one. When it's negative, the growth factor will be less than one.

Graphical Features of Exponential Functions

Graphically, in the function $f(x) = ab^x$

a is the vertical intercept of the graph

b determines the rate at which the graph grows. When a is positive,

      the function will increase if $b > 1$

      the function will decrease if $0 < b < 1$

The graph will have a horizontal asymptote at $y = 0$

The graph will be concave up if $a > 0$; concave down if $a < 0$.

The domain of the function is all real numbers

The range of the function is $(0, \infty)$

When sketching the graph of an exponential function, it can be helpful to remember that the graph will pass through the points $(0, a)$ and $(1, ab)$.

The value b will determine the function's long run behavior:

If $b > 1$, as $x \to \infty$, $f(x) \to \infty$ and as $x \to -\infty$, $f(x) \to 0$.

If $0 < b < 1$, as $x \to \infty$, $f(x) \to 0$ and as $x \to -\infty$, $f(x) \to \infty$.

Example 1

Sketch a graph of $f(x) = 4\left(\dfrac{1}{3}\right)^x$

This graph will have a vertical intercept at $(0, 4)$, and pass through the point $\left(1, \dfrac{4}{3}\right)$. Since $b < 1$, the graph will be decreasing towards zero. Since $a > 0$, the graph will be concave up.

We can also see from the graph the long run behavior: as $x \to \infty$, $f(x) \to 0$ and as $x \to -\infty$, $f(x) \to \infty$.

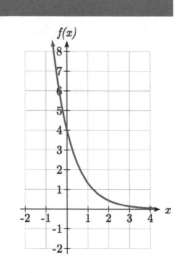

To get a better feeling for the effect of a and b on the graph, examine the sets of graphs below. The first set shows various graphs, where a remains the same and we only change the value for b.

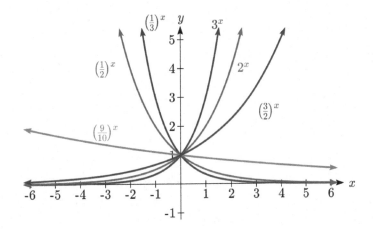

Notice that the closer the value of b is to 1, the less steep the graph will be.

In the next set of graphs, a is altered and our value for b remains the same.

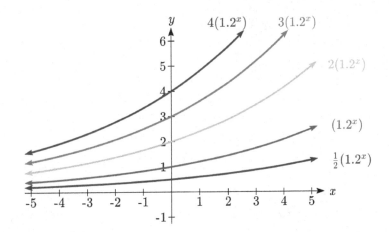

Notice that changing the value for a changes the vertical intercept. Since a is multiplying the b^x term, a acts as a vertical stretch factor, not as a shift. Notice also that the long run behavior for all of these functions is the same because the growth factor did not change and none of these a values introduced a vertical flip.

Example 2

Match each equation with its graph.

$$f(x) = 2(1.3)^x$$

$$g(x) = 2(1.8)^x$$

$$h(x) = 4(1.3)^x$$

$$k(x) = 4(0.7)^x$$

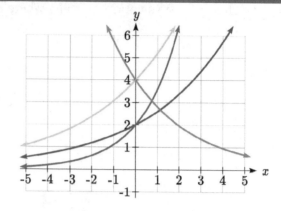

The graph of *k(x)* is the easiest to identify, since it is the only equation with a growth factor less than one, which will produce a decreasing graph. The graph of *h(x)* can be identified as the only growing exponential function with a vertical intercept at (0,4). The graphs of *f(x)* and *g(x)* both have a vertical intercept at (0,2), but since *g(x)* has a larger growth factor, we can identify it as the graph increasing faster.

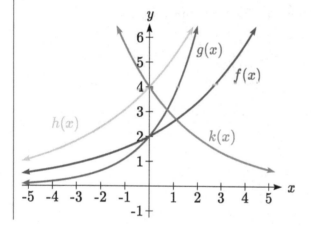

Try it Now

1. Graph the following functions on the same axis:

$$f(x) = (2)^x \; ; \; g(x) = 2(2)^x \; ; \; h(x) = 2(1/2)^x.$$

Transformations of Exponential Graphs

While exponential functions can be transformed following the same rules as any function, there are a few interesting features of transformations that can be identified. The first was seen at the beginning of the section – that a horizontal reflection is equivalent to a change in the growth factor. Likewise, since *a* is itself a stretch factor, a vertical stretch of an exponential corresponds with a change in the initial value of the function.

Next consider the effect of a horizontal shift on an exponential function. Shifting the function $f(x) = 3(2)^x$ four units to the left would give $f(x+4) = 3(2)^{x+4}$. Employing exponent rules, we could rewrite this:

$$f(x+4) = 3(2)^{x+4} = 3(2)^x(2^4) = 48(2)^x$$

Interestingly, it turns out that a horizontal shift of an exponential function corresponds with a change in initial value of the function.

Lastly, consider the effect of a vertical shift on an exponential function. Shifting $f(x) = 3(2)^x$ down 4 units would give the equation $f(x) = 3(2)^x - 4$.

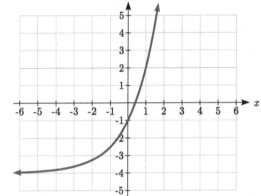

Graphing that, notice it is substantially different than the basic exponential graph. Unlike a basic exponential, this graph does not have a horizontal asymptote at $y = 0$; due to the vertical shift, the horizontal asymptote has also shifted to $y = -4$. We can see that as $x \to \infty$, $f(x) \to \infty$ and as $x \to -\infty$, $f(x) \to -4$.

We have determined that a vertical shift is the only transformation of an exponential function that changes the graph in a way that cannot be achieved by altering the parameters a and b in the basic exponential function $f(x) = ab^x$.

Transformations of Exponentials
Any transformed exponential can be written in the form $$f(x) = ab^x + c$$ where $y = c$ is the horizontal asymptote.

Note that, due to the shift, the vertical intercept is shifted to $(0, a+c)$.

Try it Now

2. Write the equation and graph the exponential function described as follows:
$f(x) = e^x$ is vertically stretched by a factor of 2, flipped across the y axis and shifted up 4 units.

Example 3

Sketch a graph of $f(x) = -3\left(\dfrac{1}{2}\right)^x + 4$.

Notice that in this exponential function, the negative in the stretch factor -3 will cause a vertical reflection, and the vertical shift up 4 will move the horizontal asymptote to

$y = 4$. Sketching this as a transformation of $g(x) = \left(\dfrac{1}{2}\right)^x$,

The basic $g(x) = \left(\dfrac{1}{2}\right)^x$ Vertically reflected and stretched by 3

  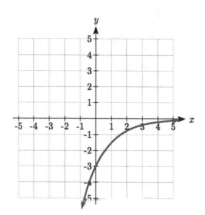

Vertically shifted up four units

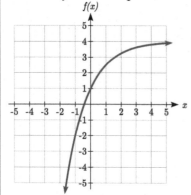

Notice that while the domain of this function is unchanged, due to the reflection and shift, the range of this function is $(-\infty, 4)$.

As $x \to \infty$, $f(x) \to 4$ and as $x \to -\infty$, $f(x) \to -\infty$.

Functions leading to graphs like the one above are common as models for learning and models of growth approaching a limit.

Example 4

Find an equation for the function graphed.

Looking at this graph, it appears to have a horizontal asymptote at $y = 5$, suggesting an equation of the form $f(x) = ab^x + 5$. To find values for a and b, we can identify two other points on the graph. It appears the graph passes through $(0,2)$ and $(-1,3)$, so we can use those points. Substituting in $(0,2)$ allows us to solve for a.

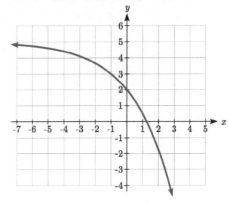

$2 = ab^0 + 5$

$2 = a + 5$

$a = -3$

Substituting in $(-1,3)$ allows us to solve for b

$3 = -3b^{-1} + 5$

$-2 = \dfrac{-3}{b}$

$-2b = -3$

$b = \dfrac{3}{2} = 1.5$

The final formula for our function is $f(x) = -3(1.5)^x + 5$.

Try it Now

3. Given the graph of the transformed exponential function, find a formula and describe the long run behavior.

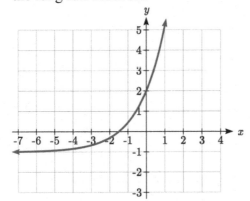

Important Topics of this Section
Graphs of exponential functions
Intercept
Growth factor
Exponential Growth
Exponential Decay
Horizontal intercepts
Long run behavior
Transformations

Try it Now Answers

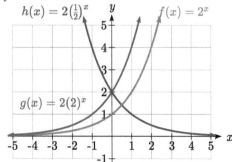

1.

2. $f(x) = -2e^x + 4$

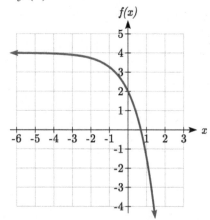

3. Horizontal asymptote at $y = -1$, so $f(x) = ab^x - 1$. Substitute (0, 2) to find $a = 3$.

    Substitute (1,5) to find $5 = 3b^1 - 1$, $b = 2$.

    $f(x) = 3(2^x) - 1$ or $f(x) = 3(.5)^{-x} - 1$

    As $x \to \infty$, $f(x) \to \infty$ and as $x \to -\infty$, $f(x) \to -1$

Section 4.2 Exercises

Match each function with one of the graphs below.

1. $f(x) = 2(0.69)^x$

2. $f(x) = 2(1.28)^x$

3. $f(x) = 2(0.81)^x$

4. $f(x) = 4(1.28)^x$

5. $f(x) = 2(1.59)^x$

6. $f(x) = 4(0.69)^x$

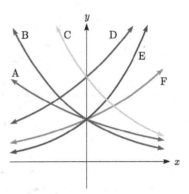

If all the graphs to the right have equations with form
$f(x) = ab^x$,

7. Which graph has the largest value for b?

8. Which graph has the smallest value for b?

9. Which graph has the largest value for a?

10. Which graph has the smallest value for a?

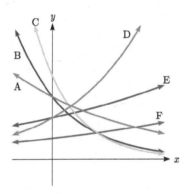

Sketch a graph of each of the following transformations of $f(x) = 2^x$

11. $f(x) = 2^{-x}$ 12. $g(x) = -2^x$

13. $h(x) = 2^x + 3$ 14. $f(x) = 2^x - 4$

15. $f(x) = 2^{x-2}$ 16. $k(x) = 2^{x-3}$

Starting with the graph of $f(x) = 4^x$, find a formula for the function that results from

17. Shifting $f(x)$ 4 units upwards

18. Shifting $f(x)$ 3 units downwards

19. Shifting $f(x)$ 2 units left

20. Shifting $f(x)$ 5 units right

21. Reflecting $f(x)$ about the x-axis

22. Reflecting $f(x)$ about the y-axis

Describe the long run behavior, as $x \to \infty$ and $x \to -\infty$ of each function

23. $f(x) = -5(4^x) - 1$

24. $f(x) = -2(3^x) + 2$

25. $f(x) = 3\left(\dfrac{1}{2}\right)^x - 2$

26. $f(x) = 4\left(\dfrac{1}{4}\right)^x + 1$

27. $f(x) = 3(4)^{-x} + 2$

28. $f(x) = -2(3)^{-x} - 1$

Find a formula for each function graphed as a transformation of $f(x) = 2^x$.

29.

30.

31.

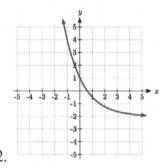

32.

Find an equation for the exponential function graphed.

33.

34.

35.

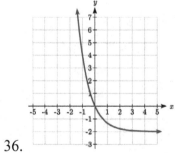

36.

Section 4.3 Logarithmic Functions

A population of 50 flies is expected to double every week, leading to a function of the form $f(x) = 50(2)^x$, where x represents the number of weeks that have passed. When will this population reach 500? Trying to solve this problem leads to:

$500 = 50(2)^x$ Dividing both sides by 50 to isolate the exponential

$10 = 2^x$

While we have set up exponential models and used them to make predictions, you may have noticed that solving exponential equations has not yet been mentioned. The reason is simple: none of the algebraic tools discussed so far are sufficient to solve exponential equations. Consider the equation $2^x = 10$ above. We know that $2^3 = 8$ and $2^4 = 16$, so it is clear that x must be some value between 3 and 4 since $g(x) = 2^x$ is increasing. We could use technology to create a table of values or graph to better estimate the solution.

From the graph, we could better estimate the solution to be around 3.3. This result is still fairly unsatisfactory, and since the exponential function is one-to-one, it would be great to have an inverse function. None of the functions we have already discussed would serve as an inverse function and so we must introduce a new function, named **log** as the inverse of an exponential function. Since exponential functions have different bases, we will define corresponding logarithms of different bases as well.

Logarithm
The logarithm (base b) function, written $\log_b(x)$, is the inverse of the exponential function (base b), b^x.

Since the logarithm and exponential are inverses, it follows that:

Properties of Logs: Inverse Properties
$\log_b(b^x) = x$
$b^{\log_b x} = x$

Recall from the definition of an inverse function that if $f(a) = c$, then $f^{-1}(c) = a$. Applying this to the exponential and logarithmic functions, we can convert between a logarithmic equation and its equivalent exponential.

> **Logarithm Equivalent to an Exponential**
>
> The statement $b^a = c$ is equivalent to the statement $\log_b(c) = a$.

Alternatively, we could show this by starting with the exponential function $c = b^a$, then taking the log base b of both sides, giving $\log_b(c) = \log_b b^a$. Using the inverse property of logs, we see that $\log_b(c) = a$.

Since log is a function, it is most correctly written as $\log_b(c)$, using parentheses to denote function evaluation, just as we would with *f(c)*. However, when the input is a single variable or number, it is common to see the parentheses dropped and the expression written as $\log_b c$.

Example 1

Write these exponential equations as logarithmic equations:

a) $2^3 = 8$ b) $5^2 = 25$ c) $10^{-4} = \dfrac{1}{10000}$

a) $2^3 = 8$ is equivalent to $\log_2(8) = 3$

b) $5^2 = 25$ is equivalent to $\log_5(25) = 2$

c) $10^{-4} = \dfrac{1}{10000}$ is equivalent to $\log_{10}\left(\dfrac{1}{10000}\right) = -4$

Example 2

Write these logarithmic equations as exponential equations:

a) $\log_6\left(\sqrt{6}\right) = \dfrac{1}{2}$ b) $\log_3(9) = 2$

a) $\log_6\left(\sqrt{6}\right) = \dfrac{1}{2}$ is equivalent to $6^{1/2} = \sqrt{6}$

b) $\log_3(9) = 2$ is equivalent to $3^2 = 9$

Try it Now

1. Write the exponential equation $4^2 = 16$ as a logarithmic equation.

By establishing the relationship between exponential and logarithmic functions, we can now solve basic logarithmic and exponential equations by rewriting.

Example 3

Solve $\log_4(x) = 2$ for x.

By rewriting this expression as an exponential, $4^2 = x$, so $x = 16$.

Example 4

Solve $2^x = 10$ for x.

By rewriting this expression as a logarithm, we get $x = \log_2(10)$.

While this does define a solution, and an exact solution at that, you may find it somewhat unsatisfying since it is difficult to compare this expression to the decimal estimate we made earlier. Also, giving an exact expression for a solution is not always useful – often we really need a decimal approximation to the solution. Luckily, this is a task calculators and computers are quite adept at. Unluckily for us, most calculators and computers will only evaluate logarithms of two bases. Happily, this ends up not being a problem, as we'll see briefly.

Common and Natural Logarithms

The **common log** is the logarithm with base 10, and is typically written $\log(x)$.

The **natural log** is the logarithm with base e, and is typically written $\ln(x)$.

Example 5

Evaluate $\log(1000)$ using the definition of the common log.

To evaluate $\log(1000)$, we can let $x = \log(1000)$, then rewrite into exponential form using the common log base of 10: $10^x = 1000$.

From this, we might recognize that 1000 is the cube of 10, so $x = 3$.

Values of the common log

number	number as exponential	log(number)
1000	10^3	3
100	10^2	2
10	10^1	1
1	10^0	0
0.1	10^{-1}	-1
0.01	10^{-2}	-2
0.001	10^{-3}	-3

We also can use the inverse property of logs to write $\log_{10}(10^3) = 3$.

Try it Now
2. Evaluate $\log(1000000)$.

Example 6

Evaluate $\ln\left(\sqrt{e}\right)$.

We can rewrite $\ln\left(\sqrt{e}\right)$ as $\ln\left(e^{1/2}\right)$. Since ln is a log base e, we can use the inverse property for logs: $\ln\left(e^{1/2}\right) = \log_e\left(e^{1/2}\right) = \dfrac{1}{2}$.

Example 7

Evaluate $\log(500)$ using your calculator or computer.

Using a computer, we can evaluate $\log(500) \approx 2.69897$

To utilize the common or natural logarithm functions to evaluate expressions like $\log_2(10)$, we need to establish some additional properties.

Properties of Logs: Exponent Property

$$\log_b\left(A^r\right) = r\log_b(A)$$

To show why this is true, we offer a proof:

Since the logarithmic and exponential functions are inverses, $b^{\log_b A} = A$.

Raising both sides to the r power, we get $A^r = \left(b^{\log_b A}\right)^r$.

Utilizing the exponential rule that states $\left(x^p\right)^q = x^{pq}$, $A^r = \left(b^{\log_b A}\right)^r = b^{r\log_b A}$

Taking the log of both sides, $\log_b\left(A^r\right) = \log_b\left(b^{r\log_b A}\right)$

Utilizing the inverse property on the right side yields the result: $\log_b\left(A^r\right) = r\log_b A$

Example 8

Rewrite $\log_3(25)$ using the exponent property for logs.

Since $25 = 5^2$,

$\log_3(25) = \log_3\left(5^2\right) = 2\log_3(5)$

Example 9

Rewrite $4\ln(x)$ using the exponent property for logs.

Using the property in reverse, $4\ln(x) = \ln\left(x^4\right)$.

Try it Now

3. Rewrite using the exponent property for logs: $\ln\left(\dfrac{1}{x^2}\right)$.

The exponent property allows us to find a method for changing the base of a logarithmic expression.

Properties of Logs: Change of Base

$$\log_b(A) = \frac{\log_c(A)}{\log_c(b)}$$

Proof:

Let $\log_b(A) = x$.

Rewriting as an exponential gives $b^x = A$.

Taking the log base c of both sides of this equation gives $\log_c b^x = \log_c A$,

Now utilizing the exponent property for logs on the left side, $x\log_c b = \log_c A$

Dividing, we obtain $x = \dfrac{\log_c A}{\log_c b}$. Replacing our original expression for x,

$\log_b A = \dfrac{\log_c A}{\log_c b}$

With this change of base formula, we can finally find a good decimal approximation to our question from the beginning of the section.

Example 10

Evaluate $\log_2(10)$ using the change of base formula.

According to the change of base formula, we can rewrite the log base 2 as a logarithm of any other base. Since our calculators can evaluate the natural log, we might choose to use the natural logarithm, which is the log base e:

$$\log_2 10 = \frac{\log_e 10}{\log_e 2} = \frac{\ln 10}{\ln 2}$$

Using our calculators to evaluate this,
$$\frac{\ln 10}{\ln 2} \approx \frac{2.30259}{0.69315} \approx 3.3219$$

This finally allows us to answer our original question – the population of flies we discussed at the beginning of the section will take 3.32 weeks to grow to 500.

Example 11

Evaluate $\log_5(100)$ using the change of base formula.

We can rewrite this expression using any other base. If our calculators are able to evaluate the common logarithm, we could rewrite using the common log, base 10.
$$\log_5(100) = \frac{\log_{10} 100}{\log_{10} 5} \approx \frac{2}{0.69897} = 2.861$$

While we can solve the basic exponential equation $2^x = 10$ by rewriting in logarithmic form and then using the change of base formula to evaluate the logarithm, the proof of the change of base formula illuminates an alternative approach to solving exponential equations.

Solving exponential equations:
1. Isolate the exponential expressions when possible
2. Take the logarithm of both sides
3. Utilize the exponent property for logarithms to pull the variable out of the exponent
4. Use algebra to solve for the variable.

Example 12

Solve $2^x = 10$ for x.

Using this alternative approach, rather than rewrite this exponential into logarithmic form, we will take the logarithm of both sides of the equation. Since we often wish to evaluate the result to a decimal answer, we will usually utilize either the common log or natural log. For this example, we'll use the natural log:

$\ln\left(2^x\right) = \ln(10)$ Utilizing the exponent property for logs,

$x\ln(2) = \ln(10)$ Now dividing by $\ln(2)$,

$x = \dfrac{\ln(10)}{\ln(2)} \approx 3.3219$

Notice that this result matches the result we found using the change of base formula.

Example 13

In the first section, we predicted the population (in billions) of India t years after 2008 by using the function $f(t) = 1.14(1+0.0134)^t$. If the population continues following this trend, when will the population reach 2 billion?

We need to solve for time t so that $f(t) = 2$.

$2 = 1.14(1.0134)^t$ Divide by 1.14 to isolate the exponential expression

$\dfrac{2}{1.14} = 1.0134^t$ Take the logarithm of both sides of the equation

$\ln\left(\dfrac{2}{1.14}\right) = \ln\left(1.0134^t\right)$ Apply the exponent property on the right side

$\ln\left(\dfrac{2}{1.14}\right) = t\ln(1.0134)$ Divide both sides by $\ln(1.0134)$

$t = \dfrac{\ln\left(\dfrac{2}{1.14}\right)}{\ln(1.0134)} \approx 42.23$ years

If this growth rate continues, the model predicts the population of India will reach 2 billion about 42 years after 2008, or approximately in the year 2050.

Try it Now
4. Solve $5(0.93)^x = 10$.

Example 14

Solve $5(1.07)^{3t} = 2$

To start, we want to isolate the exponential part of the expression, the $(1.07)^{3t}$, so it is alone on one side of the equation. Then we can use the log to solve the equation. We can use any base log; this time we'll use the common log.

$5(1.07)^{3t} = 2$ Divide both sides by 5 to isolate the exponential

$(1.07)^{3t} = \dfrac{2}{5}$ Take the log of both sides.

$\log\left((1.07)^{3t}\right) = \log\left(\dfrac{2}{5}\right)$ Use the exponent property for logs

$3t\log(1.07) = \log\left(\dfrac{2}{5}\right)$ Divide by $3\log(1.07)$ on both sides

$\dfrac{3t\log(1.07)}{3\log(1.07)} = \dfrac{\log\left(\dfrac{2}{5}\right)}{3\log(1.07)}$ Simplify and evaluate

$t = \dfrac{\log\left(\dfrac{2}{5}\right)}{3\log(1.07)} \approx -4.5143$

Note that when entering that expression on your calculator, be sure to put parentheses around the whole denominator to ensure the proper order of operations:
`log(2/5)/(3*log(1.07))`

In addition to solving exponential equations, logarithmic expressions are common in many physical situations.

Example 15

In chemistry, pH is a measure of the acidity or basicity of a liquid. The pH is related to the concentration of hydrogen ions, $[H^+]$, measured in moles per liter, by the equation $pH = -\log\left(\left[H^+\right]\right)$.

If a liquid has concentration of 0.0001 moles per liber, determine the pH.
Determine the hydrogen ion concentration of a liquid with pH of 7.

To answer the first question, we evaluate the expression $-\log(0.0001)$. While we could use our calculators for this, we do not really need them here, since we can use the inverse property of logs:
$-\log(0.0001) = -\log\left(10^{-4}\right) = -(-4) = 4$

To answer the second question, we need to solve the equation $7 = -\log\left(\left[H^+\right]\right)$. Begin by isolating the logarithm on one side of the equation by multiplying both sides by -1:
$-7 = \log\left(\left[H^+\right]\right)$. Rewriting into exponential form yields the answer:
$\left[H^+\right] = 10^{-7} = 0.0000001$ moles per liter.

Logarithms also provide us a mechanism for finding continuous growth models for exponential growth given two data points.

Example 15

A population grows from 100 to 130 in 2 weeks. Find the continuous growth rate.

Measuring t in weeks, we are looking for an equation $P(t) = ae^{rt}$ so that $P(0) = 100$ and $P(2) = 130$. Using the first pair of values,
$100 = ae^{r \cdot 0}$, so $a = 100$.

Using the second pair of values,

$130 = 100e^{r \cdot 2}$	Divide by 100
$\dfrac{130}{100} = e^{r2}$	Take the natural log of both sides
$\ln(1.3) = \ln\left(e^{r2}\right)$	Use the inverse property of logs
$\ln(1.3) = 2r$	

$$r = \frac{\ln(1.3)}{2} \approx 0.1312$$

This population is growing at a continuous rate of 13.12% per week.

In general, we can relate the standard form of an exponential with the continuous growth form by noting (using k to represent the continuous growth rate to avoid the confusion of using r in two different ways in the same formula):

$$a(1+r)^x = ae^{kx}$$
$$(1+r)^x = e^{kx}$$
$$1+r = e^k$$

Converting Between Periodic to Continuous Growth Rate

In the equation $f(x) = a(1+r)^x$, r is the **periodic growth rate**, the percent growth each time period (weekly growth, annual growth, etc.).

In the equation $f(x) = ae^{kx}$, k is the **continuous growth rate**.

You can convert between these using: $1 + r = e^k$.

Remember that the continuous growth rate k represents the nominal growth rate before accounting for the effects of continuous compounding, while r represents the actual percent increase in one time unit (one week, one year, etc.).

Example 16

A company's sales can be modeled by the function $S(t) = 5000e^{0.12t}$, with t measured in years. Find the annual growth rate.

Noting that $1 + r = e^k$, then $r = e^{0.12} - 1 = 0.1275$, so the annual growth rate is 12.75%. The sales function could also be written in the form $S(t) = 5000(1 + 0.1275)^t$.

Important Topics of this Section

The Logarithmic function as the inverse of the exponential function

Writing logarithmic & exponential expressions

Properties of logs

    Inverse properties

    Exponential properties

    Change of base

Common log

Natural log

Solving exponential equations

Converting between periodic and continuous growth rate.

Try it Now Answers

1. $\log_4(16) = 2 = \log_4 4^2 = 2\log_4 4$

2. $\log(1000000) = \log(10^6) = 6$

3. $\ln\left(\dfrac{1}{x^2}\right) = \ln(x^{-2}) = -2\ln(x)$

4. $5(0.93)^x = 10$

    $(0.93)^x = 2$

    $\ln(0.93^x) = \ln(2)$

    $x\ln(0.93) = \ln(2)$

    $\dfrac{\ln(2)}{\ln(0.93)} \approx -9.5513$

Section 4.3 Exercises

Rewrite each equation in exponential form

1. $\log_4(q) = m$ 2. $\log_3(t) = k$ 3. $\log_a(b) = c$ 4. $\log_p(z) = u$

5. $\log(v) = t$ 6. $\log(r) = s$ 7. $\ln(w) = n$ 8. $\ln(x) = y$

Rewrite each equation in logarithmic form.

9. $4^x = y$ 10. $5^y = x$ 11. $c^d = k$ 12. $n^z = L$

13. $10^a = b$ 14. $10^p = v$ 15. $e^k = h$ 16. $e^y = x$

Solve for x.

17. $\log_3(x) = 2$ 18. $\log_4(x) = 3$ 19. $\log_2(x) = -3$ 20. $\log_5(x) = -1$

21. $\log(x) = 3$ 22. $\log(x) = 5$ 23. $\ln(x) = 2$ 24. $\ln(x) = -2$

Simplify each expression using logarithm properties.

25. $\log_5(25)$ 26. $\log_2(8)$ 27. $\log_3\left(\dfrac{1}{27}\right)$ 28. $\log_6\left(\dfrac{1}{36}\right)$

29. $\log_6\left(\sqrt{6}\right)$ 30. $\log_5\left(\sqrt[3]{5}\right)$ 31. $\log(10,000)$ 32. $\log(100)$

33. $\log(0.001)$ 34. $\log(0.00001)$ 35. $\ln\left(e^{-2}\right)$ 36. $\ln\left(e^3\right)$

Evaluate using your calculator.

37. $\log(0.04)$ 38. $\log(1045)$ 39. $\ln(15)$ 40. $\ln(0.02)$

Solve each equation for the variable.

41. $5^x = 14$ 42. $3^x = 23$ 43. $7^x = \dfrac{1}{15}$ 44. $3^x = \dfrac{1}{4}$

45. $e^{5x} = 17$ 46. $e^{3x} = 12$ 47. $3^{4x-5} = 38$ 48. $4^{2x-3} = 44$

49. $1000(1.03)^t = 5000$ 50. $200(1.06)^t = 550$

51. $3(1.04)^{3t} = 8$ 52. $2(1.08)^{4t} = 7$

53. $50e^{-0.12t} = 10$ 54. $10e^{-0.03t} = 4$

55. $10 - 8\left(\dfrac{1}{2}\right)^x = 5$ 56. $100 - 100\left(\dfrac{1}{4}\right)^x = 70$

Convert the equation into continuous growth form, $f(t) = ae^{kt}$.

57. $f(t) = 300(0.91)^t$

58. $f(t) = 120(0.07)^t$

59. $f(t) = 10(1.04)^t$

60. $f(t) = 1400(1.12)^t$

Convert the equation into annual growth form, $f(t) = ab^t$.

61. $f(t) = 150e^{0.06t}$

62. $f(t) = 100e^{0.12t}$

63. $f(t) = 50e^{-0.012t}$

64. $f(t) = 80e^{-0.85t}$

65. The population of Kenya was 39.8 million in 2009 and has been growing by about 2.6% each year. If this trend continues, when will the population exceed 45 million?

66. The population of Algeria was 34.9 million in 2009 and has been growing by about 1.5% each year. If this trend continues, when will the population exceed 45 million?

67. The population of Seattle grew from 563,374 in 2000 to 608,660 in 2010. If the population continues to grow exponentially at the same rate, when will the population exceed 1 million people?

68. The median household income (adjusted for inflation) in Seattle grew from $42,948 in 1990 to $45,736 in 2000. If it continues to grow exponentially at the same rate, when will median income exceed $50,000?

69. A scientist begins with 100 mg of a radioactive substance. After 4 hours, it has decayed to 80 mg. How long after the process began will it take to decay to 15 mg?

70. A scientist begins with 100 mg of a radioactive substance. After 6 days, it has decayed to 60 mg. How long after the process began will it take to decay to 10 mg?

71. If $1000 is invested in an account earning 3% compounded monthly, how long will it take the account to grow in value to $1500?

72. If $1000 is invested in an account earning 2% compounded quarterly, how long will it take the account to grow in value to $1300?

Section 4.4 Logarithmic Properties

In the previous section, we derived two important properties of logarithms, which allowed us to solve some basic exponential and logarithmic equations.

Properties of Logs

Inverse Properties:

$$\log_b\left(b^x\right) = x$$

$$b^{\log_b x} = x$$

Exponential Property:

$$\log_b\left(A^r\right) = r\log_b\left(A\right)$$

Change of Base:

$$\log_b\left(A\right) = \frac{\log_c\left(A\right)}{\log_c\left(b\right)}$$

While these properties allow us to solve a large number of problems, they are not sufficient to solve all problems involving exponential and logarithmic equations.

Properties of Logs

Sum of Logs Property:

$$\log_b\left(A\right) + \log_b\left(C\right) = \log_b\left(AC\right)$$

Difference of Logs Property:

$$\log_b\left(A\right) - \log_b\left(C\right) = \log_b\left(\frac{A}{C}\right)$$

It's just as important to know what properties logarithms do *not* satisfy as to memorize the valid properties listed above. In particular, the logarithm is not a linear function, which means that it does not distribute: $\log(A + B) \neq \log(A) + \log(B)$.

To help in this process we offer a proof to help solidify our new rules and show how they follow from properties you've already seen.

Let $a = \log_b(A)$ and $c = \log_b(C)$.

By definition of the logarithm, $b^a = A$ and $b^c = C$.

Using these expressions, $AC = b^a b^c$

Using exponent rules on the right, $AC = b^{a+c}$

Taking the log of both sides, and utilizing the inverse property of logs,

$$\log_b(AC) = \log_b(b^{a+c}) = a + c$$

Replacing a and c with their definition establishes the result

$$\log_b(AC) = \log_b A + \log_b C$$

The proof for the difference property is very similar.

With these properties, we can rewrite expressions involving multiple logs as a single log, or break an expression involving a single log into expressions involving multiple logs.

Example 1

Write $\log_3(5) + \log_3(8) - \log_3(2)$ as a single logarithm.

Using the sum of logs property on the first two terms,

$$\log_3(5) + \log_3(8) = \log_3(5 \cdot 8) = \log_3(40)$$

This reduces our original expression to $\log_3(40) - \log_3(2)$

Then using the difference of logs property,

$$\log_3(40) - \log_3(2) = \log_3\left(\frac{40}{2}\right) = \log_3(20)$$

Example 2

Evaluate $2\log(5) + \log(4)$ without a calculator by first rewriting as a single logarithm.

On the first term, we can use the exponent property of logs to write

$$2\log(5) = \log(5^2) = \log(25)$$

With the expression reduced to a sum of two logs, $\log(25) + \log(4)$, we can utilize the sum of logs property

$$\log(25) + \log(4) = \log(4 \cdot 25) = \log(100)$$

Since $100 = 10^2$, we can evaluate this log without a calculator:

$$\log(100) = \log(10^2) = 2$$

Try it Now
1. Without a calculator evaluate by first rewriting as a single logarithm:

$$\log_2(8) + \log_2(4)$$

Example 3

Rewrite $\ln\left(\dfrac{x^4 y}{7}\right)$ as a sum or difference of logs

First, noticing we have a quotient of two expressions, we can utilize the difference property of logs to write

$$\ln\left(\frac{x^4 y}{7}\right) = \ln(x^4 y) - \ln(7)$$

Then seeing the product in the first term, we use the sum property

$$\ln(x^4 y) - \ln(7) = \ln(x^4) + \ln(y) - \ln(7)$$

Finally, we could use the exponent property on the first term

$$\ln(x^4) + \ln(y) - \ln(7) = 4\ln(x) + \ln(y) - \ln(7)$$

Interestingly, solving exponential equations was not the reason logarithms were originally developed. Historically, up until the advent of calculators and computers, the power of logarithms was that these log properties reduced multiplication, division, roots, or powers to be evaluated using addition, subtraction, division and multiplication, respectively, which are much easier to compute without a calculator. Large books were published listing the logarithms of numbers, such as in the table to the right. To find the product of two numbers, the sum of log property was used. Suppose for example we didn't know the value of 2 times 3. Using the sum property of logs:

value	log(value)
1	0.0000000
2	0.3010300
3	0.4771213
4	0.6020600
5	0.6989700
6	0.7781513
7	0.8450980
8	0.9030900
9	0.9542425
10	1.0000000

$$\log(2 \cdot 3) = \log(2) + \log(3)$$

Using the log table,
$$\log(2 \cdot 3) = \log(2) + \log(3) = 0.3010300 + 0.4771213 = 0.7781513$$

We can then use the table again in reverse, looking for 0.7781513 as an output of the logarithm. From that we can determine:
$$\log(2 \cdot 3) = 0.7781513 = \log(6).$$

By using addition and the table of logs, we were able to determine $2 \cdot 3 = 6$.

Likewise, to compute a cube root like $\sqrt[3]{8}$

$$\log(\sqrt[3]{8}) == \log\left(8^{1/3}\right) = \frac{1}{3}\log(8) = \frac{1}{3}(0.9030900) = 0.3010300 = \log(2)$$

So $\sqrt[3]{8} = 2$.

Although these calculations are simple and insignificant, they illustrate the same idea that was used for hundreds of years as an efficient way to calculate the product, quotient, roots, and powers of large and complicated numbers, either using tables of logarithms or mechanical tools called slide rules.

These properties still have other practical applications for interpreting changes in exponential and logarithmic relationships.

Example 4

Recall that in chemistry, $pH = -\log\left(\left[H^+\right]\right)$. If the concentration of hydrogen ions in a liquid is doubled, what is the affect on pH?

Suppose C is the original concentration of hydrogen ions, and P is the original pH of the liquid, so $P = -\log(C)$. If the concentration is doubled, the new concentration is $2C$. Then the pH of the new liquid is
$$pH - \log(2C)$$

Using the sum property of logs,
$$pH = -\log(2C) = -\left(\log(2) + \log(C)\right) = -\log(2) - \log(C)$$

Since $P = -\log(C)$, the new pH is
$$pH = P - \log(2) = P - 0.301$$

When the concentration of hydrogen ions is doubled, the pH decreases by 0.301.

Log properties in solving equations

The logarithm properties often arise when solving problems involving logarithms. First, we'll look at a simpler log equation.

Example 5

Solve $\log(2x - 6) = 3$.

To solve for x, we need to get it out from inside the log function. There are two ways we can approach this.

Method 1: Rewrite as an exponential.

Recall that since the common log is base 10, $\log(A) = B$ can be rewritten as the exponential $10^B = A$. Likewise, $\log(2x - 6) = 3$ can be rewritten in exponential form as $10^3 = 2x - 6$

Method 2: Exponentiate both sides.

If $A = B$, then $10^A = 10^B$. Using this idea, since $\log(2x - 6) = 3$, then $10^{\log(2x-6)} = 10^3$. Use the inverse property of logs to rewrite the left side and get $2x - 6 = 10^3$.

Using either method, we now need to solve $2x - 6 = 10^3$. Evaluate 10^3 to get

$2x - 6 = 1000$ Add 6 to both sides

$2x = 1006$ Divide both sides by 2

$x = 503$

Occasionally the solving process will result in extraneous solutions – answers that are outside the domain of the original equation. In this case, our answer looks fine.

Example 6

Solve $\log(50x + 25) - \log(x) = 2$.

In order to rewrite in exponential form, we need a single logarithmic expression on the left side of the equation. Using the difference property of logs, we can rewrite the left side:

$$\log\left(\frac{50x + 25}{x}\right) = 2$$

Rewriting in exponential form reduces this to an algebraic equation:

$\dfrac{50x + 25}{x} = 10^2 = 100$ Multiply both sides by x

$50x + 25 = 100x$ Combine like terms

$25 = 50x$ Divide by 50

$x = \dfrac{25}{50} = \dfrac{1}{2}$

Checking this answer in the original equation, we can verify there are no domain issues, and this answer is correct.

Try it Now

2. Solve $\log(x^2 - 4) = 1 + \log(x + 2)$.

294   Chapter 4

Example 7

Solve $\ln(x+2)+\ln(x+1)=\ln(4x+14)$.

$\ln(x+2)+\ln(x+1)=\ln(4x+14)$ Use the sum of logs property on the right

$\ln\big((x+2)(x+1)\big)=\ln(4x+14)$ Expand

$\ln\big(x^2+3x+2\big)=\ln(4x+14)$

We have a log on both side of the equation this time. Rewriting in exponential form would be tricky, so instead we can exponentiate both sides.

$e^{\ln(x^2+3x+2)}=e^{\ln(4x+13)}$ Use the inverse property of logs

$x^2+3x+2=4x+14$ Move terms to one side

$x^2-x-12=0$ Factor

$(x+4)(x-3)=0$

$x=-4$ or $x=3$.

Checking our answers, notice that evaluating the original equation at $x=-4$ would result in us evaluating $\ln(-2)$, which is undefined. That answer is outside the domain of the original equation, so it is an extraneous solution and we discard it. There is one solution: $x=3$.

More complex exponential equations can often be solved in more than one way. In the following example, we will solve the same problem in two ways – one using logarithm properties, and the other using exponential properties.

Example 8a

In 2008, the population of Kenya was approximately 38.8 million, and was growing by 2.64% each year, while the population of Sudan was approximately 41.3 million and growing by 2.24% each year[2]. If these trends continue, when will the population of Kenya match that of Sudan?

We start by writing an equation for each population in terms of t, the number of years after 2008.

$Kenya(t)=38.8(1+0.0264)^t$

$Sudan(t)=41.3(1+0.0224)^t$

To find when the populations will be equal, we can set the equations equal

$38.8(1.0264)^t=41.3(1.0224)^t$

---

[2] World Bank, World Development Indicators, as reported on http://www.google.com/publicdata, retrieved August 24, 2010

For our first approach, we take the log of both sides of the equation.
$$\log\left(38.8(1.0264)^t\right) = \log\left(41.3(1.0224)^t\right)$$

Utilizing the sum property of logs, we can rewrite each side,
$$\log(38.8) + \log\left(1.0264^t\right) = \log(41.3) + \log\left(1.0224^t\right)$$

Then utilizing the exponent property, we can pull the variables out of the exponent
$$\log(38.8) + t\log\left(1.0264\right) = \log(41.3) + t\log\left(1.0224\right)$$

Moving all the terms involving t to one side of the equation and the rest of the terms to the other side,
$$t\log\left(1.0264\right) - t\log\left(1.0224\right) = \log(41.3) - \log(38.8)$$

Factoring out the t on the left,
$$t\left(\log\left(1.0264\right) - \log\left(1.0224\right)\right) = \log(41.3) - \log(38.8)$$

Dividing to solve for t
$$t = \frac{\log(41.3) - \log(38.8)}{\log\left(1.0264\right) - \log\left(1.0224\right)} \approx 15.991 \text{ years until the populations will be equal.}$$

Example 8b

Solve the problem above by rewriting before taking the log.

Starting at the equation
$$38.8(1.0264)^t = 41.3(1.0224)^t$$

Divide to move the exponential terms to one side of the equation and the constants to the other side
$$\frac{1.0264^t}{1.0224^t} = \frac{41.3}{38.8}$$

Using exponent rules to group on the left,
$$\left(\frac{1.0264}{1.0224}\right)^t = \frac{41.3}{38.8}$$

Taking the log of both sides
$$\log\left(\left(\frac{1.0264}{1.0224}\right)^t\right) = \log\left(\frac{41.3}{38.8}\right)$$

Utilizing the exponent property on the left,

$$t \log\left(\frac{1.0264}{1.0224}\right) = \log\left(\frac{41.3}{38.8}\right)$$

Dividing gives

$$t = \frac{\log\left(\frac{41.3}{38.8}\right)}{\log\left(\frac{1.0264}{1.0224}\right)} \approx 15.991 \text{ years}$$

While the answer does not immediately appear identical to that produced using the previous method, note that by using the difference property of logs, the answer could be rewritten:

$$t = \frac{\log\left(\frac{41.3}{38.8}\right)}{\log\left(\frac{1.0264}{1.0224}\right)} = \frac{\log(41.3) - \log(38.8)}{\log(1.0264) - \log(1.0224)}$$

While both methods work equally well, it often requires fewer steps to utilize algebra before taking logs, rather than relying solely on log properties.

Try it Now

3. Tank A contains 10 liters of water, and 35% of the water evaporates each week. Tank B contains 30 liters of water, and 50% of the water evaporates each week. In how many weeks will the tanks contain the same amount of water?

Important Topics of this Section
Inverse
Exponential
Change of base
Sum of logs property
Difference of logs property
Solving equations using log rules

Try it Now Answers

1. $\log_2(8 \cdot 4) = \log_2(32) = \log_2(2^5) = 5$

2. $\log(x^2 - 4) = 1 + \log(x + 2)$ Move both logs to one side

$\log(x^2 - 4) - \log(x + 2) = 1$ Use the difference property of logs

$\log\left(\dfrac{x^2 - 4}{x + 2}\right) = 1$ Factor

$\log\left(\dfrac{(x + 2)(x - 2)}{x + 2}\right) = 1$ Simplify

$\log(x - 2) = 1$ Rewrite as an exponential

$10^1 = x - 2$ Add 2 to both sides

$x = 12$

3. Tank A: $A(t) = 10(1 - 0.35)^t$.   Tank B: $B(t) = 30(1 - 0.50)^t$

Solving $A(t) = B(t)$,

$10(0.65)^t = 30(0.5)^t$ Using the method from Example 8b

$\dfrac{(0.65)^t}{(0.5)^t} = \dfrac{30}{10}$ Regroup

$\left(\dfrac{0.65}{0.5}\right)^t = 3$ Simplify

$(1.3)^t = 3$ Take the log of both sides

$\log\left((1.3)^t\right) = \log(3)$ Use the exponent property of logs

$t \log(1.3) = \log(3)$ Divide and evaluate

$t = \dfrac{\log(3)}{\log(1.3)} \approx 4.1874$ weeks

Section 4.4 Exercises

Simplify to a single logarithm, using logarithm properties.

1. $\log_3(28) - \log_3(7)$

2. $\log_3(32) - \log_3(4)$

3. $-\log_3\left(\dfrac{1}{7}\right)$

4. $-\log_4\left(\dfrac{1}{5}\right)$

5. $\log_3\left(\dfrac{1}{10}\right) + \log_3(50)$

6. $\log_4(3) + \log_4(7)$

7. $\dfrac{1}{3}\log_7(8)$

8. $\dfrac{1}{2}\log_5(36)$

9. $\log(2x^4) + \log(3x^5)$

10. $\ln(4x^2) + \ln(3x^3)$

11. $\ln(6x^9) - \ln(3x^2)$

12. $\log(12x^4) - \log(4x)$

13. $2\log(x) + 3\log(x+1)$

14. $3\log(x) + 2\log(x^2)$

15. $\log(x) - \dfrac{1}{2}\log(y) + 3\log(z)$

16. $2\log(x) + \dfrac{1}{3}\log(y) - \log(z)$

Use logarithm properties to expand each expression.

17. $\log\left(\dfrac{x^{15}y^{13}}{z^{19}}\right)$

18. $\log\left(\dfrac{a^2b^3}{c^5}\right)$

19. $\ln\left(\dfrac{a^{-2}}{b^{-4}c^5}\right)$

20. $\ln\left(\dfrac{a^{-2}b^3}{c^{-5}}\right)$

21. $\log\left(\sqrt{x^3y^{-4}}\right)$

22. $\log\left(\sqrt{x^{-3}y^2}\right)$

23. $\ln\left(y\sqrt{\dfrac{y}{1-y}}\right)$

24. $\ln\left(\dfrac{x}{\sqrt{1-x^2}}\right)$

25. $\log\left(x^2y^3\sqrt[3]{x^2y^5}\right)$

26. $\log\left(x^3y^4\sqrt[7]{x^3y^9}\right)$

Solve each equation for the variable.

27. $4^{4x-7} = 3^{9x-6}$

28. $2^{2x-5} = 7^{3x-7}$

29. $17(1.14)^x = 19(1.16)^x$

30. $20(1.07)^x = 8(1.13)^x$

31. $5e^{0.12t} = 10e^{0.08t}$

32. $3e^{0.09t} = e^{0.14t}$

33. $\log_2(7x+6) = 3$

34. $\log_3(2x+4) = 2$

35. $2\ln(3x) + 3 = 1$

36. $4\ln(5x) + 5 = 2$

37. $\log(x^3) = 2$

38. $\log(x^5) = 3$

39. $\log(x) + \log(x+3) = 3$

40. $\log(x+4) + \log(x) = 9$

41. $\log(x+4) - \log(x+3) = 1$

42. $\log(x+5) - \log(x+2) = 2$

43. $\log_6(x^2) - \log_6(x+1) = 1$

44. $\log_3(x^2) - \log_3(x+2) = 5$

45. $\log(x+12) = \log(x) + \log(12)$

46. $\log(x+15) = \log(x) + \log(15)$

47. $\ln(x) + \ln(x-3) = \ln(7x)$

48. $\ln(x) + \ln(x-6) = \ln(6x)$

Section 4.5 Graphs of Logarithmic Functions

Recall that the exponential function $f(x) = 2^x$ produces this table of values

x	-3	-2	-1	0	1	2	3
$f(x)$	$\dfrac{1}{8}$	$\dfrac{1}{4}$	$\dfrac{1}{2}$	1	2	4	8

Since the logarithmic function is an inverse of the exponential, $g(x) = \log_2(x)$ produces the table of values

x	$\dfrac{1}{8}$	$\dfrac{1}{4}$	$\dfrac{1}{2}$	1	2	4	8
$g(x)$	-3	-2	-1	0	1	2	3

In this second table, notice that
1) As the input increases, the output increases.
2) As input increases, the output increases more slowly.
3) Since the exponential function only outputs positive values, the logarithm can only accept positive values as inputs, so the domain of the log function is $(0, \infty)$.
4) Since the exponential function can accept all real numbers as inputs, the logarithm can output any real number, so the range is all real numbers or $(-\infty, \infty)$.

Sketching the graph, notice that as the input approaches zero from the right, the output of the function grows very large in the negative direction, indicating a vertical asymptote at $x = 0$.
In symbolic notation we write
as $x \to 0^+, f(x) \to -\infty$, and as $x \to \infty, f(x) \to \infty$

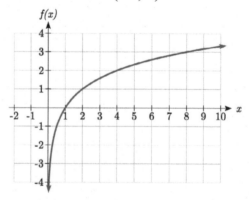

Graphical Features of the Logarithm
Graphically, in the function $g(x) = \log_b(x)$
The graph has a horizontal intercept at (1, 0)
The graph has a vertical asymptote at $x = 0$
The graph is increasing and concave down
The domain of the function is $x > 0$, or $(0, \infty)$
The range of the function is all real numbers, or $(-\infty, \infty)$

When sketching a general logarithm with base b, it can be helpful to remember that the graph will pass through the points $(1, 0)$ and $(b, 1)$.

To get a feeling for how the base affects the shape of the graph, examine the graphs below.

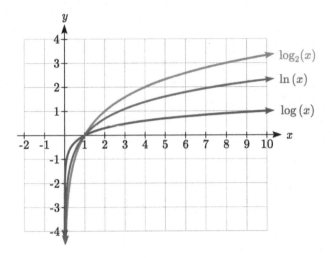

Notice that the larger the base, the slower the graph grows. For example, the common log graph, while it grows without bound, it does so very slowly. For example, to reach an output of 8, the input must be 100,000,000.

Another important observation made was the domain of the logarithm. Like the reciprocal and square root functions, the logarithm has a restricted domain which must be considered when finding the domain of a composition involving a log.

Example 1

Find the domain of the function $f(x) = \log(5 - 2x)$

The logarithm is only defined with the input is positive, so this function will only be defined when $5 - 2x > 0$. Solving this inequality,

$$-2x > -5$$

$$x < \frac{5}{2}$$

The domain of this function is $x < \dfrac{5}{2}$, or in interval notation, $\left(-\infty, \dfrac{5}{2}\right)$

Try it Now

1. Find the domain of the function $f(x) = \log(x - 5) + 2$; before solving this as an inequality, consider how the function has been transformed.

Transformations of the Logarithmic Function

Transformations can be applied to a logarithmic function using the basic transformation techniques, but as with exponential functions, several transformations result in interesting relationships.

First recall the change of base property tells us that $\log_b x = \dfrac{\log_c x}{\log_c b} = \dfrac{1}{\log_c b} \log_c x$

From this, we can see that $\log_b x$ is a vertical stretch or compression of the graph of the $\log_c x$ graph. This tells us that a vertical stretch or compression is equivalent to a change of base. For this reason, we typically represent all graphs of logarithmic functions in terms of the common or natural log functions.

Next, consider the effect of a horizontal compression on the graph of a logarithmic function. Considering $f(x) = \log(cx)$, we can use the sum property to see
$$f(x) = \log(cx) = \log(c) + \log(x)$$

Since $\log(c)$ is a constant, the effect of a horizontal compression is the same as the effect of a vertical shift.

Sketch $f(x) = \ln(x)$ and $g(x) = \ln(x) + 2$.

Graphing these,

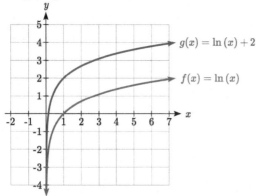

Note that this vertical shift could also be written as a horizontal compression, since
$$g(x) = \ln(x) + 2 = \ln(x) + \ln(e^2) = \ln(e^2 x).$$

While a horizontal stretch or compression can be written as a vertical shift, a horizontal reflection is unique and separate from vertical shifting.

Finally, we will consider the effect of a horizontal shift on the graph of a logarithm.

Example 3

Sketch a graph of $f(x) = \ln(x+2)$.

This is a horizontal shift to the left by 2 units. Notice that none of our logarithm rules allow us rewrite this in another form, so the effect of this transformation is unique. Shifting the graph,

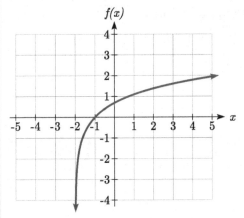

Notice that due to the horizontal shift, the vertical asymptote shifted to $x = -2$, and the domain shifted to $(-2, \infty)$.

Combining these transformations,

Example 4

Sketch a graph of $f(x) = 5\log(-x+2)$.

Factoring the inside as $f(x) = 5\log(-(x-2))$ reveals that this graph is that of the common logarithm, horizontally reflected, vertically stretched by a factor of 5, and shifted to the right by 2 units.

The vertical asymptote will be shifted to $x = 2$, and the graph will have domain $(\infty, 2)$. A rough sketch can be created by using the vertical asymptote along with a couple points on the graph, such as

$f(1) = 5\log(-1+2) = 5\log(1) = 0$

$f(-8) = 5\log(-(-8)+2) = 5\log(10) = 5$

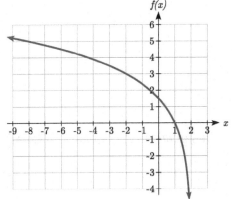

2. Sketch a graph of the function $f(x) = -3\log(x-2) + 1$.

Transformations of Logs

Any transformed logarithmic function can be written in the form

$f(x) = a\log(x-b) + k$, or $f(x) = a\log\big(-(x-b)\big) + k$ if horizontally reflected,

where $x = b$ is the vertical asymptote.

Example 5

Find an equation for the logarithmic function graphed.

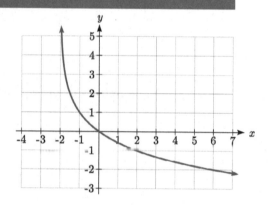

This graph has a vertical asymptote at $x = -2$ and has been vertically reflected. We do not know yet the vertical shift (equivalent to horizontal stretch) or the vertical stretch (equivalent to a change of base). We know so far that the equation will have form
$f(x) = -a\log(x+2) + k$

It appears the graph passes through the points (–1, 1) and (2, –1). Substituting in (–1, 1),
$1 = -a\log(-1+2) + k$

$1 = -a\log(1) + k$

$1 = k$

Next, substituting in (2, –1),
$-1 = -a\log(2+2) + 1$

$-2 = -a\log(4)$

$a = \dfrac{2}{\log(4)}$

This gives us the equation $f(x) = -\dfrac{2}{\log(4)}\log(x+2) + 1$.

This could also be written as $f(x) = -2\log_4(x+2) + 1$.

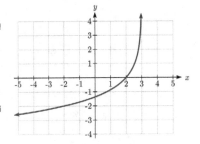

Try it Now

3. Write an equation for the function graphed here.

Flashback

4. Write the domain and range of the function graphed in Example 5, and describe its long run behavior.

Important Topics of this Section
Graph of the logarithmic function (domain and range)
Transformation of logarithmic functions
Creating graphs from equations
Creating equations from graphs

Try it Now and Flashback Answers

1. Domain: $\{x \mid x > 5\}$

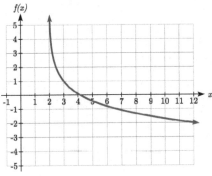

2.

3. The graph is horizontally reflected and has a vertical asymptote at $x = 3$, giving form $f(x) = a\log(-(x-3)) + k$. Substituting in the point (2,0) gives $0 = a\log(-(2-3)) + k$, simplifying to $k = 0$. Substituting in (-2,-2), $-2 = a\log(-(-2-3))$, so $\dfrac{-2}{\log(5)} = a$.

The equation is $f(x) = \dfrac{-2}{\log(5)}\log(-(x-3))$ or $f(x) = -2\log_5(-(x-3))$.

4. Domain: $\{x \mid x > -2\}$, Range: all real numbers; As $x \to -2^+, f(x) \to \infty$ and as $x \to \infty, f(x) \to -\infty$.

Section 4.5 Exercises

For each function, find the domain and the vertical asymptote.

1. $f(x) = \log(x - 5)$

2. $f(x) = \log(x + 2)$

3. $f(x) = \ln(3 - x)$

4. $f(x) = \ln(5 - x)$

5. $f(x) = \log(3x + 1)$

6. $f(x) = \log(2x + 5)$

7. $f(x) = 3\log(-x) + 2$

8. $f(x) = 2\log(-x) + 1$

Sketch a graph of each pair of functions.

9. $f(x) = \log(x), g(x) = \ln(x)$

10. $f(x) = \log_2(x), g(x) = \log_4(x)$

Sketch each transformation.

11. $f(x) = 2\log(x)$

12. $f(x) = 3\ln(x)$

13. $f(x) = \ln(-x)$

14. $f(x) = -\log(x)$

15. $f(x) = \log_2(x + 2)$

16. $f(x) = \log_3(x + 4)$

Find a formula for the transformed logarithm graph shown.

17.

18.

19.

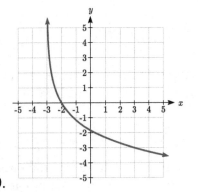

20.

Find a formula for the transformed logarithm graph shown.

21.

22.

23.

24.

Section 4.6 Exponential and Logarithmic Models

While we have explored some basic applications of exponential and logarithmic functions, in this section we explore some important applications in more depth.

Radioactive Decay

In an earlier section, we discussed radioactive decay – the idea that radioactive isotopes change over time. One of the common terms associated with radioactive decay is half-life.

Half Life

The **half-life** of a radioactive isotope is the time it takes for half the substance to decay.

Given the basic exponential growth/decay equation $h(t) = ab^t$, half-life can be found by solving for when half the original amount remains; by solving $\frac{1}{2}a = a(b)^t$, or more simply $\frac{1}{2} = b^t$. Notice how the initial amount is irrelevant when solving for half-life.

Example 1

Bismuth-210 is an isotope that decays by about 13% each day. What is the half-life of Bismuth-210?

We were not given a starting quantity, so we could either make up a value or use an unknown constant to represent the starting amount. To show that starting quantity does not affect the result, let us denote the initial quantity by the constant a. Then the decay of Bismuth-210 can be described by the equation $Q(d) = a(0.87)^d$.

To find the half-life, we want to determine when the remaining quantity is half the original: $\frac{1}{2}a$. Solving,

$$\frac{1}{2}a = a(0.87)^d \qquad \text{Divide by } a,$$

$$\frac{1}{2} = 0.87^d \qquad \text{Take the log of both sides}$$

$$\log\left(\frac{1}{2}\right) = \log\left(0.87^d\right) \qquad \text{Use the exponent property of logs}$$

$$\log\left(\frac{1}{2}\right) = d\log(0.87) \qquad \text{Divide to solve for } d$$

$$d = \frac{\log\left(\frac{1}{2}\right)}{\log(0.87)} \approx 4.977 \text{ days}$$

This tells us that the half-life of Bismuth-210 is approximately 5 days.

Example 2

Cesium-137 has a half-life of about 30 years. If you begin with 200mg of cesium-137, how much will remain after 30 years? 60 years? 90 years?

Since the half-life is 30 years, after 30 years, half the original amount, 100mg, will remain.

After 60 years, another 30 years have passed, so during that second 30 years, another half of the substance will decay, leaving 50mg.

After 90 years, another 30 years have passed, so another half of the substance will decay, leaving 25mg.

Example 3

Cesium-137 has a half-life of about 30 years. Find the annual decay rate.

Since we are looking for an annual decay rate, we will use an equation of the form $Q(t) = a(1+r)^t$. We know that after 30 years, half the original amount will remain. Using this information

$$\frac{1}{2}a = a(1+r)^{30} \qquad \text{Dividing by } a$$

$$\frac{1}{2} = (1+r)^{30} \qquad \text{Taking the 30}^{\text{th}} \text{ root of both sides}$$

$$\sqrt[30]{\frac{1}{2}} = 1+r \qquad \text{Subtracting one from both sides,}$$

$$r = \sqrt[30]{\frac{1}{2}} - 1 \approx -0.02284$$

This tells us cesium-137 is decaying at an annual rate of 2.284% per year.

1. Chlorine-36 is eliminated from the body with a biological half-life of 10 days[3]. Find the daily decay rate.

Example 4

Carbon-14 is a radioactive isotope that is present in organic materials, and is commonly used for dating historical artifacts. Carbon-14 has a half-life of 5730 years. If a bone fragment is found that contains 20% of its original carbon-14, how old is the bone?

To find how old the bone is, we first will need to find an equation for the decay of the carbon-14. We could either use a continuous or annual decay formula, but opt to use the continuous decay formula since it is more common in scientific texts. The half life tells us that after 5730 years, half the original substance remains. Solving for the rate,

$\dfrac{1}{2}a = ae^{r5730}$ Dividing by a

$\dfrac{1}{2} = e^{r5730}$ Taking the natural log of both sides

$\ln\left(\dfrac{1}{2}\right) = \ln\left(e^{r5730}\right)$ Use the inverse property of logs on the right side

$\ln\left(\dfrac{1}{2}\right) = 5730r$ Divide by 5730

$r = \dfrac{\ln\left(\dfrac{1}{2}\right)}{5730} \approx -0.000121$

Now we know the decay will follow the equation $Q(t) = ae^{-0.000121t}$. To find how old the bone fragment is that contains 20% of the original amount, we solve for t so that $Q(t) = 0.20a$.

$0.20a = ae^{-0.000121t}$

$0.20 = e^{-0.000121t}$

$\ln(0.20) = \ln\left(e^{-0.000121t}\right)$

$\ln(0.20) = -0.000121t$

$t = \dfrac{\ln(0.20)}{-0.000121} \approx 13301$ years

The bone fragment is about 13,300 years old.

---

[3] http://www.ead.anl.gov/pub/doc/chlorine.pdf

Try it Now
2. In Example 2, we learned that Cesium-137 has a half-life of about 30 years. If you begin with 200mg of cesium-137, will it take more or less than 230 years until only 1 milligram remains?

Doubling Time

For decaying quantities, we asked how long it takes for half the substance to decay. For growing quantities we might ask how long it takes for the quantity to double.

Doubling Time
The **doubling time** of a growing quantity is the time it takes for the quantity to double.

Given the basic exponential growth equation $h(t) = ab^t$, doubling time can be found by solving for when the original quantity has doubled; by solving $2a = a(b)^x$, or more simply $2 = b^x$. Like with decay, the initial amount is irrelevant when solving for doubling time.

Example 5

Cancer cells sometimes increase exponentially. If a cancerous growth contained 300 cells last month and 360 cells this month, how long will it take for the number of cancer cells to double?

Defining t to be time in months, with $t = 0$ corresponding to this month, we are given two pieces of data: this month, $(0, 360)$, and last month, $(-1, 300)$.

From this data, we can find an equation for the growth. Using the form $C(t) = ab^t$, we know immediately $a = 360$, giving $C(t) = 360b^t$. Substituting in $(-1, 300)$,

$$300 = 360b^{-1}$$

$$300 = \frac{360}{b}$$

$$b = \frac{360}{300} = 1.2$$

This gives us the equation $C(t) = 360(1.2)^t$

To find the doubling time, we look for the time when we will have twice the original amount, so when $C(t) = 2a$.

$$2a = a(1.2)^t$$
$$2 = (1.2)^t$$
$$\log(2) = \log(1.2^t)$$
$$\log(2) = t\log(1.2)$$
$$t = \frac{\log(2)}{\log(1.2)} \approx 3.802 \text{ months for the number of cancer cells to double.}$$

Example 6

Use of a new social networking website has been growing exponentially, with the number of new members doubling every 5 months. If the site currently has 120,000 users and this trend continues, how many users will the site have in 1 year?

We can use the doubling time to find a function that models the number of site users, and then use that equation to answer the question. While we could use an arbitrary a as we have before for the initial amount, in this case, we know the initial amount was 120,000.

If we use a continuous growth equation, it would look like $N(t) = 120e^{rt}$, measured in thousands of users after t months. Based on the doubling time, there would be 240 thousand users after 5 months. This allows us to solve for the continuous growth rate:
$$240 = 120e^{r5}$$
$$2 = e^{r5}$$
$$\ln 2 = 5r$$
$$r = \frac{\ln 2}{5} \approx 0.1386$$

Now that we have an equation, $N(t) = 120e^{0.1386t}$, we can predict the number of users after 12 months:
$$N(12) = 120e^{0.1386(12)} = 633.140 \text{ thousand users.}$$

So after 1 year, we would expect the site to have around 633,140 users.

Try it Now

3. If tuition at a college is increasing by 6.6% each year, how many years will it take for tuition to double?

Newton's Law of Cooling

When a hot object is left in surrounding air that is at a lower temperature, the object's temperature will decrease exponentially, leveling off towards the surrounding air temperature. This "leveling off" will correspond to a horizontal asymptote in the graph of the temperature function. Unless the room temperature is zero, this will correspond to a vertical shift of the generic exponential decay function.

Newton's Law of Cooling

The temperature of an object, T, in surrounding air with temperature T_s will behave according to the formula

$$T(t) = ae^{kt} + T_s$$

Where

t is time

a is a constant determined by the initial temperature of the object

k is a constant, the continuous rate of cooling of the object

---

While an equation of the form $T(t) = ab^t + T_s$ could be used, the continuous growth form is more common.

Example 7

A cheesecake is taken out of the oven with an ideal internal temperature of 165 degrees Fahrenheit, and is placed into a 35 degree refrigerator. After 10 minutes, the cheesecake has cooled to 150 degrees. If you must wait until the cheesecake has cooled to 70 degrees before you eat it, how long will you have to wait?

Since the surrounding air temperature in the refrigerator is 35 degrees, the cheesecake's temperature will decay exponentially towards 35, following the equation
$$T(t) = ae^{kt} + 35$$

We know the initial temperature was 165, so $T(0) = 165$. Substituting in these values,

$165 = ae^{k0} + 35$

$165 = a + 35$

$a = 130$

We were given another pair of data, $T(10) = 150$, which we can use to solve for k

$150 = 130e^{k10} + 35$

$115 = 130e^{k10}$

$\dfrac{115}{130} = e^{10k}$

$\ln\left(\dfrac{115}{130}\right) = 10k$

$k = \dfrac{\ln\left(\dfrac{115}{130}\right)}{10} = -0.0123$

Together this gives us the equation for cooling: $T(t) = 130e^{-0.0123t} + 35$.

Now we can solve for the time it will take for the temperature to cool to 70 degrees.

$70 = 130e^{-0.0123t} + 35$

$35 = 130e^{-0.0123t}$

$\dfrac{35}{130} = e^{-0.0123t}$

$\ln\left(\dfrac{35}{130}\right) = -0.0123t$

$t = \dfrac{\ln\left(\dfrac{35}{130}\right)}{-0.0123} \approx 106.68$

It will take about 107 minutes, or one hour and 47 minutes, for the cheesecake to cool. Of course, if you like your cheesecake served chilled, you'd have to wait a bit longer.

Try it Now

4. A pitcher of water at 40 degrees Fahrenheit is placed into a 70 degree room. One hour later the temperature has risen to 45 degrees. How long will it take for the temperature to rise to 60 degrees?

Logarithmic Scales

For quantities that vary greatly in magnitude, a standard scale of measurement is not always effective, and utilizing logarithms can make the values more manageable. For example, if the average distances from the sun to the major bodies in our solar system are listed, you see they vary greatly.

Planet	Distance (millions of km)
Mercury	58
Venus	108
Earth	150
Mars	228
Jupiter	779
Saturn	1430
Uranus	2880
Neptune	4500

Placed on a linear scale – one with equally spaced values – these values get bunched up.

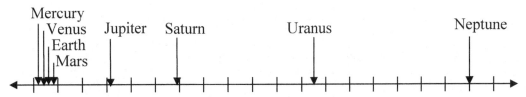

However, computing the logarithm of each value and plotting these new values on a number line results in a more manageable graph, and makes the relative distances more apparent.[4]

Planet	Distance (millions of km)	log(distance)
Mercury	58	1.76
Venus	108	2.03
Earth	150	2.18
Mars	228	2.36
Jupiter	779	2.89
Saturn	1430	3.16
Uranus	2880	3.46
Neptune	4500	3.65

---

[4] It is interesting to note the large gap between Mars and Jupiter on the log number line. The asteroid belt is located there, which scientists believe is a planet that never formed because of the effects of the gravity of Jupiter.

Sometimes, as shown above, the scale on a logarithmic number line will show the log values, but more commonly the original values are listed as powers of 10, as shown below.

Example 8

Estimate the value of point P on the log scale above

The point P appears to be half way between -2 and -1 in log value, so if V is the value of this point,

$\log(V) \approx -1.5$ Rewriting in exponential form,

$V \approx 10^{-1.5} = 0.0316$

Example 9

Place the number 6000 on a logarithmic scale.

Since $\log(6000) \approx 3.8$, this point would belong on the log scale about here:

Try it Now

5. Plot the data in the table below on a logarithmic scale[5].

Source of Sound/Noise	Approximate Sound Pressure in μPa (micro Pascals)
Launching of the Space Shuttle	2,000,000,000
Full Symphony Orchestra	2,000,000
Diesel Freight Train at High Speed at 25 m	200,000
Normal Conversation	20,000
Soft Whispering at 2 m in Library	2,000
Unoccupied Broadcast Studio	200
Softest Sound a human can hear	20

---

[5] From http://www.epd.gov.hk/epd/noise_education/web/ENG_EPD_HTML/m1/intro_5.html, retrieved Oct 2, 2010

Notice that on the log scale above Example 8, the visual distance on the scale between points A and B and between C and D is the same. When looking at the values these points correspond to, notice B is ten times the value of A, and D is ten times the value of C. A visual *linear* difference between points corresponds to a *relative* (ratio) change between the corresponding values.

Logarithms are useful for showing these relative changes. For example, comparing $1,000,000 to $10,000, the first is 100 times larger than the second.

$$\frac{1,000,000}{10,000} = 100 = 10^2$$

Likewise, comparing $1000 to $10, the first is 100 times larger than the second.

$$\frac{1,000}{10} = 100 = 10^2$$

When one quantity is roughly ten times larger than another, we say it is one **order of magnitude** larger. In both cases described above, the first number was two orders of magnitude larger than the second.

Notice that the order of magnitude can be found as the common logarithm of the ratio of the quantities. On the log scale above, B is one order of magnitude larger than A, and D is one order of magnitude larger than C.

Orders of Magnitude

Given two values A and B, to determine how many **orders of magnitude** A is greater than B,

$$\text{Difference in orders of magnitude} = \log\left(\frac{A}{B}\right)$$

Example 10

On the log scale above Example 8, how many orders of magnitude larger is C than B?

The value B corresponds to $10^2 = 100$
The value C corresponds to $10^5 = 100,000$

The relative change is $\dfrac{100,000}{100} = 1000 = \dfrac{10^5}{10^2} = 10^3$. The log of this value is 3.

C is three orders of magnitude greater than B, which can be seen on the log scale by the visual difference between the points on the scale.

Try it Now
6. Using the table from Try it Now #5, what is the difference of order of magnitude between the softest sound a human can hear and the launching of the space shuttle?

Earthquakes

An example of a logarithmic scale is the Moment Magnitude Scale (MMS) used for earthquakes. This scale is commonly and mistakenly called the Richter Scale, which was a very similar scale succeeded by the MMS.

Moment Magnitude Scale

For an earthquake with seismic moment S, a measurement of earth movement, the MMS value, or magnitude of the earthquake, is

$$M = \frac{2}{3}\log\left(\frac{S}{S_0}\right)$$

Where $S_0 = 10^{16}$ is a baseline measure for the seismic moment.

Example 11

If one earthquake has a MMS magnitude of 6.0, and another has a magnitude of 8.0, how much more powerful (in terms of earth movement) is the second earthquake?

Since the first earthquake has magnitude 6.0, we can find the amount of earth movement for that quake, which we'll denote S_1. The value of S_0 is not particularly relevant, so we will not replace it with its value.

$$6.0 = \frac{2}{3}\log\left(\frac{S_1}{S_0}\right)$$

$$6.0\left(\frac{3}{2}\right) = \log\left(\frac{S_1}{S_0}\right)$$

$$9 = \log\left(\frac{S_1}{S_0}\right)$$

$$\frac{S_1}{S_0} = 10^9$$

$$S_1 = 10^9 S_0$$

This tells us the first earthquake has about 10^9 times more earth movement than the baseline measure.

Doing the same with the second earthquake, S_2, with a magnitude of 8.0,

$$8.0 = \frac{2}{3} \log \left(\frac{S_2}{S_0} \right)$$

$$S_2 = 10^{12} S_0$$

Comparing the earth movement of the second earthquake to the first,

$$\frac{S_2}{S_1} = \frac{10^{12} S_0}{10^9 S_0} = 10^3 = 1000$$

The second value's earth movement is 1000 times as large as the first earthquake.

Example 12

One earthquake has magnitude of 3.0. If a second earthquake has twice as much earth movement as the first earthquake, find the magnitude of the second quake.

Since the first quake has magnitude 3.0,

$$3.0 = \frac{2}{3} \log \left(\frac{S}{S_0} \right)$$

Solving for S,

$$3.0 \left(\frac{3}{2} \right) = \log \left(\frac{S}{S_0} \right)$$

$$4.5 = \log \left(\frac{S}{S_0} \right)$$

$$10^{4.5} = \frac{S}{S_0}$$

$$S = 10^{4.5} S_0$$

Since the second earthquake has twice as much earth movement, for the second quake,
$$S = 2 \cdot 10^{4.5} S_0$$

Finding the magnitude,

$$M = \frac{2}{3} \log \left(\frac{2 \cdot 10^{4.5} S_0}{S_0} \right)$$

$$M = \frac{2}{3} \log \left(2 \cdot 10^{4.5} \right) \approx 3.201$$

The second earthquake with twice as much earth movement will have a magnitude of about 3.2.

In fact, using log properties, we could show that whenever the earth movement doubles, the magnitude will increase by about 0.201:

$$M = \frac{2}{3}\log\left(\frac{2S}{S_0}\right) = \frac{2}{3}\log\left(2 \cdot \frac{S}{S_0}\right)$$

$$M = \frac{2}{3}\left(\log(2) + \log\left(\frac{S}{S_0}\right)\right)$$

$$M = \frac{2}{3}\log(2) + \frac{2}{3}\log\left(\frac{S}{S_0}\right)$$

$$M = 0.201 + \frac{2}{3}\log\left(\frac{S}{S_0}\right)$$

This illustrates the most important feature of a log scale: that *multiplying* the quantity being considered will *add* to the scale value, and vice versa.

Important Topics of this Section
Radioactive decay
Half life
Doubling time
Newton's law of cooling
Logarithmic Scales
Orders of Magnitude
Moment Magnitude scale

Try it Now Answers

1. $r = \sqrt[10]{\dfrac{1}{2}} - 1 \approx -0.067$ or 6.7% is the daily rate of decay.

2. Less than 230 years, 229.3157 to be exact

3. Solving $a(1 + 0.066)^t = 2a$, it will take $t = \dfrac{\log(2)}{\log(1.066)} \approx 10.845$ years, or approximately

   11 years, for tuition to double.

4. $T(t) = ae^{kt} + 70$. Substituting (0, 40), we find $a = -30$. Substituting (1, 45), we solve

   $45 = -30e^{k(1)} + 70$ to get $k = \ln\left(\dfrac{25}{30}\right) = -0.1823$.

   Solving $60 = -30e^{-0.1823t} + 70$ gives

   $t = \dfrac{\ln(1/3)}{-0.1823} = 6.026$ hours

5.

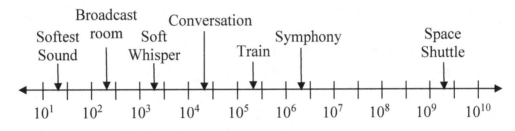

6. $\dfrac{2x10^9}{2x10^1} = 10^8$. The sound pressure in μPa created by launching the space shuttle is 8

   orders of magnitude greater than the sound pressure in μPa created by the softest
   sound a human ear can hear.

Section 4.6 Exercises

1. You go to the doctor and he injects you with 13 milligrams of radioactive dye. After 12 minutes, 4.75 milligrams of dye remain in your system. To leave the doctor's office, you must pass through a radiation detector without sounding the alarm. If the detector will sound the alarm whenever more than 2 milligrams of the dye are in your system, how long will your visit to the doctor take, assuming you were given the dye as soon as you arrived and the amount of dye decays exponentially?

2. You take 200 milligrams of a headache medicine, and after 4 hours, 120 milligrams remain in your system. If the effects of the medicine wear off when less than 80 milligrams remain, when will you need to take a second dose, assuming the amount of medicine in your system decays exponentially?

3. The half-life of Radium-226 is 1590 years. If a sample initially contains 200 mg, how many milligrams will remain after 1000 years?

4. The half-life of Fermium-253 is 3 days. If a sample initially contains 100 mg, how many milligrams will remain after 1 week?

5. The half-life of Erbium-165 is 10.4 hours. After 24 hours a sample still contains 2 mg. What was the initial mass of the sample, and how much will remain after another 3 days?

6. The half-life of Nobelium-259 is 58 minutes. After 3 hours a sample still contains10 mg. What was the initial mass of the sample, and how much will remain after another 8 hours?

7. A scientist begins with 250 grams of a radioactive substance. After 225 minutes, the sample has decayed to 32 grams. Find the half-life of this substance.

8. A scientist begins with 20 grams of a radioactive substance. After 7 days, the sample has decayed to 17 grams. Find the half-life of this substance.

9. A wooden artifact from an archeological dig contains 60 percent of the carbon-14 that is present in living trees. How long ago was the artifact made? (The half-life of carbon-14 is 5730 years.)

10. A wooden artifact from an archeological dig contains 15 percent of the carbon-14 that is present in living trees. How long ago was the artifact made? (The half-life of carbon-14 is 5730 years.)

11. A bacteria culture initially contains 1500 bacteria and doubles in size every half hour. Find the size of the population after: a) 2 hours   b) 100 minutes

12. A bacteria culture initially contains 2000 bacteria and doubles in size every half hour. Find the size of the population after: a) 3 hours   b) 80 minutes

13. The count of bacteria in a culture was 800 after 10 minutes and 1800 after 40 minutes.
    a. What was the initial size of the culture?
    b. Find the doubling time.
    c. Find the population after 105 minutes.
    d. When will the population reach 11000?

14. The count of bacteria in a culture was 600 after 20 minutes and 2000 after 35 minutes.
    a. What was the initial size of the culture?
    b. Find the doubling time.
    c. Find the population after 170 minutes.
    d. When will the population reach 12000?

15. Find the time required for an investment to double in value if invested in an account paying 3% compounded quarterly.

16. Find the time required for an investment to double in value if invested in an account paying 4% compounded monthly

17. The number of crystals that have formed after t hours is given by $n(t) = 20e^{0.013t}$. How long does it take the number of crystals to double?

18. The number of building permits in Pasco t years after 1992 roughly followed the equation $n(t) = 400e^{0.143t}$. What is the doubling time?

19. A turkey is pulled from the oven when the internal temperature is 165° Fahrenheit, and is allowed to cool in a 75° room. If the temperature of the turkey is 145° after half an hour,

    a.   What will the temperature be after 50 minutes?
    b.   How long will it take the turkey to cool to 110°?

20. A cup of coffee is poured at 190° Fahrenheit, and is allowed to cool in a 70° room. If the temperature of the coffee is 170° after half an hour,

    a.   What will the temperature be after 70 minutes?
    b.   How long will it take the coffee to cool to 120°?

21. The population of fish in a farm-stocked lake after t years could be modeled by the equation $P(t) = \dfrac{1000}{1 + 9e^{-0.6t}}$.

    a.   Sketch a graph of this equation.
    b.   What is the initial population of fish?
    c.   What will the population be after 2 years?
    d.   How long will it take for the population to reach 900?

22. The number of people in a town who have heard a rumor after t days can be modeled by the equation $N(t) = \dfrac{500}{1 + 49e^{-0.7t}}$.

    a.   Sketch a graph of this equation.
    b.   How many people started the rumor?
    c.   How many people have heard the rumor after 3 days?
    d.   How long will it take until 300 people have heard the rumor?

Find the value of the number shown on each logarithmic scale

23.

24.

25.

26.

Plot each set of approximate values on a logarithmic scale.

27. Intensity of sounds: Whisper: $10^{-10}\ W/m^2$, Vacuum: $10^{-4}W/m^2$, Jet: $10^2\ W/m^2$

28. Mass: Amoeba: $10^{-5}g$, Human: $10^5 g$, Statue of Liberty: $10^8 g$

29. The 1906 San Francisco earthquake had a magnitude of 7.9 on the MMS scale. Later there was an earthquake with magnitude 4.7 that caused only minor damage. How many times more intense was the San Francisco earthquake than the second one?

30. The 1906 San Francisco earthquake had a magnitude of 7.9 on the MMS scale. Later there was an earthquake with magnitude 6.5 that caused less damage. How many times more intense was the San Francisco earthquake than the second one?

31. One earthquake has magnitude 3.9 on the MMS scale. If a second earthquake has 750 times as much energy as the first, find the magnitude of the second quake.

32. One earthquake has magnitude 4.8 on the MMS scale. If a second earthquake has 1200 times as much energy as the first, find the magnitude of the second quake.

33. A colony of yeast cells is estimated to contain 10^6 cells at time $t = 0$. After collecting experimental data in the lab, you decide that the total population of cells at time t hours is given by the function $f(t) = 10^6 e^{0.495105t}$.   [UW]

    a.   How many cells are present after one hour?
    b.   How long does it take of the population to double?.
    c.   Cherie, another member of your lab, looks at your notebook and says: "That formula is wrong, my calculations predict the formula for the number of yeast cells is given by the function. $f(t) = 10^6 (2.042727)^{0.693147t}$." Should you be worried by Cherie's remark?
    d.   Anja, a third member of your lab working with the same yeast cells, took these two measurements: 7.246×10^6 cells after 4 hours; 16.504×10^6 cells after 6 hours. Should you be worried by Anja's results? If Anja's measurements are correct, does your model over estimate or under estimate the number of yeast cells at time t?

34. As light from the surface penetrates water, its intensity is diminished. In the clear waters of the Caribbean, the intensity is decreased by 15 percent for every 3 meters of depth. Thus, the intensity will have the form of a general exponential function. [UW]
    a.   If the intensity of light at the water's surface is I_0, find a formula for $I(d)$, the intensity of light at a depth of d meters. Your formula should depend on I_0 and d.
    b.   At what depth will the light intensity be decreased to 1% of its surface intensity?

35. Myoglobin and hemoglobin are oxygen-carrying molecules in the human body. Hemoglobin is found inside red blood cells, which flow from the lungs to the muscles through the bloodstream. Myoglobin is found in muscle cells. The function

$Y = M(p) = \dfrac{p}{1+p}$ calculates the fraction of myoglobin saturated with oxygen at a

given pressure p Torrs. For example, at a pressure of 1 Torr, $M(1) = 0.5$, which means half of the myoglobin (i.e. 50%) is oxygen saturated. (Note: More precisely, you need to use something called the "partial pressure", but the distinction is not important for

this problem.) Likewise, the function $Y = H(p) = \dfrac{p^{2.8}}{26^{2.8} + p^{2.8}}$ calculates the fraction

of hemoglobin saturated with oxygen at a given pressure p.   [UW]

a. The graphs of $M(p)$ and $H(p)$ are given here on the domain $0 \le p \le 100$; which is which?

b. If the pressure in the lungs is 100 Torrs, what is the level of oxygen saturation of the hemoglobin in the lungs?

c. The pressure in an active muscle is 20 Torrs. What is the level of oxygen saturation of myoglobin in an active muscle? What is the level of hemoglobin in an active muscle?

d. Define the efficiency of oxygen transport at a given pressure p to be $M(p) - H(p)$. What is the oxygen transport efficiency at 20 Torrs? At 40 Torrs? At 60 Torrs? Sketch the graph of $M(p) - H(p)$; are there conditions under which transport efficiency is maximized (explain)?

36. The length of some fish are modeled by a von Bertalanffy growth function. For Pacific halibut, this function has the form $L(t) = 200\left(1 - 0.957e^{-0.18t}\right)$ where $L(t)$ is the length (in centimeters) of a fish t years old. [UW]

a. What is the length of a newborn halibut at birth?

b. Use the formula to estimate the length of a 6–year–old halibut.

c. At what age would you expect the halibut to be 120 cm long?

d. What is the practical (physical) significance of the number 200 in the formula for $L(t)$?

37. A cancer cell lacks normal biological growth regulation and can divide continuously. Suppose a single mouse skin cell is cancerous and its mitotic cell cycle (the time for the cell to divide once) is 20 hours. The number of cells at time t grows according to an exponential model. [UW]
    a. Find a formula $C(t)$ for the number of cancerous skin cells after t hours.
    b. Assume a typical mouse skin cell is spherical of radius 50×10^{-4} cm. Find the combined volume of all cancerous skin cells after t hours. When will the volume of cancerous cells be 1 cm^3?

38. A ship embarked on a long voyage. At the start of the voyage, there were 500 ants in the cargo hold of the ship. One week into the voyage, there were 800 ants. Suppose the population of ants is an exponential function of time. [UW]
    a. How long did it take the population to double?
    b. How long did it take the population to triple?
    c. When were there be 10,000 ants on board?
    d. There also was an exponentially growing population of anteaters on board. At the start of the voyage there were 17 anteaters, and the population of anteaters doubled every 2.8 weeks. How long into the voyage were there 200 ants per anteater?

39. The populations of termites and spiders in a certain house are growing exponentially. The house contains 100 termites the day you move in. After 4 days, the house contains 200 termites. Three days after moving in, there are two times as many termites as spiders. Eight days after moving in, there were four times as many termites as spiders. How long (in days) does it take the population of spiders to triple? [UW]

Section 4.7 Fitting Exponential Models to Data

In the previous section, we saw number lines using logarithmic scales. It is also common to see two dimensional graphs with one or both axes using a logarithmic scale.

One common use of a logarithmic scale on the vertical axis is to graph quantities that are changing exponentially, since it helps reveal relative differences. This is commonly used in stock charts, since values historically have grown exponentially over time. Both stock charts below show the Dow Jones Industrial Average, from 1928 to 2010.

Both charts have a linear horizontal scale, but the first graph has a linear vertical scale, while the second has a logarithmic vertical scale. The first scale is the one we are more familiar with, and shows what appears to be a strong exponential trend, at least up until the year 2000.

Example 1

There were stock market drops in 1929 and 2008. Which was larger?

In the first graph, the stock market drop around 2008 looks very large, and in terms of dollar values, it was indeed a large drop. However, the second graph shows relative changes, and the drop in 2009 seems less major on this graph, and in fact the drop starting in 1929 was, percentage-wise, much more significant.

Specifically, in 2008, the Dow value dropped from about 14,000 to 8,000, a drop of 6,000. This is obviously a large value drop, and amounts to about a 43% drop. In 1929, the Dow value dropped from a high of around 380 to a low of 42 by July of 1932. While value-wise this drop of 338 is much smaller than the 2008 drop, it corresponds to a 89% drop, a much larger relative drop than in 2008. The logarithmic scale shows these relative changes.

The second graph above, in which one axis uses a linear scale and the other axis uses a logarithmic scale, is an example of a **semi-log** graph.

Semi-log and Log-log Graphs

A **semi-log** graph is a graph with one axis using a linear scale and one axis using a logarithmic scale.

A **log-log** graph is a graph with both axes using logarithmic scales.

Example 2

Plot 5 points on the graph of $f(x) = 3(2)^x$ on a semi-log graph with a logarithmic scale on the vertical axis.

To do this, we need to find 5 points on the graph, then calculate the logarithm of the output value. Arbitrarily choosing 5 input values,

x	$f(x)$	$\log(f(x))$
-3	$3(2)^{-3} = \dfrac{3}{8}$	-0.426
-1	$3(2)^{-1} = \dfrac{3}{2}$	0.176
0	$3(2)^0 = 3$	0.477
2	$3(2)^2 = 12$	1.079
5	$3(2)^5 = 96$	1.982

Plotting these values on a semi-log graph,

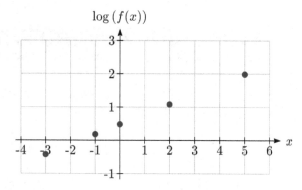

Notice that on this semi-log scale, values from the exponential function appear linear. We can show this behavior is expected by utilizing logarithmic properties. For the function $f(x) = ab^x$, finding log(f(x)) gives

$\log(f(x)) = \log(ab^x)$ Utilizing the sum property of logs,

$\log(f(x)) = \log(a) + \log(b^x)$ Now utilizing the exponent property,

$\log(f(x)) = \log(a) + x\log(b)$

This relationship is linear, with log(a) as the vertical intercept, and log(b) as the slope. This relationship can also be utilized in reverse.

Example 3

An exponential graph is plotted on semi-log axes. Find a formula for the exponential function $g(x)$ that generated this graph.

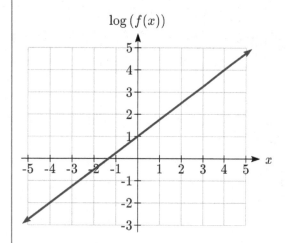

The graph is linear, with vertical intercept at (0, 1). Looking at the change between the points (0, 1) and (4, 4), we can determine the slope of the line is $\frac{3}{4}$. Since the output is log($g(x)$), this leads to the equation $\log(g(x)) = 1 + \frac{3}{4}x$.

We can solve this formula for *g(x)* by rewriting in exponential form and simplifying:

$$\log(g(x)) = 1 + \frac{3}{4}x \qquad \text{Rewriting as an exponential,}$$

$$g(x) = 10^{1+\frac{3}{4}x} \qquad \text{Breaking this apart using exponent rules,}$$

$$g(x) = 10^1 \cdot 10^{\frac{3}{4}x} \qquad \text{Using exponent rules to group the second factor,}$$

$$g(x) = 10^1 \cdot \left(10^{\frac{3}{4}}\right)^x \qquad \text{Evaluating the powers of 10,}$$

$$g(x) = 10(5.623)^x$$

Try it Now

1. An exponential graph is plotted on a semi-log graph below. Find a formula for the exponential function *g(x)* that generated this graph.

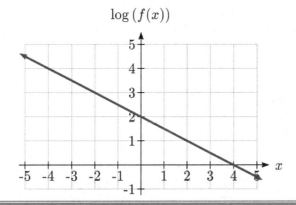

Fitting Exponential Functions to Data

Some technology options provide dedicated functions for finding exponential functions that fit data, but many only provide functions for fitting linear functions to data. The semi-log scale provides us with a method to fit an exponential function to data by building upon the techniques we have for fitting linear functions to data.

To fit an exponential function to a set of data using linearization
1. Find the log of the data output values
2. Find the linear equation that fits the (input, log(output)) pairs. This equation will be of the form $\log(f(x)) = b + mx$.
3. Solve this equation for the exponential function *f(x)*

Example 4

The table below shows the cost in dollars per megabyte of storage space on computer hard drives from 1980 to 2004, and the data is shown on a standard graph to the right, with the input changed to years after 1980.

Year	Cost per MB
1980	192.31
1984	87.86
1988	15.98
1992	4
1996	0.173
2000	0.006849
2004	0.001149

This data appears to be decreasing exponentially. To find a function that models this decay, we would start by finding the log of the costs.

Year	t	Cost per MB	log(Cost)
1980	0	192.31	2.284002
1984	4	87.86	1.943791
1988	8	15.98	1.203577
1992	12	4	0.60206
1996	16	0.173	-0.76195
2000	20	0.006849	-2.16437
2004	24	0.001149	-2.93952

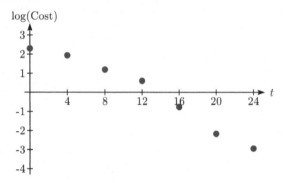

As hoped, the graph of the log of costs appears fairly linear, suggesting an exponential function will fit the original data will fit reasonably well. Using technology, we can find a linear equation to fit the log(Cost) values. Using t as years after 1980, linear regression gives the equation:

$$\log(C(t)) = 2.794 - 0.231t$$

Solving for $C(t)$,

$$C(t) = 10^{2.794 - 0.231t}$$

$$C(t) = 10^{2.794} \cdot 10^{-0.231t}$$

$$C(t) = 10^{2.794} \cdot \left(10^{-0.231}\right)^{t}$$

$$C(t) = 622 \cdot (0.5877)^{t}$$

This equation suggests that the cost per megabyte for storage on computer hard drives is decreasing by about 41% each year.

Using this function, we could predict the cost of storage in the future. Predicting the cost in the year 2020 ($t = 40$):

$C(40) = 622(0.5877)^{40} \approx 0.000000364$ dollars per megabyte, a really small number. That is equivalent to $0.36 per terabyte of hard drive storage.

Comparing the values predicted by this model to the actual data, we see the model matches the original data in order of magnitude, but the specific values appear quite different. This is, unfortunately, the best exponential model that can fit the data. It is possible that a non-exponential model would fit the data better, or there could just be wide enough variability in the data that no relatively simple model would fit the data any better.

Year	Actual Cost per MB	Cost predicted by model
1980	192.31	622.3
1984	87.86	74.3
1988	15.98	8.9
1992	4	1.1
1996	0.173	0.13
2000	0.006849	0.015
2004	0.001149	0.0018

Try it Now

2. The table below shows the value V, in billions of dollars, of US imports from China t years after 2000.

year	2000	2001	2002	2003	2004	2005
t	0	1	2	3	4	5
V	100	102.3	125.2	152.4	196.7	243.5

This data appears to be growing exponentially. Linearize this data and build a model to predict how many billions of dollars of imports were expected in 2011.

Important Topics of this Section
Semi-log graph
Log-log graph
Linearizing exponential functions
Fitting an exponential equation to data

Try it Now Answers

1. $g(x) = 10^{2-0.5x} = 10^2 (10^{-0.5})^x$. $f(x) = 100(0.3162)^x$

2. $V(t) = 90.545(1.2078)^t$.   Predicting in 2011, $V(11) = 722.45$ billion dollars

Section 4.7 Exercises

Graph each function on a semi-log scale, then find a formula for the linearized function in the form $\log(f(x)) = mx + b$.

1. $f(x) = 4(1.3)^x$

2. $f(x) = 2(1.5)^x$

3. $f(x) = 10(0.2)^x$

4. $f(x) = 30(0.7)^x$

The graph below is on a semi-log scale, as indicated. Find a formula for the exponential function $y(x)$.

5.

6.

7.

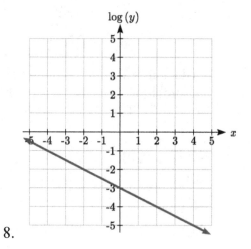

8.

Use regression to find an exponential function that best fits the data given.

9.
x	1	2	3	4	5	6
y	1125	1495	2310	3294	4650	6361

10.
x	1	2	3	4	5	6
y	643	829	920	1073	1330	1631

11.
x	1	2	3	4	5	6
y	555	383	307	210	158	122

12.
x	1	2	3	4	5	6
y	699	701	695	668	683	712

13. Total expenditures (in billions of dollars) in the US for nursing home care are shown below. Use regression to find an exponential function that models the data. What does the model predict expenditures will be in 2015?

Year	1990	1995	2000	2003	2005	2008
Expenditure	53	74	95	110	121	138

14. Light intensity as it passes through water decreases exponentially with depth. The data below shows the light intensity (in lumens) at various depths. Use regression to find an function that models the data. What does the model predict the intensity will be at 25 feet?

Depth (ft)	3	6	9	12	15	18
Lumen	11.5	8.6	6.7	5.2	3.8	2.9

15. The average price of electricity (in cents per kilowatt hour) from 1990 through 2008 is given below. Determine if a linear or exponential model better fits the data, and use the better model to predict the price of electricity in 2014.

Year	1990	1992	1994	1996	1998	2000	2002	2004	2006	2008
Cost	7.83	8.21	8.38	8.36	8.26	8.24	8.44	8.95	10.40	11.26

16. The average cost of a loaf of white bread from 1986 through 2008 is given below. Determine if a linear or exponential model better fits the data, and use the better model to predict the price of a loaf of bread in 2016.

Year	1986	1988	1990	1995	1997	2000	2002	2004	2006	2008
Cost	0.57	0.66	0.70	0.84	0.88	0.99	1.03	0.97	1.14	1.42

Chapter 5: Trigonometric Functions of Angles

In the previous chapters, we have explored a variety of functions which could be combined to form a variety of shapes. In this discussion, one common shape has been missing: the circle. We already know certain things about the circle, like how to find area and circumference, and the relationship between radius and diameter, but now, in this chapter, we explore the circle and its unique features that lead us into the rich world of trigonometry.

Section 5.1 Circles

To begin, we need to find distances. Starting with the Pythagorean Theorem, which relates the sides of a right triangle, we can find the distance between two points.

Pythagorean Theorem

The Pythagorean Theorem states that the sum of the squares of the legs of a right triangle will equal the square of the hypotenuse of the triangle.

In graphical form, given the triangle shown, $a^2 + b^2 = c^2$.

We can use the Pythagorean Theorem to find the distance between two points on a graph.

Example 1

Find the distance between the points (-3, 2) and (2, 5).

By plotting these points on the plane, we can then draw a right triangle with these points at each end of the hypotenuse. We can calculate horizontal width of the triangle to be 5 and the vertical height to be 3.

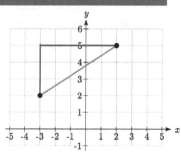

From these we can find the distance between the points using the Pythagorean Theorem:

$$dist^2 = 5^2 + 3^2 = 34$$

$$dist = \sqrt{34}$$

Notice that the width of the triangle was calculated using the difference between the *x* (input) values of the two points, and the height of the triangle was found using the difference between the *y* (output) values of the two points. Generalizing this process gives us the distance formula.

Distance Formula

The distance between two points (x_1, y_1) and (x_2, y_2) can be calculated as

$$dist = \sqrt{(x_2 - x_1)^2 + (y_2 - y_1)^2}$$

Try it Now
1. Find the distance between the points (1, 6) and (3, -5).

Circles

If we wanted to find an equation to represent a circle with a radius of *r* centered at a point (*h, k*), we notice that the distance between any point (*x, y*) on the circle and the center point is always the same: *r*. Noting this, we can use our distance formula to write an equation for the radius:

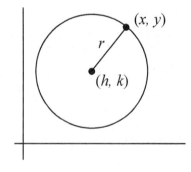

$$r = \sqrt{(x - h)^2 + (y - k)^2}$$

Squaring both sides of the equation gives us the standard equation for a circle.

Equation of a Circle

The **equation of a circle** centered at the point (*h, k*) with radius *r* can be written as

$$(x - h)^2 + (y - k)^2 = r^2$$

Notice that a circle does not pass the vertical line test. It is not possible to write *y* as a function of *x* or vice versa.

Example 2

Write an equation for a circle centered at the point (-3, 2) with radius 4.

Using the equation from above, $h = -3$, $k = 2$, and the radius $r = 4$. Using these in our formula,

$(x-(-3))^2 + (y-2)^2 = 4^2$ simplified, this gives

$(x+3)^2 + (y-2)^2 = 16$

Example 3

Write an equation for the circle graphed here.

This circle is centered at the origin, the point (0, 0). By measuring horizontally or vertically from the center out to the circle, we can see the radius is 3. Using this information in our formula gives:

$(x-0)^2 + (y-0)^2 = 3^2$ simplified, this gives

$x^2 + y^2 = 9$

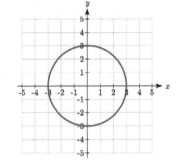

Try it Now

2. Write an equation for a circle centered at (4, -2) with radius 6.

Notice that, relative to a circle centered at the origin, horizontal and vertical shifts of the circle are revealed in the values of h and k, which are the coordinates for the center of the circle.

Points on a Circle

As noted earlier, an equation for a circle cannot be written so that y is a function of x or vice versa. To find coordinates on the circle given only the x or y value, we must solve algebraically for the unknown values.

Example 4

Find the points on a circle of radius 5 centered at the origin with an x value of 3.

We begin by writing an equation for the circle centered at the origin with a radius of 5.

$x^2 + y^2 = 25$

Substituting in the desired x value of 3 gives an equation we can solve for y.

$$3^2 + y^2 = 25$$
$$y^2 = 25 - 9 = 16$$
$$y = \pm\sqrt{16} = \pm 4$$

There are two points on the circle with an x value of 3: (3, 4) and (3, -4).

Example 5

Find the x intercepts of a circle with radius 6 centered at the point (2, 4).

We can start by writing an equation for the circle.
$$(x-2)^2 + (y-4)^2 = 36$$

To find the x intercepts, we need to find the points where $y = 0$. Substituting in zero for y, we can solve for x.
$$(x-2)^2 + (0-4)^2 = 36$$
$$(x-2)^2 + 16 = 36$$
$$(x-2)^2 = 20$$
$$x - 2 = \pm\sqrt{20}$$
$$x = 2 \pm \sqrt{20} = 2 \pm 2\sqrt{5}$$

The x intercepts of the circle are $\left(2 + 2\sqrt{5}, 0\right)$ and $\left(2 - 2\sqrt{5}, 0\right)$

Example 6

In a town, Main Street runs east to west, and Meridian Road runs north to south. A pizza store is located on Meridian 2 miles south of the intersection of Main and Meridian. If the store advertises that it delivers within a 3-mile radius, how much of Main Street do they deliver to?

This type of question is one in which introducing a coordinate system and drawing a picture can help us solve the problem. We could either place the origin at the intersection of the two streets, or place the origin at the pizza store itself. It is often easier to work with circles centered at the origin, so we'll place the origin at the pizza store, though either approach would work fine.

Placing the origin at the pizza store, the delivery area with radius 3 miles can be described as the region inside the circle described by $x^2 + y^2 = 9$.

Main Street, located 2 miles north of the pizza store and running east to west, can be described by the equation $y = 2$.

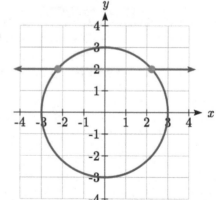

To find the portion of Main Street the store will deliver to, we first find the boundary of their delivery region by looking for where the delivery circle intersects Main Street. To find the intersection, we look for the points on the circle where $y = 2$. Substituting $y = 2$ into the circle equation lets us solve for the corresponding x values.

$x^2 + 2^2 = 9$

$x^2 = 9 - 4 = 5$

$x = \pm\sqrt{5} \approx \pm 2.236$

This means the pizza store will deliver 2.236 miles down Main Street east of Meridian and 2.236 miles down Main Street west of Meridian. We can conclude that the pizza store delivers to a 4.472 mile long segment of Main St.

In addition to finding where a vertical or horizontal line intersects the circle, we can also find where an arbitrary line intersects a circle.

Example 7

Find where the line $f(x) = 4x$ intersects the circle $(x - 2)^2 + y^2 = 16$.

Normally, to find an intersection of two functions $f(x)$ and $g(x)$ we would solve for the x value that would make the functions equal by solving the equation $f(x) = g(x)$. In the case of a circle, it isn't possible to represent the equation as a function, but we can utilize the same idea.

The output value of the line determines the y value: $y = f(x) = 4x$. We want the y value of the circle to equal the y value of the line, which is the output value of the function. To do this, we can substitute the expression for y from the line into the circle equation.

$(x - 2)^2 + y^2 = 16$ replace y with the line formula: $y = 4x$

$(x - 2)^2 + (4x)^2 = 16$ expand

$x^2 - 4x + 4 + 16x^2 = 16$ simplify

$17x^2 - 4x + 4 = 16$ since this equation is quadratic, we arrange one side to be 0

$17x^2 - 4x - 12 = 0$

Since this quadratic doesn't appear to be easily factorable, we can use the quadratic formula to solve for x:

$$x = \frac{-(-4) \pm \sqrt{(-4)^2 - 4(17)(-12)}}{2(17)} = \frac{4 \pm \sqrt{832}}{34}, \text{ or approximately } x \approx 0.966 \text{ or } -0.731$$

From these x values we can use either equation to find the corresponding y values.

Since the line equation is easier to evaluate, we might choose to use it:
$y = f(0.966) = 4(0.966) = 3.864$
$y = f(-0.731) = 4(-0.731) = -2.923$

The line intersects the circle at the points (0.966, 3.864) and (-0.731, -2.923).

Try it Now

3. A small radio transmitter broadcasts in a 50 mile radius. If you drive along a straight line from a city 60 miles north of the transmitter to a second city 70 miles east of the transmitter, during how much of the drive will you pick up a signal from the transmitter?

Important Topics of This Section
Distance formula
Equation of a Circle
Finding the x coordinate of a point on the circle given the y coordinate or vice versa
Finding the intersection of a circle and a line

Try it Now Answers

1. $5\sqrt{5}$

2. $(x-4)^2 + (y+2)^2 = 36$

3. The circle can be represented by $x^2 + y^2 = 50^2$.

Finding a line from (0,60) to (70,0) gives $y = 60 - \dfrac{60}{70}x$.

Substituting the line equation into the circle gives $x^2 + \left(60 - \dfrac{60}{70}x\right)^2 = 50^2$.

Solving this equation, we find $x = 14$ or $x = 45.29$, corresponding to points (14, 48) and (45.29, 21.18).
The distance between these points is 41.21 miles.

Section 5.1 Exercises

1. Find the distance between the points (5,3) and (-1,-5).

2. Find the distance between the points (3,3) and (-3,-2).

3. Write an equation of the circle centered at (8 , -10) with radius 8.

4. Write an equation of the circle centered at (-9, 9) with radius 16.

5. Write an equation of the circle centered at (7, -2) that passes through (-10, 0).

6. Write an equation of the circle centered at (3, -7) that passes through (15, 13).

7. Write an equation for a circle where the points (2, 6) and (8, 10) lie along a diameter.

8. Write an equation for a circle where the points (-3, 3) and (5, 7) lie along a diameter.

9. Sketch a graph of $(x-2)^2 + (y+3)^2 = 9$.

10. Sketch a graph of $(x+1)^2 + (y-2)^2 = 16$.

11. Find the y intercept(s) of the circle with center (2, 3) with radius 3.

12. Find the x intercept(s) of the circle with center (2, 3) with radius 4.

13. At what point in the first quadrant does the line with equation $y=2x+5$ intersect a circle with radius 3 and center (0, 5)?

14. At what point in the first quadrant does the line with equation $y=x+2$ intersect the circle with radius 6 and center (0, 2)?

15. At what point in the second quadrant does the line with equation $y=2x+5$ intersect a circle with radius 3 and center (-2, 0)?

16. At what point in the first quadrant does the line with equation $y=x+2$ intersect the circle with radius 6 and center (-1,0)?

17. A small radio transmitter broadcasts in a 53 mile radius. If you drive along a straight line from a city 70 miles north of the transmitter to a second city 74 miles east of the transmitter, during how much of the drive will you pick up a signal from the transmitter?

18. A small radio transmitter broadcasts in a 44 mile radius. If you drive along a straight line from a city 56 miles south of the transmitter to a second city 53 miles west of the transmitter, during how much of the drive will you pick up a signal from the transmitter?

19. A tunnel connecting two portions of a space
    station has a circular cross-section of radius 15
    feet. Two walkway decks are constructed in the
    tunnel. Deck A is along a horizontal diameter
    and another parallel Deck B is 2 feet below Deck
    A. Because the space station is in a weightless
    environment, you can walk vertically upright
    along Deck A, or vertically upside down along
    Deck B. You have been assigned to paint "safety
    stripes" on each deck level, so that a 6 foot
    person can safely walk upright along either deck.
    Determine the width of the "safe walk zone" on
    each deck. [UW]

(a) Cross-section of tunnel.

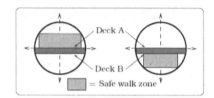

(b) Walk zones.

20. A crawling tractor sprinkler is
    located as pictured here, 100 feet
    south of a sidewalk. Once the water
    is turned on, the sprinkler waters a
    circular disc of radius 20 feet and
    moves north along the hose at the
    rate of ½ inch/second. The hose is
    perpendicular to the 10 ft. wide
    sidewalk. Assume there is grass on
    both sides of the sidewalk. [UW]

    a) Impose a coordinate system.
        Describe the initial coordinates
        of the sprinkler and find
        equations of the lines forming and find equations of the lines forming the north
        and south boundaries of the sidewalk.
    b) When will the water first strike the sidewalk?
    c) When will the water from the sprinkler fall completely north of the sidewalk?
    d) Find the total amount of time water from the sprinkler falls on the sidewalk.
    e) Sketch a picture of the situation after 33 minutes. Draw an accurate picture of the
        watered portion of the sidewalk.
    f) Find the area of grass watered after one hour.

21. Erik's disabled sailboat is floating anchored 3 miles East and 2 miles north of Kingsford. A ferry leaves Kingsford heading toward Eaglerock at 12 mph. Eaglerock is 6 miles due east of Kingsford. After 20 minutes the ferry turns, heading due south. Bander is 8 miles south and 1 mile west of Eaglerock. Impose coordinates with Bander as the origin. [UW]

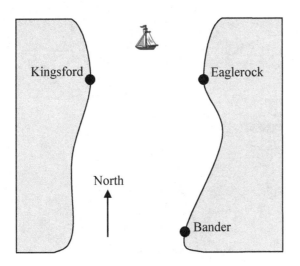

a) Find equations for the lines along which the ferry is moving and draw in these lines.

b) The sailboat has a radar scope that will detect any object within 3 miles of the sailboat. Looking down from above, as in the picture, the radar region looks like a circular disk. The boundary is the "edge" or circle around this disk, the interior is everything inside of the circle, and the exterior is everything outside of the circle. Give the mathematical description (an equation or inequality) of the boundary, interior and exterior of the radar zone. Sketch an accurate picture of the radar zone by determining where the line connecting Kingsford and Eaglerock would cross the radar zone.

c) When does the ferry enter the radar zone?

d) Where and when does the ferry exit the radar zone?

e) How long does the ferry spend inside the radar zone?

22. Nora spends part of her summer driving a combine during the wheat harvest. Assume she starts at the indicated position heading east at 10 ft/sec toward a circular wheat field of radius 200 ft. The combine cuts a swath 20 feet wide and begins when the corner of the machine labeled "a" is 60 feet north and 60 feet west of the western-most edge of the field. [UW]

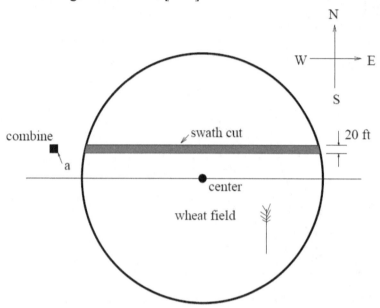

a) When does Nora's combine first start cutting the wheat?
b) When does Nora's combine first start cutting a swath 20 feet wide?
c) Find the total amount of time wheat is being cut during this pass across the field.
d) Estimate the area of the swath cut during this pass across the field.

23. The vertical cross-section of a drainage ditch is pictured to the right. Here, R indicates in each case the radius of a circle with $R = 10$ feet, where all of the indicated circle centers lie along a horizontal line 10 feet above and parallel to the ditch bottom. Assume that water is flowing into the ditch so that the level above the bottom is rising at a rate of 2 inches per minute. [UW]

3D−view of ditch

vertical cross-section

a) When will the ditch be completely full?
b) Find a piecewise defined function that models the vertical cross-section of the ditch.
c) What is the width of the filled portion of the ditch after 1 hour and 18 minutes?
d) When will the filled portion of the ditch be 42 feet wide? 50 feet wide? 73 feet wide?

Section 5.2 Angles

Because many applications involving circles also involve a rotation of the circle, it is natural to introduce a measure for the rotation, or angle, between two rays (line segments) emanating from the center of a circle. The angle measurement you are most likely familiar with is degrees, so we'll begin there.

Measure of an Angle

The **measure of an angle** is a measurement between two intersecting lines, line segments or rays, starting at the **initial side** and ending at the **terminal side**. It is a rotational measure, not a linear measure.

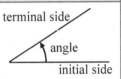

Measuring Angles

Degrees

A **degree** is a measurement of angle. One full rotation around the circle is equal to 360 degrees, so one degree is 1/360 of a circle.

An angle measured in degrees should always include the unit "degrees" after the number, or include the degree symbol °. For example, 90 degrees = 90°.

Standard Position

When measuring angles on a circle, unless otherwise directed, we measure angles in **standard position**: starting at the positive horizontal axis and with counter-clockwise rotation.

Example 1

Give the degree measure of the angle shown on the circle.

The vertical and horizontal lines divide the circle into quarters. Since one full rotation is 360 degrees= 360°, each quarter rotation is 360/4 = 90° or 90 degrees.

Example 2

Show an angle of 30° on the circle.

An angle of 30° is 1/3 of 90°, so by dividing a quarter rotation into thirds, we can sketch a line at 30°.

Going Greek

When representing angles using variables, it is traditional to use Greek letters. Here is a list of commonly encountered Greek letters.

θ	φ or ϕ	α	β	γ
theta	phi	alpha	beta	gamma

Working with Angles in Degrees

Notice that since there are 360 degrees in one rotation, an angle greater than 360 degrees would indicate more than 1 full rotation. Shown on a circle, the resulting direction in which this angle's terminal side points would be the same as for another angle between 0 and 360 degrees.   These angles would be called **coterminal**.

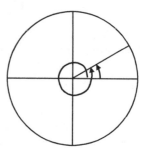

Coterminal Angles

After completing their full rotation based on the given angle, two angles are **coterminal** if they terminate in the same position, so their terminal sides coincide (point in the same direction).

Example 3

Find an angle θ that is coterminal with $800°$, where $0° \leq \theta < 360°$

Since adding or subtracting a full rotation, 360 degrees, would result in an angle with terminal side pointing in the same direction, we can find coterminal angles by adding or subtracting 360 degrees.   An angle of 800 degrees is coterminal with an angle of 800-360 = 440 degrees. It would also be coterminal with an angle of 440-360 = 80 degrees.

The angle $\theta = 80°$ is coterminal with $800°$.

By finding the coterminal angle between 0 and 360 degrees, it can be easier to see which direction the terminal side of an angle points in.

Try it Now

1. Find an angle α that is coterminal with $870°$, where $0° \leq \alpha < 360°$.

On a number line a positive number is measured to the right and a negative number is measured in the opposite direction (to the left). Similarly a positive angle is measured counterclockwise and a negative angle is measured in the opposite direction (clockwise).

Show the angle $-45°$ on the circle and find a positive angle α that is coterminal and $0° \le \alpha < 360°$.

Since 45 degrees is half of 90 degrees, we can start at the positive horizontal axis and measure clockwise half of a 90 degree angle.

Since we can find coterminal angles by adding or subtracting a full rotation of 360 degrees, we can find a positive coterminal angle here by adding 360 degrees:
$-45° + 360° = 315°$

Try it Now

2. Find an angle β coterminal with $-300°$ where $0° \le \beta < 360°$.

It can be helpful to have a familiarity with the frequently encountered angles in one rotation of a circle. It is common to encounter multiples of 30, 45, 60, and 90 degrees. These values are shown to the right. Memorizing these angles and understanding their properties will be very useful as we study the properties associated with angles

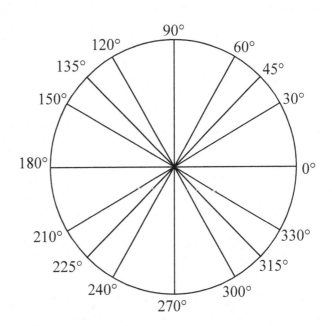

Angles in Radians

While measuring angles in degrees may be familiar, doing so often complicates matters since the units of measure can get in the way of calculations. For this reason, another measure of angles is commonly used. This measure is based on the distance around a circle.

Arclength
Arclength is the length of an arc, s, along a circle of radius r subtended (drawn out) by an angle θ.   It is the portion of the circumference between the initial and terminal sides of the angle.

The length of the arc around an entire circle is called the circumference of a circle. The circumference of a circle is $C = 2\pi r$. The ratio of the circumference to the radius, produces the constant 2π. Regardless of the radius, this ratio is always the same, just as how the degree measure of an angle is independent of the radius.

To elaborate on this idea, consider two circles, one with radius 2 and one with radius 3. Recall the circumference (perimeter) of a circle is $C = 2\pi r$, where r is the radius of the circle. The smaller circle then has circumference $2\pi(2) = 4\pi$ and the larger has circumference $2\pi(3) = 6\pi$.

Drawing a 45 degree angle on the two circles, we might be interested in the length of the arc of the circle that the angle indicates. In both cases, the 45 degree angle draws out an arc that is 1/8th of the full circumference, so for the smaller circle, the

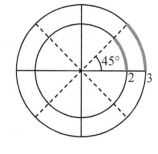

arclength $= \dfrac{1}{8}(4\pi) = \dfrac{1}{2}\pi$, and for the larger circle, the length of the

arc or arclength $= \dfrac{1}{8}(6\pi) = \dfrac{3}{4}\pi$.

Notice what happens if we find the *ratio* of the arclength divided by the radius of the circle:

Smaller circle: $\dfrac{\frac{1}{2}\pi}{2} = \dfrac{1}{4}\pi$

Larger circle: $\dfrac{\frac{3}{4}\pi}{3} = \dfrac{1}{4}\pi$

The ratio is the same regardless of the radius of the circle – it only depends on the angle. This property allows us to define a measure of the angle based on arclength.

Radians

The **radian measure** of an angle is the ratio of the length of the circular arc subtended by the angle to the radius of the circle.

In other words, if s is the length of an arc of a circle, and r is the radius of the circle, then

$$\text{radian measure} = \frac{s}{r}$$

If the circle has radius 1, then the radian measure corresponds to the length of the arc.

Because radian measure is the ratio of two lengths, it is a **unitless measure**. It is not necessary to write the label "radians" after a radian measure, and if you see an angle that is not labeled with "degrees" or the degree symbol, you should assume that it is a radian measure.

Considering the most basic case, the unit circle (a circle with radius 1), we know that 1 rotation equals 360 degrees, $360°$. We can also track one rotation around a circle by finding the circumference, $C = 2\pi r$, and for the unit circle $C = 2\pi$. These two different ways to rotate around a circle give us a way to convert from degrees to radians.

1 rotation = $360° = 2\pi$ radians
½ rotation = $180°$ = π radians
¼ rotation = $90° = \dfrac{\pi}{2}$ radians

Example 5

Find the radian measure of one third of a full rotation.

For any circle, the arclength along such a rotation would be one third of the circumference, $C = \dfrac{1}{3}(2\pi r) = \dfrac{2\pi r}{3}$. The radian measure would be the arclength divided by the radius:

$$\text{Radian measure} = \frac{2\pi r}{3r} = \frac{2\pi}{3}.$$

Converting Between Radians and Degrees

1 degree $= \dfrac{\pi}{180}$ radians

or: to convert from degrees to radians, multiply by $\dfrac{\pi \text{ radians}}{180°}$

1 radian $= \dfrac{180}{\pi}$ degrees

or: to convert from radians to degrees, multiply by $\dfrac{180°}{\pi \text{ radians}}$

Example 6

Convert $\dfrac{\pi}{6}$ radians to degrees.

Since we are given a problem in radians and we want degrees, we multiply by $\dfrac{180°}{\pi}$.

Remember radians are a unitless measure, so we don't need to write "radians."

$\dfrac{\pi}{6}$ radians $= \dfrac{\pi}{6} \cdot \dfrac{180°}{\pi} = 30$ degrees

Example 7

Convert 15 degrees to radians.

In this example, we start with degrees and want radians so we use the other conversion $\dfrac{\pi}{180°}$ so that the degree units cancel and we are left with the unitless measure of radians.

15 degrees $= 15° \cdot \dfrac{\pi}{180°} = \dfrac{\pi}{12}$

Try it Now

3. Convert $\dfrac{7\pi}{10}$ radians to degrees.

Just as we listed all the common angles in degrees on a circle, we should also list the corresponding radian values for the common measures of a circle corresponding to degree multiples of 30, 45, 60, and 90 degrees. As with the degree measurements, it would be advisable to commit these to memory.

We can work with the radian measures of an angle the same way we work with degrees.

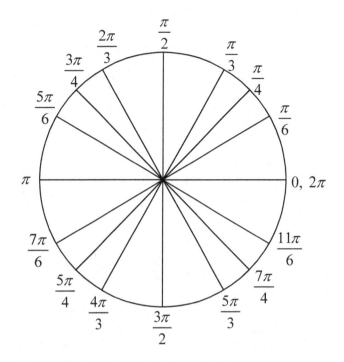

Example 8

Find an angle β that is coterminal with $\dfrac{19\pi}{4}$, where $0 \le \beta < 2\pi$.

When working in degrees, we found coterminal angles by adding or subtracting 360 degrees, a full rotation. Likewise, in radians, we can find coterminal angles by adding or subtracting full rotations of 2π radians.

$$\frac{19\pi}{4} - 2\pi = \frac{19\pi}{4} - \frac{8\pi}{4} = \frac{11\pi}{4}$$

The angle $\dfrac{11\pi}{4}$ is coterminal, but not less than 2π, so we subtract another rotation.

$$\frac{11\pi}{4} - 2\pi = \frac{11\pi}{4} - \frac{8\pi}{4} = \frac{3\pi}{4}$$

The angle $\dfrac{3\pi}{4}$ is coterminal with $\dfrac{19\pi}{4}$.

Try it Now

4. Find an angle ϕ that is coterminal with $-\dfrac{17\pi}{6}$ where $0 \le \phi < 2\pi$.

Arclength and Area of a Sector

Recall that the radian measure of an angle was defined as the ratio of the arclength of a circular arc to the radius of the circle, $\theta = \dfrac{s}{r}$. From this relationship, we can find arclength along a circle given an angle.

Arclength on a Circle

The length of an arc, s, along a circle of radius r subtended by angle θ in radians is

$s = r\theta$

Example 9

Mercury orbits the sun at a distance of approximately 36 million miles. In one Earth day, it completes 0.0114 rotation around the sun. If the orbit was perfectly circular, what distance through space would Mercury travel in one Earth day?

To begin, we will need to convert the decimal rotation value to a radian measure.

Since one rotation = 2π radians,
0.0114 rotation = $2\pi(0.0114) = 0.0716$ radians.

Combining this with the given radius of 36 million miles, we can find the arclength:
$s = (36)(0.0716) = 2.578$ million miles travelled through space.

Try it Now
5. Find the arclength along a circle of radius 10 subtended by an angle of 215 degrees.

In addition to arclength, we can also use angles to find the area of a sector of a circle. A sector is a portion of a circle contained between two lines from the center, like a slice of pizza or pie.

Recall that the area of a circle with radius r can be found using the formula $A = \pi r^2$. If a sector is cut out by an angle of θ, measured in radians, then the fraction of full circle that angle has cut out is $\dfrac{\theta}{2\pi}$, since 2π is one full rotation. Thus, the area of the sector would be this fraction of the whole area:

$\text{Area of sector} = \left(\dfrac{\theta}{2\pi} \right) \pi r^2 = \dfrac{\theta \pi r^2}{2\pi} = \dfrac{1}{2}\theta r^2$

Area of a Sector

The **area of a sector** of a circle with radius r subtended by an angle θ, measured in radians, is

Area of sector $= \dfrac{1}{2}\theta r^2$

Example 10

An automatic lawn sprinkler sprays a distance of 20 feet while rotating 30 degrees. What is the area of the sector of grass the sprinkler waters?

First, we need to convert the angle measure into radians. Since 30 degrees is one of our common angles, you ideally should already know the equivalent radian measure, but if not we can convert:

30 degrees $= 30 \cdot \dfrac{\pi}{180} = \dfrac{\pi}{6}$ radians.

The area of the sector is then Area $= \dfrac{1}{2}\left(\dfrac{\pi}{6}\right)(20)^2 = 104.72$ ft^2

Try it Now

6. In central pivot irrigation, a large irrigation pipe on wheels rotates around a center point, as pictured here[1]. A farmer has a central pivot system with a radius of 400 meters. If water restrictions only allow her to water 150 thousand square meters a day, what angle should she set the system to cover?

Linear and Angular Velocity

When your car drives down a road, it makes sense to describe its speed in terms of miles per hour or meters per second. These are measures of speed along a line, also called linear velocity. When a point on a circle rotates, we would describe its angular velocity, or rotational speed, in radians per second, rotations per minute, or degrees per hour.

---

[1]

Angular and Linear Velocity

As a point moves along a circle of radius r, its **angular velocity, ω,** can be found as the angular rotation θ per unit time, t.

$$\omega = \frac{\theta}{t}$$

The **linear velocity**, v, of the point can be found as the distance travelled, arclength s, per unit time, t.

$$v = \frac{s}{t}$$

Example 11

A water wheel completes 1 rotation every 5 seconds. Find the angular velocity in radians per second.[2]

The wheel completes 1 rotation = 2π radians in 5 seconds, so the angular velocity would be $\omega = \dfrac{2\pi}{5} \approx 1.257$ radians per second.

Combining the definitions above with the arclength equation, $s = r\theta$, we can find a relationship between angular and linear velocities. The angular velocity equation can be solved for θ, giving $\theta = \omega t$. Substituting this into the arclength equation gives $s = r\theta = r\omega t$.

Substituting this into the linear velocity equation gives

$$v = \frac{s}{t} = \frac{r\omega t}{t} = r\omega$$

Relationship Between Linear and Angular Velocity

When the angular velocity is measured in radians per unit time, linear velocity and angular velocity are related by the equation

$$v = r\omega$$

---

[2] http://en.wikipedia.org/wiki/File:R%C3%B6mische_S%C3%A4gem%C3%BChle.svg

Example 12

A bicycle has wheels 28 inches in diameter. A tachometer determines the wheels are rotating at 180 RPM (revolutions per minute). Find the speed the bicycle is travelling down the road.

Here we have an angular velocity and need to find the corresponding linear velocity, since the linear speed of the outside of the tires is the speed at which the bicycle travels down the road.

We begin by converting from rotations per minute to radians per minute. It can be helpful to utilize the units to make this conversion

$$180 \frac{\text{rotations}}{\text{minute}} \cdot \frac{2\pi \text{ radians}}{\text{rotation}} = 360\pi \frac{\text{radians}}{\text{minute}}$$

Using the formula from above along with the radius of the wheels, we can find the linear velocity

$$v = (14 \text{ inches})\left(360\pi \frac{\text{radians}}{\text{minute}}\right) = 5040\pi \frac{\text{inches}}{\text{minute}}$$

You may be wondering where the "radians" went in this last equation. Remember that radians are a unitless measure, so it is not necessary to include them.

Finally, we may wish to convert this linear velocity into a more familiar measurement, like miles per hour.

$$5040\pi \frac{\text{inches}}{\text{minute}} \cdot \frac{1 \text{ feet}}{12 \text{ inches}} \cdot \frac{1 \text{ mile}}{5280 \text{ feet}} \cdot \frac{60 \text{ minutes}}{1 \text{ hour}} = 14.99 \text{ miles per hour (mph)}.$$

Try it Now

7. A satellite is rotating around the earth at 27,934 kilometers per minute at an altitude of 242 km above the earth. If the radius of the earth is 6378 kilometers, find the angular velocity of the satellite.

Important Topics of This Section
Degree measure of angle
Radian measure of angle
Conversion between degrees and radians
Common angles in degrees and radians
Coterminal angles
Arclength
Area of a sector
Linear and angular velocity

Try it Now Answers

1. $\alpha = 870 - 360 - 360 = 150°$

2. $\beta = -300 + 360 = 60°$

3. $\dfrac{7\pi}{10} \cdot \dfrac{180°}{\pi} = 126°$

4. $-\dfrac{17\pi}{6} + 2\pi + 2\pi = -\dfrac{17\pi}{6} + \dfrac{12\pi}{6} + \dfrac{12\pi}{6} = \dfrac{7\pi}{6}$

5. $215° = \dfrac{215\pi}{180}$ radians. $s = 10 \cdot \dfrac{215\pi}{180} = \dfrac{215\pi}{18} \approx 37.525$

6. $\dfrac{1}{2}\theta(400)^2 = 150{,}000$. $\theta = 1.875$, or $107.43°$

7. $v = 27934$. $r = 6378 + 242 = 6620$. $\omega = \dfrac{v}{r} = \dfrac{27934}{6620} = 4.2196$ radians per hour.

Section 5.2 Exercises

1. Indicate each angle on a circle: $30°$, $300°$, $-135°$, $70°$, $\dfrac{2\pi}{3}$, $\dfrac{7\pi}{4}$

2. Indicate each angle on a circle: $30°$, $315°$, $-135°$, $80°$, $\dfrac{7\pi}{6}$, $\dfrac{3\pi}{4}$

3. Convert the angle $180°$ to radians.

4. Convert the angle $30°$ to radians.

5. Convert the angle $\dfrac{5\pi}{6}$ from radians to degrees.

6. Convert the angle $\dfrac{11\pi}{6}$ from radians to degrees.

7. Find the angle between $0°$ and $360°$ that is coterminal with a $685°$ angle.

8. Find the angle between $0°$ and $360°$ that is coterminal with a $451°$ angle.

9. Find the angle between $0°$ and $360°$ that is coterminal with a $-1746°$ angle.

10. Find the angle between $0°$ and $360°$ that is coterminal with a $-1400°$ angle.

11. Find the angle between 0 and 2π in radians that is coterminal with the angle $\dfrac{26\pi}{9}$.

12. Find the angle between 0 and 2π in radians that is coterminal with the angle $\dfrac{17\pi}{3}$.

13. Find the angle between 0 and 2π in radians that is coterminal with the angle $-\dfrac{3\pi}{2}$.

14. Find the angle between 0 and 2π in radians that is coterminal with the angle $-\dfrac{7\pi}{6}$.

15. On a circle of radius 7 miles, find the length of the arc that subtends a central angle of 5 radians.

16. On a circle of radius 6 feet, find the length of the arc that subtends a central angle of 1 radian.

17. On a circle of radius 12 cm, find the length of the arc that subtends a central angle of 120 degrees.

18. On a circle of radius 9 miles, find the length of the arc that subtends a central angle of 800 degrees.

19. Find the distance along an arc on the surface of the Earth that subtends a central angle of 5 minutes (1 minute = 1/60 degree). The radius of the Earth is 3960 miles.

20. Find the distance along an arc on the surface of the Earth that subtends a central angle of 7 minutes (1 minute = 1/60 degree). The radius of the Earth is 3960 miles.

21. On a circle of radius 6 feet, what angle in degrees would subtend an arc of length 3 feet?

22. On a circle of radius 5 feet, what angle in degrees would subtend an arc of length 2 feet?

23. A sector of a circle has a central angle of 45°. Find the area of the sector if the radius of the circle is 6 cm.

24. A sector of a circle has a central angle of 30°. Find the area of the sector if the radius of the circle is 20 cm.

25. A truck with 32-in.-diameter wheels is traveling at 60 mi/h. Find the angular speed of the wheels in rad/min. How many revolutions per minute do the wheels make?

26. A bicycle with 24-in.-diameter wheels is traveling at 15 mi/h. Find the angular speed of the wheels in rad/min. How many revolutions per minute do the wheels make?

27. A wheel of radius 8 in. is rotating 15°/sec. What is the linear speed v, the angular speed in RPM, and the angular speed in rad/sec?

28. A wheel of radius 14 in. is rotating 0.5 rad/sec. What is the linear speed v, the angular speed in RPM, and the angular speed in deg/sec?

29. A CD has diameter of 120 millimeters. When playing audio, the angular speed varies to keep the linear speed constant where the disc is being read. When reading along the outer edge of the disc, the angular speed is about 200 RPM (revolutions per minute). Find the linear speed.

30. When being burned in a writable CD-R drive, the angular speed of a CD is often much faster than when playing audio, but the angular speed still varies to keep the linear speed constant where the disc is being written. When writing along the outer edge of the disc, the angular speed of one drive is about 4800 RPM (revolutions per minute). Find the linear speed.

31. You are standing on the equator of the Earth (radius 3960 miles). What is your linear and angular speed?

32. The restaurant in the Space Needle in Seattle rotates at the rate of one revolution every 47 minutes. [UW]
    a) Through how many radians does it turn in 100 minutes?
    b) How long does it take the restaurant to rotate through 4 radians?
    c) How far does a person sitting by the window move in 100 minutes if the radius of the restaurant is 21 meters?

Section 5.3 Points on Circles Using Sine and Cosine

While it is convenient to describe the location of a point on a circle using an angle or a distance along the circle, relating this information to the x and y coordinates and the circle equation we explored in Section 5.1 is an important application of trigonometry.

A distress signal is sent from a sailboat during a storm, but the transmission is unclear and the rescue boat sitting at the marina cannot determine the sailboat's location. Using high powered radar, they determine the distress signal is coming from a distance of 20 miles at an angle of 225 degrees from the marina. How many miles east/west and north/south of the rescue boat is the stranded sailboat?

In a general sense, to investigate this, we begin by drawing a circle centered at the origin with radius r, and marking the point on the circle indicated by some angle θ. This point has coordinates (x, y).

If we drop a line segment vertically down from this point to the x axis, we would form a right triangle inside of the circle.

No matter which quadrant our angle θ puts us in we can draw a triangle by dropping a perpendicular line segment to the x axis, keeping in mind that the values of x and y may be positive or negative, depending on the quadrant.

Additionally, if the angle θ puts us on an axis, we simply measure the radius as the x or y with the other value being 0, again ensuring we have appropriate signs on the coordinates based on the quadrant.

Triangles obtained from different radii will all be similar triangles, meaning corresponding sides scale proportionally. While the lengths of the sides may change, as we saw in the last section, the *ratios* of the side lengths will always remain constant for any given angle.

$$\frac{y_1}{r_1} = \frac{y_2}{r_2}$$

$$\frac{x_1}{r_1} = \frac{x_2}{r_2}$$

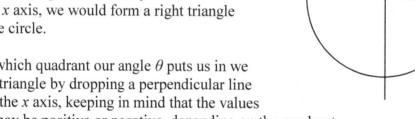

To be able to refer to these ratios more easily, we will give them names. Since the ratios depend on the angle, we will write them as functions of the angle θ.

Sine and Cosine

For the point (x, y) on a circle of radius r at an angle of θ, we can define two important functions as the ratios of the sides of the corresponding triangle:

The **sine** function: $\sin(\theta) = \dfrac{y}{r}$

The **cosine** function: $\cos(\theta) = \dfrac{x}{r}$

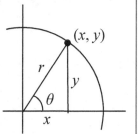

In this chapter, we will explore these functions using both circles and right triangles. In the next chapter, we will take a closer look at the behavior and characteristics of the sine and cosine functions.

Example 1

The point $(3, 4)$ is on the circle of radius 5 at some angle θ. Find $\cos(\theta)$ and $\sin(\theta)$.

Knowing the radius of the circle and coordinates of the point, we can evaluate the cosine and sine functions as the ratio of the sides.

$$\cos(\theta) = \frac{x}{r} = \frac{3}{5} \qquad\qquad \sin(\theta) = \frac{y}{r} = \frac{4}{5}$$

There are a few cosine and sine values which we can determine fairly easily because the corresponding point on the circle falls on the x or y axis.

Example 2

Find $\cos(90°)$ and $\sin(90°)$

On any circle, the terminal side of a 90 degree angle points straight up, so the coordinates of the corresponding point on the circle would be $(0, r)$. Using our definitions of cosine and sine,

$$\cos(90°) = \frac{x}{r} = \frac{0}{r} = 0$$

$$\sin(90°) = \frac{y}{r} = \frac{r}{r} = 1$$

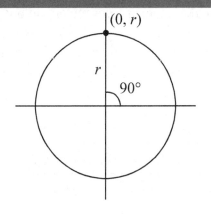

Try it Now

Try it Now
1. Find cosine and sine of the angle π.

Notice that the definitions above can also be stated as:

> ### Coordinates of the Point on a Circle at a Given Angle
>
> On a circle of radius r at an angle of θ, we can find the **coordinates of the point** (x, y) at that angle using
>
> $x = r\cos(\theta)$
>
> $y = r\sin(\theta)$
>
> On a unit circle, a circle with radius 1, $x = \cos(\theta)$ and $y = \sin(\theta)$.

Utilizing the basic equation for a circle centered at the origin, $x^2 + y^2 = r^2$, combined with the relationships above, we can establish a new identity.

$x^2 + y^2 = r^2$	substituting the relations above,
$(r\cos(\theta))^2 + (r\sin(\theta))^2 = r^2$	simplifying,
$r^2(\cos(\theta))^2 + r^2(\sin(\theta))^2 = r^2$	dividing by r^2
$(\cos(\theta))^2 + (\sin(\theta))^2 = 1$	or using shorthand notation
$\cos^2(\theta) + \sin^2(\theta) = 1$	

Here $\cos^2(\theta)$ is a commonly used shorthand notation for $(\cos(\theta))^2$. Be aware that many calculators and computers do not understand the shorthand notation.

In Section 5.1 we related the Pythagorean Theorem $a^2 + b^2 = c^2$ to the basic equation of a circle $x^2 + y^2 = r^2$, which we have now used to arrive at the Pythagorean Identity.

> ### Pythagorean Identity
>
> The **Pythagorean Identity**. For any angle θ, $\cos^2(\theta) + \sin^2(\theta) = 1$.

One use of this identity is that it helps us to find a cosine value of an angle if we know the sine value of that angle or vice versa. However, since the equation will yield two possible values, we will need to utilize additional knowledge of the angle to help us find the desired value.

Example 3

If $\sin(\theta) = \dfrac{3}{7}$ and θ is in the second quadrant, find $\cos(\theta)$.

Substituting the known value for sine into the Pythagorean identity,

$$\cos^2(\theta) + \left(\frac{3}{7}\right)^2 = 1$$

$$\cos^2(\theta) + \frac{9}{49} = 1$$

$$\cos^2(\theta) = \frac{40}{49}$$

$$\cos(\theta) = \pm\sqrt{\frac{40}{49}} = \pm\frac{\sqrt{40}}{7} = \pm\frac{2\sqrt{10}}{7}$$

Since the angle is in the second quadrant, we know the x value of the point would be negative, so the cosine value should also be negative. Using this additional information, we can conclude that $\cos(\theta) = -\dfrac{2\sqrt{10}}{7}$.

Values for Sine and Cosine

At this point, you may have noticed that we haven't found any cosine or sine values from angles not on an axis. To do this, we will need to utilize our knowledge of triangles.

First, consider a point on a circle at an angle of 45 degrees, or $\dfrac{\pi}{4}$.

At this angle, the x and y coordinates of the corresponding point on the circle will be equal because 45 degrees divides the first quadrant in half. Since the x and y values will be the same, the sine and cosine values will also be equal. Utilizing the Pythagorean Identity,

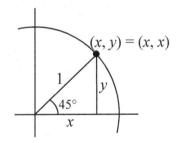

$$\cos^2\left(\frac{\pi}{4}\right) + \sin^2\left(\frac{\pi}{4}\right) = 1 \qquad \text{since the sine and cosine are equal, we can}$$

substitute sine with cosine

$$\cos^2\left(\frac{\pi}{4}\right) + \cos^2\left(\frac{\pi}{4}\right) = 1 \qquad \text{add like terms}$$

$$2\cos^2\left(\frac{\pi}{4}\right) = 1 \qquad \text{divide}$$

$$\cos^2\left(\frac{\pi}{4}\right) = \frac{1}{2} \qquad \text{since the } x \text{ value is positive, we'll keep the positive root}$$

$$\cos\left(\frac{\pi}{4}\right) = \sqrt{\frac{1}{2}}$$
often this value is written with a rationalized denominator

Remember, to rationalize the denominator we multiply by a term equivalent to 1 to get rid of the radical in the denominator.

$$\cos\left(\frac{\pi}{4}\right) = \sqrt{\frac{1}{2}}\sqrt{\frac{2}{2}} = \sqrt{\frac{2}{4}} = \frac{\sqrt{2}}{2}$$

Since the sine and cosine are equal, $\sin\left(\frac{\pi}{4}\right) = \frac{\sqrt{2}}{2}$ as well. The (x, y) coordinates for a point on a circle of radius 1 at an angle of 45 degrees are $\left(\frac{\sqrt{2}}{2}, \frac{\sqrt{2}}{2}\right)$.

Example 4

Find the coordinates of the point on a circle of radius 6 at an angle of $\frac{\pi}{4}$.

Using our new knowledge that $\sin\left(\frac{\pi}{4}\right) = \frac{\sqrt{2}}{2}$ and $\cos\left(\frac{\pi}{4}\right) = \frac{\sqrt{2}}{2}$, along with our relationships that stated $x = r\cos(\theta)$ and $y = r\sin(\theta)$, we can find the coordinates of the point desired:

$$x = 6\cos\left(\frac{\pi}{4}\right) = 6\left(\frac{\sqrt{2}}{2}\right) = 3\sqrt{2}$$

$$y = 6\sin\left(\frac{\pi}{4}\right) = 6\left(\frac{\sqrt{2}}{2}\right) = 3\sqrt{2}$$

Try it Now

2. Find the coordinates of the point on a circle of radius 3 at an angle of $90°$.

Next, we will find the cosine and sine at an angle of 30 degrees, or $\frac{\pi}{6}$. To do this, we will first draw a triangle inside a circle with one side at an angle of 30 degrees, and another at an angle of -30 degrees. If the resulting two right triangles are combined into one

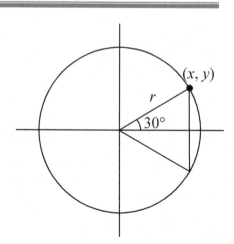

large triangle, notice that all three angles of this larger triangle will be 60 degrees. Since all the angles are equal, the sides will all be equal as well. The vertical line has length $2y$, and since the sides are all equal we can conclude that $2y = r$, or $y = \dfrac{r}{2}$. Using this, we can find the sine value:

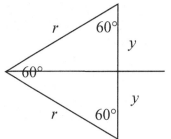

$$\sin\left(\frac{\pi}{6}\right) = \frac{y}{r} = \frac{r/2}{r} = \frac{r}{2} \cdot \frac{1}{r} = \frac{1}{2}$$

Using the Pythagorean Identity, we can find the cosine value:

$$\cos^2\left(\frac{\pi}{6}\right) + \sin^2\left(\frac{\pi}{6}\right) = 1$$

$$\cos^2\left(\frac{\pi}{6}\right) + \left(\frac{1}{2}\right)^2 = 1$$

$$\cos^2\left(\frac{\pi}{6}\right) = \frac{3}{4} \qquad \text{since the } x \text{ value is positive, we'll keep the positive root}$$

$$\cos\left(\frac{\pi}{6}\right) = \sqrt{\frac{3}{4}} = \frac{\sqrt{3}}{2}$$

The (x, y) coordinates for the point on a circle of radius 1 at an angle of 30 degrees are $\left(\dfrac{\sqrt{3}}{2}, \dfrac{1}{2}\right)$.

By drawing a the triangle inside the unit circle with a 30 degree angle and reflecting it over the line $y = x$, we can find the cosine and sine for 60 degrees, or $\dfrac{\pi}{3}$, without any additional work.

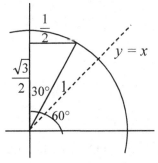

By this symmetry, we can see the coordinates of the point on the unit circle at an angle of 60 degrees will be $\left(\dfrac{1}{2}, \dfrac{\sqrt{3}}{2}\right)$, giving

$$\cos\left(\frac{\pi}{3}\right) = \frac{1}{2} \quad \text{and} \quad \sin\left(\frac{\pi}{3}\right) = \frac{\sqrt{3}}{2}$$

We have now found the cosine and sine values for all the commonly encountered angles in the first quadrant of the unit circle.

Angle	0	$\frac{\pi}{6}$, or 30°	$\frac{\pi}{4}$, or 45°	$\frac{\pi}{3}$, or 60°	$\frac{\pi}{2}$, or 90°
Cosine	1	$\frac{\sqrt{3}}{2}$	$\frac{\sqrt{2}}{2}$	$\frac{1}{2}$	0
Sine	0	$\frac{1}{2}$	$\frac{\sqrt{2}}{2}$	$\frac{\sqrt{3}}{2}$	1

For any given angle in the first quadrant, there will be an angle in another quadrant with the same sine value, and yet another angle in yet another quadrant with the same cosine value. Since the sine value is the y coordinate on the unit circle, the other angle with the same sine will share the same y value, but have the opposite x value. Likewise, the angle with the same cosine will share the same x value, but have the opposite y value.

As shown here, angle α has the same sine value as angle θ; the cosine values would be opposites. The angle β has the same cosine value as the angle θ; the sine values would be opposites.

$\sin(\theta) = \sin(\alpha)$ and $\cos(\theta) = -\cos(\alpha)$ $\qquad$ $\sin(\theta) = -\sin(\beta)$ and $\cos(\theta) = \cos(\beta)$

 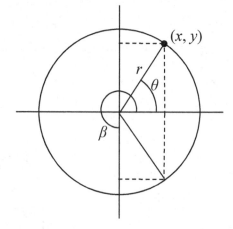

It is important to notice the relationship between the angles. If, from the angle, you measured the smallest angle to the horizontal axis, all would have the same measure in absolute value. We say that all these angles have a **reference angle** of θ.

Reference Angle

An angle's **reference angle** is the size of the smallest angle to the horizontal axis.

A reference angle is always an angle between 0 and 90 degrees, or 0 and $\frac{\pi}{2}$ radians.

Angles share the same cosine and sine values as their reference angles, except for signs (positive or negative) which can be determined from the quadrant of the angle.

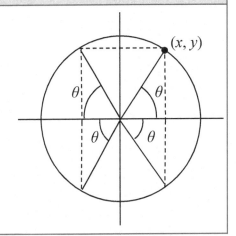

Example 5

Find the reference angle of 150 degrees. Use it to find $\cos(150°)$ and $\sin(150°)$.

150 degrees is located in the second quadrant. It is 30 degrees short of the horizontal axis at 180 degrees, so the reference angle is 30 degrees.

This tells us that 150 degrees has the same sine and cosine values as 30 degrees, except for sign. We know that $\sin(30°) = \frac{1}{2}$ and $\cos(30°) = \frac{\sqrt{3}}{2}$. Since 150 degrees is in the second quadrant, the x coordinate of the point on the circle would be negative, so the cosine value will be negative. The y coordinate is positive, so the sine value will be positive.

$$\sin(150°) = \frac{1}{2} \text{ and } \cos(150°) = -\frac{\sqrt{3}}{2}$$

The (x, y) coordinates for the point on a unit circle at an angle of $150°$ are $\left(\frac{-\sqrt{3}}{2}, \frac{1}{2} \right)$.

Using symmetry and reference angles, we can fill in cosine and sine values at the rest of the special angles on the unit circle. Take time to learn the (x, y) coordinates of all the major angles in the first quadrant!

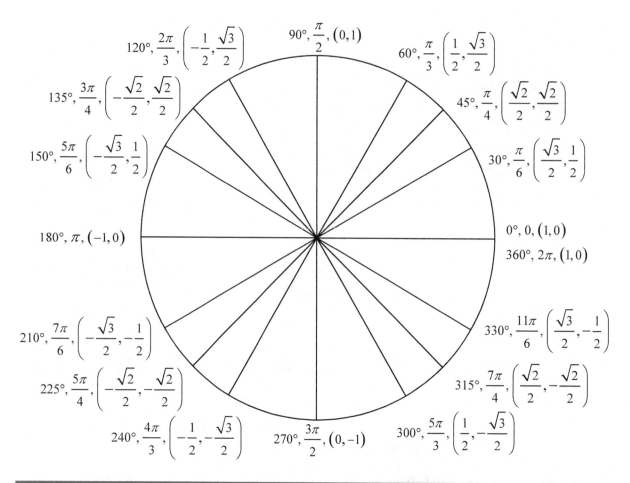

Example 6

Find the coordinates of the point on a circle of radius 12 at an angle of $\dfrac{7\pi}{6}$.

Note that this angle is in the third quadrant, where both x and y are negative. Keeping this in mind can help you check your signs of the sine and cosine function.

$$x = 12\cos\left(\frac{7\pi}{6}\right) = 12\left(\frac{-\sqrt{3}}{2}\right) = -6\sqrt{3}$$

$$y = 12\sin\left(\frac{7\pi}{6}\right) = 12\left(\frac{-1}{2}\right) = -6$$

The coordinates of the point are $(-6\sqrt{3}, -6)$.

Try it Now

3. Find the coordinates of the point on a circle of radius 5 at an angle of $\dfrac{5\pi}{3}$.

Example 7

We now have the tools to return to the sailboat question posed at the beginning of this section.

A distress signal is sent from a sailboat during a storm, but the transmission is unclear and the rescue boat sitting at the marina cannot determine the sailboat's location. Using high powered radar, they determine the distress signal is coming from a point 20 miles away at an angle of 225 degrees from the marina. How many miles east/west and north/south of the rescue boat is the stranded sailboat?

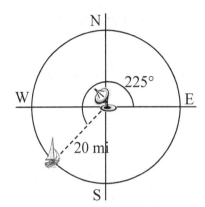

We can now answer the question by finding the coordinates of the point on a circle with a radius of 20 miles at an angle of 225 degrees.

$$x = 20\cos(225°) = 20\left(\frac{-\sqrt{2}}{2}\right) \approx -14.142 \text{ miles}$$

$$y = 20\sin(225°) = 20\left(\frac{-\sqrt{2}}{2}\right) \approx -14.142 \text{ miles}$$

The sailboat is located 14.142 miles west and 14.142 miles south of the marina.

The special values of sine and cosine in the first quadrant are very useful to know, since knowing them allows you to quickly evaluate the sine and cosine of very common angles without needing to look at a reference or use your calculator. However, scenarios do come up where we need to know the sine and cosine of other angles.

To find the cosine and sine of any other angle, we turn to a computer or calculator. **Be aware**: most calculators can be set into "degree" or "radian" mode, which tells the calculator the units for the input value. When you evaluate "cos(30)" on your calculator, it will evaluate it as the cosine of 30 degrees if the calculator is in degree mode, or the cosine of 30 radians if the calculator is in radian mode. Most computer software with cosine and sine functions only operates in radian mode.

Example 8

Evaluate the cosine of 20 degrees using a calculator or computer.

On a calculator that can be put in degree mode, you can evaluate this directly to be approximately 0.939693.

On a computer or calculator without degree mode, you would first need to convert the angle to radians, or equivalently evaluate the expression $\cos\left(20 \cdot \dfrac{\pi}{180}\right)$.

Important Topics of This Section

The sine function

The cosine function

Pythagorean Identity

Unit Circle values

Reference angles

Using technology to find points on a circle

Try it Now Answers

1. $\cos(\pi) = -1 \quad \sin(\pi) = 0$

2.
$$x = 3\cos\left(\frac{\pi}{2}\right) = 3 \cdot 0 = 0$$
$$y = 3\sin\left(\frac{\pi}{2}\right) = 3 \cdot 1 = 3$$

3. $\left(5\cos\left(\dfrac{5\pi}{3}\right), 5\sin\left(\dfrac{5\pi}{3}\right)\right) = \left(\dfrac{5}{2}, \dfrac{-5\sqrt{3}}{2}\right)$

Section 5.3 Exercises

1. Find the quadrant in which the terminal point determined by t lies if
   a. $\sin(t) < 0$ and $\cos(t) < 0$ b. $\sin(t) > 0$ and $\cos(t) < 0$

2. Find the quadrant in which the terminal point determined by t lies if
   a. $\sin(t) < 0$ and $\cos(t) > 0$ b. $\sin(t) > 0$ and $\cos(t) > 0$

3. The point P is on the unit circle. If the y-coordinate of P is $\dfrac{3}{5}$, and P is in quadrant II, find the x coordinate.

4. The point P is on the unit circle. If the x-coordinate of P is $\dfrac{1}{5}$, and P is in quadrant IV, find the y coordinate.

5. If $\cos(\theta) = \dfrac{1}{7}$ and θ is in the 4th quadrant, find $\sin(\theta)$.

6. If $\cos(\theta) = \dfrac{2}{9}$ and θ is in the 1st quadrant, find $\sin(\theta)$.

7. If $\sin(\theta) = \dfrac{3}{8}$ and θ is in the 2nd quadrant, find $\cos(\theta)$.

8. If $\sin(\theta) = -\dfrac{1}{4}$ and θ is in the 3rd quadrant, find $\cos(\theta)$.

9. For each of the following angles, find the reference angle and which quadrant the angle lies in. Then compute sine and cosine of the angle.
   a. $225°$ b. $300°$ c. $135°$ d. $210°$

10. For each of the following angles, find the reference angle and which quadrant the angle lies in. Then compute sine and cosine of the angle.
    a. $120°$ b. $315°$ c. $250°$ d. $150°$

11. For each of the following angles, find the reference angle and which quadrant the angle lies in. Then compute sine and cosine of the angle.
    a. $\dfrac{5\pi}{4}$ b. $\dfrac{7\pi}{6}$ c. $\dfrac{5\pi}{3}$ d. $\dfrac{3\pi}{4}$

12. For each of the following angles, find the reference angle and which quadrant the angle lies in. Then compute sine and cosine of the angle.
    a. $\dfrac{4\pi}{3}$ b. $\dfrac{2\pi}{3}$ c. $\dfrac{5\pi}{6}$ d. $\dfrac{7\pi}{4}$

13. Give exact values for $\sin(\theta)$ and $\cos(\theta)$ for each of these angles.

a. $-\dfrac{3\pi}{4}$ b. $\dfrac{23\pi}{6}$ c. $-\dfrac{\pi}{2}$ d. 5π

14. Give exact values for $\sin(\theta)$ and $\cos(\theta)$ for each of these angles.

a. $-\dfrac{2\pi}{3}$ b. $\dfrac{17\pi}{4}$ c. $-\dfrac{\pi}{6}$ d. 10π

15. Find an angle θ with $0 < \theta < 360°$ or $0 < \theta < 2\pi$ that has the same sine value as:

a. $\dfrac{\pi}{3}$ b. $80°$ c. $140°$ d. $\dfrac{4\pi}{3}$ e. $305°$

16. Find an angle θ with $0 < \theta < 360°$ or $0 < \theta < 2\pi$ that has the same sine value as:

a. $\dfrac{\pi}{4}$ b. $15°$ c. $160°$ d. $\dfrac{7\pi}{6}$ e. $340°$

17. Find an angle θ with $0 < \theta < 360°$ or $0 < \theta < 2\pi$ that has the same cosine value as:

a. $\dfrac{\pi}{3}$ b. $80°$ c. $140°$ d. $\dfrac{4\pi}{3}$ e. $305°$

18. Find an angle θ with $0 < \theta < 360°$ or $0 < \theta < 2\pi$ that has the same cosine value as:

a. $\dfrac{\pi}{4}$ b. $15°$ c. $160°$ d. $\dfrac{7\pi}{6}$ e. $340°$

19. Find the coordinates of the point on a circle with radius 15 corresponding to an angle of 220°.

20. Find the coordinates of the point on a circle with radius 20 corresponding to an angle of 280°.

21. Marla is running clockwise around a circular track. She runs at a constant speed of 3 meters per second. She takes 46 seconds to complete one lap of the track. From her starting point, it takes her 12 seconds to reach the northernmost point of the track. Impose a coordinate system with the center of the track at the origin, and the northernmost point on the positive y-axis. [UW]
   a) Give Marla's coordinates at her starting point.
   b) Give Marla's coordinates when she has been running for 10 seconds.
   c) Give Marla's coordinates when she has been running for 901.3 seconds.

Section 5.4 The Other Trigonometric Functions

In the previous section, we defined the sine and cosine functions as ratios of the sides of a right triangle in a circle. Since the triangle has 3 sides there are 6 possible combinations of ratios. While the sine and cosine are the two prominent ratios that can be formed, there are four others, and together they define the 6 trigonometric functions.

Tangent, Secant, Cosecant, and Cotangent Functions

For the point (x, y) on a circle of radius r at an angle of θ, we can define four additional important functions as the ratios of the sides of the corresponding triangle:

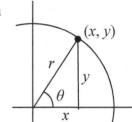

The **tangent** function: $\tan(\theta) = \dfrac{y}{x}$

The **secant** function: $\sec(\theta) = \dfrac{r}{x}$

The **cosecant** function: $\csc(\theta) = \dfrac{r}{y}$

The **cotangent** function: $\cot(\theta) = \dfrac{x}{y}$

Geometrically, notice that the definition of tangent corresponds with the slope of the line segment between the origin $(0, 0)$ and the point (x, y). This relationship can be very helpful in thinking about tangent values.

You may also notice that the ratios defining the secant, cosecant, and cotangent are the reciprocals of the ratios defining the cosine, sine, and tangent functions, respectively. Additionally, notice that using our results from the last section,

$$\tan(\theta) = \frac{y}{x} = \frac{r\sin(\theta)}{r\cos(\theta)} = \frac{\sin(\theta)}{\cos(\theta)}$$

Applying this concept to the other trig functions we can state the reciprocal identities.

Identities

The other four trigonometric functions can be related back to the sine and cosine functions using these basic relationships:

$$\tan(\theta) = \frac{\sin(\theta)}{\cos(\theta)} \qquad \sec(\theta) = \frac{1}{\cos(\theta)} \qquad \csc(\theta) = \frac{1}{\sin(\theta)} \qquad \cot(\theta) = \frac{1}{\tan(\theta)} = \frac{\cos(\theta)}{\sin(\theta)}$$

These relationships are called **identities**. Identities are statements that are true for all values of the input on which they are defined. Identities are usually something that can be derived from definitions and relationships we already know, similar to how the identities above were derived from the circle relationships of the six trig functions. The Pythagorean Identity we learned earlier was derived from the Pythagorean Theorem and the definitions of sine and cosine. We will discuss the role of identities more after an example.

Example 1

Evaluate $\tan(45°)$ and $\sec\left(\dfrac{5\pi}{6}\right)$.

Since we know the sine and cosine values for these angles, it makes sense to relate the tangent and secant values back to the sine and cosine values.

$$\tan(45°) = \frac{\sin(45°)}{\cos(45°)} = \frac{\sqrt{2}/2}{\sqrt{2}/2} = 1$$

Notice this result is consistent with our interpretation of the tangent value as the slope of the line passing through the origin at the given angle: a line at 45 degrees would indeed have a slope of 1.

$$\sec\left(\frac{5\pi}{6}\right) = \frac{1}{\cos\left(\dfrac{5\pi}{6}\right)} = \frac{1}{-\sqrt{3}/2} = \frac{-2}{\sqrt{3}}, \text{ which could also be written as } \frac{-2\sqrt{3}}{3}.$$

Try it Now

1. Evaluate $\csc\left(\dfrac{7\pi}{6}\right)$.

Just as we often need to simplify algebraic expressions, it is often also necessary or helpful to simplify trigonometric expressions. To do so, we utilize the definitions and identities we have established.

Example 2

Simplify $\dfrac{\sec(\theta)}{\tan(\theta)}$.

We can simplify this by rewriting both functions in terms of sine and cosine

$\dfrac{\sec(\theta)}{\tan(\theta)} = \dfrac{{}^{1}\!/\!{}_{\cos(\theta)}}{{}^{\sin(\theta)}\!/\!{}_{\cos(\theta)}}$ To divide the fractions we could invert and multiply

$= \dfrac{1}{\cos(\theta)} \dfrac{\cos(\theta)}{\sin(\theta)}$ cancelling the cosines,

$= \dfrac{1}{\sin(\theta)} = \csc(\theta)$ simplifying and using the identity

By showing that $\dfrac{\sec(\theta)}{\tan(\theta)}$ can be simplified to $\csc(\theta)$, we have, in fact, established a new

identity: that $\dfrac{\sec(\theta)}{\tan(\theta)} = \csc(\theta)$.

Occasionally a question may ask you to "prove the identity" or "establish the identity." This is the same idea as when an algebra book asks a question like "show that $(x-1)^2 = x^2 - 2x + 1$." In this type of question, we must show the algebraic manipulations that demonstrate that the left and right side of the equation are in fact equal. You can think of a "prove the identity" problem as a simplification problem where you *know the answer*: you know what the end goal of the simplification should be, and just need to show the steps to get there.

To prove an identity, in most cases you will start with the expression on one side of the identity and manipulate it using algebra and trigonometric identities until you have simplified it to the expression on the other side of the equation. **Do not** treat the identity like an equation to solve – it isn't! The proof is establishing *if* the two expressions are equal, so we must take care to work with one side at a time rather than applying an operation simultaneously to both sides of the equation.

Example 3

Prove the identity $\dfrac{1 + \cot(\alpha)}{\csc(\alpha)} = \sin(\alpha) + \cos(\alpha)$.

Since the left side seems a bit more complicated, we will start there and simplify the expression until we obtain the right side. We can use the right side as a guide for what might be good steps to make. In this case, the left side involves a fraction while the right side doesn't, which suggests we should look to see if the fraction can be reduced.

Additionally, since the right side involves sine and cosine and the left does not, it suggests that rewriting the cotangent and cosecant using sine and cosine might be a good idea.

$$\frac{1+\cot(\alpha)}{\csc(\alpha)}$$ Rewriting the cotangent and cosecant

$$= \frac{1 + \dfrac{\cos(\alpha)}{\sin(\alpha)}}{\dfrac{1}{\sin(\alpha)}}$$ To divide the fractions, we invert and multiply

$$= \left(1 + \frac{\cos(\alpha)}{\sin(\alpha)}\right)\frac{\sin(\alpha)}{1}$$ Distributing,

$$= 1 \cdot \frac{\sin(\alpha)}{1} + \frac{\cos(\alpha)}{\sin(\alpha)} \cdot \frac{\sin(\alpha)}{1}$$ Simplifying the fractions,

$$= \sin(\alpha) + \cos(\alpha)$$ Establishing the identity.

Notice that in the second step, we could have combined the 1 and $\dfrac{\cos(\alpha)}{\sin(\alpha)}$ before inverting and multiplying. It is very common when proving or simplifying identities for there to be more than one way to obtain the same result.

We can also utilize identities we have previously learned, like the Pythagorean Identity, while simplifying or proving identities.

Example 4

Establish the identity $\dfrac{\cos^2(\theta)}{1+\sin(\theta)} = 1 - \sin(\theta)$.

Since the left side of the identity is more complicated, it makes sense to start there. To simplify this, we will have to reduce the fraction, which would require the numerator to have a factor in common with the denominator. Additionally, we notice that the right side only involves sine. Both of these suggest that we need to convert the cosine into something involving sine.

Recall the Pythagorean Identity told us $\cos^2(\theta) + \sin^2(\theta) = 1$. By moving one of the trig functions to the other side, we can establish:

$$\sin^2(\theta) = 1 - \cos^2(\theta) \qquad \text{and} \qquad \cos^2(\theta) = 1 - \sin^2(\theta)$$

Utilizing this, we now can establish the identity. We start on one side and manipulate:

$$\frac{\cos^2(\theta)}{1+\sin(\theta)}$$ Utilizing the Pythagorean Identity

$$=\frac{1-\sin^2(\theta)}{1+\sin(\theta)}$$ Factoring the numerator

$$=\frac{(1-\sin(\theta))(1+\sin(\theta))}{1+\sin(\theta)}$$ Cancelling the like factors

$$=1-\sin(\theta)$$ Establishing the identity

We can also build new identities from previously established identities. For example, if we divide both sides of the Pythagorean Identity by cosine squared (which is allowed since we've already shown the identity is true),

$$\frac{\cos^2(\theta)+\sin^2(\theta)}{\cos^2(\theta)}=\frac{1}{\cos^2(\theta)}$$ Splitting the fraction on the left,

$$\frac{\cos^2(\theta)}{\cos^2(\theta)}+\frac{\sin^2(\theta)}{\cos^2(\theta)}=\frac{1}{\cos^2(\theta)}$$ Simplifying and using the definitions of tan and sec

$$1+\tan^2(\theta)=\sec^2(\theta).$$

Try it Now

2. Use a similar approach to establish that $\cot^2(\theta)+1=\csc^2(\theta)$.

Identities

Alternate forms of the Pythagorean Identity

$1+\tan^2(\theta)=\sec^2(\theta)$

$\cot^2(\theta)+1=\csc^2(\theta)$

Example 5

If $\tan(\theta)=\dfrac{2}{7}$ and θ is in the 3rd quadrant, find $\cos(\theta)$.

There are two approaches to this problem, both of which work equally well.

Approach 1

Since $\tan(\theta) = \dfrac{y}{x}$ and the angle is in the third quadrant, we can imagine a triangle in a

circle of some radius so that the point on the circle is (-7, -2), so $\dfrac{y}{x} = \dfrac{-2}{-7} = \dfrac{2}{7}$.

Using the Pythagorean Theorem, we can find the radius of the circle:
$(-7)^2 + (-2)^2 = r^2$, so $r = \sqrt{53}$.

Now we can find the cosine value:
$$\cos(\theta) = \frac{x}{r} = \frac{-7}{\sqrt{53}}$$

Approach 2

Using the $1 + \tan^2(\theta) = \sec^2(\theta)$ form of the Pythagorean Identity with the known
tangent value,
$$1 + \tan^2(\theta) = \sec^2(\theta)$$
$$1 + \left(\frac{2}{7}\right)^2 = \sec^2(\theta)$$
$$\frac{53}{49} = \sec^2(\theta)$$
$$\sec(\theta) = \pm\sqrt{\frac{53}{49}} = \pm\frac{\sqrt{53}}{7}$$

Since the angle is in the third quadrant, the cosine value will be negative so the secant
value will also be negative. Keeping the negative result, and using definition of secant,
$$\sec(\theta) = -\frac{\sqrt{53}}{7}$$
$$\frac{1}{\cos(\theta)} = -\frac{\sqrt{53}}{7} \qquad \text{Inverting both sides}$$
$$\cos(\theta) = -\frac{7}{\sqrt{53}} = -\frac{7\sqrt{53}}{53}$$

Try it Now

3. If $\sec(\phi) = -\dfrac{7}{3}$ and $\dfrac{\pi}{2} < \phi < \pi$, find $\tan(\phi)$ and $\sin(\phi)$.

Important Topics of This Section
6 Trigonometric Functions:
Sine
Cosine
Tangent
Cosecant
Secant
Cotangent
Trig identities

Try it Now Answers

1. $\csc\left(\dfrac{7\pi}{6}\right) = \dfrac{1}{\sin\left(\dfrac{7\pi}{6}\right)} = \dfrac{1}{-\frac{1}{2}} = -2$

2.

$$\frac{\cos^2(\theta) + \sin^2(\theta)}{\sin^2\theta} = 1$$

$$\frac{\cos^2(\theta)}{\sin^2(\theta)} + \frac{\sin^2(\theta)}{\sin^2(\theta)} = \frac{1}{\sin^2(\theta)}$$

$$\cot^2(\theta) + 1 = \csc^2(\theta)$$

3. $\sec(\phi) = -\dfrac{7}{3}$. By definition, $\dfrac{1}{\cos(\phi)} = -\dfrac{7}{3}$, so $\cos(\phi) = -\dfrac{3}{7}$.

Using Pythagorean Identity with the sec, $1 + \tan^2(\phi) = \left(-\dfrac{7}{3}\right)^2$. Solving gives

$\tan(\phi) = \dfrac{\sqrt{40}}{-3}$. We use the negative square root since an angle in the second quadrant would have a negative tangent.

Using Pythagorean Identity with the cos, $\sin^2(\phi) + \left(-\dfrac{3}{7}\right)^2 = 1$. Solving,

$\sin(\phi) = \dfrac{\sqrt{40}}{7}$.

382 Chapter 5

Section 5.4 Exercises

1. If $\theta = \dfrac{\pi}{4}$, find exact values for $\sec(\theta), \csc(\theta), \tan(\theta), \cot(\theta)$.

2. If $\theta = \dfrac{7\pi}{4}$, find exact values for $\sec(\theta), \csc(\theta), \tan(\theta), \cot(\theta)$.

3. If $\theta = \dfrac{5\pi}{6}$, find exact values for $\sec(\theta), \csc(\theta), \tan(\theta), \cot(\theta)$.

4. If $\theta = \dfrac{\pi}{6}$, find exact values for $\sec(\theta), \csc(\theta), \tan(\theta), \cot(\theta)$.

5. If $\theta = \dfrac{2\pi}{3}$, find exact values for $\sec(\theta), \csc(\theta), \tan(\theta), \cot(\theta)$.

6. If $\theta = \dfrac{4\pi}{3}$, find exact values for $\sec(\theta), \csc(\theta), \tan(\theta), \cot(\theta)$.

7. Evaluate: a. $\sec(135°)$ b. $\csc(210°)$ c. $\tan(60°)$ d. $\cot(225°)$

8. Evaluate: a. $\sec(30°)$ b. $\csc(315°)$ c. $\tan(135°)$ d. $\cot(150°)$

9. If $\sin(\theta) = \dfrac{3}{4}$, and θ is in quadrant II, find $\cos(\theta), \sec(\theta), \csc(\theta), \tan(\theta), \cot(\theta)$.

10. If $\sin(\theta) = \dfrac{2}{7}$, and θ is in quadrant II, find $\cos(\theta), \sec(\theta), \csc(\theta), \tan(\theta), \cot(\theta)$.

11. If $\cos(\theta) = -\dfrac{1}{3}$, and θ is in quadrant III, find
    $\sin(\theta), \sec(\theta), \csc(\theta), \tan(\theta), \cot(\theta)$.

12. If $\cos(\theta) = \dfrac{1}{5}$, and θ is in quadrant I, find $\sin(\theta), \sec(\theta), \csc(\theta), \tan(\theta), \cot(\theta)$.

13. If $\tan(\theta) = \dfrac{12}{5}$, and $0 \le \theta < \dfrac{\pi}{2}$, find $\sin(\theta), \cos(\theta), \sec(\theta), \csc(\theta), \cot(\theta)$.

14. If $\tan(\theta) = 4$, and $0 \le \theta < \dfrac{\pi}{2}$, find $\sin(\theta), \cos(\theta), \sec(\theta), \csc(\theta), \cot(\theta)$.

15. Use a calculator to find sine, cosine, and tangent of the following values:

    a. 0.15   b. 4   c. 70°   d. 283°

16. Use a calculator to find sine, cosine, and tangent of the following values:

    a. 0.5   b. 5.2   c. 10°   d. 195°

Simplify each of the following to an expression involving a single trig function with no fractions.

17. $\csc(t)\tan(t)$

18. $\cos(t)\csc(t)$

19. $\dfrac{\sec(t)}{\csc(t)}$

20. $\dfrac{\cot(t)}{\csc(t)}$

21. $\dfrac{\sec(t)-\cos(t)}{\sin(t)}$

22. $\dfrac{\tan(t)}{\sec(t)-\cos(t)}$

23. $\dfrac{1+\cot(t)}{1+\tan(t)}$

24. $\dfrac{1+\sin(t)}{1+\csc(t)}$

25. $\dfrac{\sin^2(t)+\cos^2(t)}{\cos^2(t)}$

26. $\dfrac{1-\sin^2(t)}{\sin^2(t)}$

Prove the identities.

27. $\dfrac{\sin^2(\theta)}{1+\cos(\theta)} = 1 - \cos(\theta)$

28. $\tan^2(t) = \dfrac{1}{\cos^2(t)} - 1$

29. $\sec(a) - \cos(a) = \sin(a)\tan(a)$

30. $\dfrac{1+\tan^2(b)}{\tan^2(b)} = \csc^2(b)$

31. $\dfrac{\csc^2(x) - \sin^2(x)}{\csc(x) + \sin(x)} = \cos(x)\cot(x)$

32. $\dfrac{\sin(\theta) - \cos(\theta)}{\sec(\theta) - \csc(\theta)} = \sin(\theta)\cos(\theta)$

33. $\dfrac{\csc^2(\alpha) - 1}{\csc^2(\alpha) - \csc(\alpha)} = 1 + \sin(\alpha)$

34. $1 + \cot(x) = \cos(x)\big(\sec(x) + \csc(x)\big)$

35. $\dfrac{1+\cos(u)}{\sin(u)} = \dfrac{\sin(u)}{1-\cos(u)}$

36. $2\sec^2(t) = \dfrac{1-\sin(t)}{\cos^2(t)} + \dfrac{1}{1-\sin(t)}$

37. $\dfrac{\sin^4(\gamma) - \cos^4(\gamma)}{\sin(\gamma) - \cos(\gamma)} = \sin(\gamma) + \cos(\gamma)$

38. $\dfrac{\big(1+\cos(A)\big)\big(1-\cos(A)\big)}{\sin(A)} = \sin(A)$

Section 5.5 Right Triangle Trigonometry

In section 5.3 we were introduced to the sine and cosine function as ratios of the sides of a triangle drawn inside a circle, and spent the rest of that section discussing the role of those functions in finding points on the circle. In this section, we return to the triangle, and explore the applications of the trigonometric functions to right triangles where circles may not be involved.

Recall that we defined sine and cosine as

$$\sin(\theta) = \frac{y}{r}$$

$$\cos(\theta) = \frac{x}{r}$$

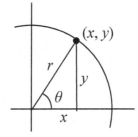

Separating the triangle from the circle, we can make equivalent but more general definitions of the sine, cosine, and tangent on a right triangle. On the right triangle, we will label the hypotenuse as well as the side opposite the angle and the side adjacent (next to) the angle.

Right Triangle Relationships
Given a right triangle with an angle of θ $\sin(\theta) = \dfrac{\text{opposite}}{\text{hypotenuse}}$ $\cos(\theta) = \dfrac{\text{adjacent}}{\text{hypotenuse}}$ $\tan(\theta) = \dfrac{\text{opposite}}{\text{adjacent}}$

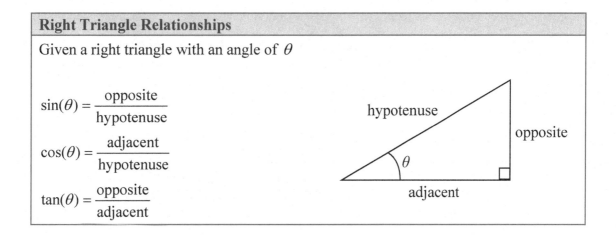

A common mnemonic for remembering these relationships is SohCahToa, formed from the first letters of "Sine is opposite over hypotenuse, Cosine is adjacent over hypotenuse, Tangent is opposite over adjacent."

Example 1

Given the triangle shown, find the value for $\cos(\alpha)$.

The side adjacent to the angle is 15, and the hypotenuse of the triangle is 17, so

$$\cos(\alpha) = \frac{\text{adjacent}}{\text{hypotenuse}} = \frac{15}{17}$$

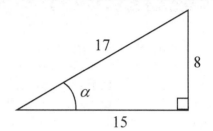

When working with general right triangles, the same rules apply regardless of the orientation of the triangle. In fact, we can evaluate the sine and cosine of either of the two acute angles in the triangle.

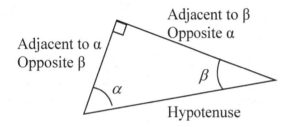

Example 2

Using the triangle shown, evaluate $\cos(\alpha)$, $\sin(\alpha)$, $\cos(\beta)$, and $\sin(\beta)$.

$$\cos(\alpha) = \frac{\text{adjacent to } \alpha}{\text{hypotenuse}} = \frac{3}{5}$$

$$\sin(\alpha) = \frac{\text{opposite } \alpha}{\text{hypotenuse}} = \frac{4}{5}$$

$$\cos(\beta) = \frac{\text{adjacent to } \beta}{\text{hypotenuse}} = \frac{4}{5}$$

$$\sin(\beta) = \frac{\text{opposite } \beta}{\text{hypotenuse}} = \frac{3}{5}$$

Try it Now

1. A right triangle is drawn with angle α opposite a side with length 33, angle β opposite a side with length 56, and hypotenuse 65. Find the sine and cosine of α and β.

You may have noticed that in the above example that $\cos(\alpha) = \sin(\beta)$ and $\cos(\beta) = \sin(\alpha)$. This makes sense since the side opposite α is also adjacent to β. Since the three angles in a triangle need to add to π, or 180 degrees, then the other two angles must add to $\dfrac{\pi}{2}$, or 90 degrees, so $\beta = \dfrac{\pi}{2} - \alpha$, and $\alpha = \dfrac{\pi}{2} - \beta$. Since $\cos(\alpha) = \sin(\beta)$, then $\cos(\alpha) = \sin\left(\dfrac{\pi}{2} - \alpha\right)$.

Cofunction Identities

The **cofunction identities** for sine and cosine are:

$$\cos(\theta) = \sin\left(\frac{\pi}{2} - \theta\right) \qquad\qquad \sin(\theta) = \cos\left(\frac{\pi}{2} - \theta\right)$$

In the previous examples, we evaluated the sine and cosine on triangles where we knew all three sides of the triangle. Right triangle trigonometry becomes powerful when we start looking at triangles in which we know an angle but don't know all the sides.

Example 3

Find the unknown sides of the triangle pictured here.

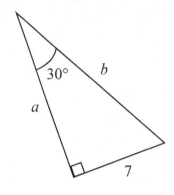

Since $\sin(\theta) = \dfrac{\text{opposite}}{\text{hypotenuse}}$, $\sin(30°) = \dfrac{7}{b}$.

From this, we can solve for the side b.
$b\sin(30°) = 7$

$b = \dfrac{7}{\sin(30°)}$

To obtain a value, we can evaluate the sine and simplify

$b = \dfrac{7}{\frac{1}{2}} = 14$

To find the value for side a, we could use the cosine, or simply apply the Pythagorean Theorem:
$a^2 + 7^2 = b^2$
$a^2 + 7^2 = 14^2$
$a = \sqrt{147}$

Notice that if we know at least one of the non-right angles of a right triangle and one side, we can find the rest of the sides and angles.

Try it Now

2. A right triangle has one angle of $\frac{\pi}{3}$ and a hypotenuse of 20. Find the unknown sides and angles of the triangle.

Example 4

To find the height of a tree, a person walks to a point 30 feet from the base of the tree, and measures the angle from the ground to the top of the tree to be 57 degrees. Find the height of the tree.

We can introduce a variable, h, to represent the height of the tree. The two sides of the triangle that are most important to us are the side opposite the angle, the height of the tree we are looking for, and the adjacent side, the side we are told is 30 feet long.

The trigonometric function which relates the side opposite of the angle and the side adjacent to the angle is the tangent.

$\tan(57°) = \dfrac{\text{opposite}}{\text{adjacent}} = \dfrac{h}{30}$ Solving for h,

$h = 30\tan(57°)$ Using technology, we can approximate a value

$h = 30\tan(57°) \approx 46.2$ feet

The tree is approximately 46 feet tall.

Example 5

A person standing on the roof of a 100 foot tall building is looking towards a skyscraper a few blocks away, wondering how tall it is. She measures the angle of declination from the roof of the building to the base of the skyscraper to be 20 degrees and the angle of inclination to the top of the skyscraper to be 42 degrees.

To approach this problem, it would be good to start with a picture. Although we are interested in the height, *h*, of the skyscraper, it can be helpful to also label other unknown quantities in the picture – in this case the horizontal distance *x* between the buildings and *a*, the height of the skyscraper above the person.

To start solving this problem, notice we have two right triangles. In the top triangle, we know one angle is 42 degrees, but we don't know any of the sides of the triangle, so we don't yet know enough to work with this triangle.

In the lower right triangle, we know one angle is 20 degrees, and we know the vertical height measurement of 100 ft. Since we know these two pieces of information, we can solve for the unknown distance *x*.

$$\tan(20°) = \frac{\text{opposite}}{\text{adjacent}} = \frac{100}{x}$$ Solving for *x*

$$x\tan(20°) = 100$$

$$x = \frac{100}{\tan(20°)}$$

Now that we have found the distance *x*, we know enough information to solve the top right triangle.

$$\tan(42°) = \frac{\text{opposite}}{\text{adjacent}} = \frac{a}{x} = \frac{a}{100/\tan(20°)}$$

$$\tan(42°) = \frac{a\tan(20°)}{100}$$

$$100\tan(42°) = a\tan(20°)$$

$$\frac{100\tan(42°)}{\tan(20°)} = a$$

Approximating a value,

$$a = \frac{100\tan(42°)}{\tan(20°)} \approx 247.4 \text{ feet}$$

Adding the height of the first building, we determine that the skyscraper is about 347 feet tall.

Important Topics of This Section
SOH CAH TOA
Cofunction identities
Applications with right triangles

Try it Now Answers

1. $\sin(\alpha) = \dfrac{33}{65}$ $\cos(\alpha) = \dfrac{56}{65}$ $\sin(\beta) = \dfrac{56}{65}$ $\cos(\beta) = \dfrac{33}{65}$

2. $\cos\left(\dfrac{\pi}{3}\right) = \dfrac{\text{adjacent}}{\text{hypoteuse}} = \dfrac{\text{Adj}}{20}$ so, $\text{adjacent} = 20\cos\left(\dfrac{\pi}{3}\right) = 20\left(\dfrac{1}{2}\right) = 10$

$\sin\left(\dfrac{\pi}{3}\right) = \dfrac{\text{Opposite}}{\text{hypoteuse}} = \dfrac{\text{Opp}}{20}$ so, $\text{opposite} = 20\sin\left(\dfrac{\pi}{3}\right) = 20\left(\dfrac{\sqrt{3}}{2}\right) = 10\sqrt{3}$

Missing angle = 180-90-60 = 30 degrees or $\pi/6$.

Section 5.5 Exercises

Note: pictures may not be drawn to scale.

In each of the triangles below, find $\sin(A), \cos(A), \tan(A), \sec(A), \csc(A), \cot(A)$.

1.

2.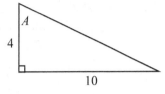

In each of the following triangles, solve for the unknown sides and angles.

3.

4.

5.

6.

7.

8.

9. A 33-ft ladder leans against a building so that the angle between the ground and the ladder is 80°. How high does the ladder reach up the side of the building?

10. A 23-ft ladder leans against a building so that the angle between the ground and the ladder is 80°. How high does the ladder reach up the side of the building?

11. The angle of elevation to the top of a building in New York is found to be 9 degrees from the ground at a distance of 1 mile from the base of the building. Using this information, find the height of the building.

12. The angle of elevation to the top of a building in Seattle is found to be 2 degrees from the ground at a distance of 2 miles from the base of the building. Using this information, find the height of the building.

13. A radio tower is located 400 feet from a building. From a window in the building, a person determines that the angle of elevation to the top of the tower is 36° and that the angle of depression to the bottom of the tower is 23°. How tall is the tower?

14. A radio tower is located 325 feet from a building. From a window in the building, a person determines that the angle of elevation to the top of the tower is 43° and that the angle of depression to the bottom of the tower is 31°. How tall is the tower?

15. A 200 foot tall monument is located in the distance. From a window in a building, a person determines that the angle of elevation to the top of the monument is 15° and that the angle of depression to the bottom of the tower is 2°. How far is the person from the monument?

16. A 400 foot tall monument is located in the distance. From a window in a building, a person determines that the angle of elevation to the top of the monument is 18° and that the angle of depression to the bottom of the tower is 3°. How far is the person from the monument?

17. There is an antenna on the top of a building. From a location 300 feet from the base of the building, the angle of elevation to the top of the building is measured to be 40°. From the same location, the angle of elevation to the top of the antenna is measured to be 43°. Find the height of the antenna.

18. There is lightning rod on the top of a building. From a location 500 feet from the base of the building, the angle of elevation to the top of the building is measured to be 36°. From the same location, the angle of elevation to the top of the lightning rod is measured to be 38°. Find the height of the lightning rod.

19. Find the length x.

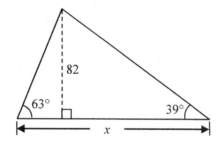

20. Find the length x.

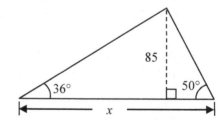

21. Find the length x.

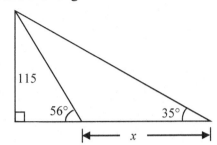

22. Find the length x.

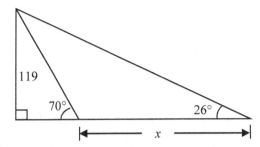

23. A plane is flying 2000 feet above sea level toward a mountain. The pilot observes the top of the mountain to be $18°$ above the horizontal, then immediately flies the plane at an angle of $20°$ above horizontal. The airspeed of the plane is 100 mph. After 5 minutes, the plane is directly above the top of the mountain. How high is the plane above the top of the mountain (when it passes over)? What is the height of the mountain? [UW]

24. Three airplanes depart SeaTac Airport. A United flight is heading in a direction $50°$ counterclockwise from east, an Alaska flight is heading $115°$ counterclockwise from east and a Delta flight is heading $20°$ clockwise from east. [UW]
    a. Find the location of the United flight when it is 20 miles north of SeaTac.
    b. Find the location of the Alaska flight when it is 50 miles west of SeaTac.
    c. Find the location of the Delta flight when it is 30 miles east of SeaTac.

(a) The flight paths of three airplanes

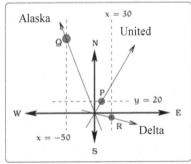

(b) Modeling the paths of each flight

25. The crew of a helicopter needs to
land temporarily in a forest and spot
a flat piece of ground (a clearing in
the forest) as a potential landing site,
but are uncertain whether it is wide
enough. They make two
measurements from A (see picture)
finding $\alpha = 25°$ and $\beta = 54°$. They
rise vertically 100 feet to B and
measure $\gamma = 47°$. Determine the width of the clearing to the nearest foot. [UW]

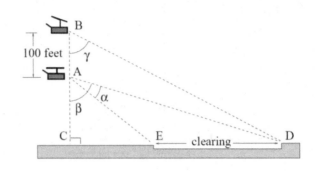

26. A Forest Service helicopter needs to determine
the width of a deep canyon. While hovering,
they measure the angle $\gamma = 48°$ at position B
(see picture), then descend 400 feet to position
A and make two measurements: $\alpha = 13°$ (the
measure of $\angle EAD$), $\beta = 53°$ (the measure of
$\angle CAD$). Determine the width of the canyon
to the nearest foot. [UW]

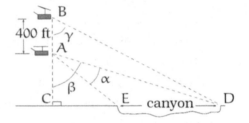

Chapter 6: Periodic Functions

In the previous chapter, the trigonometric functions were introduced as ratios of sides of a right triangle, and related to points on a circle. We noticed how the x and y values of the points did not change with repeated revolutions around the circle by finding coterminal angles. In this chapter, we will take a closer look at the important characteristics and applications of these types of functions, and begin solving equations involving them.

Section 6.1 Sinusoidal Graphs

The London Eye[1] is a huge Ferris wheel 135 meters (394 feet) tall in London, England, which completes one rotation every 30 minutes. When we look at the behavior of this Ferris wheel it is clear that it completes 1 cycle, or 1 revolution, and then repeats this revolution over and over again.

This is an example of a periodic function, because the Ferris wheel repeats its revolution or one cycle every 30 minutes, and so we say it has a period of 30 minutes.

In this section, we will work to sketch a graph of a rider's height above the ground over time and express this height as a function of time.

Periodic Functions

A **periodic function** is a function for which a specific horizontal shift, P, results in the original function: $f(x + P) = f(x)$ for all values of x. When this occurs we call the smallest such horizontal shift with $P > 0$ the **period** of the function.

---

[1] London Eye photo by authors, 2010, CC-BY

This chapter is part of *Precalculus: An Investigation of Functions* © Lippman & Rasmussen 2017.

You might immediately guess that there is a connection here to finding points on a circle, since the height above ground would correspond to the y value of a point on the circle. We can determine the y value by using the sine function. To get a better sense of this function's behavior, we can create a table of values we know, and use them to sketch a graph of the sine and cosine functions.

Listing some of the values for sine and cosine on a unit circle,

θ	0	$\dfrac{\pi}{6}$	$\dfrac{\pi}{4}$	$\dfrac{\pi}{3}$	$\dfrac{\pi}{2}$	$\dfrac{2\pi}{3}$	$\dfrac{3\pi}{4}$	$\dfrac{5\pi}{6}$	π
cos	1	$\dfrac{\sqrt{3}}{2}$	$\dfrac{\sqrt{2}}{2}$	$\dfrac{1}{2}$	0	$-\dfrac{1}{2}$	$-\dfrac{\sqrt{2}}{2}$	$-\dfrac{\sqrt{3}}{2}$	-1
sin	0	$\dfrac{1}{2}$	$\dfrac{\sqrt{2}}{2}$	$\dfrac{\sqrt{3}}{2}$	1	$\dfrac{\sqrt{3}}{2}$	$\dfrac{\sqrt{2}}{2}$	$\dfrac{1}{2}$	0

Here you can see how for each angle, we use the y value of the point on the circle to determine the output value of the sine function.

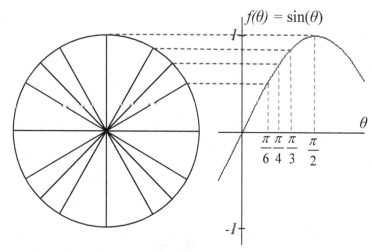

Plotting more points gives the full shape of the sine and cosine functions.

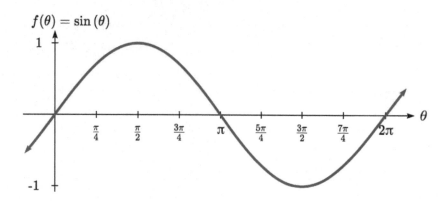

Notice how the sine values are positive between 0 and π, which correspond to the values of sine in quadrants 1 and 2 on the unit circle, and the sine values are negative between π and 2π, corresponding to quadrants 3 and 4.

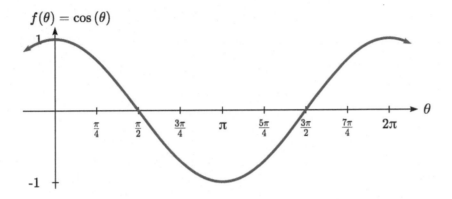

Like the sine function we can track the value of the cosine function through the 4 quadrants of the unit circle as we place it on a graph.

Both of these functions are defined for all real numbers, since we can evaluate the sine and cosine of any angle. By thinking of sine and cosine as coordinates of points on a unit circle, it becomes clear that the range of both functions must be the interval $[-1, 1]$.

Domain and Range of Sine and Cosine
The domain of sine and cosine is all real numbers, $(-\infty, \infty)$.
The range of sine and cosine is the interval [-1, 1].

Both these graphs are called **sinusoidal** graphs.

In both graphs, the shape of the graph begins repeating after 2π. Indeed, since any coterminal angles will have the same sine and cosine values, we could conclude that $\sin(\theta + 2\pi) = \sin(\theta)$ and $\cos(\theta + 2\pi) = \cos(\theta)$.

In other words, if you were to shift either graph horizontally by 2π, the resulting shape would be identical to the original function. Sinusoidal functions are a specific type of periodic function.

Period of Sine and Cosine
The periods of the sine and cosine functions are both 2π.

Looking at these functions on a domain centered at the vertical axis helps reveal symmetries.

sine   cosine

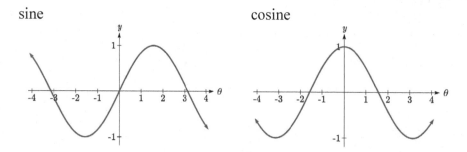

The sine function is symmetric about the origin, the same symmetry the cubic function has, making it an odd function. The cosine function is clearly symmetric about the y axis, the same symmetry as the quadratic function, making it an even function.

Negative Angle Identities

The sine is an odd function, symmetric about the *origin*, so $\sin(-\theta) = -\sin(\theta)$.

The cosine is an even function, symmetric about the y-axis, so $\cos(-\theta) = \cos(\theta)$.

These identities can be used, among other purposes, for helping with simplification and proving identities.

You may recall the cofunction identity from last chapter, $\sin(\theta) = \cos\left(\dfrac{\pi}{2} - \theta\right)$.

Graphically, this tells us that the sine and cosine graphs are horizontal transformations of each other. We can prove this by using the cofunction identity and the negative angle identity for cosine.

$$\sin(\theta) = \cos\left(\frac{\pi}{2} - \theta\right) = \cos\left(-\theta + \frac{\pi}{2}\right) = \cos\left(-\left(\theta - \frac{\pi}{2}\right)\right) = \cos\left(\theta - \frac{\pi}{2}\right)$$

Now we can clearly see that if we horizontally shift the cosine function to the right by $\pi/2$ we get the sine function.

Remember this shift is not representing the period of the function. It only shows that the cosine and sine function are transformations of each other.

Example 1

Simplify $\dfrac{\sin(-\theta)}{\tan(\theta)}$.

We start by using the negative angle identity for sine.

$\dfrac{-\sin(\theta)}{\tan(\theta)}$ Rewriting the tangent

$\dfrac{-\sin(\theta)}{\sin(\theta)\Big/\cos(\theta)}$ Inverting and multiplying

$-\sin(\theta)\cdot\dfrac{\cos(\theta)}{\sin(\theta)}$ Simplifying we get

$-\cos(\theta)$

Transforming Sine and Cosine

Example 2

A point rotates around a circle of radius 3. Sketch a graph of the y coordinate of the point.

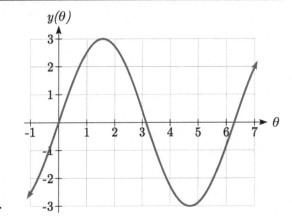

Recall that for a point on a circle of radius r, the y coordinate of the point is $y = r\sin(\theta)$, so in this case, we get the equation $y(\theta) = 3\sin(\theta)$.

The constant 3 causes a vertical stretch of the y values of the function by a factor of 3.

Notice that the period of the function does not change.

Since the outputs of the graph will now oscillate between -3 and 3, we say that the **amplitude** of the sine wave is 3.

Try it Now

1. What is the amplitude of the function $f(\theta) = 7\cos(\theta)$? Sketch a graph of this function.

Example 3

A circle with radius 3 feet is mounted with its center 4 feet off the ground. The point closest to the ground is labeled P. Sketch a graph of the height above ground of the point P as the circle is rotated, then find a function that gives the height in terms of the angle of rotation.

Sketching the height, we note that it will start 1 foot above the ground, then increase up to 7 feet above the ground, and continue to oscillate 3 feet above and below the center value of 4 feet.

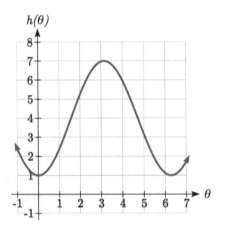

Although we could use a transformation of either the sine or cosine function, we start by looking for characteristics that would make one function easier to use than the other.

We decide to use a cosine function because it starts at the highest or lowest value, while a sine function starts at the middle value. A standard cosine starts at the highest value, and this graph starts at the lowest value, so we need to incorporate a vertical reflection.

Second, we see that the graph oscillates 3 above and below the center, while a basic cosine has an amplitude of one, so this graph has been vertically stretched by 3, as in the last example.

Finally, to move the center of the circle up to a height of 4, the graph has been vertically shifted up by 4. Putting these transformations together,

$$h(\theta) = -3\cos(\theta) + 4$$

Midline

The center value of a sinusoidal function, the value that the function oscillates above and below, is called the **midline** of the function, corresponding to a vertical shift.

The function $f(\theta) = \cos(\theta) + k$ has midline at $y = k$.

Try it Now

2. What is the midline of the function $f(\theta) = 3\cos(\theta) - 4$? Sketch a graph of the function.

To answer the Ferris wheel problem at the beginning of the section, we need to be able to express our sine and cosine functions at inputs of time. To do so, we will utilize composition. Since the sine function takes an input of an angle, we will look for a function that takes time as an input and outputs an angle. If we can find a suitable $\theta(t)$ function, then we can compose this with our $f(\theta) = \cos(\theta)$ function to obtain a sinusoidal function of time: $f(t) = \cos(\theta(t))$.

Example 4

A point completes 1 revolution every 2 minutes around a circle of radius 5. Find the x coordinate of the point as a function of time, if it starts at (5, 0).

Normally, we would express the x coordinate of a point on a unit circle using $x = r\cos(\theta)$, here we write the function $x(\theta) = 5\cos(\theta)$.

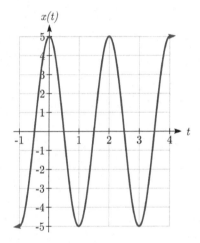

The rotation rate of 1 revolution every 2 minutes is an angular velocity. We can use this rate to find a formula for the angle as a function of time. The point begins at an angle of 0. Since the point rotates 1 revolution = 2π radians every 2 minutes, it rotates π radians every minute. After t minutes, it will have rotated:

$\theta(t) = \pi t$ radians

Composing this with the cosine function, we obtain a function of time.

$x(t) = 5\cos(\theta(t)) = 5\cos(\pi t)$

Notice that this composition has the effect of a horizontal compression, changing the period of the function.

To see how the period relates to the stretch or compression coefficient B in the equation $f(t) = \sin(Bt)$, note that the period will be the time it takes to complete one full revolution of a circle. If a point takes P minutes to complete 1 revolution, then the angular velocity is $\dfrac{2\pi \text{ radians}}{P \text{ minutes}}$. Then $\theta(t) = \dfrac{2\pi}{P}t$. Composing with a sine function,

$f(t) = \sin(\theta(t)) = \sin\left(\dfrac{2\pi}{P}t\right)$

From this, we can determine the relationship between the coefficient B and the period:

$B = \dfrac{2\pi}{P}$.

Notice that the stretch or compression coefficient B is a ratio of the "normal period of a sinusoidal function" to the "new period."   If we know the stretch or compression coefficient B, we can solve for the "new period": $P = \dfrac{2\pi}{B}$.

Summarizing our transformations so far:

Transformations of Sine and Cosine

Given an equation in the form $f(t) = A\sin(Bt) + k$ or $f(t) = A\cos(Bt) + k$

A is the vertical stretch, and is the **amplitude** of the function.

B is the horizontal stretch/compression, and is related to the **period, P,** by $P = \dfrac{2\pi}{B}$.

k is the vertical shift and determines the **midline** of the function.

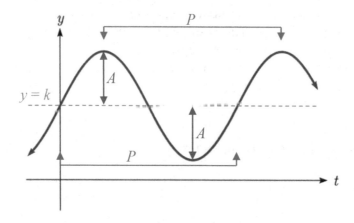

Example 5

What is the period of the function $f(t) = \sin\left(\dfrac{\pi}{6}t\right)$?

Using the relationship above, the stretch/compression factor is $B = \dfrac{\pi}{6}$, so the period will be $P = \dfrac{2\pi}{B} = \dfrac{2\pi}{\dfrac{\pi}{6}} = 2\pi \cdot \dfrac{6}{\pi} = 12$.

While it is common to compose sine or cosine with functions involving time, the composition can be done so that the input represents any reasonable quantity.

Example 6

A bicycle wheel with radius 14 inches has the bottom-most point on the wheel marked in red. The wheel then begins rolling down the street. Write a formula for the height above ground of the red point after the bicycle has travelled x inches.

The height of the point begins at the lowest value, 0, increases to the highest value of 28 inches, and continues to oscillate above and below a center height of 14 inches. In terms of the angle of rotation, θ:
$$h(\theta) = -14\cos(\theta) + 14$$

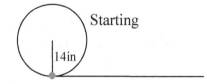

In this case, x is representing a linear distance the wheel has travelled, corresponding to an arclength along the circle. Since arclength and angle can be related by $s = r\theta$, in this case we can write $x = 14\theta$, which allows us to express the angle in terms of x:

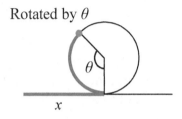

$$\theta(x) = \frac{x}{14}$$

Composing this with our cosine-based function from above,
$$h(x) = h(\theta(x)) = -14\cos\left(\frac{x}{14}\right) + 14 = -14\cos\left(\frac{1}{14}x\right) + 14$$

The period of this function would be $P = \dfrac{2\pi}{B} = \dfrac{2\pi}{\dfrac{1}{14}} = 2\pi \cdot 14 = 28\pi$, the circumference

of the circle. This makes sense – the wheel completes one full revolution after the bicycle has travelled a distance equivalent to the circumference of the wheel.

Example 7

Determine the midline, amplitude, and period of the function $f(t) = 3\sin(2t) + 1$.

The amplitude is 3

The period is $P = \dfrac{2\pi}{B} = \dfrac{2\pi}{2} = \pi$

The midline is at $y = 1$

Amplitude, midline, and period, when combined with vertical flips, allow us to write equations for a variety of sinusoidal situations.

Try it Now
3. If a sinusoidal function starts on the midline at point (0,3), has an amplitude of 2, and a period of 4, write a formula for the function

Example 8

Find a formula for the sinusoidal function graphed here.

The graph oscillates from a low of -1 to a high of 3, putting the midline at $y = 1$, halfway between.

The amplitude will be 2, the distance from the midline to the highest value (or lowest value) of the graph.

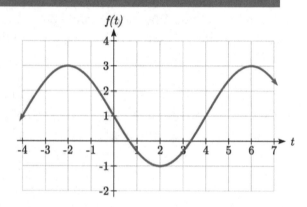

The period of the graph is 8. We can measure this from the first peak at $x = -2$ to the second at $x = 6$. Since the period is 8, the stretch/compression factor we will use will be

$$B = \frac{2\pi}{P} = \frac{2\pi}{8} = \frac{\pi}{4}$$

At $x = 0$, the graph is at the midline value, which tells us the graph can most easily be represented as a sine function. Since the graph then decreases, this must be a vertical reflection of the sine function. Putting this all together,

$$f(t) = -2\sin\left(\frac{\pi}{4}t\right) + 1$$

With these transformations, we are ready to answer the Ferris wheel problem from the beginning of the section.

Example 9

The London Eye is a huge Ferris wheel in London, England, which completes one rotation every 30 minutes. The diameter of the wheel is 120 meters, but the passenger capsules sit outside the wheel. Suppose the diameter at the capsules is 130 meters, and riders board from a platform 5 meters above the ground. Express a rider's height above ground as a function of time in minutes.

It can often help to sketch a graph of the situation before trying to find the equation.

With a diameter of 130 meters, the wheel has a radius of 65 meters. The height will oscillate with amplitude of 65 meters above and below the center.

Passengers board 5 meters above ground level, so the center of the wheel must be located $65 + 5 = 70$ meters above ground level. The midline of the oscillation will be at 70 meters.

The wheel takes 30 minutes to complete 1 revolution, so the height will oscillate with period of 30 minutes.

Lastly, since the rider boards at the lowest point, the height will start at the smallest value and increase, following the shape of a flipped cosine curve.
Putting these together:
Amplitude: 65
Midline: 70

Period: 30, so $B = \dfrac{2\pi}{30} = \dfrac{\pi}{15}$

Shape: negative cosine

An equation for the rider's height would be

$$h(t) = -65\cos\left(\frac{\pi}{15}t\right) + 70$$

Try it Now
4. The Ferris wheel at the Puyallup Fair[2] has a diameter of about 70 feet and takes 3 minutes to complete a full rotation. Passengers board from a platform 10 feet above the ground. Write an equation for a rider's height above ground over time.

While these transformations are sufficient to represent many situations, occasionally we encounter a sinusoidal function that does not have a vertical intercept at the lowest point, highest point, or midline. In these cases, we need to use horizontal shifts. Since we are combining horizontal shifts with horizontal stretches, we need to be careful. Recall that when the inside of the function is factored, it reveals the horizontal shift.

Horizontal Shifts of Sine and Cosine

Given an equation in the form $f(t) = A\sin(B(t - h)) + k$ or $f(t) = A\cos(B(t - h)) + k$

h is the horizontal shift of the function

Example 10

Sketch a graph of $f(t) = 3\sin\left(\dfrac{\pi}{4}t - \dfrac{\pi}{4}\right)$.

To reveal the horizontal shift, we first need to factor inside the function:

$$f(t) = 3\sin\left(\dfrac{\pi}{4}(t - 1)\right)$$

This graph will have the shape of a sine function, starting at the midline and increasing, with an amplitude of 3. The period of the graph will be $P = \dfrac{2\pi}{B} = \dfrac{2\pi}{\dfrac{\pi}{4}} = 2\pi \cdot \dfrac{4}{\pi} = 8$.

Finally, the graph will be shifted to the right by 1.

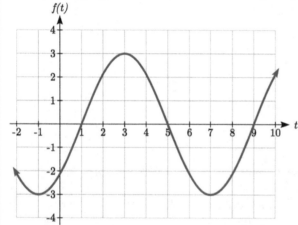

In some physics and mathematics books, you will hear the horizontal shift referred to as **phase shift**. In other physics and mathematics books, they would say the phase shift of the equation above is $\dfrac{\pi}{4}$, the value in the unfactored form. Because of this ambiguity, we will not use the term phase shift any further, and will only talk about the horizontal shift.

Example 11

Find a formula for the function graphed here.

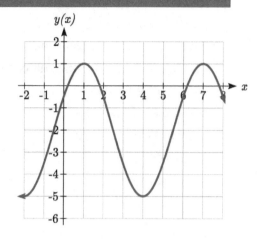

With highest value at 1 and lowest value at -5, the midline will be halfway between at -2.

The distance from the midline to the highest or lowest value gives an amplitude of 3.

The period of the graph is 6, which can be measured from the peak at $x = 1$ to the next peak at $x = 7$, or from the distance between the lowest points. This gives $B = \dfrac{2\pi}{P} = \dfrac{2\pi}{6} = \dfrac{\pi}{3}$.

For the shape and shift, we have more than one option. We could either write this as:
A cosine shifted 1 to the right
A negative cosine shifted 2 to the left
A sine shifted ½ to the left
A negative sine shifted 2.5 to the right

While any of these would be fine, the cosine shifts are easier to work with than the sine shifts in this case, because they involve integer values. Writing these:

$$y(x) = 3\cos\left(\frac{\pi}{3}(x-1)\right) - 2 \qquad \text{or}$$

$$y(x) = -3\cos\left(\frac{\pi}{3}(x+2)\right) - 2$$

Again, these functions are equivalent, so both yield the same graph.

Try it Now

5.   Write a formula for the function graphed here.

408 Chapter 6

Important Topics of This Section
Periodic functions
Sine and cosine function from the unit circle
Domain and range of sine and cosine functions
Sinusoidal functions
Negative angle identity
Simplifying expressions
Transformations
Amplitude
Midline
Period
Horizontal shifts

Try it Now Answers

1. 7

2. -4

3. $f(x) = 2\sin\left(\dfrac{\pi}{2}x\right) + 3$

4. $h(t) = -35\cos\left(\dfrac{2\pi}{3}t\right) + 45$

5. Two possibilities: $f(x) = 4\cos\left(\dfrac{\pi}{5}(x - 3.5)\right) + 4$ or $f(x) = 4\sin\left(\dfrac{\pi}{5}(x - 1)\right) + 4$

Section 6.1 Exercises

1. Sketch a graph of $f(x) = -3\sin(x)$.

2. Sketch a graph of $f(x) = 4\sin(x)$.

3. Sketch a graph of $f(x) = 2\cos(x)$.

4. Sketch a graph of $f(x) = -4\cos(x)$.

For the graphs below, determine the amplitude, midline, and period, then find a formula for the function.

5.

6.

7.

8.

9.

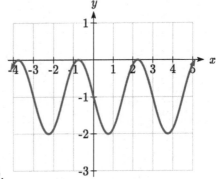

10.

For each of the following equations, find the amplitude, period, horizontal shift, and midline.

11. $y = 3\sin(8(x+4)) + 5$

12. $y = 4\sin\left(\dfrac{\pi}{2}(x-3)\right) + 7$

13. $y = 2\sin(3x - 21) + 4$

14. $y = 5\sin(5x + 20) - 2$

15. $y = \sin\left(\dfrac{\pi}{6}x + \pi\right) - 3$

16. $y = 8\sin\left(\dfrac{7\pi}{6}x + \dfrac{7\pi}{2}\right) + 6$

Find a formula for each of the functions graphed below.

17.

18.

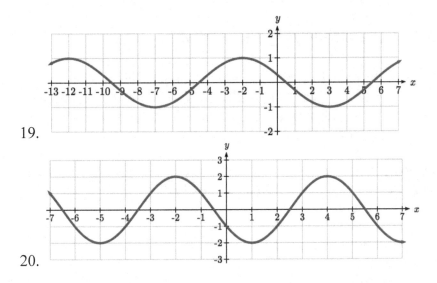

19.

20.

21. Outside temperature over the course of a day can be modeled as a sinusoidal function. Suppose you know the temperature is 50 degrees at midnight and the high and low temperature during the day are 57 and 43 degrees, respectively. Assuming t is the number of hours since midnight, find a function for the temperature, D, in terms of t.

22. Outside temperature over the course of a day can be modeled as a sinusoidal function. Suppose you know the temperature is 68 degrees at midnight and the high and low temperature during the day are 80 and 56 degrees, respectively. Assuming t is the number of hours since midnight, find a function for the temperature, D, in terms of t.

23. A Ferris wheel is 25 meters in diameter and boarded from a platform that is 1 meters above the ground. The six o'clock position on the Ferris wheel is level with the loading platform. The wheel completes 1 full revolution in 10 minutes. The function $h(t)$ gives your height in meters above the ground t minutes after the wheel begins to turn.

    a. Find the amplitude, midline, and period of $h(t)$.

    b. Find a formula for the height function $h(t)$.

    c. How high are you off the ground after 5 minutes?

24. A Ferris wheel is 35 meters in diameter and boarded from a platform that is 3 meters above the ground. The six o'clock position on the Ferris wheel is level with the loading platform. The wheel completes 1 full revolution in 8 minutes. The function $h(t)$ gives your height in meters above the ground t minutes after the wheel begins to turn.

    a. Find the amplitude, midline, and period of $h(t)$.

    b. Find a formula for the height function $h(t)$.

    c. How high are you off the ground after 4 minutes?

Section 6.2 Graphs of the Other Trig Functions

In this section, we will explore the graphs of the other four trigonometric functions. We'll begin with the tangent function. Recall that in Chapter 5 we defined tangent as y/x or sine/cosine, so you can think of the tangent as the slope of a line through the origin making the given angle with the positive x axis.

At an angle of 0, the line would be horizontal with a slope of zero. As the angle increases towards $\pi/2$, the slope increases more and more. At an angle of $\pi/2$, the line would be vertical and the slope would be undefined.

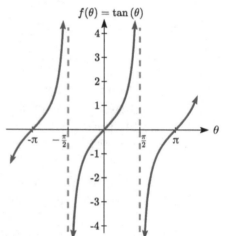

Immediately past $\pi/2$, the line would have a steep negative slope, giving a large negative tangent value. There is a break in the function at $\pi/2$, where the tangent value jumps from large positive to large negative.

We can use these ideas along with the definition of tangent to sketch a graph. Since tangent is defined as sine/cosine, we can determine that tangent will be zero when sine is zero: at $-\pi$, 0, π, and so on. Likewise, tangent will be undefined when cosine is zero: at $-\pi/2$, $\pi/2$, and so on.

The tangent is positive from 0 to $\pi/2$ and π to $3\pi/2$, corresponding to quadrants 1 and 3 of the unit circle.

Using technology, we can obtain a graph of tangent on a standard grid.

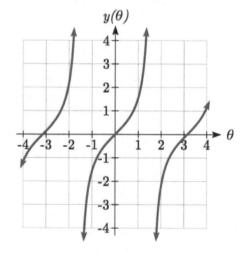

Notice that the graph appears to repeat itself. For any angle on the circle, there is a second angle with the same slope and tangent value halfway around the circle, so the graph repeats itself with a period of π; we can see one continuous cycle from $-\pi/2$ to $\pi/2$, before it jumps and repeats itself.

The graph has vertical asymptotes and the tangent is undefined wherever a line at that angle would be vertical: at $\pi/2$, $3\pi/2$, and so on. While the domain of the function is limited in this way, the range of the function is all real numbers.

Features of the Graph of Tangent

The graph of the tangent function $m(\theta) = \tan(\theta)$

The **period** of the tangent function is π

The **domain** of the tangent function is $\theta \neq \dfrac{\pi}{2} + k\pi$, where k is an integer

The **range** of the tangent function is all real numbers, $(-\infty, \infty)$

With the tangent function, like the sine and cosine functions, horizontal stretches/compressions are distinct from vertical stretches/compressions. The horizontal stretch can typically be determined from the period of the graph. With tangent graphs, it is often necessary to determine a vertical stretch using a point on the graph.

Example 1

Find a formula for the function graphed here.

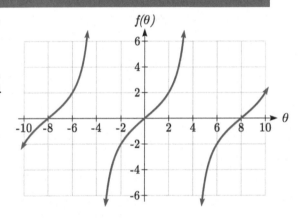

The graph has the shape of a tangent function, however the period appears to be 8. We can see one full continuous cycle from -4 to 4, suggesting a horizontal stretch. To stretch π to 8, the input values would have to be multiplied by $\dfrac{8}{\pi}$. Since the constant k in $f(\theta) = a\tan(k\theta)$ is the reciprocal of the horizontal stretch $\dfrac{8}{\pi}$, the equation must have form

$$f(\theta) = a\tan\left(\frac{\pi}{8}\theta\right).$$

We can also think of this the same way we did with sine and cosine. The period of the tangent function is π but it has been transformed and now it is 8; remember the ratio of the "normal period" to the "new period" is $\dfrac{\pi}{8}$ and so this becomes the value on the inside of the function that tells us how it was horizontally stretched.

To find the vertical stretch a, we can use a point on the graph. Using the point (2, 2)

$$2 = a\tan\left(\frac{\pi}{8} \cdot 2\right) = a\tan\left(\frac{\pi}{4}\right). \quad \text{Since } \tan\left(\frac{\pi}{4}\right) = 1, \quad a = 2.$$

This function would have a formula $f(\theta) = 2\tan\left(\dfrac{\pi}{8}\theta\right).$

1. Sketch a graph of $f(\theta) = 3\tan\left(\dfrac{\pi}{6}\theta\right)$.

For the graph of secant, we remember the reciprocal identity where $\sec(\theta) = \dfrac{1}{\cos(\theta)}$.

Notice that the function is undefined when the cosine is 0, leading to a vertical asymptote in the graph at $\pi/2$, $3\pi/2$, etc. Since the cosine is always no more than one in absolute value, the secant, being the reciprocal, will always be no less than one in absolute value. Using technology, we can generate the graph. The graph of the cosine is shown dashed so you can see the relationship.

$$f(\theta) = \sec(\theta) = \dfrac{1}{\cos(\theta)}$$

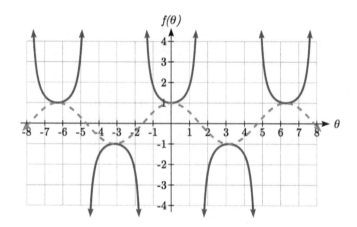

The graph of cosecant is similar. In fact, since $\sin(\theta) = \cos\left(\dfrac{\pi}{2} - \theta\right)$, it follows that $\csc(\theta) = \sec\left(\dfrac{\pi}{2} - \theta\right)$, suggesting the cosecant graph is a horizontal shift of the secant graph. This graph will be undefined where sine is 0. Recall from the unit circle that this occurs at 0, π, 2π, etc. The graph of sine is shown dashed along with the graph of the cosecant.

$$f(\theta) = \csc(\theta) = \dfrac{1}{\sin(\theta)}$$

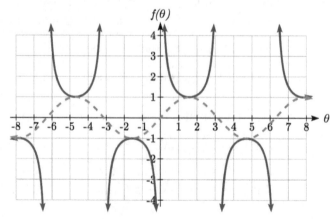

Features of the Graph of Secant and Cosecant

The secant and cosecant graphs have period 2π like the sine and cosine functions.

Secant has domain $\theta \neq \dfrac{\pi}{2} + k\pi$, where k is an integer

Cosecant has domain $\theta \neq k\pi$, where k is an integer

Both secant and cosecant have range of $(-\infty, -1] \cup [1, \infty)$

Example 2

Sketch a graph of $f(\theta) = 2\csc\left(\dfrac{\pi}{2}\theta\right) + 1$. What is the domain and range of this function?

The basic cosecant graph has vertical asymptotes at the integer multiples of π. Because of the factor $\dfrac{\pi}{2}$ inside the cosecant, the graph will be compressed by $\dfrac{2}{\pi}$, so the vertical asymptotes will be compressed to $\theta = \dfrac{2}{\pi} \cdot k\pi = 2k$. In other words, the graph will have vertical asymptotes at the integer multiples of 2, and the domain will correspondingly be $\theta \neq 2k$, where k is an integer.

The basic sine graph has a range of [-1, 1]. The vertical stretch by 2 will stretch this to [-2, 2], and the vertical shift up 1 will shift the range of this function to [-1, 3].

The basic cosecant graph has a range of $(-\infty, -1] \cup [1, \infty)$. The vertical stretch by 2 will stretch this to $(-\infty, -2] \cup [2, \infty)$, and the vertical shift up 1 will shift the range of this function to $(-\infty, -1] \cup [3, \infty)$.

The resulting graph is shown to the right.

Notice how the graph of the transformed cosecant relates to the graph of

$f(\theta) = 2\sin\left(\dfrac{\pi}{2}\theta\right) + 1$ shown dashed.

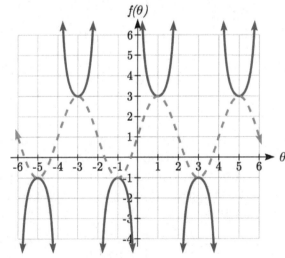

2. Given the graph of $f(\theta) = 2\cos\left(\dfrac{\pi}{2}\theta\right) + 1$ shown, sketch the

    graph of $g(\theta) = 2\sec\left(\dfrac{\pi}{2}\theta\right) + 1$ on the same axes.

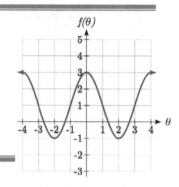

Finally, we'll look at the graph of cotangent. Based on its definition as the ratio of cosine to sine, it will be undefined when the sine is zero: at at 0, π, 2π, etc. The resulting graph is similar to that of the tangent. In fact, it is a horizontal flip and shift of the tangent function, as we'll see shortly in the next example.

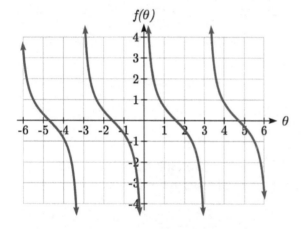

Features of the Graph of Cotangent
The cotangent graph has period π
Cotangent has domain $\theta \neq k\pi$, where k is an integer
Cotangent has range of all real numbers, $(-\infty, \infty)$

In Section 6.1 we determined that the sine function was an odd function and the cosine was an even function by observing the graph and establishing the negative angle identities for cosine and sine. Similarly, you may notice from its graph that the tangent function appears to be odd. We can verify this using the negative angle identities for sine and cosine:

$$\tan(-\theta) = \frac{\sin(-\theta)}{\cos(-\theta)} = \frac{-\sin(\theta)}{\cos(\theta)} = -\tan(\theta)$$

The secant, like the cosine it is based on, is an even function, while the cosecant, like the sine, is an odd function.

Negative Angle Identities Tangent, Cotangent, Secant and Cosecant	
$\tan(-\theta) = -\tan(\theta)$	$\cot(-\theta) = -\cot(\theta)$
$\sec(-\theta) = \sec(\theta)$	$\csc(-\theta) = -\csc(\theta)$

Example 3

Prove that $\tan(\theta) = -\cot\left(\theta - \dfrac{\pi}{2}\right)$

$\tan(\theta)$ Using the definition of tangent

$= \dfrac{\sin(\theta)}{\cos(\theta)}$ Using the cofunction identities

$= \dfrac{\cos\left(\dfrac{\pi}{2} - \theta\right)}{\sin\left(\dfrac{\pi}{2} - \theta\right)}$ Using the definition of cotangent

$= \cot\left(\dfrac{\pi}{2} - \theta\right)$ Factoring a negative from the inside

$= \cot\left(-\left(\theta - \dfrac{\pi}{2}\right)\right)$ Using the negative angle identity for cot

$= -\cot\left(\theta - \dfrac{\pi}{2}\right)$

Important Topics of This Section
The tangent and cotangent functions
Period
Domain
Range
The secant and cosecant functions
Period
Domain
Range
Transformations
Negative Angle identities

Try it Now Answers

1.

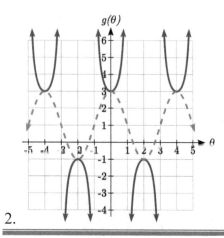

2.

Section 6.2 Exercises

Match each trigonometric function with one of the graphs.

1. $f(x) = \tan(x)$ 2. $f(x) = \sec(x)$

3. $f(x) = \csc(x)$ 4. $f(x) = \cot(x)$

I

II

III

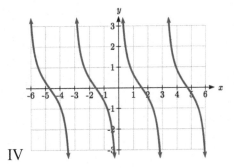

IV

Find the period and horizontal shift of each of the following functions.

5. $f(x) = 2\tan(4x - 32)$

6. $g(x) = 3\tan(6x + 42)$

7. $h(x) = 2\sec\left(\dfrac{\pi}{4}(x+1)\right)$

8. $k(x) = 3\sec\left(2\left(x + \dfrac{\pi}{2}\right)\right)$

9. $m(x) = 6\csc\left(\dfrac{\pi}{3}x + \pi\right)$

10. $n(x) = 4\csc\left(\dfrac{5\pi}{3}x - \dfrac{20\pi}{3}\right)$

11. Sketch a graph of #7 above.
12. Sketch a graph of #8 above.
13. Sketch a graph of #9 above.
14. Sketch a graph of #10 above.

15. Sketch a graph of $j(x) = \tan\left(\dfrac{\pi}{2}x\right)$.

16. Sketch a graph of $p(t) = 2\tan\left(t - \dfrac{\pi}{2}\right)$.

Find a formula for each function graphed below.

17.

18.

19.

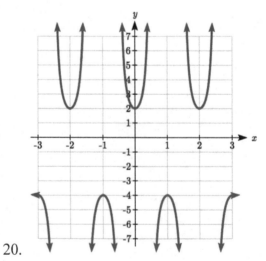

20.

21. If $\tan x = -1.5$, find $\tan(-x)$.

22. If $\tan x = 3$, find $\tan(-x)$.

23. If $\sec x = 2$, find $\sec(-x)$.

24. If $\sec x = -4$, find $\sec(-x)$.

25. If $\csc x = -5$, find $\csc(-x)$.

26. If $\csc x = 2$, find $\csc(-x)$.

Simplify each of the following expressions completely.

27. $\cot(-x)\cos(-x) + \sin(-x)$

28. $\cos(-x) + \tan(-x)\sin(-x)$

Section 6.3 Inverse Trig Functions

In previous sections, we have evaluated the trigonometric functions at various angles, but at times we need to know what angle would yield a specific sine, cosine, or tangent value. For this, we need inverse functions. Recall that for a one-to-one function, if $f(a) = b$, then an inverse function would satisfy $f^{-1}(b) = a$.

You probably are already recognizing an issue – that the sine, cosine, and tangent functions are not one-to-one functions. To define an inverse of these functions, we will need to restrict the domain of these functions to yield a new function that is one-to-one. We choose a domain for each function that includes the angle zero.

Sine, limited to $\left[-\dfrac{\pi}{2}, \dfrac{\pi}{2}\right]$ Cosine, limited to $[0, \pi]$ Tangent, limited to $\left(-\dfrac{\pi}{2}, \dfrac{\pi}{2}\right)$

    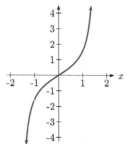

On these restricted domains, we can define the inverse sine, inverse cosine, and inverse tangent functions.

Inverse Sine, Cosine, and Tangent Functions

For angles in the interval $\left[-\dfrac{\pi}{2}, \dfrac{\pi}{2}\right]$, if $\sin(\theta) = a$, then $\sin^{-1}(a) = \theta$

For angles in the interval $[0, \pi]$, if $\cos(\theta) = a$, then $\cos^{-1}(a) = \theta$

For angles in the interval $\left(-\dfrac{\pi}{2}, \dfrac{\pi}{2}\right)$, if $\tan(\theta) = a$, then $\tan^{-1}(a) = \theta$

$\sin^{-1}(x)$ has domain [-1, 1] and range $\left[-\dfrac{\pi}{2}, \dfrac{\pi}{2}\right]$

$\cos^{-1}(x)$ has domain [-1, 1] and range $[0, \pi]$

$\tan^{-1}(x)$ has domain of all real numbers and range $\left(-\dfrac{\pi}{2}, \dfrac{\pi}{2}\right)$

The $\sin^{-1}(x)$ is sometimes called the **arcsine** function, and notated $\arcsin(a)$.

The $\cos^{-1}(x)$ is sometimes called the **arccosine** function, and notated $\arccos(a)$.

The $\tan^{-1}(x)$ is sometimes called the **arctangent** function, and notated $\arctan(a)$.

The graphs of the inverse functions are shown here:

$\sin^{-1}(x)$ $\cos^{-1}(x)$ $\tan^{-1}(x)$

          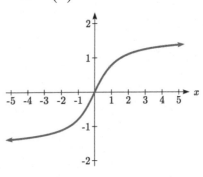

Notice that the output of each of these inverse functions is an *angle*.

Example 1

Evaluate

a) $\sin^{-1}\left(\dfrac{1}{2}\right)$ b) $\sin^{-1}\left(-\dfrac{\sqrt{2}}{2}\right)$ c) $\cos^{-1}\left(-\dfrac{\sqrt{3}}{2}\right)$ d) $\tan^{-1}(1)$

a) Evaluating $\sin^{-1}\left(\dfrac{1}{2}\right)$ is the same as asking what angle would have a sine value of $\dfrac{1}{2}$.

In other words, what angle θ would satisfy $\sin(\theta)=\dfrac{1}{2}$?

There are multiple angles that would satisfy this relationship, such as $\dfrac{\pi}{6}$ and $\dfrac{5\pi}{6}$, but

we know we need the angle in the range of $\sin^{-1}(x)$, the interval $\left[-\dfrac{\pi}{2},\dfrac{\pi}{2}\right]$, so the

answer will be $\sin^{-1}\left(\dfrac{1}{2}\right)=\dfrac{\pi}{6}$.

Remember that the inverse is a *function* so for each input, we will get exactly one output.

b) Evaluating $\sin^{-1}\left(-\dfrac{\sqrt{2}}{2}\right)$, we know that $\dfrac{5\pi}{4}$ and $\dfrac{7\pi}{4}$ both have a sine value of

$-\dfrac{\sqrt{2}}{2}$, but neither is in the interval $\left[-\dfrac{\pi}{2},\dfrac{\pi}{2}\right]$. For that, we need the negative angle

coterminal with $\dfrac{7\pi}{4}$. $\sin^{-1}\left(-\dfrac{\sqrt{2}}{2}\right)=-\dfrac{\pi}{4}$.

c) Evaluating $\cos^{-1}\left(-\dfrac{\sqrt{3}}{2}\right)$, we are looking for an angle in the interval $[0,\pi]$ with a

cosine value of $-\dfrac{\sqrt{3}}{2}$. The angle that satisfies this is $\cos^{-1}\left(-\dfrac{\sqrt{3}}{2}\right)=\dfrac{5\pi}{6}$.

d) Evaluating $\tan^{-1}(1)$, we are looking for an angle in the interval $\left(-\dfrac{\pi}{2},\dfrac{\pi}{2}\right)$ with a

tangent value of 1. The correct angle is $\tan^{-1}(1)=\dfrac{\pi}{4}$.

Try It Now
1. Evaluate

a) $\sin^{-1}(-1)$ b) $\tan^{-1}(-1)$ c) $\cos^{-1}(-1)$ d) $\cos^{-1}\left(\dfrac{1}{2}\right)$

Example 2
Evaluate $\sin^{-1}(0.97)$ using your calculator.

Since the output of the inverse function is an angle, your calculator will give you a degree value if in degree mode, and a radian value if in radian mode.

In radian mode, $\sin^{-1}(0.97)\approx 1.3252$ In degree mode, $\sin^{-1}(0.97)\approx 75.93°$

Try it Now
2. Evaluate $\cos^{-1}(-0.4)$ using your calculator.

In Section 5.5, we worked with trigonometry on a right triangle to solve for the sides of a triangle given one side and an additional angle. Using the inverse trig functions, we can solve for the angles of a right triangle given two sides.

Example 3

Solve the triangle for the angle θ.

Since we know the hypotenuse and the side adjacent to the angle, it makes sense for us to use the cosine function.

$\cos(\theta) = \dfrac{9}{12}$ Using the definition of the inverse,

$\theta = \cos^{-1}\left(\dfrac{9}{12}\right)$ Evaluating

$\theta \approx 0.7227$, or about $41.4096°$

There are times when we need to compose a trigonometric function with an inverse trigonometric function. In these cases, we can find exact values for the resulting expressions

---

Example 4

Evaluate $\sin^{-1}\left(\cos\left(\dfrac{13\pi}{6}\right)\right)$.

a) Here, we can directly evaluate the inside of the composition.

$\cos\left(\dfrac{13\pi}{6}\right) = \dfrac{\sqrt{3}}{2}$

Now, we can evaluate the inverse function as we did earlier.

$\sin^{-1}\left(\dfrac{\sqrt{3}}{2}\right) = \dfrac{\pi}{3}$

Try it Now

3. Evaluate $\cos^{-1}\left(\sin\left(-\dfrac{11\pi}{4}\right)\right)$.

Example 5

Find an exact value for $\sin\left(\cos^{-1}\left(\dfrac{4}{5}\right)\right)$.

Beginning with the inside, we can say there is some angle so $\theta = \cos^{-1}\left(\dfrac{4}{5}\right)$, which means $\cos(\theta) = \dfrac{4}{5}$, and we are looking for $\sin(\theta)$. We can use the Pythagorean identity to do this.

$\sin^2(\theta) + \cos^2(\theta) = 1$ Using our known value for cosine

$\sin^2(\theta) + \left(\dfrac{4}{5}\right)^2 = 1$ Solving for sine

$\sin^2(\theta) = 1 - \dfrac{16}{25}$

$\sin(\theta) = \pm\sqrt{\dfrac{9}{25}} = \pm\dfrac{3}{5}$

Since we know that the inverse cosine always gives an angle on the interval $[0, \pi]$, we know that the sine of that angle must be positive, so $\sin\left(\cos^{-1}\left(\dfrac{4}{5}\right)\right) = \sin(\theta) = \dfrac{3}{5}$

Example 6

Find an exact value for $\sin\left(\tan^{-1}\left(\dfrac{7}{4}\right)\right)$.

While we could use a similar technique as in the last example, we will demonstrate a different technique here. From the inside, we know there is an angle so $\tan(\theta) = \dfrac{7}{4}$. We can envision this as the opposite and adjacent sides on a right triangle.

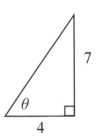

Using the Pythagorean Theorem, we can find the hypotenuse of this triangle:

$4^2 + 7^2 = hypotenuse^2$

$hypotenuse = \sqrt{65}$

Now, we can represent the sine of the angle as opposite side divided by hypotenuse.

$$\sin(\theta) = \frac{7}{\sqrt{65}}$$

This gives us our desired composition

$$\sin\left(\tan^{-1}\left(\frac{7}{4}\right)\right) = \sin(\theta) = \frac{7}{\sqrt{65}}.$$

Try it Now

4. Evaluate $\cos\left(\sin^{-1}\left(\frac{7}{9}\right)\right)$.

We can also find compositions involving algebraic expressions

Example 7

Find a simplified expression for $\cos\left(\sin^{-1}\left(\frac{x}{3}\right)\right)$, for $-3 \le x \le 3$.

We know there is an angle θ so that $\sin(\theta) = \frac{x}{3}$. Using the Pythagorean Theorem,

$$\sin^2(\theta) + \cos^2(\theta) = 1 \qquad \text{Using our known expression for sine}$$

$$\left(\frac{x}{3}\right)^2 + \cos^2(\theta) = 1 \qquad \text{Solving for cosine}$$

$$\cos^2(\theta) = 1 - \frac{x^2}{9}$$

$$\cos(\theta) = \pm\sqrt{\frac{9 - x^2}{9}} = \pm\frac{\sqrt{9 - x^2}}{3}$$

Since we know that the inverse sine must give an angle on the interval $\left[-\frac{\pi}{2}, \frac{\pi}{2}\right]$, we can deduce that the cosine of that angle must be positive. This gives us

$$\cos\left(\sin^{-1}\left(\frac{x}{3}\right)\right) = \frac{\sqrt{9 - x^2}}{3}$$

Try it Now

5. Find a simplified expression for $\sin\left(\tan^{-1}(4x)\right)$, for $-\dfrac{1}{4} \le x \le \dfrac{1}{4}$.

Important Topics of This Section

Inverse trig functions: arcsine, arccosine and arctangent
Domain restrictions
Evaluating inverses using unit circle values and the calculator
Simplifying numerical expressions involving the inverse trig functions
Simplifying algebraic expressions involving the inverse trig functions

---

Try it Now Answers

1. a) $-\dfrac{\pi}{2}$ b) $-\dfrac{\pi}{4}$ c) π d) $\dfrac{\pi}{3}$

2. 1.9823 or 113.578°

3. $\sin\left(-\dfrac{11\pi}{4}\right) = -\dfrac{\sqrt{2}}{2}$. $\cos^{-1}\left(-\dfrac{\sqrt{2}}{2}\right) = \dfrac{3\pi}{4}$

4. Let $\theta = \sin^{-1}\left(\dfrac{7}{9}\right)$ so $\sin(\theta) = \dfrac{7}{9}$. .

   Using Pythagorean Identity, $\sin^2\theta + \cos^2\theta = 1$, so $\left(\dfrac{7}{9}\right)^2 + \cos^2\theta = 1$.

   Solving, $\cos\left(\sin^{-1}\left(\dfrac{7}{9}\right)\right) = \cos(\theta) = \dfrac{4\sqrt{2}}{9}$.

5. Let $\theta = \tan^{-1}(4x)$, so $\tan(\theta) = 4x$. We can represent this on a
   triangle as $\tan(\theta) = \dfrac{4x}{1}$.

   The hypotenuse of the triangle would be $\sqrt{(4x)^2 + 1}$.

   $\sin\left(\tan^{-1}(4x)\right) = \sin(\theta) = \dfrac{4x}{\sqrt{16x^2 + 1}}$

Section 6.3 Exercises

Evaluate the following expressions, giving the answer in radians.

1. $\sin^{-1}\left(\dfrac{\sqrt{2}}{2}\right)$

2. $\sin^{-1}\left(\dfrac{\sqrt{3}}{2}\right)$

3. $\sin^{-1}\left(-\dfrac{1}{2}\right)$

4. $\sin^{-1}\left(-\dfrac{\sqrt{2}}{2}\right)$

5. $\cos^{-1}\left(\dfrac{1}{2}\right)$

6. $\cos^{-1}\left(\dfrac{\sqrt{2}}{2}\right)$

7. $\cos^{-1}\left(-\dfrac{\sqrt{2}}{2}\right)$

8. $\cos^{-1}\left(-\dfrac{\sqrt{3}}{2}\right)$

9. $\tan^{-1}(1)$

10. $\tan^{-1}\left(\sqrt{3}\right)$

11. $\tan^{-1}\left(-\sqrt{3}\right)$

12. $\tan^{-1}(-1)$

Use your calculator to evaluate each expression, giving the answer in radians.

13. $\cos^{-1}(-0.4)$

14. $\cos^{-1}(0.8)$

15. $\sin^{-1}(-0.8)$

16. $\tan^{-1}(6)$

Find the angle θ in degrees.

17.

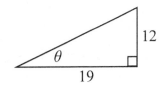

18.

Evaluate the following expressions.

19. $\sin^{-1}\left(\cos\left(\dfrac{\pi}{4}\right)\right)$

20. $\cos^{-1}\left(\sin\left(\dfrac{\pi}{6}\right)\right)$

21. $\sin^{-1}\left(\cos\left(\dfrac{4\pi}{3}\right)\right)$

22. $\cos^{-1}\left(\sin\left(\dfrac{5\pi}{4}\right)\right)$

23. $\cos\left(\sin^{-1}\left(\dfrac{3}{7}\right)\right)$

24. $\sin\left(\cos^{-1}\left(\dfrac{4}{9}\right)\right)$

25. $\cos\left(\tan^{-1}(4)\right)$

26. $\tan\left(\sin^{-1}\left(\dfrac{1}{3}\right)\right)$

Find a simplified expression for each of the following.

27. $\sin\left(\cos^{-1}\left(\dfrac{x}{5}\right)\right)$, for $-5 \le x \le 5$

28. $\tan\left(\cos^{-1}\left(\dfrac{x}{2}\right)\right)$, for $-2 \le x \le 2$

29. $\sin\left(\tan^{-1}(3x)\right)$

30. $\cos\left(\tan^{-1}(4x)\right)$

Section 6.4 Solving Trig Equations

In Section 6.1, we determined the height of a rider on the London Eye Ferris wheel could be determined by the equation $h(t) = -65\cos\left(\dfrac{\pi}{15}t\right) + 70$.

If we wanted to know length of time during which the rider is more than 100 meters above ground, we would need to solve equations involving trig functions.

Solving using known values

In the last chapter, we learned sine and cosine values at commonly encountered angles. We can use these to solve sine and cosine equations involving these common angles.

Example 1

Solve $\sin(t) = \dfrac{1}{2}$ for all possible values of t.

Notice this is asking us to identify all angles, t, that have a sine value of $\dfrac{1}{2}$. While evaluating a function always produces one result, solving for an input can yield multiple solutions. Two solutions should immediately jump to mind from the last chapter: $t = \dfrac{\pi}{6}$ and $t = \dfrac{5\pi}{6}$ because they are the common angles on the unit circle with a sin of $\dfrac{1}{2}$.

Looking at a graph confirms that there are more than these two solutions. While eight are seen on this graph, there are an infinite number of solutions!

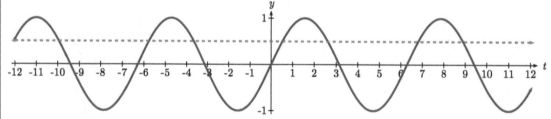

Remember that any coterminal angle will also have the same sine value, so any angle coterminal with these our first two solutions is also a solution. Coterminal angles can be found by adding full rotations of 2π, so we can write the full set of solutions:

$t = \dfrac{\pi}{6} + 2\pi k$ where k is an integer, and $t = \dfrac{5\pi}{6} + 2\pi k$ where k is an integer.

Example 2

A circle of radius $5\sqrt{2}$ intersects the line $x = -5$ at two points. Find the angles θ on the interval $0 \le \theta < 2\pi$, where the circle and line intersect.

The x coordinate of a point on a circle can be found as $x = r\cos(\theta)$, so the x coordinate of points on this circle would be $x = 5\sqrt{2}\cos(\theta)$. To find where the line $x = -5$ intersects the circle, we can solve for where the x value on the circle would be -5.

$-5 = 5\sqrt{2}\cos(\theta)$ Isolating the cosine

$\dfrac{-1}{\sqrt{2}} = \cos(\theta)$ Recall that $\dfrac{-1}{\sqrt{2}} = \dfrac{-\sqrt{2}}{2}$, so we are solving

$\cos(\theta) = \dfrac{-\sqrt{2}}{2}$

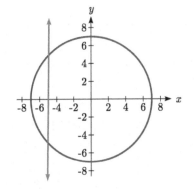

We can recognize this as one of our special cosine values from our unit circle, and it corresponds with angles

$\theta = \dfrac{3\pi}{4}$ and $\theta = \dfrac{5\pi}{4}$.

Try it Now

1. Solve $\tan(t) = 1$ for all possible values of t.

Example 3

The depth of water at a dock rises and falls with the tide, following the equation $f(t) = 4\sin\left(\dfrac{\pi}{12}t\right) + 7$, where t is measured in hours after midnight. A boat requires a depth of 9 feet to tie up at the dock. Between what times will the depth be 9 feet?

To find when the depth is 9 feet, we need to solve $f(t) = 9$.

$4\sin\left(\dfrac{\pi}{12}t\right) + 7 = 9$ Isolating the sine

$4\sin\left(\dfrac{\pi}{12}t\right) = 2$ Dividing by 4

$\sin\left(\dfrac{\pi}{12}t\right) = \dfrac{1}{2}$ We know $\sin(\theta) = \dfrac{1}{2}$ when $\theta = \dfrac{\pi}{6}$ or $\theta = \dfrac{5\pi}{6}$

While we know what angles have a sine value of $\frac{1}{2}$, because of the horizontal stretch/compression it is less clear how to proceed.

To deal with this, we can make a substitution, defining a new temporary variable u to be $u = \frac{\pi}{12}t$, so our equation $\sin\left(\frac{\pi}{12}t\right) = \frac{1}{2}$ becomes

$$\sin(u) = \frac{1}{2}$$

From earlier, we saw the solutions to this equation were

$u = \frac{\pi}{6} + 2\pi k$ where k is an integer, and

$u = \frac{5\pi}{6} + 2\pi k$ where k is an integer

To undo our substitution, we replace the u in the solutions with $u = \frac{\pi}{12}t$ and solve for t.

$\frac{\pi}{12}t = \frac{\pi}{6} + 2\pi k$ where k is an integer, and $\frac{\pi}{12}t = \frac{5\pi}{6} + 2\pi k$ where k is an integer.

Dividing by $\pi/12$, we obtain solutions

$t = 2 + 24k$ where k is an integer, and
$t = 10 + 24k$ where k is an integer.

The depth will be 9 feet and the boat will be able to approach the dock between 2am and 10am.

Notice how in both scenarios, the $24k$ shows how every 24 hours the cycle will be repeated.

In the previous example, looking back at the original simplified equation $\sin\left(\frac{\pi}{12}t\right) = \frac{1}{2}$, we can use the ratio of the "normal period" to the stretch factor to find the period: $\frac{2\pi}{\left(\frac{\pi}{12}\right)} = 2\pi\left(\frac{12}{\pi}\right) = 24$. Notice that the sine function has a period of 24, which is reflected in the solutions: there were two unique solutions on one full cycle of the sine function, and additional solutions were found by adding multiples of a full period.

2. Solve $4\sin(5t) - 1 = 1$ for all possible values of t.

Solving using the inverse trig functions

Not all equations involve the "special" values of the trig functions to we have learned. To find the solutions to these equations, we need to use the inverse trig functions.

Example 4

Use the inverse sine function to find one solution to $\sin(\theta) = 0.8$.

Since this is not a known unit circle value, calculating the inverse, $\theta = \sin^{-1}(0.8)$. This requires a calculator and we must approximate a value for this angle. If your calculator is in degree mode, your calculator will give you an angle in degrees as the output. If your calculator is in radian mode, your calculator will give you an angle in radians. In radians, $\theta = \sin^{-1}(0.8) \approx 0.927$, or in degrees, $\theta = \sin^{-1}(0.8) \approx 53.130°$.

If you are working with a composed trig function and you are not solving for an angle, you will want to ensure that you are working in radians. In calculus, we will almost always want to work with radians since they are unit-less.

Notice that the inverse trig functions do exactly what you would expect of any function – for each input they give exactly one output. While this is necessary for these to be a function, it means that to find *all* the solutions to an equation like $\sin(\theta) = 0.8$, we need to do more than just evaluate the inverse function.

To find additional solutions, it is good to remember four things:
- The sine is the y-value of a point on the unit circle
- The cosine is the x-value of a point on the unit circle
- The tangent is the slope of a line at a given angle
- Other angles with the same sin/cos/tan will have the same reference angle

Example 5

Find all solutions to $\sin(\theta) = 0.8$.

We would expect two unique angles on one cycle to have this sine value. In the previous example, we found one solution to be $\theta = \sin^{-1}(0.8) \approx 0.927$. To find the other, we need to answer the question "what other angle has the same sine value as an angle of 0.927?"

We can think of this as finding all the angles where the y-value on the unit circle is 0.8. Drawing a picture of the circle helps how the symmetry.
On a unit circle, we would recognize that the second angle would have the same reference angle and reside in the second quadrant. This second angle would be located at
$\theta = \pi - \sin^{-1}(0.8)$, or approximately $\theta \approx \pi - 0.927 = 2.214$.

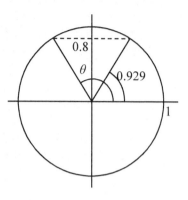

To find more solutions we recall that angles coterminal with these two would have the same sine value, so we can add full cycles of 2π.

$\theta = \sin^{-1}(0.8) + 2\pi k$ and $\theta = \pi - \sin^{-1}(0.8) + 2\pi k$ where k is an integer, or approximately, $\theta = 0.927 + 2\pi k$ and $\theta = 2.214 + 2\pi k$ where k is an integer.

Example 6

Find all solutions to $\sin(x) = -\dfrac{8}{9}$ on the interval $0° \leq x < 360°$.

We are looking for the angles with a y-value of -8/9 on the unit circle. Immediately we can see the solutions will be in the third and fourth quadrants.

First, we will turn our calculator to degree mode. Using the inverse, we can find one solution $x = \sin^{-1}\left(-\dfrac{8}{9}\right) \approx -62.734°$.

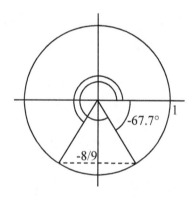

While this angle satisfies the equation, it does not lie in the domain we are looking for. To find the angles in the desired domain, we start looking for additional solutions.

First, an angle coterminal with $-62.734°$ will have the same sine. By adding a full rotation, we can find an angle in the desired domain with the same sine.
$x = -62.734° + 360° = 297.266°$

There is a second angle in the desired domain that lies in the third quadrant. Notice that $62.734°$ is the reference angle for all solutions, so this second solution would be $62.734°$ past $180°$
$x = 62.734° + 180° = 242.734°$

The two solutions on $0° \leq x < 360°$ are $x = 297.266°$ and $x = 242.734°$

Example 7

Find all solutions to $\tan(x) = 3$ on $0 \le x < 2\pi$.

Using the inverse tangent function, we can find one solution $x = \tan^{-1}(3) \approx 1.249$. Unlike the sine and cosine, the tangent function only attains any output value once per cycle, so there is no second solution in any one cycle.

By adding π, a full period of tangent function, we can find a second angle with the same tangent value. Notice this gives another angle where the line has the same slope.

If additional solutions were desired, we could continue to add multiples of π, so all solutions would take on the form $x = 1.249 + k\pi$, however we are only interested in $0 \le x < 2\pi$.
$x = 1.249 + \pi = 4.391$

The two solutions on $0 \le x < 2\pi$ are $x = 1.249$ and $x = 4.391$.

Try it Now

3. Find all solutions to $\tan(x) = 0.7$ on $0° \le x < 360°$.

Example 8

Solve $3\cos(t) + 4 = 2$ for all solutions on one cycle, $0 \le t < 2\pi$

$3\cos(t) + 4 = 2$ Isolating the cosine
$3\cos(t) = -2$

$\cos(t) = -\dfrac{2}{3}$ Using the inverse, we can find one solution

$t = \cos^{-1}\left(-\dfrac{2}{3}\right) \approx 2.301$

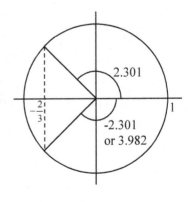

We're looking for two angles where the x-coordinate on a unit circle is -2/3. A second angle with the same cosine would be located in the third quadrant. Notice that the location of this angle could be represented as $t = -2.301$. To represent this as a positive angle we could find a coterminal angle by adding a full cycle.
$t = -2.301 + 2\pi = 3.982$

The equation has two solutions between 0 and 2π, at $t = 2.301$ and $t = 3.982$.

Example 9

Solve $\cos(3t) = 0.2$ for all solutions on two cycles, $0 \le t < \dfrac{4\pi}{3}$.

As before, with a horizontal compression it can be helpful to make a substitution, $u = 3t$. Making this substitution simplifies the equation to a form we have already solved.

$\cos(u) = 0.2$

$u = \cos^{-1}(0.2) \approx 1.369$

A second solution on one cycle would be located in the fourth quadrant with the same reference angle.

$u = 2\pi - 1.369 = 4.914$

In this case, we need all solutions on two cycles, so we need to find the solutions on the second cycle. We can do this by adding a full rotation to the previous two solutions.

$u = 1.369 + 2\pi = 7.653$

$u = 4.914 + 2\pi = 11.197$

Undoing the substitution, we obtain our four solutions:
$3t = 1.369$, so $t = 0.456$
$3t = 4.914$ so $t = 1.638$
$3t = 7.653$, so $t = 2.551$
$3t = 11.197$, so $t = 3.732$

Example 10

Solve $3\sin(\pi t) = -2$ for all solutions.

$3\sin(\pi t) = -2$ 　　　　　　　　　Isolating the sine

$\sin(\pi t) = -\dfrac{2}{3}$ 　　　　　　　　We make the substitution $u = \pi t$

$\sin(u) = -\dfrac{2}{3}$ 　　　　　　　　Using the inverse, we find one solution

$u = \sin^{-1}\left(-\dfrac{2}{3}\right) \approx -0.730$

This angle is in the fourth quadrant. A second angle with the same sine would be in the third quadrant with 0.730 as a reference angle:

$u = \pi + 0.730 = 3.871$

We can write all solutions to the equation $\sin(u) = -\dfrac{2}{3}$ as

$u = -0.730 + 2\pi k$ or $u = 3.871 + 2\pi k$, where k is an integer.

Undoing our substitution, we can replace u in our solutions with $u = \pi t$ and solve for t

$$\pi t = -0.730 + 2\pi k \quad \text{or} \qquad \pi t = 3.871 + 2\pi k \qquad\qquad \text{Divide by } \pi$$
$$t = -0.232 + 2k \quad \text{or} \qquad t = 1.232 + 2k$$

Try it Now

4. Solve $5\sin\left(\dfrac{\pi}{2}t\right) + 3 = 0$ for all solutions on one cycle, $0 \le t < 4$.

Solving Trig Equations

1) Isolate the trig function on one side of the equation
2) Make a substitution for the inside of the sine, cosine, or tangent (or other trig function)
3) Use inverse trig functions to find one solution
4) Use symmetries to find a second solution on one cycle (when a second exists)
5) Find additional solutions if needed by adding full periods
6) Undo the substitution

We now can return to the question we began the section with.

Example 11

The height of a rider on the London Eye Ferris wheel can be determined by the equation $h(t) = -65\cos\left(\dfrac{\pi}{15}t\right) + 70$. How long is the rider more than 100 meters above ground?

To find how long the rider is above 100 meters, we first find the times at which the rider is at a height of 100 meters by solving $h(t) = 100$.

$$100 = -65\cos\left(\frac{\pi}{15}t\right) + 70 \qquad\qquad \text{Isolating the cosine}$$

$$30 = -65\cos\left(\frac{\pi}{15}t\right)$$

$$\frac{30}{-65} = \cos\left(\frac{\pi}{15}t\right) \qquad\qquad \text{We make the substitution } u = \frac{\pi}{15}t$$

$$\frac{30}{-65} = \cos(u) \qquad\qquad \text{Using the inverse, we find one solution}$$

$$u = \cos^{-1}\left(\frac{30}{-65}\right) \approx 2.051$$

This angle is in the second quadrant. A second angle with the same cosine would be symmetric in the third quadrant. This angle could be represented as $u = -2.051$, but we need a coterminal positive angle, so we add 2π:
$u = 2\pi - 2.051 \approx 4.230$

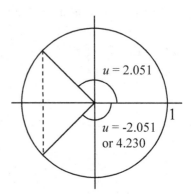

Now we can undo the substitution to solve for t

$\frac{\pi}{15}t = 2.051$ so $t = 9.793$ minutes after the start of the ride

$\frac{\pi}{15}t = 4.230$ so $t = 20.197$ minutes after the start of the ride

A rider will be at 100 meters after 9.793 minutes, and again after 20.197 minutes. From the behavior of the height graph, we know the rider will be above 100 meters between these times. A rider will be above 100 meters for 20.197 - 9.793 = 10.404 minutes of the ride.

Important Topics of This Section
Solving trig equations using known values
Using substitution to solve equations
Finding answers in one cycle or period vs. finding all possible solutions
Method for solving trig equations

Try it Now Answers

1. From our special angles, we know one answer is $t = \frac{\pi}{4}$. Tangent equations only have one unique solution per cycle or period, so additional solutions can be found by adding multiples of a full period, π. $t = \frac{\pi}{4} + \pi k$.

2. $4\sin(5t) - 1 = 1$

    $\sin(5t) = \frac{1}{2}$. Let $u = 5t$ so this becomes $\sin(u) = \frac{1}{2}$, which has solutions

    $u = \frac{\pi}{6} + 2\pi k, \frac{5\pi}{6} + 2\pi k$. Solving $5t = u = \frac{\pi}{6} + 2\pi k, \frac{5\pi}{6} + 2\pi k$ gives the solutions

    $t = \frac{\pi}{30} + \frac{2\pi}{5}k \qquad t = \frac{\pi}{6} + \frac{2\pi}{5}k$

3. The first solution is $x = \tan^{-1}(0.7) \approx 34.992°$.

    For a standard tangent, the second solution can be found by adding a full period, $180°$, giving $x = 180° + 34.99° = 214.992°$.

4. $\sin\left(\dfrac{\pi}{2}t\right) = -\dfrac{3}{5}$. Let $u = \dfrac{\pi}{2}t$, so this becomes $\sin(u) = -\dfrac{3}{5}$.

    Using the inverse, $u = \sin^{-1}\left(-\dfrac{3}{5}\right) \approx -0.6435$. Since we want positive solutions, we can find the coterminal solution by adding a full cycle: $u = -0.6435 + 2\pi = 5.6397$.

    Another angle with the same sin would be in the third quadrant with the reference angle 0.6435. $u = \pi + 0.6435 = 3.7851$.

    Solving for t, $u = \dfrac{\pi}{2}t = 5.6397$, so $t = 5.6397\left(\dfrac{2}{\pi}\right) = 3.5903$

    and $u = \dfrac{\pi}{2}t = 3.7851$, so $t = 3.7851\left(\dfrac{2}{\pi}\right) = 2.4097$.

    $t = 2.4097$ or $t = 3.5903$.

Section 6.4 Exercises

Give all answers in radians unless otherwise indicated.

Find all solutions on the interval $0 \le \theta < 2\pi$.

1. $2\sin(\theta) = -\sqrt{2}$ 2. $2\sin(\theta) = \sqrt{3}$ 3. $2\cos(\theta) = 1$ 4. $2\cos(\theta) = -\sqrt{2}$

5. $\sin(\theta) = 1$ 6. $\sin(\theta) = 0$ 7. $\cos(\theta) = 0$ 8. $\cos(\theta) = -1$

Find all solutions.

9. $2\cos(\theta) = \sqrt{2}$ 10. $2\cos(\theta) = -1$ 11. $2\sin(\theta) = -1$ 12. $2\sin(\theta) = -\sqrt{3}$

Find all solutions.

13. $2\sin(3\theta) = 1$ 14. $2\sin(2\theta) = \sqrt{3}$ 15. $2\sin(3\theta) = -\sqrt{2}$

16. $2\sin(3\theta) = -1$ 17. $2\cos(2\theta) = 1$ 18. $2\cos(2\theta) = \sqrt{3}$

19. $2\cos(3\theta) = -\sqrt{2}$ 20. $2\cos(2\theta) = -1$ 21. $\cos\left(\dfrac{\pi}{4}\theta\right) = -1$

22. $\sin\left(\dfrac{\pi}{3}\theta\right) = -1$ 23. $2\sin(\pi\theta) = 1$.   24. $2\cos\left(\dfrac{\pi}{5}\theta\right) = \sqrt{3}$

Find all solutions on the interval $0 \le x < 2\pi$.

25. $\sin(x) = 0.27$ 26. $\sin(x) = 0.48$ 27. $\sin(x) = -0.58$ 28. $\sin(x) = -0.34$

29. $\cos(x) = -0.55$ 30. $\sin(x) = 0.28$ 31. $\cos(x) = 0.71$ 32. $\cos(x) = -0.07$

Find the first two positive solutions.

33. $7\sin(6x) = 2$ 34. $7\sin(5x) = 6$ 35. $5\cos(3x) = -3$ 36. $3\cos(4x) = 2$

37. $3\sin\left(\dfrac{\pi}{4}x\right) = 2$ 38. $7\sin\left(\dfrac{\pi}{5}x\right) = 6$ 39. $5\cos\left(\dfrac{\pi}{3}x\right) = 1$ 40. $3\cos\left(\dfrac{\pi}{2}x\right) = -2$

Section 6.5 Modeling with Trigonometric Functions

Solving right triangles for angles

In Section 5.5, we used trigonometry on a right triangle to solve for the sides of a triangle given one side and an additional angle. Using the inverse trig functions, we can solve for the angles of a right triangle given two sides.

Example 1

An airplane needs to fly to an airfield located 300 miles east and 200 miles north of its current location. At what heading should the airplane fly? In other words, if we ignore air resistance or wind speed, how many degrees north of east should the airplane fly?

We might begin by drawing a picture and labeling all of the known information. Drawing a triangle, we see we are looking for the angle α. In this triangle, the side opposite the angle α is 200 miles and the side adjacent is 300 miles. Since we know the values for the opposite and adjacent sides, it makes sense to use the tangent function.

$$\tan(\alpha) = \frac{200}{300} \qquad \text{Using the inverse,}$$

$$\alpha = \tan^{-1}\left(\frac{200}{300}\right) \approx 0.588 \text{, or equivalently about 33.7 degrees.}$$

The airplane needs to fly at a heading of 33.7 degrees north of east.

Example 2

OSHA safety regulations require that the base of a ladder be placed 1 foot from the wall for every 4 feet of ladder length[3]. Find the angle such a ladder forms with the ground.

For any length of ladder, the base needs to be one quarter of the distance the foot of the ladder is away from the wall. Equivalently, if the base is a feet from the wall, the ladder can be $4a$ feet long. Since a is the side adjacent to the angle and $4a$ is the hypotenuse, we use the cosine function.

$$\cos(\theta) = \frac{a}{4a} = \frac{1}{4} \qquad \text{Using the inverse}$$

$$\theta = \cos^{-1}\left(\frac{1}{4}\right) \approx 75.52 \text{ degrees}$$

The ladder forms a 75.52 degree angle with the ground.

---

[3] http://www.osha.gov/SLTC/etools/construction/falls/4ladders.html

1. A cable that anchors the center of the London Eye Ferris wheel to the ground must be replaced. The center of the Ferris wheel is 70 meters above the ground and the second anchor on the ground is 23 meters from the base of the wheel. What is the angle from the ground up to the center of the Ferris wheel and how long is the cable?

Example 3

In a video game design, a map shows the location of other characters relative to the player, who is situated at the origin, and the direction they are facing. A character currently shows on the map at coordinates (-3, 5). If the player rotates counterclockwise by 20 degrees, then the objects in the map will correspondingly rotate 20 degrees clockwise. Find the new coordinates of the character.

To rotate the position of the character, we can imagine it as a point on a circle, and we will change the angle of the point by 20 degrees. To do so, we first need to find the radius of this circle and the original angle.

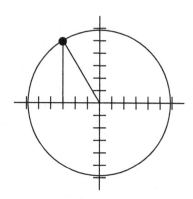

Drawing a right triangle inside the circle, we can find the radius using the Pythagorean Theorem:

$$(-3)^2 + 5^2 = r^2$$

$$r = \sqrt{9 + 25} = \sqrt{34}$$

To find the angle, we need to decide first if we are going to find the acute angle of the triangle, the reference angle, or if we are going to find the angle measured in standard position. While either approach will work, in this case we will do the latter. Since for any point on a circle we know $x = r\cos(\theta)$, using our given information we get

$$-3 = \sqrt{34}\cos(\theta)$$

$$\frac{-3}{\sqrt{34}} = \cos(\theta)$$

$$\theta = \cos^{-1}\left(\frac{-3}{\sqrt{34}}\right) \approx 120.964°$$

While there are two angles that have this cosine value, the angle of 120.964 degrees is in the second quadrant as desired, so it is the angle we were looking for.

Rotating the point clockwise by 20 degrees, the angle of the point will decrease to 100.964 degrees. We can then evaluate the coordinates of the rotated point

$$x = \sqrt{34}\cos(100.964°) \approx -1.109$$

$$y = \sqrt{34}\sin(100.964°) \approx 5.725$$

The coordinates of the character on the rotated map will be (-1.109, 5.725).

Modeling with sinusoidal functions

Many modeling situations involve functions that are periodic. Previously we learned that sinusoidal functions are a special type of periodic function. Problems that involve quantities that oscillate can often be modeled by a sine or cosine function and once we create a suitable model for the problem we can use that model to answer various questions.

Example 4

The hours of daylight in Seattle oscillate from a low of 8.5 hours in January to a high of 16 hours in July[4]. When should you plant a garden if you want to do it during a month where there are 14 hours of daylight?

To model this, we first note that the hours of daylight oscillate with a period of 12 months. $B = \dfrac{2\pi}{12} = \dfrac{\pi}{6}$ corresponds to the horizontal stretch, found by using the ratio of the original period to the new period.

With a low of 8.5 and a high of 16, the midline will be halfway between these values, at $\dfrac{16 + 8.5}{2} = 12.25$.

The amplitude will be half the difference between the highest and lowest values: $\dfrac{16 - 8.5}{2} = 3.75$, or equivalently the distance from the midline to the high or low value, 16-12.25=3.75.

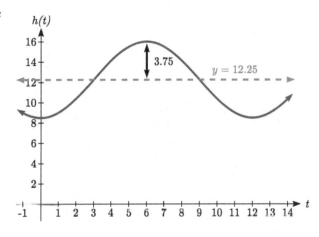

Letting January be $t = 0$, the graph starts at the lowest value, so it can be modeled as a flipped cosine graph. Putting this together, we get a model:

$$h(t) = -3.75\cos\left(\frac{\pi}{6}t\right) + 12.25$$

$h(t)$ is our model for hours of day light t months after January.

To find when there will be 14 hours of daylight, we solve $h(t) = 14$.

$$14 = -3.75\cos\left(\frac{\pi}{6}t\right) + 12.25 \qquad \text{Isolating the cosine}$$

---

[4] http://www.mountaineers.org/seattle/climbing/Reference/DaylightHrs.html

$$1.75 = -3.75\cos\left(\frac{\pi}{6}t\right) \qquad \text{Subtracting 12.25 and dividing by -3.75}$$

$$-\frac{1.75}{3.75} = \cos\left(\frac{\pi}{6}t\right) \qquad \text{Using the inverse}$$

$$\frac{\pi}{6}t = \cos^{-1}\left(-\frac{1.75}{3.75}\right) \approx 2.0563 \qquad \text{multiplying by the reciprocal}$$

$$t = 2.0563 \cdot \frac{6}{\pi} = 3.927 \qquad t=3.927 \text{ months past January}$$

There will be 14 hours of daylight 3.927 months into the year, or near the end of April.

While there would be a second time in the year when there are 14 hours of daylight, since we are planting a garden, we would want to know the first solution, in spring, so we do not need to find the second solution in this case.

Try it Now

2. The author's monthly gas usage (in therms) is shown here. Find a function to model the data.

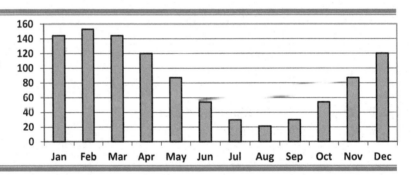

Example 6

An object is connected to the wall with a spring that has a natural length of 20 cm. The object is pulled back 8 cm past the natural length and released. The object oscillates 3 times per second. Find an equation for the horizontal position of the object ignoring the effects of friction. How much time during each cycle is the object more than 27 cm from the wall?

If we use the distance from the wall, *x*, as the desired output, then the object will oscillate equally on either side of the spring's natural length of 20, putting the midline of the function at 20 cm.

If we release the object 8 cm past the natural length, the amplitude of the oscillation will be 8 cm.

We are beginning at the largest value and so this function can most easily be modeled using a cosine function.

Since the object oscillates 3 times per second, it has a frequency of 3 and the period of one oscillation is 1/3 of second. Using this we find the horizontal compression using the ratios of the periods: $\dfrac{2\pi}{1/3} = 6\pi$.

Using all this, we can build our model:
$$x(t) = 8\cos(6\pi t) + 20$$

To find when the object is 27 cm from the wall, we can solve $x(t) = 27$

$27 = 8\cos(6\pi t) + 20$ \qquad\qquad Isolating the cosine

$7 = 8\cos(6\pi t)$

$\dfrac{7}{8} = \cos(6\pi t)$ \qquad\qquad Using the inverse

$6\pi t = \cos^{-1}\left(\dfrac{7}{8}\right) \approx 0.505$

$t = \dfrac{0.505}{6\pi} = 0.0268$

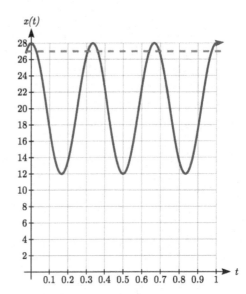

Based on the shape of the graph, we can conclude that the object will spend the first 0.0268 seconds more than 27 cm from the wall. Based on the symmetry of the function, the object will spend another 0.0268 seconds more than 27 cm from the wall at the end of the cycle. Altogether, the object spends 0.0536 seconds each cycle at a distance greater than 27 cm from the wall.

In some problems, we can use trigonometric functions to model behaviors more complicated than the basic sinusoidal function.

Example 7

A rigid rod with length 10 cm is attached to a circle of radius 4cm at point A as shown here. The point B is able to freely move along the horizontal axis, driving a piston[5]. If the wheel rotates counterclockwise at 5 revolutions per second, find the location of point B as a function of time. When will the point B be 12 cm from the center of the circle?

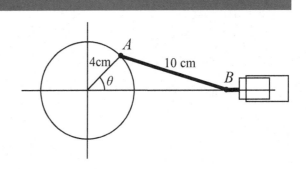

---

[5] For an animation of this situation, see http://www.mathdemos.org/mathdemos/sinusoidapp/engine1.gif

To find the position of point B, we can begin by finding the coordinates of point A. Since it is a point on a circle with radius 4, we can express its coordinates as $(4\cos(\theta), 4\sin(\theta))$, where θ is the angle shown.

The angular velocity is 5 revolutions per second, or equivalently 10π radians per second. After t seconds, the wheel will rotate by $\theta = 10\pi t$ radians. Substituting this, we can find the coordinates of A in terms of t.
$(4\cos(10\pi t), 4\sin(10\pi t))$

Notice that this is the same value we would have obtained by observing that the period of the rotation is 1/5 of a second and calculating the stretch/compression factor:
$$\frac{"original"}{"new"} \frac{2\pi}{1/5} = 10\pi \,.$$

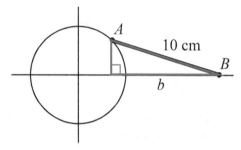

Now that we have the coordinates of the point A, we can relate this to the point B. By drawing a vertical line segment from A to the horizontal axis, we can form a right triangle. The height of the triangle is the y coordinate of the point A:
$4\sin(10\pi t)$.

Using the Pythagorean Theorem, we can find the base length of the triangle:
$$(4\sin(10\pi t))^2 + b^2 = 10^2$$
$$b^2 = 100 - 16\sin^2(10\pi t)$$
$$b = \sqrt{100 - 16\sin^2(10\pi t)}$$

Looking at the x coordinate of the point A, we can see that the triangle we drew is shifted to the right of the y axis by $4\cos(10\pi t)$. Combining this offset with the length of the base of the triangle gives the x coordinate of the point B:
$$x(t) = 4\cos(10\pi t) + \sqrt{100 - 16\sin^2(10\pi t)}$$

To solve for when the point B will be 12 cm from the center of the circle, we need to solve $x(t) = 12$.

$12 = 4\cos(10\pi t) + \sqrt{100 - 16\sin^2(10\pi t)}$	Isolate the square root
$12 - 4\cos(10\pi t) = \sqrt{100 - 16\sin^2(10\pi t)}$	Square both sides
$(12 - 4\cos(10\pi t))^2 = 100 - 16\sin^2(10\pi t)$	Expand the left side
$144 - 96\cos(10\pi t) + 16\cos^2(10\pi t) = 100 - 16\sin^2(10\pi t)$	Move all terms to the left
$44 - 96\cos(10\pi t) + 16\cos^2(10\pi t) + 16\sin^2(10\pi t) = 0$	Factor out 16
$44 - 96\cos(10\pi t) + 16\left(\cos^2(10\pi t) + \sin^2(10\pi t)\right) = 0$	

At this point, we can utilize the Pythagorean Identity, which tells us that
$\cos^2(10\pi t) + \sin^2(10\pi t) = 1$.

Using this identity, our equation simplifies to

$44 - 96\cos(10\pi t) + 16 = 0$	Combine the constants and move to the right side
$-96\cos(10\pi t) = -60$	Divide
$\cos(10\pi t) = \dfrac{60}{96}$	Make a substitution
$\cos(u) = \dfrac{60}{96}$	
$u = \cos^{-1}\left(\dfrac{60}{96}\right) \approx 0.896$	By symmetry we can find a second solution
$u = 2\pi - 0.896 = 5.388$	Undoing the substitution

$10\pi t = 0.896$, so $t = 0.0285$
$10\pi t = 5.388$, so $t = 0.1715$

The point B will be 12 cm from the center of the circle 0.0285 seconds after the process begins, 0.1715 seconds after the process begins, and every 1/5 of a second after each of those values.

Important Topics of This Section
Modeling with trig equations
Modeling with sinusoidal functions
Solving right triangles for angles in degrees and radians

Try it Now Answers
1. Angle of elevation for the cable is 71.81 degrees and the cable is 73.68 m long

2. Approximately $G(t) = 66\cos\left(\dfrac{\pi}{6}(t-1)\right) + 87$

Section 6.5 Exercises

In each of the following triangles, solve for the unknown side and angles.

1.

2.

3.

4.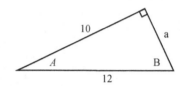

Find a possible formula for the trigonometric function whose values are in the following tables.

5.

x	0	1	2	3	4	5	6
y	-2	4	10	4	-2	4	10

6.

x	0	1	2	3	4	5	6
y	1	-3	-7	-3	1	-3	-7

7. Outside temperature over the course of a day can be modeled as a sinusoidal function. Suppose you know the high temperature for the day is 63 degrees and the low temperature of 37 degrees occurs at 5 AM. Assuming t is the number of hours since midnight, find an equation for the temperature, D, in terms of t.

8. Outside temperature over the course of a day can be modeled as a sinusoidal function. Suppose you know the high temperature for the day is 92 degrees and the low temperature of 78 degrees occurs at 4 AM. Assuming t is the number of hours since midnight, find an equation for the temperature, D, in terms of t.

9. A population of rabbits oscillates 25 above and below an average of 129 during the year, hitting the lowest value in January ($t = 0$).
    a. Find an equation for the population, P, in terms of the months since January, t.
    b. What if the lowest value of the rabbit population occurred in April instead?

10. A population of elk oscillates 150 above and below an average of 720 during the year, hitting the lowest value in January ($t = 0$).
    a. Find an equation for the population, P, in terms of the months since January, t.
    b. What if the lowest value of the rabbit population occurred in March instead?

11. Outside temperature over the course of a day can be modeled as a sinusoidal function. Suppose you know the high temperature of 105 degrees occurs at 5 PM and the average temperature for the day is 85 degrees. Find the temperature, to the nearest degree, at 9 AM.

12. Outside temperature over the course of a day can be modeled as a sinusoidal function. Suppose you know the high temperature of 84 degrees occurs at 6 PM and the average temperature for the day is 70 degrees. Find the temperature, to the nearest degree, at 7 AM.

13. Outside temperature over the course of a day can be modeled as a sinusoidal function. Suppose you know the temperature varies between 47 and 63 degrees during the day and the average daily temperature first occurs at 10 AM. How many hours after midnight does the temperature first reach 51 degrees?

14. Outside temperature over the course of a day can be modeled as a sinusoidal function. Suppose you know the temperature varies between 64 and 86 degrees during the day and the average daily temperature first occurs at 12 AM. How many hours after midnight does the temperature first reach 70 degrees?

15. A Ferris wheel is 20 meters in diameter and boarded from a platform that is 2 meters above the ground. The six o'clock position on the Ferris wheel is level with the loading platform. The wheel completes 1 full revolution in 6 minutes. How many minutes of the ride are spent higher than 13 meters above the ground?

16. A Ferris wheel is 45 meters in diameter and boarded from a platform that is 1 meter above the ground. The six o'clock position on the Ferris wheel is level with the loading platform. The wheel completes 1 full revolution in 10 minutes. How many minutes of the ride are spent higher than 27 meters above the ground?

17. The sea ice area around the North Pole fluctuates between about 6 million square kilometers in September to 14 million square kilometers in March. Assuming sinusoidal fluctuation, during how many months are there less than 9 million square kilometers of sea ice?

18. The sea ice area around the South Pole fluctuates between about 18 million square kilometers in September to 3 million square kilometers in March. Assuming sinusoidal fluctuation, during how many months are there more than 15 million square kilometers of sea ice?

19. A respiratory ailment called "Cheyne-Stokes Respiration" causes the volume per breath to increase and decrease in a sinusoidal manner, as a function of time. For one particular patient with this condition, a machine begins recording a plot of volume per breath versus time (in seconds). Let $b(t)$ be a function of time t that tells us the volume (in liters) of a breath that starts at time t. During the test, the smallest volume per breath is 0.6 liters and this first occurs for a breath that starts 5 seconds into the test. The largest volume per breath is 1.8 liters and this first occurs for a breath beginning 55 seconds into the test. [UW]
    a. Find a formula for the function $b(t)$ whose graph will model the test data for this patient.
    b. If the patient begins a breath every 5 seconds, what are the breath volumes during the first minute of the test?

20. Suppose the high tide in Seattle occurs at 1:00 a.m. and 1:00 p.m, at which time the water is 10 feet above the height of low tide. Low tides occur 6 hours after high tides. Suppose there are two high tides and two low tides every day and the height of the tide varies sinusoidally. [UW]
    a. Find a formula for the function $y=h(t)$ that computes the height of the tide above low tide at time t. (In other words, $y = 0$ corresponds to low tide.)
    b. What is the tide height at 11:00 a.m.?

21. A communications satellite orbits the earth t miles above the surface. Assume the radius of the earth is 3,960 miles. The satellite can only "see" a portion of the earth's surface, bounded by what is called a horizon circle. This leads to a two-dimensional cross-sectional picture we can use to study the size of the horizon slice: [UW]

CROSS-SECTION

    a. Find a formula for α in terms of t.
    b. If $t = 30,000$ miles, what is α? What percentage of the circumference of the earth is covered by the satellite? What would be the minimum number of such satellites required to cover the circumference?
    c. If $t = 1,000$ miles, what is α? What percentage of the circumference of the earth is covered by the satellite? What would be the minimum number of such satellites required to cover the circumference?
    d. Suppose you wish to place a satellite into orbit so that 20% of the circumference is covered by the satellite. What is the required distance t?

22. Tiffany is a model rocket enthusiast. She has been working on a pressurized rocket filled with nitrous oxide. According to her design, if the atmospheric pressure exerted on the rocket is less than 10 pounds/sq.in., the nitrous oxide chamber inside the rocket will explode. Tiff worked from a formula $p = 14.7e^{-h/10}$ pounds/sq.in. for the atmospheric pressure h miles above sea level. Assume that the rocket is launched at an angle of α above level ground at sea level with an initial speed of 1400 feet/sec. Also, assume the height (in feet) of the rocket at time t seconds is given by the equation $y(t) = -16t^2 + 1400\sin(\alpha)t$.   [UW]

   a. At what altitude will the rocket explode?
   b. If the angle of launch is $\alpha = 12°$, determine the minimum atmospheric pressure exerted on the rocket during its flight. Will the rocket explode in midair?
   c. If the angle of launch is $\alpha = 82°$, determine the minimum atmospheric pressure exerted on the rocket during its flight. Will the rocket explode in midair?
   d. Find the largest launch angle α so that the rocket will not explode.

Chapter 7: Trigonometric Equations and Identities

In the last two chapters we have used basic definitions and relationships to simplify trigonometric expressions and solve trigonometric equations. In this chapter we will look at more complex relationships. By conducting a deeper study of trigonometric identities we can learn to simplify complicated expressions, allowing us to solve more interesting applications.

Section 7.1 Solving Trigonometric Equations with Identities

In the last chapter, we solved basic trigonometric equations. In this section, we explore the techniques needed to solve more complicated trig equations. Building from what we already know makes this a much easier task.

Consider the function $f(x) = 2x^2 + x$. If you were asked to solve $f(x) = 0$, it requires simple algebra:

$2x^2 + x = 0$ Factor

$x(2x + 1) = 0$ Giving solutions

$x = 0$ or $x = -\dfrac{1}{2}$

Similarly, for $g(t) = \sin(t)$, if we asked you to solve $g(t) = 0$, you can solve this using unit circle values:

$\sin(t) = 0$ for $t = 0, \pi, 2\pi$ and so on.

Using these same concepts, we consider the composition of these two functions:

$f(g(t)) = 2(\sin(t))^2 + (\sin(t)) = 2\sin^2(t) + \sin(t)$

This creates an equation that is a polynomial trig function. With these types of functions, we use algebraic techniques like factoring and the quadratic formula, along with trigonometric identities and techniques, to solve equations.

As a reminder, here are some of the essential trigonometric identities that we have learned so far:

Identities

Pythagorean Identities

$$\cos^2(t) + \sin^2(t) = 1 \qquad 1 + \cot^2(t) = \csc^2(t) \qquad 1 + \tan^2(t) = \sec^2(t)$$

Negative Angle Identities

$$\sin(-t) = -\sin(t) \qquad \cos(-t) = \cos(t) \qquad \tan(-t) = -\tan(t)$$
$$\csc(-t) = -\csc(t) \qquad \sec(-t) = \sec(t) \qquad \cot(-t) = -\cot(t)$$

Reciprocal Identities

$$\sec(t) = \frac{1}{\cos(t)} \qquad \csc(t) = \frac{1}{\sin(t)} \qquad \tan(t) = \frac{\sin(t)}{\cos(t)} \qquad \cot(t) = \frac{1}{\tan(t)}$$

Example 1

Solve $2\sin^2(t) + \sin(t) = 0$ for all solutions with $0 \le t < 2\pi$.

This equation kind of looks like a quadratic equation, but with $\sin(t)$ in place of an algebraic variable (we often call such an equation "quadratic in sine"). As with all quadratic equations, we can use factoring techniques or the quadratic formula. This expression factors nicely, so we proceed by factoring out the common factor of $\sin(t)$:

$$\sin(t)\big(2\sin(t) + 1\big) = 0$$

Using the zero product theorem, we know that the product on the left will equal zero if either factor is zero, allowing us to break this equation into two cases:

$$\sin(t) = 0 \qquad \text{or} \qquad 2\sin(t) + 1 = 0$$

We can solve each of these equations independently, using our knowledge of special angles.

$$\sin(t) = 0 \qquad\qquad 2\sin(t) + 1 = 0$$

$$t = 0 \text{ or } t = \pi \qquad \sin(t) = -\frac{1}{2}$$

$$t = \frac{7\pi}{6} \text{ or } t = \frac{11\pi}{6}$$

Together, this gives us four solutions to the equation on $0 \le t < 2\pi$:

$$t = 0, \pi, \frac{7\pi}{6}, \frac{11\pi}{6}$$

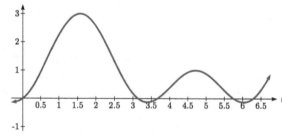

We could check these answers are reasonable by graphing the function and comparing the zeros.

Example 2

Solve $3\sec^2(t) - 5\sec(t) - 2 = 0$ for all solutions with $0 \le t < 2\pi$.

Since the left side of this equation is quadratic in secant, we can try to factor it, and hope it factors nicely.

If it is easier to for you to consider factoring without the trig function present, consider using a substitution $u = \sec(t)$, resulting in $3u^2 - 5u - 2 = 0$, and then try to factor:
$$3u^2 - 5u - 2 = (3u + 1)(u - 2)$$

Undoing the substitution,
$(3\sec(t) + 1)(\sec(t) - 2) = 0$

Since we have a product equal to zero, we break it into the two cases and solve each separately.

$3\sec(t) + 1 = 0$ Isolate the secant

$\sec(t) = -\dfrac{1}{3}$ Rewrite as a cosine

$\dfrac{1}{\cos(t)} = -\dfrac{1}{3}$ Invert both sides

$\cos(t) = -3$

Since the cosine has a range of [-1, 1], the cosine will never take on an output of -3. There are no solutions to this case.

Continuing with the second case,

$\sec(t) - 2 = 0$ Isolate the secant
$\sec(t) = 2$ Rewrite as a cosine

$\dfrac{1}{\cos(t)} = 2$ Invert both sides

$\cos(t) = \dfrac{1}{2}$ This gives two solutions

$t = \dfrac{\pi}{3}$ or $t = \dfrac{5\pi}{3}$

These are the only two solutions on the interval. By utilizing technology to graph $f(t) = 3\sec^2(t) - 5\sec(t) - 2$, a look at a graph confirms there are only two zeros for this function on the interval [0, 2π), which assures us that we didn't miss anything.

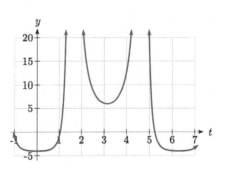

---

1. Solve $2\sin^2(t) + 3\sin(t) + 1 = 0$ for all solutions with $0 \le t < 2\pi$.

When solving some trigonometric equations, it becomes necessary to first rewrite the equation using trigonometric identities. One of the most common is the Pythagorean Identity, $\sin^2(\theta) + \cos^2(\theta) = 1$ which allows you to rewrite $\sin^2(\theta)$ in terms of $\cos^2(\theta)$ or vice versa,

Identities
Alternate Forms of the Pythagorean Identity
$\sin^2(\theta) = 1 - \cos^2(\theta)$ $\cos^2(\theta) = 1 - \sin^2(\theta)$

These identities become very useful whenever an equation involves a combination of sine and cosine functions.

Example 3

Solve $2\sin^2(t) - \cos(t) = 1$ for all solutions with $0 \le t < 2\pi$.

Since this equation has a mix of sine and cosine functions, it becomes more complicated to solve. It is usually easier to work with an equation involving only one trig function. This is where we can use the Pythagorean Identity.

$$2\sin^2(t) - \cos(t) = 1 \qquad \text{Using } \sin^2(\theta) = 1 - \cos^2(\theta)$$
$$2\left(1 - \cos^2(t)\right) - \cos(t) = 1 \qquad \text{Distributing the 2}$$
$$2 - 2\cos^2(t) - \cos(t) = 1$$

Since this is now quadratic in cosine, we rearrange the equation so one side is zero and factor.

$$-2\cos^2(t) - \cos(t) + 1 = 0 \qquad \text{Multiply by -1 to simplify the factoring}$$
$$2\cos^2(t) + \cos(t) - 1 = 0 \qquad \text{Factor}$$
$$\left(2\cos(t) - 1\right)\left(\cos(t) + 1\right) = 0$$

This product will be zero if either factor is zero, so we can break this into two separate cases and solve each independently.

$$2\cos(t) - 1 = 0 \qquad \text{or} \qquad \cos(t) + 1 = 0$$

$$\cos(t) = \frac{1}{2} \qquad \text{or} \qquad \cos(t) = -1$$

$$t = \frac{\pi}{3} \text{ or } t = \frac{5\pi}{3} \qquad \text{or} \qquad t = \pi$$

Try it Now

2. Solve $2\sin^2(t) = 3\cos(t)$ for all solutions with $0 \le t < 2\pi$.

In addition to the Pythagorean Identity, it is often necessary to rewrite the tangent, secant, cosecant, and cotangent as part of solving an equation.

Example 4

Solve $\tan(x) = 3\sin(x)$ for all solutions with $0 \le x < 2\pi$.

With a combination of tangent and sine, we might try rewriting tangent
$$\tan(x) = 3\sin(x)$$

$$\frac{\sin(x)}{\cos(x)} = 3\sin(x) \qquad\qquad \text{Multiplying both sides by cosine}$$

$$\sin(x) = 3\sin(x)\cos(x)$$

At this point, you may be tempted to divide both sides of the equation by $\sin(x)$. **Resist the urge.** When we divide both sides of an equation by a quantity, we are assuming the quantity is never zero. In this case, when $\sin(x) = 0$ the equation is satisfied, so we'd lose those solutions if we divided by the sine.

To avoid this problem, we can rearrange the equation so that one side is zero[1].
$$\sin(x) - 3\sin(x)\cos(x) = 0 \qquad \text{Factoring out } \sin(x) \text{ from both parts}$$
$$\sin(x)\big(1 - 3\cos(x)\big) = 0$$

From here, we can see we get solutions when $\sin(x) = 0$ or $1 - 3\cos(x) = 0$.

Using our knowledge of the special angles of the unit circle,
$\sin(x) = 0$ when $x = 0$ or $x = \pi$.

---

[1] You technically *can* divide by $\sin(x)$, as long as you separately consider the case where $\sin(x) = 0$. Since it is easy to forget this step, the factoring approach used in the example is recommended.

For the second equation, we will need the inverse cosine.

$1 - 3\cos(x) = 0$

$\cos(x) = \dfrac{1}{3}$ Using our calculator or technology

$x = \cos^{-1}\left(\dfrac{1}{3}\right) \approx 1.231$ Using symmetry to find a second solution

$x = 2\pi - 1.231 = 5.052$

We have four solutions on $0 \le x < 2\pi$:

$x = 0,\ 1.231,\ \pi,\ 5.052$

Try it Now

3. Solve $\sec(\theta) = 2\cos(\theta)$ to find the first four positive solutions.

Example 5

Solve $\dfrac{4}{\sec^2(\theta)} + 3\cos(\theta) = 2\cot(\theta)\tan(\theta)$ for all solutions with $0 \le \theta < 2\pi$.

$\dfrac{4}{\sec^2(\theta)} + 3\cos(\theta) = 2\cot(\theta)\tan(\theta)$ Using the reciprocal identities

$4\cos^2(\theta) + 3\cos(\theta) = 2\dfrac{1}{\tan(\theta)}\tan(\theta)$ Simplifying

$4\cos^2(\theta) + 3\cos(\theta) = 2$ Subtracting 2 from each side

$4\cos^2(\theta) + 3\cos(\theta) - 2 = 0$

This does not appear to factor nicely so we use the quadratic formula, remembering that we are solving for $\cos(\theta)$.

$\cos(\theta) = \dfrac{-3 \pm \sqrt{3^2 - 4(4)(-2)}}{2(4)} = \dfrac{-3 \pm \sqrt{41}}{8}$

Using the negative square root first,

$\cos(\theta) = \dfrac{-3 - \sqrt{41}}{8} = -1.175$

This has no solutions, since the cosine can't be less than -1.

Using the positive square root,

$$\cos(\theta) = \frac{-3+\sqrt{41}}{8} = 0.425$$

$$\theta = \cos^{-1}(0.425) = 1.131 \qquad \text{By symmetry, a second solution can be found}$$

$$\theta = 2\pi - 1.131 = 5.152$$

Important Topics of This Section
Review of Trig Identities
Solving Trig Equations
By Factoring
Using the Quadratic Formula
Utilizing Trig Identities to simplify

Try it Now Answers

1. Factor as $(2\sin(t)+1)(\sin(t)+1)=0$

$2\sin(t)+1=0$ at $t = \dfrac{7\pi}{6}, \dfrac{11\pi}{6}$

$\sin(t)+1=0$ at $t = \dfrac{3\pi}{2}$

$t = \dfrac{7\pi}{6}, \dfrac{3\pi}{2}, \dfrac{11\pi}{6}$

2. $2\left(1-\cos^2(t)\right) = 3\cos(t)$

$2\cos^2(t) + 3\cos(t) - 2 = 0$

$(2\cos(t)-1)(\cos(t)+2) = 0$

$\cos(t)+2=0$ has no solutions

$2\cos(t)-1=0$ at $t = \dfrac{\pi}{3}, \dfrac{5\pi}{3}$

3. $\dfrac{1}{\cos(\theta)} = 2\cos(\theta)$

$\dfrac{1}{2} = \cos^2(\theta)$

$\cos(\theta) = \pm\sqrt{\dfrac{1}{2}} = \pm\dfrac{\sqrt{2}}{2}$

$\theta = \dfrac{\pi}{4}, \dfrac{3\pi}{4}, \dfrac{5\pi}{4}, \dfrac{7\pi}{4}$

Section 7.1 Exercises

Find all solutions on the interval $0 \le \theta < 2\pi$.

1. $2\sin(\theta) = -1$ 2. $2\sin(\theta) = \sqrt{3}$ 3. $2\cos(\theta) = 1$ 4. $2\cos(\theta) = -\sqrt{2}$

Find all solutions.

5. $2\sin\left(\dfrac{\pi}{4}x\right) = 1$ 6. $2\sin\left(\dfrac{\pi}{3}x\right) = \sqrt{2}$ 7. $2\cos(2t) = -\sqrt{3}$ 8. $2\cos(3t) = -1$

9. $3\cos\left(\dfrac{\pi}{5}x\right) = 2$ 10. $8\cos\left(\dfrac{\pi}{2}x\right) = 6$ 11. $7\sin(3t) = -2$ 12. $4\sin(4t) = 1$

Find all solutions on the interval $[0, 2\pi)$.

13. $10\sin(x)\cos(x) = 6\cos(x)$ 14. $-3\sin(t) = 15\cos(t)\sin(t)$

15. $\csc(2x) - 9 = 0$ 16. $\sec(2\theta) = 3$

17. $\sec(x)\sin(x) - 2\sin(x) = 0$ 18. $\tan(x)\sin(x) - \sin(x) = 0$

19. $\sin^2 x = \dfrac{1}{4}$ 20. $\cos^2\theta = \dfrac{1}{2}$

21. $\sec^2 x = 7$ 22. $\csc^2 t = 3$

23. $2\sin^2 w + 3\sin w + 1 = 0$ 24. $8\sin^2 x + 6\sin(x) + 1 = 0$

25. $2\cos^2 t + \cos(t) = 1$ 26. $8\cos^2(\theta) = 3 - 2\cos(\theta)$

27. $4\cos^2(x) - 4 = 15\cos(x)$ 28. $9\sin(w) - 2 = 4\sin^2(w)$

29. $12\sin^2(t) + \cos(t) - 6 = 0$ 30. $6\cos^2(x) + 7\sin(x) - 8 = 0$

31. $\cos^2\phi = -6\sin\phi$ 32. $\sin^2 t = \cos t$

33. $\tan^3(x) = 3\tan(x)$ 34. $\cos^3(t) = -\cos(t)$

35. $\tan^5(x) = \tan(x)$ 36. $\tan^5(x) - 9\tan(x) = 0$

37. $4\sin(x)\cos(x) + 2\sin(x) - 2\cos(x) - 1 = 0$

38. $2\sin(x)\cos(x) - \sin(x) + 2\cos(x) - 1 = 0$

39. $\tan(x) - 3\sin(x) = 0$ 40. $3\cos(x) = \cot(x)$

41. $2\tan^2(t) = 3\sec(t)$ 42. $1 - 2\tan(w) = \tan^2(w)$

Section 7.2 Addition and Subtraction Identities

In this section, we begin expanding our repertoire of trigonometric identities.

Identities
The sum and difference identities
$\cos(\alpha - \beta) = \cos(\alpha)\cos(\beta) + \sin(\alpha)\sin(\beta)$
$\cos(\alpha + \beta) = \cos(\alpha)\cos(\beta) - \sin(\alpha)\sin(\beta)$
$\sin(\alpha + \beta) = \sin(\alpha)\cos(\beta) + \cos(\alpha)\sin(\beta)$
$\sin(\alpha - \beta) = \sin(\alpha)\cos(\beta) - \cos(\alpha)\sin(\beta)$

We will prove the difference of angles identity for cosine. The rest of the identities can be derived from this one.

<u>Proof of the difference of angles identity for cosine</u>
Consider two points on a unit circle:
P at an angle of α from the positive x axis
with coordinates $(\cos(\alpha), \sin(\alpha))$, and Q at
an angle of β with coordinates
$(\cos(\beta), \sin(\beta))$.

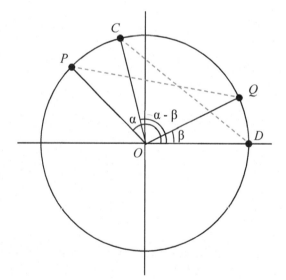

Notice the measure of angle POQ is $\alpha - \beta$.
Label two more points:
C at an angle of $\alpha - \beta$, with coordinates
$(\cos(\alpha - \beta), \sin(\alpha - \beta))$,
D at the point $(1, 0)$.

Notice that the distance from C to D is the
same as the distance from P to Q because
triangle COD is a rotation of triangle POQ.

Using the distance formula to find the distance from P to Q yields

$$\sqrt{(\cos(\alpha) - \cos(\beta))^2 + (\sin(\alpha) - \sin(\beta))^2}$$

Expanding this

$$\sqrt{\cos^2(\alpha) - 2\cos(\alpha)\cos(\beta) + \cos^2(\beta) + \sin^2(\alpha) - 2\sin(\alpha)\sin(\beta) + \sin^2(\beta)}$$

Applying the Pythagorean Identity and simplifying

$$\sqrt{2-2\cos(\alpha)\cos(\beta)-2\sin(\alpha)\sin(\beta)}$$

Similarly, using the distance formula to find the distance from C to D

$$\sqrt{(\cos(\alpha-\beta)-1)^2 + (\sin(\alpha-\beta)-0)^2}$$

Expanding this

$$\sqrt{\cos^2(\alpha-\beta)-2\cos(\alpha-\beta)+1+\sin^2(\alpha-\beta)}$$

Applying the Pythagorean Identity and simplifying

$$\sqrt{-2\cos(\alpha-\beta)+2}$$

Since the two distances are the same we set these two formulas equal to each other and simplify

$$\sqrt{2-2\cos(\alpha)\cos(\beta)-2\sin(\alpha)\sin(\beta)} = \sqrt{-2\cos(\alpha-\beta)+2}$$
$$2-2\cos(\alpha)\cos(\beta)-2\sin(\alpha)\sin(\beta) = -2\cos(\alpha-\beta)+2$$
$$\cos(\alpha)\cos(\beta)+\sin(\alpha)\sin(\beta) = \cos(\alpha-\beta)$$

This establishes the identity.

Try it Now

1. By writing $\cos(\alpha+\beta)$ as $\cos(\alpha-(-\beta))$, show the sum of angles identity for cosine follows from the difference of angles identity proven above.

The sum and difference of angles identities are often used to rewrite expressions in other forms, or to rewrite an angle in terms of simpler angles.

Example 1

Find the exact value of $\cos(75°)$.

Since $75° = 30° + 45°$, we can evaluate $\cos(75°)$ as

$\cos(75°) = \cos(30° + 45°)$ Apply the cosine sum of angles identity

$= \cos(30°)\cos(45°) - \sin(30°)\sin(45°)$ Evaluate

$= \dfrac{\sqrt{3}}{2}\cdot\dfrac{\sqrt{2}}{2} - \dfrac{1}{2}\cdot\dfrac{\sqrt{2}}{2}$ Simply

$= \dfrac{\sqrt{6}-\sqrt{2}}{4}$

Try it Now

2. Find the exact value of $\sin\left(\dfrac{\pi}{12}\right)$.

Example 2

Rewrite $\sin\left(x - \dfrac{\pi}{4}\right)$ in terms of $\sin(x)$ and $\cos(x)$.

$\sin\left(x - \dfrac{\pi}{4}\right)$ Use the difference of angles identity for sine

$= \sin(x)\cos\left(\dfrac{\pi}{4}\right) - \cos(x)\sin\left(\dfrac{\pi}{4}\right)$ Evaluate the cosine and sine and rearrange

$= \dfrac{\sqrt{2}}{2}\sin(x) - \dfrac{\sqrt{2}}{2}\cos(x)$

Additionally, these identities can be used to simplify expressions or prove new identities

Example 3

Prove $\dfrac{\sin(a + b)}{\sin(a - b)} = \dfrac{\tan(a) + \tan(b)}{\tan(a) - \tan(b)}$.

As with any identity, we need to first decide which side to begin with. Since the left side involves sum and difference of angles, we might start there

$\dfrac{\sin(a + b)}{\sin(a - b)}$ Apply the sum and difference of angle identities

$= \dfrac{\sin(a)\cos(b) + \cos(a)\sin(b)}{\sin(a)\cos(b) - \cos(a)\sin(b)}$

Since it is not immediately obvious how to proceed, we might start on the other side, and see if the path is more apparent.

$\dfrac{\tan(a) + \tan(b)}{\tan(a) - \tan(b)}$ Rewriting the tangents using the tangent identity

$$= \frac{\dfrac{\sin(a)}{\cos(a)} + \dfrac{\sin(b)}{\cos(b)}}{\dfrac{\sin(a)}{\cos(a)} - \dfrac{\sin(b)}{\cos(b)}}$$ Multiplying the top and bottom by $\cos(a)\cos(b)$

$$= \frac{\left(\dfrac{\sin(a)}{\cos(a)} + \dfrac{\sin(b)}{\cos(b)}\right)\cos(a)\cos(b)}{\left(\dfrac{\sin(a)}{\cos(a)} - \dfrac{\sin(b)}{\cos(b)}\right)\cos(a)\cos(b)}$$ Distributing and simplifying

$$= \frac{\sin(a)\cos(b) + \sin(b)\cos(a)}{\sin(a)\cos(b) - \sin(b)\cos(a)}$$ From above, we recognize this

$$= \frac{\sin(a+b)}{\sin(a-b)}$$ Establishing the identity

These identities can also be used to solve equations.

Example 4

Solve $\sin(x)\sin(2x) + \cos(x)\cos(2x) = \dfrac{\sqrt{3}}{2}$.

By recognizing the left side of the equation as the result of the difference of angles identity for cosine, we can simplify the equation

$$\sin(x)\sin(2x) + \cos(x)\cos(2x) = \frac{\sqrt{3}}{2}$$ Apply the difference of angles identity

$$\cos(x - 2x) = \frac{\sqrt{3}}{2}$$

$$\cos(-x) = \frac{\sqrt{3}}{2}$$ Use the negative angle identity

$$\cos(x) = \frac{\sqrt{3}}{2}$$

Since this is a special cosine value we recognize from the unit circle, we can quickly write the answers:

$$x = \frac{\pi}{6} + 2\pi k$$
$$x = \frac{11\pi}{6} + 2\pi k$$
, where k is an integer

Combining Waves of Equal Period

A sinusoidal function of the form $f(x) = A\sin(Bx + C)$ can be rewritten using the sum of angles identity.

> **Example 5**
>
> Rewrite $f(x) = 4\sin\left(3x + \dfrac{\pi}{3}\right)$ as a sum of sine and cosine.
>
> $4\sin\left(3x + \dfrac{\pi}{3}\right)$ Using the sum of angles identity
>
> $= 4\left(\sin(3x)\cos\left(\dfrac{\pi}{3}\right) + \cos(3x)\sin\left(\dfrac{\pi}{3}\right)\right)$ Evaluate the sine and cosine
>
> $= 4\left(\sin(3x)\cdot\dfrac{1}{2} + \cos(3x)\cdot\dfrac{\sqrt{3}}{2}\right)$ Distribute and simplify
>
> $= 2\sin(3x) + 2\sqrt{3}\cos(3x)$

Notice that the result is a stretch of the sine added to a different stretch of the cosine, but both have the same horizontal compression, which results in the same period.

We might ask now whether this process can be reversed – can a combination of a sine and cosine of the same period be written as a single sinusoidal function? To explore this, we will look in general at the procedure used in the example above.

$f(x) = A\sin(Bx + C)$ Use the sum of angles identity
$= A\big(\sin(Bx)\cos(C) + \cos(Bx)\sin(C)\big)$ Distribute the A
$= A\sin(Bx)\cos(C) + A\cos(Bx)\sin(C)$ Rearrange the terms a bit
$= A\cos(C)\sin(Bx) + A\sin(C)\cos(Bx)$

Based on this result, if we have an expression of the form $m\sin(Bx) + n\cos(Bx)$, we could rewrite it as a single sinusoidal function if we can find values A and C so that $m\sin(Bx) + n\cos(Bx) = A\cos(C)\sin(Bx) + A\sin(C)\cos(Bx)$, which will require that:

$m = A\cos(C)$ which can be rewritten as $\dfrac{m}{A} = \cos(C)$
$n = A\sin(C)$ $\dfrac{n}{A} = \sin(C)$

To find A,

$$m^2 + n^2 = \left(A\cos(C)\right)^2 + \left(A\sin(C)\right)^2$$
$$= A^2\cos^2(C) + A^2\sin^2(C)$$
$$= A^2\left(\cos^2(C) + \sin^2(C)\right) \qquad \text{Apply the Pythagorean Identity and simplify}$$
$$= A^2$$

Rewriting a Sum of Sine and Cosine as a Single Sine

To rewrite $m\sin(Bx) + n\cos(Bx)$ as $A\sin(Bx + C)$

$$A^2 = m^2 + n^2, \ \cos(C) = \frac{m}{A}, \ \text{and } \sin(C) = \frac{n}{A}$$

You can use either of the last two equations to solve for possible values of C. Since there will usually be two possible solutions, we will need to look at both to determine which quadrant C is in and determine which solution for C satisfies both equations.

Example 6

Rewrite $4\sqrt{3}\sin(2x) - 4\cos(2x)$ as a single sinusoidal function.

Using the formulas above, $A^2 = \left(4\sqrt{3}\right)^2 + \left(-4\right)^2 = 16\cdot 3 + 16 = 64$, so $A = 8$.

Solving for C,
$$\cos(C) = \frac{4\sqrt{3}}{8} = \frac{\sqrt{3}}{2}, \text{ so } C = \frac{\pi}{6} \text{ or } C = \frac{11\pi}{6}.$$
However, notice $\sin(C) = \frac{-4}{8} = -\frac{1}{2}$. Sine is negative in the third and fourth quadrant,

so the angle that works for both is $C = \frac{11\pi}{6}$.

Combining these results gives us the expression
$$8\sin\left(2x + \frac{11\pi}{6}\right)$$

Try it Now

3. Rewrite $-3\sqrt{2}\sin(5x) + 3\sqrt{2}\cos(5x)$ as a single sinusoidal function.

Rewriting a combination of sine and cosine of equal periods as a single sinusoidal function provides an approach for solving some equations.

Example 7

Solve $3\sin(2x) + 4\cos(2x) = 1$ to find two positive solutions.

Since the sine and cosine have the same period, we can rewrite them as a single sinusoidal function.
$A^2 = (3)^2 + (4)^2 = 25$, so $A = 5$

$\cos(C) = \dfrac{3}{5}$, so $C = \cos^{-1}\left(\dfrac{3}{5}\right) \approx 0.927$ or $C = 2\pi - 0.927 = 5.356$

Since $\sin(C) = \dfrac{4}{5}$, a positive value, we need the angle in the first quadrant, $C = 0.927$.

Using this, our equation becomes

$5\sin(2x + 0.927) = 1$	Divide by 5
$\sin(2x + 0.927) = \dfrac{1}{5}$	Make the substitution $u = 2x + 0.927$
$\sin(u) = \dfrac{1}{5}$	The inverse gives a first solution
$u = \sin^{-1}\left(\dfrac{1}{5}\right) \approx 0.201$	By symmetry, the second solution is
$u = \pi - 0.201 = 2.940$	A third solution would be
$u = 2\pi + 0.201 = 6.485$	

Undoing the substitution, we can find two positive solutions for x.

$2x + 0.927 = 0.201$ or	$2x + 0.927 = 2.940$ or	$2x + 0.927 = 6.485$
$2x = -0.726$	$2x = 2.013$	$2x = 5.558$
$x = -0.363$	$x = 1.007$	$x = 2.779$

Since the first of these is negative, we eliminate it and keep the two positive solutions, $x = 1.007$ and $x = 2.779$.

The Product-to-Sum and Sum-to-Product Identities

Identities

The Product-to-Sum Identities

$$\sin(\alpha)\cos(\beta) = \frac{1}{2}\left(\sin(\alpha + \beta) + \sin(\alpha - \beta)\right)$$

$$\sin(\alpha)\sin(\beta) = \frac{1}{2}\left(\cos(\alpha - \beta) - \cos(\alpha + \beta)\right)$$

$$\cos(\alpha)\cos(\beta) = \frac{1}{2}\left(\cos(\alpha + \beta) + \cos(\alpha - \beta)\right)$$

We will prove the first of these, using the sum and difference of angles identities from the beginning of the section. The proofs of the other two identities are similar and are left as an exercise.

Proof of the product-to-sum identity for sin(α)cos(β)

Recall the sum and difference of angles identities from earlier
$$\sin(\alpha + \beta) = \sin(\alpha)\cos(\beta) + \cos(\alpha)\sin(\beta)$$
$$\sin(\alpha - \beta) = \sin(\alpha)\cos(\beta) - \cos(\alpha)\sin(\beta)$$

Adding these two equations, we obtain
$$\sin(\alpha + \beta) + \sin(\alpha - \beta) = 2\sin(\alpha)\cos(\beta)$$

Dividing by 2, we establish the identity
$$\sin(\alpha)\cos(\beta) = \frac{1}{2}\left(\sin(\alpha + \beta) + \sin(\alpha - \beta)\right)$$

Example 8

Write $\sin(2t)\sin(4t)$ as a sum or difference.

Using the product-to-sum identity for a product of sines
$$\sin(2t)\sin(4t) = \frac{1}{2}\left(\cos(2t - 4t) - \cos(2t + 4t)\right)$$

$$= \frac{1}{2}\left(\cos(-2t) - \cos(6t)\right) \qquad \text{If desired, apply the negative angle identity}$$

$$= \frac{1}{2}\left(\cos(2t) - \cos(6t)\right) \qquad \text{Distribute}$$

$$= \frac{1}{2}\cos(2t) - \frac{1}{2}\cos(6t)$$

Try it Now

4. Evaluate $\cos\left(\dfrac{11\pi}{12}\right)\cos\left(\dfrac{\pi}{12}\right)$.

Identities

The Sum-to-Product Identities

$$\sin(u) + \sin(v) = 2\sin\left(\frac{u+v}{2}\right)\cos\left(\frac{u-v}{2}\right)$$

$$\sin(u) - \sin(v) = 2\sin\left(\frac{u-v}{2}\right)\cos\left(\frac{u+v}{2}\right)$$

$$\cos(u) + \cos(v) = 2\cos\left(\frac{u+v}{2}\right)\cos\left(\frac{u-v}{2}\right)$$

$$\cos(u) - \cos(v) = -2\sin\left(\frac{u+v}{2}\right)\sin\left(\frac{u-v}{2}\right)$$

We will again prove one of these and leave the rest as an exercise.

<u>Proof of the sum-to-product identity for sine functions</u>
We define two new variables:
$$u = \alpha + \beta$$
$$v = \alpha - \beta$$

Adding these equations yields $u + v = 2\alpha$, giving $\alpha = \dfrac{u+v}{2}$

Subtracting the equations yields $u - v = 2\beta$, or $\beta = \dfrac{u-v}{2}$

Substituting these expressions into the product-to-sum identity
$$\sin(\alpha)\cos(\beta) = \frac{1}{2}\left(\sin(\alpha+\beta) + \sin(\alpha-\beta)\right) \text{ gives}$$

$$\sin\left(\frac{u+v}{2}\right)\cos\left(\frac{u-v}{2}\right) = \frac{1}{2}\left(\sin(u) + \sin(v)\right) \qquad \text{Multiply by 2 on both sides}$$

$$2\sin\left(\frac{u+v}{2}\right)\cos\left(\frac{u-v}{2}\right) = \sin(u) + \sin(v) \qquad \text{Establishing the identity}$$

Try it Now

5. Notice that, using the negative angle identity, $\sin(u) - \sin(v) = \sin(u) + \sin(-v)$. Use this along with the sum of sines identity to prove the sum-to-product identity for $\sin(u) - \sin(v)$.

Example 9

Evaluate $\cos(15°) - \cos(75°)$.

Using the sum-to-product identity for the difference of cosines,

$\cos(15°) - \cos(75°)$

$= -2\sin\left(\dfrac{15° + 75°}{2}\right)\sin\left(\dfrac{15° - 75°}{2}\right)$ Simplify

$= -2\sin(45°)\sin(-30°)$ Evaluate

$= -2 \cdot \dfrac{\sqrt{2}}{2} \cdot \dfrac{-1}{2} = \dfrac{\sqrt{2}}{2}$

Example 10

Prove the identity $\dfrac{\cos(4t) - \cos(2t)}{\sin(4t) + \sin(2t)} = -\tan(t)$.

Since the left side seems more complicated, we can start there and simplify.

$\dfrac{\cos(4t) - \cos(2t)}{\sin(4t) + \sin(2t)}$ Use the sum-to-product identities

$= \dfrac{-2\sin\left(\dfrac{4t + 2t}{2}\right)\sin\left(\dfrac{4t - 2t}{2}\right)}{2\sin\left(\dfrac{4t + 2t}{2}\right)\cos\left(\dfrac{4t - 2t}{2}\right)}$ Simplify

$= \dfrac{-2\sin(3t)\sin(t)}{2\sin(3t)\cos(t)}$ Simplify further

$= \dfrac{-\sin(t)}{\cos(t)}$ Rewrite as a tangent

$= -\tan(t)$ Establishing the identity

Example 11

Solve $\sin(\pi t) + \sin(3\pi t) = \cos(\pi t)$ for all solutions with $0 \le t < 2$.

In an equation like this, it is not immediately obvious how to proceed. One option would be to combine the two sine functions on the left side of the equation. Another would be to move the cosine to the left side of the equation, and combine it with one of the sines. For no particularly good reason, we'll begin by combining the sines on the left side of the equation and see how things work out.

$\sin(\pi t) + \sin(3\pi t) = \cos(\pi t)$ Apply the sum to product identity on the left

$2\sin\left(\dfrac{\pi t + 3\pi t}{2}\right)\cos\left(\dfrac{\pi t - 3\pi t}{2}\right) = \cos(\pi t)$ Simplify

$2\sin(2\pi t)\cos(-\pi t) = \cos(\pi t)$ Apply the negative angle identity

$2\sin(2\pi t)\cos(\pi t) = \cos(\pi t)$ Rearrange the equation to be 0 on one side

$2\sin(2\pi t)\cos(\pi t) - \cos(\pi t) = 0$ Factor out the cosine

$\cos(\pi t)\big(2\sin(2\pi t) - 1\big) = 0$

Using the Zero Product Theorem we know that at least one of the two factors must be zero. The first factor, $\cos(\pi t)$, has period $P = \dfrac{2\pi}{\pi} = 2$, so the solution interval of $0 \le t < 2$ represents one full cycle of this function.

$\cos(\pi t) = 0$ Substitute $u = \pi t$

$\cos(u) = 0$ On one cycle, this has solutions

$u = \dfrac{\pi}{2}$ or $u = \dfrac{3\pi}{2}$ Undo the substitution

$\pi t = \dfrac{\pi}{2}$, so $t = \dfrac{1}{2}$

$\pi t = \dfrac{3\pi}{2}$, so $t = \dfrac{3}{2}$

The second factor, $2\sin(2\pi t) - 1$, has period of $P = \dfrac{2\pi}{2\pi} = 1$, so the solution interval $0 \le t < 2$ contains two complete cycles of this function.

$2\sin(2\pi t) - 1 = 0$ Isolate the sine

$\sin(2\pi t) = \dfrac{1}{2}$ Substitute $u = 2\pi t$

$$\sin(u) = \frac{1}{2}$$ On one cycle, this has solutions

$$u = \frac{\pi}{6} \text{ or } u = \frac{5\pi}{6}$$ On the second cycle, the solutions are

$$u = 2\pi + \frac{\pi}{6} = \frac{13\pi}{6} \text{ or } u = 2\pi + \frac{5\pi}{6} = \frac{17\pi}{6}$$ Undo the substitution

$$2\pi t = \frac{\pi}{6}, \text{ so } t = \frac{1}{12}$$

$$2\pi t = \frac{5\pi}{6}, \text{ so } t = \frac{5}{12}$$

$$2\pi t = \frac{13\pi}{6}, \text{ so } t = \frac{13}{12}$$

$$2\pi t = \frac{17\pi}{6}, \text{ so } t = \frac{17}{12}$$

Altogether, we found six solutions on $0 \le t < 2$, which we can confirm by looking at the graph.

$$t = \frac{1}{12}, \frac{5}{12}, \frac{1}{2}, \frac{13}{12}, \frac{3}{2}, \frac{17}{12}$$

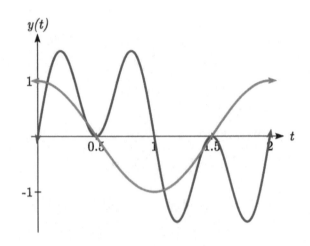

Important Topics of This Section
The sum and difference identities
Combining waves of equal periods
Product-to-sum identities
Sum-to-product identities
Completing proofs

Try it Now Answers

1.
$$\cos(\alpha + \beta) = \cos(\alpha - (-\beta))$$
$$\cos(\alpha)\cos(-\beta) + \sin(\alpha)\sin(-\beta)$$
$$\cos(\alpha)\cos(\beta) + \sin(\alpha)(-\sin(\beta))$$
$$\cos(\alpha)\cos(\beta) - \sin(\alpha)\sin(\beta)$$

2. $\sin\left(\dfrac{\pi}{12}\right) = \sin\left(\dfrac{\pi}{3} - \dfrac{\pi}{4}\right) = \sin\left(\dfrac{\pi}{3}\right)\cos\left(\dfrac{\pi}{4}\right) - \cos\left(\dfrac{\pi}{3}\right)\sin\left(\dfrac{\pi}{4}\right)$

$= \dfrac{\sqrt{3}}{2}\dfrac{\sqrt{2}}{2} - \dfrac{1}{2}\dfrac{\sqrt{2}}{2}\dfrac{\sqrt{6}-\sqrt{2}}{4}$

3. $A^2 = \left(-3\sqrt{2}\right)^2 + \left(3\sqrt{2}\right)^2 = 36$. $A = 6$

$\cos(C) = \dfrac{-3\sqrt{2}}{6} = \dfrac{-\sqrt{2}}{2}$, $\sin(C) = \dfrac{3\sqrt{2}}{6} = \dfrac{\sqrt{2}}{2}$. $C = \dfrac{3\pi}{4}$

$6\sin\left(5x + \dfrac{3\pi}{4}\right)$

4. $\cos\left(\dfrac{11\pi}{12}\right)\cos\left(\dfrac{\pi}{12}\right) = \dfrac{1}{2}\left(\cos\left(\dfrac{11\pi}{12} + \dfrac{\pi}{12}\right) + \cos\left(\dfrac{11\pi}{12} - \dfrac{\pi}{12}\right)\right)$

$= \dfrac{1}{2}\left(\cos(\pi) + \cos\left(\dfrac{5\pi}{6}\right)\right) = \dfrac{1}{2}\left(-1 - \dfrac{\sqrt{3}}{2}\right)$

$= \dfrac{-2-\sqrt{3}}{4}$

5. $\sin(u) - \sin(v)$ Use negative angle identity for sine

$\sin(u) + \sin(-v)$ Use sum-to-product identity for sine

$2\sin\left(\dfrac{u+(-v)}{2}\right)\cos\left(\dfrac{u-(-v)}{2}\right)$ Eliminate the parenthesis

$2\sin\left(\dfrac{u-v}{2}\right)\cos\left(\dfrac{u+v}{2}\right)$ Establishing the identity

Section 7.2 Exercises

Find an exact value for each of the following.

1. $\sin(75°)$ 2. $\sin(195°)$ 3. $\cos(165°)$ 4. $\cos(345°)$

5. $\cos\left(\dfrac{7\pi}{12}\right)$ 6. $\cos\left(\dfrac{\pi}{12}\right)$ 7. $\sin\left(\dfrac{5\pi}{12}\right)$ 8. $\sin\left(\dfrac{11\pi}{12}\right)$

Rewrite in terms of $\sin(x)$ and $\cos(x)$.

9. $\sin\left(x+\dfrac{11\pi}{6}\right)$ 10. $\sin\left(x-\dfrac{3\pi}{4}\right)$ 11. $\cos\left(x-\dfrac{5\pi}{6}\right)$ 12. $\cos\left(x+\dfrac{2\pi}{3}\right)$

Simplify each expression.

13. $\csc\left(\dfrac{\pi}{2}-t\right)$ 14. $\sec\left(\dfrac{\pi}{2}-w\right)$ 15. $\cot\left(\dfrac{\pi}{2}-x\right)$ 16. $\tan\left(\dfrac{\pi}{2}-x\right)$

Rewrite the product as a sum.

17. $16\sin(16x)\sin(11x)$ 18. $20\cos(36t)\cos(6t)$

19. $2\sin(5x)\cos(3x)$ 20. $10\cos(5x)\sin(10x)$

Rewrite the sum as a product.

21. $\cos(6t)+\cos(4t)$ 22. $\cos(6u)+\cos(4u)$

23. $\sin(3x)+\sin(7x)$ 24. $\sin(h)+\sin(3h)$

25. Given $\sin(a)=\dfrac{2}{3}$ and $\cos(b)=-\dfrac{1}{4}$, with a and b both in the interval $\left[\dfrac{\pi}{2},\pi\right)$:

     a. Find $\sin(a+b)$ b. Find $\cos(a-b)$

26. Given $\sin(a)=\dfrac{4}{5}$ and $\cos(b)=\dfrac{1}{3}$, with a and b both in the interval $\left[0,\dfrac{\pi}{2}\right)$:

     a. Find $\sin(a-b)$ b. Find $\cos(a+b)$

Solve each equation for all solutions.

27. $\sin(3x)\cos(6x)-\cos(3x)\sin(6x)=-0.9$

28. $\sin(6x)\cos(11x)-\cos(6x)\sin(11x)=-0.1$

29. $\cos(2x)\cos(x)+\sin(2x)\sin(x)=1$

30. $\cos(5x)\cos(3x)-\sin(5x)\sin(3x)=\dfrac{\sqrt{3}}{2}$

Solve each equation for all solutions.

31. $\cos(5x) = -\cos(2x)$

32. $\sin(5x) = \sin(3x)$

33. $\cos(6\theta) - \cos(2\theta) = \sin(4\theta)$

34. $\cos(8\theta) - \cos(2\theta) = \sin(5\theta)$

Rewrite as a single function of the form $A\sin(Bx + C)$.

35. $4\sin(x) - 6\cos(x)$ 36. $-\sin(x) - 5\cos(x)$

37. $5\sin(3x) + 2\cos(3x)$ 38. $-3\sin(5x) + 4\cos(5x)$

Solve for the first two positive solutions.

39. $-5\sin(x) + 3\cos(x) = 1$ 40. $3\sin(x) + \cos(x) = 2$

41. $3\sin(2x) - 5\cos(2x) = 3$ 42. $-3\sin(4x) - 2\cos(4x) = 1$

Simplify.

43. $\dfrac{\sin(7t) + \sin(5t)}{\cos(7t) + \cos(5t)}$ 44. $\dfrac{\sin(9t) - \sin(3t)}{\cos(9t) + \cos(3t)}$

Prove the identity.

44. $\tan\left(x + \dfrac{\pi}{4}\right) = \dfrac{\tan(x) + 1}{1 - \tan(x)}$

45. $\tan\left(\dfrac{\pi}{4} - t\right) = \dfrac{1 - \tan(t)}{1 + \tan(t)}$

46. $\cos(a + b) + \cos(a - b) = 2\cos(a)\cos(b)$

47. $\dfrac{\cos(a + b)}{\cos(a - b)} = \dfrac{1 - \tan(a)\tan(b)}{1 + \tan(a)\tan(b)}$

48. $\dfrac{\tan(a + b)}{\tan(a - b)} = \dfrac{\sin(a)\cos(a) + \sin(b)\cos(b)}{\sin(a)\cos(a) - \sin(b)\cos(b)}$

49. $2\sin(a + b)\sin(a - b) = \cos(2b) - \cos(2a)$

50. $\dfrac{\sin(x) + \sin(y)}{\cos(x) + \cos(y)} = \tan\left(\dfrac{1}{2}(x + y)\right)$

Prove the identity.

51. $\dfrac{\cos(a+b)}{\cos(a)\cos(b)} = 1 - \tan(a)\tan(b)$

52. $\cos(x+y)\cos(x-y) = \cos^2 x - \sin^2 y$

53. Use the sum and difference identities to establish the product-to-sum identity

$$\sin(\alpha)\sin(\beta) = \frac{1}{2}\left(\cos(\alpha-\beta) - \cos(\alpha+\beta)\right)$$

54. Use the sum and difference identities to establish the product-to-sum identity

$$\cos(\alpha)\cos(\beta) = \frac{1}{2}\left(\cos(\alpha+\beta) + \cos(\alpha-\beta)\right)$$

55. Use the product-to-sum identities to establish the sum-to-product identity

$$\cos(u) + \cos(v) = 2\cos\left(\frac{u+v}{2}\right)\cos\left(\frac{u-v}{2}\right)$$

56. Use the product-to-sum identities to establish the sum-to-product identity

$$\cos(u) - \cos(v) = -2\sin\left(\frac{u+v}{2}\right)\sin\left(\frac{u-v}{2}\right)$$

Section 7.3 Double Angle Identities

Two special cases of the sum of angles identities arise often enough that we choose to state these identities separately.

Identities
The double angle identities
$\sin(2\alpha) = 2\sin(\alpha)\cos(\alpha)$
$\cos(2\alpha) \quad = \quad \cos^2(\alpha) - \sin^2(\alpha)$
$\qquad\qquad = \qquad 1 - 2\sin^2(\alpha)$
$\qquad\qquad = \qquad 2\cos^2(\alpha) - 1$

These identities follow from the sum of angles identities.

<u>Proof of the sine double angle identity</u>
$\sin(2\alpha)$

$= \sin(\alpha + \alpha)$	Apply the sum of angles identity
$= \sin(\alpha)\cos(\alpha) + \cos(\alpha)\sin(\alpha)$	Simplify
$= 2\sin(\alpha)\cos(\alpha)$	Establishing the identity

Try it Now

1. Show $\cos(2\alpha) = \cos^2(\alpha) - \sin^2(\alpha)$ by using the sum of angles identity for cosine.

For the cosine double angle identity, there are three forms of the identity stated because the basic form, $\cos(2\alpha) = \cos^2(\alpha) - \sin^2(\alpha)$, can be rewritten using the Pythagorean Identity. Rearranging the Pythagorean Identity results in the equality $\cos^2(\alpha) = 1 - \sin^2(\alpha)$, and by substituting this into the basic double angle identity, we obtain the second form of the double angle identity.

$\cos(2\alpha) = \cos^2(\alpha) - \sin^2(\alpha)$	Substituting using the Pythagorean identity
$\cos(2\alpha) = 1 - \sin^2(\alpha) - \sin^2(\alpha)$	Simplifying
$\cos(2\alpha) = 1 - 2\sin^2(\alpha)$	

Example 1

If $\sin(\theta) = \dfrac{3}{5}$ and θ is in the second quadrant, find exact values for $\sin(2\theta)$ and $\cos(2\theta)$.

To evaluate $\cos(2\theta)$, since we know the value for $\sin(\theta)$ we can use the version of the double angle that only involves sine.

$$\cos(2\theta) = 1 - 2\sin^2(\theta) = 1 - 2\left(\frac{3}{5}\right)^2 = 1 - \frac{18}{25} = \frac{7}{25}$$

Since the double angle for sine involves both sine and cosine, we'll need to first find $\cos(\theta)$, which we can do using the Pythagorean Identity.

$$\sin^2(\theta) + \cos^2(\theta) = 1$$

$$\left(\frac{3}{5}\right)^2 + \cos^2(\theta) = 1$$

$$\cos^2(\theta) = 1 - \frac{9}{25}$$

$$\cos(\theta) = \pm\sqrt{\frac{16}{25}} = \pm\frac{4}{5}$$

Since θ is in the second quadrant, we know that $\cos(\theta) < 0$, so

$$\cos(\theta) = -\frac{4}{5}$$

Now we can evaluate the sine double angle

$$\sin(2\theta) = 2\sin(\theta)\cos(\theta) = 2\left(\frac{3}{5}\right)\left(-\frac{4}{5}\right) = -\frac{24}{25}$$

Example 2

Simplify the expressions

a) $2\cos^2(12°) - 1$ b) $8\sin(3x)\cos(3x)$

a) Notice that the expression is in the same form as one version of the double angle identity for cosine: $\cos(2\theta) = 2\cos^2(\theta) - 1$. Using this,

$$2\cos^2(12°) - 1 = \cos(2 \cdot 12°) = \cos(24°)$$

b) This expression looks similar to the result of the double angle identity for sine.

$8\sin(3x)\cos(3x)$ Factoring a 4 out of the original expression

$4 \cdot 2\sin(3x)\cos(3x)$ Applying the double angle identity

$4\sin(6x)$

We can use the double angle identities to simplify expressions and prove identities.

Example 2

Simplify $\dfrac{\cos(2t)}{\cos(t) - \sin(t)}$.

With three choices for how to rewrite the double angle, we need to consider which will be the most useful. To simplify this expression, it would be great if the denominator would cancel with something in the numerator, which would require a factor of $\cos(t) - \sin(t)$ in the numerator, which is most likely to occur if we rewrite the numerator with a mix of sine and cosine.

$\dfrac{\cos(2t)}{\cos(t) - \sin(t)}$ Apply the double angle identity

$= \dfrac{\cos^2(t) - \sin^2(t)}{\cos(t) - \sin(t)}$ Factor the numerator

$= \dfrac{\big(\cos(t) - \sin(t)\big)\big(\cos(t) + \sin(t)\big)}{\cos(t) - \sin(t)}$ Cancelling the common factor

$= \cos(t) + \sin(t)$ Resulting in the most simplified form

Example 3

Prove $\sec(2\alpha) = \dfrac{\sec^2(\alpha)}{2 - \sec^2(\alpha)}$.

Since the right side seems a bit more complicated than the left side, we begin there.

$\dfrac{\sec^2(\alpha)}{2 - \sec^2(\alpha)}$ Rewrite the secants in terms of cosine

$= \dfrac{\dfrac{1}{\cos^2(\alpha)}}{2 - \dfrac{1}{\cos^2(\alpha)}}$

At this point, we could rewrite the bottom with common denominators, subtract the terms, invert and multiply, then simplify. Alternatively, we can multiple both the top and bottom by $\cos^2(\alpha)$, the common denominator:

$= \dfrac{\dfrac{1}{\cos^2(\alpha)} \cdot \cos^2(\alpha)}{\left(2 - \dfrac{1}{\cos^2(\alpha)}\right) \cdot \cos^2(\alpha)}$ Distribute on the bottom

$$= \frac{\dfrac{\cos^2(\alpha)}{\cos^2(\alpha)}}{2\cos^2(\alpha) - \dfrac{\cos^2(\alpha)}{\cos^2(\alpha)} \cdot}$$ 　Simplify

$$= \frac{1}{2\cos^2(\alpha) - 1}$$ 　Rewrite the denominator as a double angle

$$= \frac{1}{\cos(2\alpha)}$$ 　Rewrite as a secant

$$= \sec(2\alpha)$$ 　Establishing the identity

Try it Now

2. Use an identity to find the exact value of $\cos^2(75°) - \sin^2(75°)$.

As with other identities, we can also use the double angle identities for solving equations.

Example 4

Solve $\cos(2t) = \cos(t)$ for all solutions with $0 \le t < 2\pi$.

In general when solving trig equations, it makes things more complicated when we have a mix of sines and cosines and when we have a mix of functions with different periods. In this case, we can use a double angle identity to rewrite the cos(2t). When choosing which form of the double angle identity to use, we notice that we have a cosine on the right side of the equation. We try to limit our equation to one trig function, which we can do by choosing the version of the double angle formula for cosine that only involves cosine.

$\cos(2t) = \cos(t)$ 　Apply the double angle identity

$2\cos^2(t) - 1 = \cos(t)$ 　This is quadratic in cosine, so make one side 0

$2\cos^2(t) - \cos(t) - 1 = 0$ 　Factor

$(2\cos(t) + 1)(\cos(t) - 1) = 0$ 　Break this apart to solve each part separately

$2\cos(t) + 1 = 0$ 　or 　$\cos(t) - 1 = 0$

$\cos(t) = -\dfrac{1}{2}$ 　or 　$\cos(t) = 1$

$t = \dfrac{2\pi}{3}$ or $t = \dfrac{4\pi}{3}$ 　or 　$t = 0$

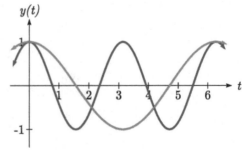

Looking at a graph of cos(2t) and cos(t) shown together, we can verify that these three

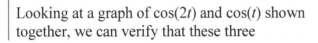

solutions on $[0, 2\pi)$ seem reasonable.

Example 5

A cannonball is fired with velocity of 100 meters per second. If it is launched at an angle of θ, the vertical component of the velocity will be $100\sin(\theta)$ and the horizontal component will be $100\cos(\theta)$. Ignoring wind resistance, the height of the cannonball will follow the equation $h(t) = -4.9t^2 + 100\sin(\theta)t$ and horizontal position will follow the equation $x(t) = 100\cos(\theta)t$. If you want to hit a target 900 meters away, at what angle should you aim the cannon?

To hit the target 900 meters away, we want $x(t) = 900$ at the time when the cannonball hits the ground, when $h(t) = 0$. To solve this problem, we will first solve for the time, t, when the cannonball hits the ground. Our answer will depend upon the angle θ.

$h(t) = 0$

$-4.9t^2 + 100\sin(\theta)t = 0$ Factor

$t(-4.9t + 100\sin(\theta)) = 0$ Break this apart to find two solutions

$t = 0$ or $-4.9t + 100\sin(\theta) = 0$ Solve for t

$\qquad\qquad\qquad -4.9t = -100\sin(\theta)$

$\qquad\qquad\qquad t = \dfrac{100\sin(\theta)}{4.9}$

This shows that the height is 0 twice, once at $t = 0$ when the cannonball is fired, and again when the cannonball hits the ground after flying through the air. This second value of t gives the time when the ball hits the ground in terms of the angle θ. We want the horizontal distance $x(t)$ to be 900 when the ball hits the ground, in other words when $t = \dfrac{100\sin(\theta)}{4.9}$.

Since the target is 900 m away we start with

$x(t) = 900$ Use the formula for $x(t)$

$100\cos(\theta)t = 900$ Substitute the desired time, t from above

$100\cos(\theta)\dfrac{100\sin(\theta)}{4.9} = 900$ Simplify

$\dfrac{100^2}{4.9}\cos(\theta)\sin(\theta) = 900$ Isolate the cosine and sine product

$\cos(\theta)\sin(\theta) = \dfrac{900(4.9)}{100^2}$

The left side of this equation almost looks like the result of the double angle identity for sine: $\sin(2\theta) = 2\sin(\theta)\cos(\theta)$.

Multiplying both sides of our equation by 2,

$$2\cos(\theta)\sin(\theta) = \frac{2(900)(4.9)}{100^2}$$ Using the double angle identity on the left

$$\sin(2\theta) = \frac{2(900)(4.9)}{100^2}$$ Use the inverse sine

$$2\theta = \sin^{-1}\left(\frac{2(900)(4.9)}{100^2}\right) \approx 1.080$$ Divide by 2

$$\theta = \frac{1.080}{2} = 0.540 \text{, or about 30.94 degrees}$$

Power Reduction and Half Angle Identities

Another use of the cosine double angle identities is to use them in reverse to rewrite a squared sine or cosine in terms of the double angle. Starting with one form of the cosine double angle identity:

$$\cos(2\alpha) = 2\cos^2(\alpha) - 1$$ Isolate the cosine squared term

$$\cos(2\alpha) + 1 = 2\cos^2(\alpha)$$ Add 1

$$\cos^2(\alpha) = \frac{\cos(2\alpha) + 1}{2}$$ Divide by 2

$$\cos^2(\alpha) = \frac{\cos(2\alpha) + 1}{2}$$ This is called a **power reduction identity**

Try it Now

3. Use another form of the cosine double angle identity to prove the identity

$$\sin^2(\alpha) = \frac{1 - \cos(2\alpha)}{2}.$$

The cosine double angle identities can also be used in reverse for evaluating angles that are half of a common angle. Building from our formula $\cos^2(\alpha) = \frac{\cos(2\alpha) + 1}{2}$, if we let $\theta = 2\alpha$, then $\alpha = \frac{\theta}{2}$ this identity becomes $\cos^2\left(\frac{\theta}{2}\right) = \frac{\cos(\theta) + 1}{2}$. Taking the square root, we obtain

$$\cos\left(\frac{\theta}{2}\right) = \pm\sqrt{\frac{\cos(\theta) + 1}{2}}\text{ , where the sign is determined by the quadrant.}$$

This is called a **half-angle identity**.

Try it Now

4. Use your results from the last Try it Now to prove the identity

$$\sin\left(\frac{\theta}{2}\right) = \pm\sqrt{\frac{1-\cos(\theta)}{2}}$$.

Identities

Half-Angle Identities

$$\cos\left(\frac{\theta}{2}\right) = \pm\sqrt{\frac{\cos(\theta)+1}{2}} \qquad \sin\left(\frac{\theta}{2}\right) = \pm\sqrt{\frac{1-\cos(\theta)}{2}}$$

Power Reduction Identities

$$\cos^2(\alpha) = \frac{\cos(2\alpha)+1}{2} \qquad \sin^2(\alpha) = \frac{1-\cos(2\alpha)}{2}$$

Since these identities are easy to derive from the double-angle identities, the power reduction and half-angle identities are not ones you should need to memorize separately.

Example 6

Rewrite $\cos^4(x)$ without any powers.

$$\cos^4(x) = \left(\cos^2(x)\right)^2 \qquad\qquad \text{Using the power reduction formula}$$

$$= \left(\frac{\cos(2x)+1}{2}\right)^2 \qquad\qquad \text{Square the numerator and denominator}$$

$$= \frac{\left(\cos(2x)+1\right)^2}{4} \qquad\qquad \text{Expand the numerator}$$

$$= \frac{\cos^2(2x)+2\cos(2x)+1}{4} \qquad\qquad \text{Split apart the fraction}$$

$$= \frac{\cos^2(2x)}{4} + \frac{2\cos(2x)}{4} + \frac{1}{4} \qquad\qquad \text{Apply the formula above to } \cos^2(2x)$$

$$\cos^2(2x) = \frac{\cos(2\cdot2x)+1}{2}$$

$$= \frac{\left(\dfrac{\cos(4x)+1}{2}\right)}{4} + \frac{2\cos(2x)}{4} + \frac{1}{4} \qquad \text{Simplify}$$

$$= \frac{\cos(4x)}{8} + \frac{1}{8} + \frac{1}{2}\cos(2x) + \frac{1}{4} \qquad \text{Combine the constants}$$

$$= \frac{\cos(4x)}{8} + \frac{1}{2}\cos(2x) + \frac{3}{8}$$

Example 7

Find an exact value for $\cos(15°)$.

Since 15 degrees is half of 30 degrees, we can use our result from above:

$$\cos(15°) = \cos\left(\frac{30°}{2}\right) = \pm\sqrt{\frac{\cos(30°)+1}{2}}$$

We can evaluate the cosine. Since 15 degrees is in the first quadrant, we need the positive result.

$$\sqrt{\frac{\cos(30°)+1}{2}} = \sqrt{\frac{\dfrac{\sqrt{3}}{2}+1}{2}}$$

$$= \sqrt{\frac{\sqrt{3}}{4} + \frac{1}{2}}$$

Important Topics of This Section
Double angle identity
Power reduction identity
Half angle identity
Using identities
Simplify equations
Prove identities
Solve equations

Try it Now Answers

$$\cos(2\alpha) = \cos(\alpha + \alpha)$$

1. $\cos(\alpha)\cos(\alpha) - \sin(\alpha)\sin(\alpha)$

    $\cos^2(\alpha) - \sin^2(\alpha)$

2. $\cos^2(75°) - \sin^2(75°) = \cos(2 \cdot 75°) = \cos(150°) = \dfrac{-\sqrt{3}}{2}$

3. $\dfrac{1 - \cos(2\alpha)}{2}$

    $\dfrac{1 - \left(\cos^2(\alpha) - \sin^2(\alpha)\right)}{2}$

    $\dfrac{1 - \cos^2(\alpha) + \sin^2(\alpha)}{2}$

    $\dfrac{\sin^2(\alpha) + \sin^2(\alpha)}{2}$

    $\dfrac{2\sin^2(\alpha)}{2} = \sin^2(\alpha)$

4. $\sin^2(\alpha) = \dfrac{1 - \cos(2\alpha)}{2}$

    $\sin(\alpha) = \pm\sqrt{\dfrac{1 - \cos(2\alpha)}{2}}$

    $\alpha = \dfrac{\theta}{2}$

    $\sin\left(\dfrac{\theta}{2}\right) = \pm\sqrt{\dfrac{1 - \cos\left(2\left(\dfrac{\theta}{2}\right)\right)}{2}}$

    $\sin\left(\dfrac{\theta}{2}\right) = \pm\sqrt{\dfrac{1 - \cos(\theta)}{2}}$

Section 7.3 Exercises

1. If $\sin(x) = \dfrac{1}{8}$ and x is in quadrant I, then find exact values for (without solving for x):

    a. $\sin(2x)$ b. $\cos(2x)$ c. $\tan(2x)$

2. If $\cos(x) = \dfrac{2}{3}$ and x is in quadrant I, then find exact values for (without solving for x):

    a. $\sin(2x)$ b. $\cos(2x)$ c. $\tan(2x)$

Simplify each expression.

3. $\cos^2(28°) - \sin^2(28°)$ 4. $2\cos^2(37°) - 1$

5. $1 - 2\sin^2(17°)$ 6. $\cos^2(37°) - \sin^2(37°)$

7. $\cos^2(9x) - \sin^2(9x)$ 8. $\cos^2(6x) - \sin^2(6x)$

9. $4\sin(8x)\cos(8x)$ 10. $6\sin(5x)\cos(5x)$

Solve for all solutions on the interval $[0, 2\pi)$.

11. $6\sin(2t) + 9\sin(t) = 0$ 12. $2\sin(2t) + 3\cos(t) = 0$

13. $9\cos(2\theta) = 9\cos^2(\theta) - 4$ 14. $8\cos(2\alpha) = 8\cos^2(\alpha) - 1$

15. $\sin(2t) = \cos(t)$ 16. $\cos(2t) = \sin(t)$

17. $\cos(6x) - \cos(3x) = 0$ 18. $\sin(4x) - \sin(2x) = 0$

Use a double angle, half angle, or power reduction formula to rewrite without exponents.

19. $\cos^2(5x)$ 20. $\cos^2(6x)$

21. $\sin^4(8x)$ 22. $\sin^4(3x)$

23. $\cos^2 x \sin^4 x$ 24. $\cos^4 x \sin^2 x$

25. If $\csc(x) = 7$ and $90° < x < 180°$, then find exact values for (without solving for x):

    a. $\sin\left(\dfrac{x}{2}\right)$ b. $\cos\left(\dfrac{x}{2}\right)$ c. $\tan\left(\dfrac{x}{2}\right)$

26. If $\sec(x) = 4$ and $270° < x < 360°$, then find exact values for (without solving for x):

    a. $\sin\left(\dfrac{x}{2}\right)$ b. $\cos\left(\dfrac{x}{2}\right)$ c. $\tan\left(\dfrac{x}{2}\right)$

Prove the identity.

27. $\left(\sin t - \cos t\right)^2 = 1 - \sin\left(2t\right)$

28. $\left(\sin^2 x - 1\right)^2 = \cos\left(2x\right) + \sin^4 x$

29. $\sin\left(2x\right) = \dfrac{2\tan\left(x\right)}{1 + \tan^2\left(x\right)}$

30. $\tan\left(2x\right) = \dfrac{2\sin\left(x\right)\cos\left(x\right)}{2\cos^2\left(x\right) - 1}$

31. $\cot\left(x\right) - \tan\left(x\right) = 2\cot\left(2x\right)$

32. $\dfrac{\sin\left(2\theta\right)}{1 + \cos\left(2\theta\right)} = \tan\left(\theta\right)$

33. $\cos\left(2\alpha\right) = \dfrac{1 - \tan^2\left(\alpha\right)}{1 + \tan^2\left(\alpha\right)}$

34. $\dfrac{1 + \cos\left(2t\right)}{\sin\left(2t\right) - \cos\left(t\right)} = \dfrac{2\cos\left(t\right)}{2\sin\left(t\right) - 1}$

35. $\sin\left(3x\right) = 3\sin\left(x\right)\cos^2\left(x\right) - \sin^3\left(x\right)$.

36. $\cos\left(3x\right) = \cos^3\left(x\right) - 3\sin^2\left(x\right)\cos\left(x\right)$

Section 7.4 Modeling Changing Amplitude and Midline

While sinusoidal functions can model a variety of behaviors, it is often necessary to combine sinusoidal functions with linear and exponential curves to model real applications and behaviors. We begin this section by looking at changes to the midline of a sinusoidal function. Recall that the midline describes the middle, or average value, of the sinusoidal function.

Changing Midlines

Example 1

A population of elk currently averages 2000 elk, and that average has been growing by 4% each year. Due to seasonal fluctuation, the population oscillates from 50 below average in the winter up to 50 above average in the summer. Find a function that models the number of elk after t years, starting in the winter.

There are two components to the behavior of the elk population: the changing average, and the oscillation. The average is an exponential growth, starting at 2000 and growing by 4% each year. Writing a formula for this:

$$average = initial(1+r)^t = 2000(1+0.04)^t$$

For the oscillation, since the population oscillates 50 above and below average, the amplitude will be 50. Since it takes one year for the population to cycle, the period is 1.

We find the value of the horizontal stretch coefficient $B = \dfrac{\text{original period}}{\text{new period}} = \dfrac{2\pi}{1} = 2\pi$.

The function starts in winter, so the shape of the function will be a negative cosine, since it starts at the lowest value.

Putting it all together, the equation would be:

$$P(t) = -50\cos(2\pi t) + midline$$

Since the midline represents the average population, we substitute in the exponential function into the population equation to find our final equation:

$$P(t) = -50\cos(2\pi t) + 2000(1+0.04)^t$$

This is an example of changing midline – in this case an exponentially changing midline.

Changing Midline

A function of the form $f(t) = A\sin(Bt) + g(t)$ will oscillate above and below the average given by the function *g(t)*.

Changing midlines can be exponential, linear, or any other type of function. Here are some examples:

Linear midline

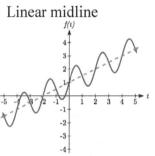

$$f(t) = A\sin(Bt) + (mt + b)$$

Exponential midline

$$f(t) = A\sin(Bt) + (ab^t)$$

Quadratic midline

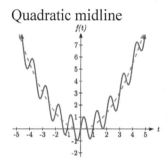

$$f(t) = A\sin(Bt) + (at^2)$$

Example 2

Find a function with linear midline of the form $f(t) = A\sin\left(\dfrac{\pi}{2}t\right) + mt + b$ that will pass through the points given below.

t	0	1	2	3
f(t)	5	10	9	8

Since we are given the value of the horizontal compression coefficient we can calculate

the period of this function: new period $= \dfrac{\text{original period}}{B} = \dfrac{2\pi}{\pi/2} = 4$.

Since the sine function is at the midline at the beginning of a cycle and halfway through a cycle, we would expect this function to be at the midline at $t = 0$ and $t = 2$, since 2 is half the full period of 4. Based on this, we expect the points (0, 5) and (2, 9) to be points on the midline. We can clearly see that this is not a constant function and so we use the two points to calculate a linear function: *midline* $= mt + b$. From these two points we can calculate a slope:

$$m = \frac{9-5}{2-0} = \frac{4}{2} = 2$$

Combining this with the initial value of 5, we have the midline: *midline* $= 2t + 5$.

The full function will have form $f(t) = A\sin\left(\dfrac{\pi}{2}t\right) + 2t + 5$. To find the amplitude, we can plug in a point we haven't already used, such as (1, 10).

$10 = A\sin\left(\dfrac{\pi}{2}(1)\right) + 2(1) + 5$ Evaluate the sine and combine like terms

$10 = A + 7$
$A = 3$

A function of the form given fitting the data would be

$f(t) = 3\sin\left(\dfrac{\pi}{2}t\right) + 2t + 5$

Alternative Approach
Notice we could have taken an alternate approach by plugging points (0, 5) and (2, 9) into the original equation. Substituting (0, 5),

$5 = A\sin\left(\dfrac{\pi}{2}(0)\right) + m(0) + b$ Evaluate the sine and simplify

$5 = b$

Substituting (2, 9)

$9 = A\sin\left(\dfrac{\pi}{2}(2)\right) + m(2) + 5$ Evaluate the sine and simplify

$9 = 2m + 5$
$4 = 2m$
$m = 2$, as we found above. Now we can proceed to find A the same way we did before.

Example 3

The number of tourists visiting a ski and hiking resort averages 4000 people annually and oscillates seasonally, 1000 above and below the average. Due to a marketing campaign, the average number of tourists has been increasing by 200 each year. Write an equation for the number of tourists after t years, beginning at the peak season.

Again there are two components to this problem: the oscillation and the average. For the oscillation, the number of tourists oscillates 1000 above and below average, giving an amplitude of 1000. Since the oscillation is seasonal, it has a period of 1 year. Since we are given a starting point of "peak season", we will model this scenario with a cosine function. So far, this gives an equation in the form $N(t) = 1000\cos(2\pi t) + midline$.

The average is currently 4000, and is increasing by 200 each year. This is a constant rate of change, so this is linear growth, $average = 4000 + 200t$. This function will act as the midline.

Combining these two pieces gives a function for the number of tourists:
$N(t) = 1000\cos(2\pi t) + 4000 + 200t$

Try it Now

1. Given the function $g(x) = (x^2 - 1) + 8\cos(x)$, describe the midline and amplitude using words.

Changing Amplitude

There are also situations in which the amplitude of a sinusoidal function does not stay constant. Back in Chapter 6, we modeled the motion of a spring using a sinusoidal function, but had to ignore friction in doing so. If there were friction in the system, we would expect the amplitude of the oscillation to decrease over time. In the equation $f(t) = A\sin(Bt) + k$, A gives the amplitude of the oscillation, we can allow the amplitude to change by replacing this constant A with a function $A(t)$.

> **Changing Amplitude**
>
> A function of the form $f(t) = A(t)\sin(Bt) + k$ will oscillate above and below the midline with an amplitude given by $A(t)$.

Here are some examples:

Linear amplitude

$f(t) = (mt + b)\sin(Bt) + k$

Exponential amplitude

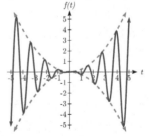

$f(t) = (ab^t)\sin(Bt) + k$

Quadratic amplitude

$f(t) = (at^2)\sin(Bt) + k$

When thinking about a spring with amplitude decreasing over time, it is tempting to use the simplest tool for the job – a linear function. But if we attempt to model the amplitude with a decreasing linear function, such as $A(t) = 10 - t$, we quickly see the problem when we graph the equation $f(t) = (10 - t)\sin(4t)$.

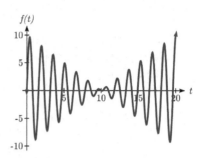

While the amplitude decreases at first as intended, the amplitude hits zero at $t = 10$, then continues past the intercept, increasing in absolute value, which is not the expected behavior. This behavior and function may model the situation on a restricted domain and we might try to chalk the rest of it up to model breakdown, but in fact springs just don't behave like this.

A better model, as you will learn later in physics and calculus, would show the amplitude decreasing by a fixed *percentage* each second, leading to an exponential decay model for the amplitude.

Damped Harmonic Motion

Damped harmonic motion, exhibited by springs subject to friction, follows a model of the form

$$f(t) = ab^t \sin(Bt) + k \quad \text{or} \quad f(t) = ae^{rt} \sin(Bt) + k.$$

Example 4

A spring with natural length of feet inches is pulled back 6 feet and released. It oscillates once every 2 seconds. Its amplitude decreases by 20% each second. Find a function that models the position of the spring t seconds after being released.

Since the spring will oscillate on either side of the natural length, the midline will be at 20 feet. The oscillation has a period of 2 seconds, and so the horizontal compression coefficient is $B = \pi$. Additionally, it begins at the furthest distance from the wall, indicating a cosine model.

Meanwhile, the amplitude begins at 6 feet, and decreases by 20% each second, giving an amplitude function of $A(t) = 6(1 - 0.20)^t$.

Combining this with the sinusoidal information gives a function for the position of the spring:

$$f(t) = 6(0.80)^t \cos(\pi t) + 20$$

Example 5

A spring with natural length of 30 cm is pulled out 10 cm and released. It oscillates 4 times per second. After 2 seconds, the amplitude has decreased to 5 cm. Find a function that models the position of the spring.

The oscillation has a period of $\frac{1}{4}$ second, so $B = \frac{2\pi}{\frac{1}{4}} = 8\pi$. Since the spring will oscillate on either side of the natural length, the midline will be at 30 cm. It begins at the furthest distance from the wall, suggesting a cosine model. Together, this gives $f(t) = A(t)\cos(8\pi t) + 30$.

For the amplitude function, we notice that the amplitude starts at 10 cm, and decreases to 5 cm after 2 seconds. This gives two points (0, 10) and (2, 5) that must be satisfied by an exponential function: $A(0) = 10$ and $A(2) = 5$. Since the function is exponential, we can use the form $A(t) = ab^t$. Substituting the first point, $10 = ab^0$, so $a = 10$. Substituting in the second point,

$5 = 10b^2$ Divide by 10

$\frac{1}{2} = b^2$ Take the square root

$b = \sqrt{\frac{1}{2}} \approx 0.707$

This gives an amplitude function of $A(t) = 10(0.707)^t$. Combining this with the oscillation,

$f(t) = 10(0.707)^t \cos(8\pi t) + 30$

Try it Now

2. A certain stock started at a high value of $7 per share, oscillating monthly above and below the average value, with the oscillation decreasing by 2% per year. However, the average value started at $4 per share and has grown linearly by 50 cents per year.
   a. Find a formula for the midline and the amplitude.
   b. Find a function S(t) that models the value of the stock after t years.

Example 6

In AM (Amplitude Modulated) radio, a carrier wave with a high frequency is used to transmit music or other signals by applying the to-be-transmitted signal as the amplitude of the carrier signal. A musical note with frequency 110 Hz (Hertz = cycles per second) is to be carried on a wave with frequency of 2 KHz (KiloHertz = thousands of cycles per second). If the musical wave has an amplitude of 3, write a function describing the broadcast wave.

The carrier wave, with a frequency of 2000 cycles per second, would have period $\dfrac{1}{2000}$ of a second, giving an equation of the form $\sin(4000\pi t)$. Our choice of a sine function here was arbitrary – it would have worked just was well to use a cosine.

The musical tone, with a frequency of 110 cycles per second, would have a period of $\dfrac{1}{110}$ of a second. With an amplitude of 3, this would correspond to a function of the form $3\sin(220\pi t)$. Again our choice of using a sine function is arbitrary.

The musical wave is acting as the amplitude of the carrier wave, so we will multiply the musical tone's function by the carrier wave function, resulting in the function
$$f(t) = 3\sin(220\pi t)\sin(4000\pi t)$$

Important Topics of This Section
Changing midline
Changing amplitude
Linear Changes
Exponential Changes
Damped Harmonic Motion

Try it Now Answers

1. The midline follows the path of the quadratic $x^2 - 1$ and the amplitude is a constant value of 8.

2.
$$m(t) = 4 + 0.5t$$
$$A(t) = 7(0.98)^t$$
$$S(t) = 7(0.98)^t \cos(24\pi t) + 4 + 0.5t$$

Section 7.4 Exercises

Find a possible formula for the trigonometric function whose values are given in the following tables.

1.

x	0	3	6	9	12	15	18
y	-4	-1	2	-1	-4	-1	2

2.

x	0	2	4	6	8	10	12
y	5	1	-3	1	5	1	-3

3. The displacement $h(t)$, in centimeters, of a mass suspended by a spring is modeled by the function $h(t) = 8\sin(6\pi t)$, where t is measured in seconds. Find the amplitude, period, and frequency of this displacement.

4. The displacement $h(t)$, in centimeters, of a mass suspended by a spring is modeled by the function $h(t) = 11\sin(12\pi t)$, where t is measured in seconds. Find the amplitude, period, and frequency of this displacement.

5. A population of rabbits oscillates 19 above and below average during the year, reaching the lowest value in January. The average population starts at 650 rabbits and increases by 160 each year. Find a function that models the population, P, in terms of the months since January, t.

6. A population of deer oscillates 15 above and below average during the year, reaching the lowest value in January. The average population starts at 800 deer and increases by 110 each year. Find a function that models the population, P, in terms of the months since January, t.

7. A population of muskrats oscillates 33 above and below average during the year, reaching the lowest value in January. The average population starts at 900 muskrats and increases by 7% each month. Find a function that models the population, P, in terms of the months since January, t.

8. A population of fish oscillates 40 above and below average during the year, reaching the lowest value in January. The average population starts at 800 fish and increases by 4% each month. Find a function that models the population, P, in terms of the months since January, t.

9. A spring is attached to the ceiling and pulled 10 cm down from equilibrium and released. The amplitude decreases by 15% each second. The spring oscillates 18 times each second. Find a function that models the distance, D, the end of the spring is below equilibrium in terms of seconds, t, since the spring was released.

10. A spring is attached to the ceiling and pulled 7 cm down from equilibrium and released. The amplitude decreases by 11% each second. The spring oscillates 20 times each second. Find a function that models the distance, D, the end of the spring is below equilibrium in terms of seconds, t, since the spring was released.

11. A spring is attached to the ceiling and pulled 17 cm down from equilibrium and released. After 3 seconds the amplitude has decreased to 13 cm. The spring oscillates 14 times each second. Find a function that models the distance, D the end of the spring is below equilibrium in terms of seconds, t, since the spring was released.

12. A spring is attached to the ceiling and pulled 19 cm down from equilibrium and released. After 4 seconds the amplitude has decreased to 14 cm. The spring oscillates 13 times each second. Find a function that models the distance, D the end of the spring is below equilibrium in terms of seconds, t, since the spring was released.

Match each equation form with one of the graphs.

13. a. $ab^x + \sin(5x)$ b. $\sin(5x) + mx + b$

14. a. $ab^x \sin(5x)$ b. $(mx + b)\sin(5x)$

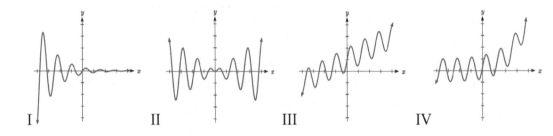

I   II   III   IV

Find a function of the form $y = ab^x + c\sin\left(\dfrac{\pi}{2}x\right)$ that fits the data given.

15.

x	0	1	2	3
y	6	29	96	379

16.

x	0	1	2	3
y	6	34	150	746

Find a function of the form $y = a\sin\left(\dfrac{\pi}{2}x\right) + m + bx$ that fits the data given.

17.

x	0	1	2	3
y	7	6	11	16

18.

x	0	1	2	3
y	-2	6	4	2

Find a function of the form $y = ab^x \cos\left(\dfrac{\pi}{2}x\right) + c$ that fits the data given.

19.

x	0	1	2	3
y	11	3	1	3

20.

x	0	1	2	3
y	4	1	-11	1

Chapter 8: Further Applications of Trigonometry

In this chapter, we will explore additional applications of trigonometry. We will begin with an extension of the right triangle trigonometry we explored in Chapter 5 to situations involving non-right triangles. We will explore the polar coordinate system and parametric equations as new ways of describing curves in the plane. In the process, we will introduce vectors and an alternative way of writing complex numbers, two important mathematical tools we use when analyzing and modeling the world around us.

Section 8.1 Non-Right Triangles: Laws of Sines and Cosines

Although right triangles allow us to solve many applications, it is more common to find scenarios where the triangle we are interested in does not have a right angle.

Two radar stations located 20 miles apart both detect a UFO located between them. The angle of elevation measured by the first station is 35 degrees. The angle of elevation measured by the second station is 15 degrees. What is the altitude of the UFO?

We see that the triangle formed by the UFO and the two stations is not a right triangle. Of course, in any triangle we could draw an **altitude**, a perpendicular line from one vertex to the opposite side, forming two right triangles, but it would be nice to have methods for working directly with non-right triangles. In this section, we will expand upon the right triangle trigonometry we learned in Chapter 5, and adapt it to non-right triangles.

Law of Sines

Given an arbitrary non-right triangle, we can drop an altitude, which we temporarily label h, to create two right triangles.

Using the right triangle relationships,
$$\sin(\alpha) = \frac{h}{b} \text{ and } \sin(\beta) = \frac{h}{a}.$$

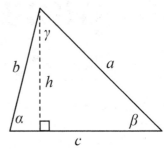

Solving both equations for h, we get $b\sin(\alpha) = h$ and $a\sin(\beta) = h$. Since the h is the same in both equations, we establish $b\sin(\alpha) = a\sin(\beta)$. Dividing both sides by ab, we conclude that
$$\frac{\sin(\alpha)}{a} = \frac{\sin(\beta)}{b}$$

Had we drawn the altitude to be perpendicular to side b or a, we could similarly establish
$$\frac{\sin(\alpha)}{a} = \frac{\sin(\gamma)}{c} \text{ and } \frac{\sin(\beta)}{b} = \frac{\sin(\gamma)}{c}$$

Collectively, these relationships are called the Law of Sines.

Law of Sines

Given a triangle with angles and sides opposite labeled as shown, the ratio of sine of angle to length of the opposite side will always be equal, or, symbolically,

$$\frac{\sin(\alpha)}{a} = \frac{\sin(\beta)}{b} = \frac{\sin(\gamma)}{c}$$

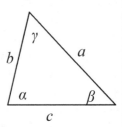

For clarity, we call side a the corresponding side of angle α.

Similarly, we call angle α, the corresponding angle of side a.

Likewise for side b and angle β, and for side c and angle γ.

When we use the law of sines, we use any pair of ratios as an equation. In the most straightforward case, we know two angles and one of the corresponding sides.

Example 1

In the triangle shown here, solve for the unknown sides and angle.

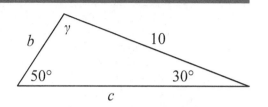

Solving for the unknown angle is relatively easy, since the three angles must add to 180 degrees.

From this, we can determine that
$\gamma = 180° - 50° - 30° = 100°$.

To find an unknown side, we need to know the corresponding angle, and we also need another known ratio.

Since we know the angle 50° and its corresponding side, we can use this for one of the two ratios. To look for side b, we would use its corresponding angle, 30°.

$$\frac{\sin(50°)}{10} = \frac{\sin(30°)}{b}$$ Multiply both sides by b

$$b\frac{\sin(50°)}{10} = \sin(30°)$$ Divide, or multiply by the reciprocal, to solve for b

$$b = \sin(30°)\frac{10}{\sin(50°)} \approx 6.527$$

Similarly, to solve for side c, we set up the equation
$$\frac{\sin(50°)}{10} = \frac{\sin(100°)}{c}$$
$$c = \sin(100°)\frac{10}{\sin(50°)} \approx 12.856$$

Example 2

Find the elevation of the UFO from the beginning of the section.

To find the elevation of the UFO, we first find the distance from one station to the UFO, such as the side a in the picture, then use right triangle relationships to find the height of the UFO, h.

Since the angles in the triangle add to 180 degrees, the unknown angle of the triangle must be $180° - 15° - 35° = 130°$. This angle is opposite the side of length 20, allowing us to set up a Law of Sines relationship:

$$\frac{\sin(130°)}{20} = \frac{\sin(35°)}{a}$$ Multiply by a

$$a\frac{\sin(130°)}{20} = \sin(35°)$$ Divide, or multiply by the reciprocal, to solve for a

$$a = \frac{20\sin(35°)}{\sin(130°)} \approx 14.975$$ Simplify

The distance from one station to the UFO is about 15 miles. Now that we know a, we can use right triangle relationships to solve for h.

$$\sin(15°) = \frac{opposite}{hypotenuse} = \frac{h}{a} = \frac{h}{14.975}$$ Solve for h

$$h = 14.975\sin(15°) \approx 3.876$$

The UFO is at an altitude of 3.876 miles.

In addition to solving triangles in which two angles are known, the law of sines can be used to solve for an angle when two sides and one corresponding angle are known.

Example 3

In the triangle shown here, solve for the unknown sides and angles.

In choosing which pair of ratios from the Law of Sines to use, we always want to pick a pair where we know three of the four pieces of information in the equation. In this case, we know the angle 85° and its corresponding side, so we will use that ratio. Since our only other known information is the side with length 9, we will use that side and solve for its corresponding angle.

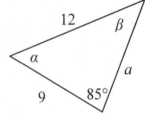

$$\frac{\sin(85°)}{12} = \frac{\sin(\beta)}{9}$$ Isolate the unknown

$$\frac{9\sin(85°)}{12} = \sin(\beta)$$ Use the inverse sine to find a first solution

Remember when we use the inverse function that there are two possible answers.

$$\beta = \sin^{-1}\left(\frac{9\sin(85°)}{12}\right) \approx 48.3438°$$ By symmetry we find the second possible solution

$$\beta = 180° - 48.3438° = 131.6562°$$

In this second case, if $\beta \approx 132°$, then α would be $\alpha = 180° - 85° - 132° = -37°$, which doesn't make sense, so the only possibility for this triangle is $\beta = 48.3438°$.

With a second angle, we can now easily find the third angle, since the angles must add to 180°, so $\alpha = 180° - 85° - 48.3438° = 46.6562°$.

Now that we know α, we can proceed as in earlier examples to find the unknown side a.

$$\frac{\sin(85°)}{12} = \frac{\sin(46.6562°)}{a}$$

$$a = \frac{12\sin(46.6562°)}{\sin(85°)} \approx 8.7603$$

Notice that in the problem above, when we use Law of Sines to solve for an unknown angle, there can be two possible solutions. This is called the **ambiguous case**, and can arise when we know two sides and a non-included angle. In the ambiguous case we may find that a particular set of given information can lead to 2, 1 or no solution at all. However, when an accurate picture of the triangle or suitable context is available, we can determine which angle is desired.

Try it Now

1. Given $\alpha = 80°, a = 120$, and $b = 121$, find the corresponding and missing side and angles. If there is more than one possible solution, show both.

Example 4

Find all possible triangles if one side has length 4 opposite an angle of 50° and a second side has length 10.

Using the given information, we can look for the angle opposite the side of length 10.

$$\frac{\sin(50°)}{4} = \frac{\sin(\alpha)}{10}$$

$$\sin(\alpha) = \frac{10\sin(50°)}{4} \approx 1.915$$

Since the range of the sine function is [-1, 1], it is impossible for the sine value to be 1.915. There are no triangles that can be drawn with the provided dimensions.

Example 5

Find all possible triangles if one side has length 6 opposite an angle of 50° and a second side has length 4.

Using the given information, we can look for the angle opposite the side of length 4.

$$\frac{\sin(50°)}{6} = \frac{\sin(\alpha)}{4}$$

$\sin(\alpha) = \dfrac{4\sin(50°)}{6} \approx 0.511$ Use the inverse to find one solution

$\alpha = \sin^{-1}(0.511) \approx 30.710°$ By symmetry there is a second possible solution

$\alpha = 180° - 30.710° = 149.290°$

If we use the angle $30.710°$, the third angle would be $180° - 50° - 30.710° = 99.290°$. We can then use Law of Sines again to find the third side.

$$\frac{\sin(50°)}{6} = \frac{\sin(99.290°)}{c}$$ Solve for c

$c = 7.730$

If we used the angle $\alpha = 149.290°$, the third angle would be $180° - 50° - 149.290° = -19.29°$, which is impossible, so the previous triangle is the only possible one.

Try it Now

2. Given $\alpha = 80°, a = 100$, and $b = 10$ find the missing side and angles. If there is more than one possible solution, show both.

Law of Cosines

Suppose a boat leaves port, travels 10 miles, turns 20 degrees, and travels another 8 miles. How far from port is the boat?

Unfortunately, while the Law of Sines lets us address many non-right triangle cases, it does not allow us to address triangles where the one known angle is included between two known sides, which means it is not a corresponding angle for a known side. For this, we need another tool.

Given an arbitrary non-right triangle, we can drop an altitude, which we temporarily label h, to create two right triangles. We will divide the base b into two pieces, one of which we will temporarily label x.

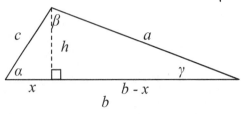

From this picture, we can establish the right triangle relationship

$\cos(\alpha) = \dfrac{x}{c}$, or equivalently, $x = c\cos(\alpha)$

Using the Pythagorean Theorem, we can establish

$(b-x)^2 + h^2 = a^2$ and $x^2 + h^2 = c^2$

Both of these equations can be solved for h^2

$h^2 = a^2 - (b-x)^2$ and $h^2 = c^2 - x^2$

Since the left side of each equation is h^2, the right sides must be equal

$c^2 - x^2 = a^2 - (b-x)^2$	Multiply out the right
$c^2 - x^2 = a^2 - (b^2 - 2bx + x^2)$	Simplify
$c^2 - x^2 = a^2 - b^2 + 2bx - x^2$	
$c^2 = a^2 - b^2 + 2bx$	Isolate a^2
$a^2 = c^2 + b^2 - 2bx$	Substitute in $c\cos(\alpha) = x$ from above
$a^2 = c^2 + b^2 - 2bc\cos(\alpha)$	

This result is called the Law of Cosines. Depending upon which side we dropped the altitude down from, we could have established this relationship using any of the angles. The important thing to note is that the right side of the equation involves an angle and the sides adjacent to that angle – the left side of the equation involves the side opposite that angle.

Law of Cosines

Given a triangle with angles and opposite sides labeled as shown,

$a^2 = c^2 + b^2 - 2bc\cos(\alpha)$

$b^2 = a^2 + c^2 - 2ac\cos(\beta)$

$c^2 = a^2 + b^2 - 2ab\cos(\gamma)$

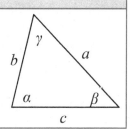

Notice that if one of the angles of the triangle is 90 degrees, $\cos(90°) = 0$, so the formula
$c^2 = a^2 + b^2 - 2ab\cos(90°)$ Simplifies to
$c^2 = a^2 + b^2$

You should recognize this as the Pythagorean Theorem. Indeed, the Law of Cosines is sometimes called the **Generalized Pythagorean Theorem**, since it extends the Pythagorean Theorem to non-right triangles.

Example 6

Returning to our question from earlier, suppose a boat leaves port, travels 10 miles, turns 20 degrees, and travels another 8 miles. How far from port is the boat?

The boat turned 20 degrees, so the obtuse angle of the non-right triangle shown in the picture is the supplemental angle, 180° - 20° = 160°.

With this, we can utilize the Law of Cosines to find the missing side of the obtuse triangle – the distance from the boat to port.

$x^2 = 8^2 + 10^2 - 2(8)(10)\cos(160°)$ Evaluate the cosine and simplify

$x^2 = 314.3508$ Square root both sides

$x = \sqrt{314.3508} = 17.730$

The boat is 17.73 miles from port.

Example 7

Find the unknown side and angles of this triangle.

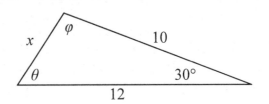

Notice that we don't have both pieces of any side/angle pair, so the Law of Sines would not work with this triangle.

Since we have the angle included between the two known sides, we can turn to Law of Cosines.

Since the left side of any of the Law of Cosines equations involves the side opposite the known angle, the left side in this situation will involve the side x. The other two sides can be used in either order.

$x^2 = 10^2 + 12^2 - 2(10)(12)\cos(30°)$ Evaluate the cosine

$x^2 = 10^2 + 12^2 - 2(10)(12)\dfrac{\sqrt{3}}{2}$ Simplify

$x^2 = 244 - 120\sqrt{3}$ Take the square root

$x = \sqrt{244 - 120\sqrt{3}} \approx 6.013$

Now that we know an angle and its corresponding side, we can use the Law of Sines to fill in the remaining angles of the triangle. Solving for angle θ,

$$\left|\frac{\sin(30°)}{6.013} = \frac{\sin(\theta)}{10}\right.$$

$$\sin(\theta) = \frac{10\sin(30°)}{6.013} \qquad\qquad \text{Use the inverse sine}$$

$$\theta = \sin^{-1}\left(\frac{10\sin(30°)}{6.013}\right) \approx 56.256°$$

The other possibility for θ would be $\theta = 180° - 56.256° = 123.744°$. In the original picture, θ is an acute angle, so $123.744°$ doesn't make sense if we assume the picture is drawn to scale.

Proceeding with $\theta = 56.256°$, we can then find the third angle of the triangle:
$\varphi = 180° - 30° - 56.256° = 93.744°$.

In addition to solving for the missing side opposite one known angle, the Law of Cosines allows us to find the angles of a triangle when we know all three sides.

Example 8

Solve for the angle α in the triangle shown.

Using the Law of Cosines,

$20^2 = 18^2 + 25^2 - 2(18)(25)\cos(\alpha) \qquad$ Simplify

$400 = 949 - 900\cos(\alpha)$

$-549 = -900\cos(\alpha)$

$\dfrac{-549}{-900} = \cos(\alpha)$

$\alpha = \cos^{-1}\left(\dfrac{-549}{-900}\right) \approx 52.410°$

Try it Now

3. Given $\alpha = 25°, b = 10$, and $c = 20$ find the missing side and angles.

Notice that since the inverse cosine can return any angle between 0 and 180 degrees, there will not be any ambiguous cases when using Law of Cosines to find an angle.

Example 9

On many cell phones with GPS, an approximate location can be given before the GPS signal is received. This is done by a process called triangulation, which works by using the distance from two known points. Suppose there are two cell phone towers within range of you, located 6000 feet apart along a straight highway that runs east to west, and you know you are north of the highway. Based on the signal delay, it can be determined you are 5050 feet from the first tower, and 2420 feet from the second. Determine your position north and east of the first tower, and determine how far you are from the highway.

For simplicity, we start by drawing a picture and labeling our given information. Using the Law of Cosines, we can solve for the angle θ.

$2420^2 = 6000^2 + 5050^2 - 2(5050)(6000)\cos(\theta)$

$5856400 = 61501500 - 60600000\cos(\theta)$

$-554646100 = -60600000\cos(\theta)$

$\cos(\theta) = \dfrac{-554646100}{-60600000} = 0.9183$

$\theta = \cos^{-1}(0.9183) = 23.328°$

Using this angle, we could then use right triangles to find the position of the cell phone relative to the western tower.

$\cos(23.328°) = \dfrac{x}{5050}$

$x = 5050\cos(23.328°) \approx 4637.2 \text{ feet}$

$\sin(23.328°) = \dfrac{y}{5050}$

$y = 5050\sin(23.328°) \approx 1999.8 \text{ feet}$

You are 5050 ft from the tower and $23.328°$ north of east (or, equivalently, $66.672°$ east of north). Specifically, you are about 4637 feet east and 2000 feet north of the first tower.

Note that if you didn't know whether you were north or south of the towers, our calculations would have given two possible locations, one north of the highway and one south. To resolve this ambiguity in real world situations, locating a position using triangulation requires a signal from a third tower.

Example 10

To measure the height of a hill, a woman measures the angle of elevation to the top of the hill to be 24 degrees. She then moves back 200 feet and measures the angle of elevation to be 22 degrees. Find the height of the hill.

As with many problems of this nature, it will be helpful to draw a picture.

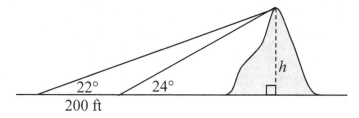

Notice there are three triangles formed here – the right triangle including the height h and the 22 degree angle, the right triangle including the height h and the 24 degree angle, and the (non-right) obtuse triangle including the 200 ft side. Since this is the triangle we have the most information for, we will begin with it. It may seem odd to work with this triangle since it does not include the desired side h, but we don't have enough information to work with either of the right triangles yet.

We can find the obtuse angle of the triangle, since it and the angle of 24 degrees complete a straight line – a 180 degree angle. The obtuse angle must be $180° - 24° = 156°$. From this, we can determine that the third angle is $2°$. We know one side is 200 feet, and its corresponding angle is $2°$, so by introducing a temporary variable x for one of the other sides (as shown below), we can use Law of Sines to solve for this length x.

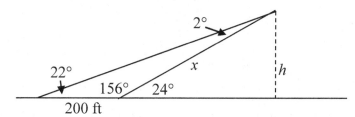

$$\frac{x}{\sin(22°)} = \frac{200}{\sin(2°)} \qquad \text{Setting up the Law of Sines}$$

$$x = \sin(22°)\frac{200}{\sin(2°)} \qquad \text{isolating the } x \text{ value}$$

$$x = 2146.77 \text{ ft}$$

Now that we know x, we can use right triangle properties to solve for h.

$$\sin(24°) = \frac{\text{opposite}}{\text{hypotenuse}} = \frac{h}{x} = \frac{h}{2146.77}$$

$$h = 2146.77\sin(24°) = 873.17 \text{ ft}. \quad \text{The hill is 873 feet high.}$$

Important Topics of This Section
Law of Sines
Solving for sides
Solving for angles
Ambiguous case, 0, 1 or 2 solutions
Law of Cosines
Solving for sides
Solving for angles
Generalized Pythagorean Theorem

Try it Now Answers

1. $\dfrac{\sin(80°)}{120} = \dfrac{\sin(\beta)}{121}$

                      $\beta = 83.2°$ $\beta = 96.8°$

     1st possible solution $\gamma = 16.8°$ 2nd solution $\gamma = 3.2°$

                      $c = 35.2$ $c = 6.9$

     If we were given a picture of the triangle it may be possible to eliminate one of these

2. $\dfrac{\sin(80°)}{120} = \dfrac{\sin(\beta)}{10}$. $\beta = 5.65°$ or $\beta = 174.35°$; only the first is reasonable.

     $\gamma = 180° - 5.65° - 80° = 94.35°$

     $\dfrac{\sin(80°)}{120} = \dfrac{\sin(94.35°)}{c}$

     $\beta = 5.65°, \gamma = 94.35°, c = 101.25$

3. $a^2 = 10^2 + 20^2 - 2(10)(20)\cos(25°)$. $a = 11.725$

     $\dfrac{\sin(25°)}{11.725} = \dfrac{\sin(\beta)}{10}$. $\beta = 21.1°$ or $\beta = 158.9°$;

          only the first is reasonable since $25° + 158.9°$ would exceed $180°$.

     $\gamma = 180° - 21.1° - 25° = 133.9°$

     $\beta = 21.1°, \quad \gamma = 133.9°, \quad a = 11.725$

Section 8.1 Exercises

Solve for the unknown sides and angles of the triangles shown.

1.

2.

3.

4.

5.

6.

7.

8.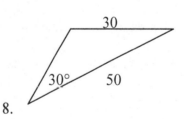

Assume α is opposite side a, β is opposite side b, and γ is opposite side c. Solve each triangle for the unknown sides and angles if possible. If there is more than one possible solution, give both.

9. $\alpha = 43°, \gamma = 69°, b = 20$

10. $\alpha = 35°, \gamma = 73°, b = 19$

11. $\alpha = 119°, a = 26, b = 14$

12. $\gamma = 113°, b = 10, c = 32$

13. $\beta = 50°, a = 105, b = 45$

14. $\beta = 67°, a = 49, b = 38$

15. $\alpha = 43.1°, a = 184.2, b = 242.8$

16. $\alpha = 36.6°, a = 186.2, b = 242.2$

Solve for the unknown sides and angles of the triangles shown.

17.

18.

19.

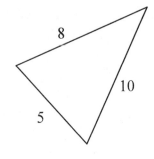

20.

Assume α is opposite side a, β is opposite side b, and γ is opposite side c. Solve each triangle for the unknown sides and angles if possible. If there is more than one possible solution, give both.

21. $\gamma = 41.2°, a = 2.49, b = 3.13$

22. $\beta = 58.7°, a = 10.6, c = 15.7$

23. $\alpha = 120°, b = 6, c = 7$

24. $\gamma = 115°, a = 18, b = 23$

25. Find the area of a triangle with sides of length 18, 21, and 32.

26. Find the area of a triangle with sides of length 20, 26, and 37.

27. To find the distance across a small lake, a surveyor has taken the measurements shown. Find the distance across the lake.

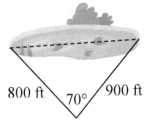

28. To find the distance between two cities, a satellite calculates the distances and angle shown (*not to scale*). Find the distance between the cities.

29. To determine how far a boat is from shore, two radar
stations 500 feet apart determine the angles out to the
boat, as shown. Find the distance of the boat from the
station A, and the distance of the boat from shore.

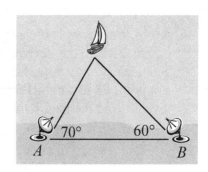

30. The path of a satellite orbiting the earth causes it to
pass directly over two tracking stations A and B,
which are 69 mi apart. When the satellite is on one
side of the two stations, the angles of elevation at A
and B are measured to be 86.2° and 83.9°,
respectively. How far is the satellite from station A
and how high is the satellite above the ground?

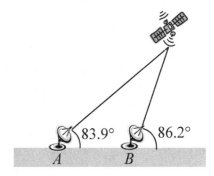

31. A communications tower is located at the top of
a steep hill, as shown. The angle of inclination of
the hill is 67°. A guy-wire is to be attached to the
top of the tower and to the ground, 165 m
downhill from the base of the tower. The angle
formed by the guy-wire and the hill is 16°. Find
the length of the cable required for the guy wire.

32. The roof of a house is at a 20° angle. An 8 foot
solar panel is to be mounted on the roof, and
should be angled 38° relative to the horizontal
for optimal results. How long does the vertical
support holding up the back of the panel need to
be?

33. A 127 foot tower is located on a hill that is
inclined 38° to the horizontal. A guy-wire is to
be attached to the top of the tower and anchored
at a point 64 feet downhill from the base of the
tower. Find the length of wire needed.

34. A 113 foot tower is located on a hill that is inclined 34° to the horizontal. A guy-wire is to be attached to the top of the tower and anchored at a point 98 feet uphill from the base of the tower. Find the length of wire needed.

35. A pilot is flying over a straight highway. He determines the angles of depression to two mileposts, 6.6 km apart, to be 37° and 44°, as shown in the figure. Find the distance of the plane from point A, and the elevation of the plane.

36. A pilot is flying over a straight highway. He determines the angles of depression to two mileposts, 4.3 km apart, to be 32° and 56°, as shown in the figure. Find the distance of the plane from point A, and the elevation of the plane.

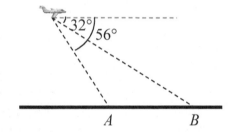

37. To estimate the height of a building, two students find the angle of elevation from a point (at ground level) down the street from the building to the top of the building is 39°. From a point that is 300 feet closer to the building, the angle of elevation (at ground level) to the top of the building is 50°. If we assume that the street is level, use this information to estimate the height of the building.

38. To estimate the height of a building, two students find the angle of elevation from a point (at ground level) down the street from the building to the top of the building is 35°. From a point that is 300 feet closer to the building, the angle of elevation (at ground level) to the top of the building is 53°. If we assume that the street is level, use this information to estimate the height of the building.

39. A pilot flies in a straight path for 1 hour 30 min. She then makes a course correction, heading 10 degrees to the right of her original course, and flies 2 hours in the new direction. If she maintains a constant speed of 680 miles per hour, how far is she from her starting position?

40. Two planes leave the same airport at the same time. One flies at 20 degrees east of north at 500 miles per hour. The second flies at 30 east of south at 600 miles per hour. How far apart are the planes after 2 hours?

41. The four sequential sides of a quadrilateral have lengths 4.5 cm, 7.9 cm, 9.4 cm, and 12.9 cm. The angle between the two smallest sides is 117°. What is the area of this quadrilateral?

42. The four sequential sides of a quadrilateral have lengths 5.7 cm, 7.2 cm, 9.4 cm, and 12.8 cm. The angle between the two smallest sides is 106°. What is the area of this quadrilateral?

43. Three circles with radii 6, 7, and 8, all touch as shown. Find the shaded area bounded by the three circles.

44. A rectangle is inscribed in a circle of radius 10 cm as shown. Find the shaded area, inside the circle but outside the rectangle.

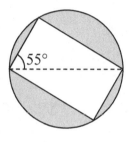

Section 8.2 Polar Coordinates

The coordinate system we are most familiar with is called the Cartesian coordinate system, a rectangular plane divided into four quadrants by horizontal and vertical axes.

In earlier chapters, we often found the Cartesian coordinates of a point on a circle at a given angle from the positive horizontal axis. Sometimes that angle, along with the point's distance from the origin, provides a more useful way of describing the point's location than conventional Cartesian coordinates.

Polar Coordinates

Polar coordinates of a point consist of an ordered pair, (r, θ), where r is the distance from the point to the origin, and θ is the angle measured in standard position.

Notice that if we were to "grid" the plane for polar coordinates, it would look like the graph to the right, with circles at incremental radii, and rays drawn at incremental angles.

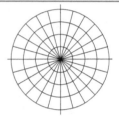

Example 1

Plot the polar point $\left(3, \dfrac{5\pi}{6}\right)$.

This point will be a distance of 3 from the origin, at an angle of $\dfrac{5\pi}{6}$. Plotting this,

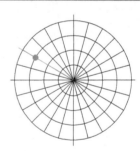

Example 2

Plot the polar point $\left(-2, \dfrac{\pi}{4}\right)$.

Typically we use positive r values, but occasionally we run into cases where r is negative. On a regular number line, we measure positive values to the right and negative values to the left. We will plot this point similarly. To start, we rotate to an angle of $\dfrac{\pi}{4}$.

Moving this direction, into the first quadrant, would be positive r values. For negative r values, we move the opposite direction, into the third quadrant. Plotting this:

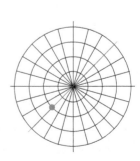

Note the resulting point is the same as the polar point $\left(2, \dfrac{5\pi}{4}\right)$. In fact, any Cartesian

point can be represented by an infinite number of different polar coordinates by adding or subtracting full rotations to these points. For example, same point could also be

represented as $\left(2, \dfrac{13\pi}{4}\right)$.

Try it Now
1. Plot the following points given in polar coordinates and label them.

   a. $A = \left(3, \dfrac{\pi}{6}\right)$ b. $B = \left(-2, \dfrac{\pi}{3}\right)$ c. $C = \left(4, \dfrac{3\pi}{4}\right)$

Converting Points

To convert between polar coordinates and Cartesian coordinates, we recall the relationships we developed back in Chapter 5.

Converting Between Polar and Cartesian Coordinates

To convert between polar (r, θ) and Cartesian (x, y) coordinates, we use the relationships

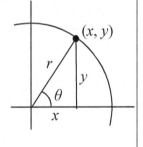

$$\cos(\theta) = \dfrac{x}{r} \qquad x = r\cos(\theta)$$

$$\sin(\theta) = \dfrac{y}{r} \qquad y = r\sin(\theta)$$

$$\tan(\theta) = \dfrac{y}{x} \qquad x^2 + y^2 = r^2$$

From these relationship and our knowledge of the unit circle, if $r = 1$ and $\theta = \dfrac{\pi}{3}$, the

polar coordinates would be $(r, \theta) = \left(1, \dfrac{\pi}{3}\right)$, and the corresponding Cartesian coordinates

$(x, y) = \left(\dfrac{1}{2}, \dfrac{\sqrt{3}}{2}\right)$.

Remembering your unit circle values will come in very handy as you convert between Cartesian and polar coordinates.

Example 3

Find the Cartesian coordinates of a point with polar coordinates $(r, \theta) = \left(5, \frac{2\pi}{3}\right)$.

To find the x and y coordinates of the point,

$$x = r\cos(\theta) = 5\cos\left(\frac{2\pi}{3}\right) = 5\left(-\frac{1}{2}\right) = -\frac{5}{2}$$

$$y = r\sin(\theta) = 5\sin\left(\frac{2\pi}{3}\right) = 5\left(\frac{\sqrt{3}}{2}\right) = \frac{5\sqrt{3}}{2}$$

The Cartesian coordinates are $\left(-\frac{5}{2}, \frac{5\sqrt{3}}{2}\right)$.

Example 4

Find the polar coordinates of the point with Cartesian coordinates $(-3, -4)$.

We begin by finding the distance r using the Pythagorean relationship $x^2 + y^2 = r^2$

$$(-3)^2 + (-4)^2 = r^2$$
$$9 + 16 = r^2$$
$$r^2 = 25$$
$$r = 5$$

Now that we know the radius, we can find the angle using any of the three trig relationships. Keep in mind that any of the relationships will produce two solutions on the circle, and we need to consider the quadrant to determine which solution to accept. Using the cosine, for example:

$$\cos(\theta) = \frac{x}{r} = \frac{-3}{5}$$

$$\theta = \cos^{-1}\left(\frac{-3}{5}\right) \approx 2.214 \qquad \text{By symmetry, there is a second possibility at}$$

$$\theta = 2\pi - 2.214 = 4.069$$

Since the point (-3, -4) is located in the 3rd quadrant, we can determine that the second angle is the one we need. The polar coordinates of this point are $(r, \theta) = (5, 4.069)$.

Try it Now

2. Convert the following.
  a. Convert polar coordinates $(r, \theta) = (2, \pi)$ to (x, y).
  b. Convert Cartesian coordinates $(x, y) = (0, -4)$ to (r, θ).

Polar Equations

Just as a Cartesian equation like $y = x^2$ describes a relationship between x and y values on a Cartesian grid, a polar equation can be written describing a relationship between r and θ values on the polar grid.

Example 5

Sketch a graph of the polar equation $r = \theta$.

The equation $r = \theta$ describes all the points for which the radius r is equal to the angle. To visualize this relationship, we can create a table of values.

θ	0	$\pi/4$	$\pi/2$	$3\pi/4$	π	$5\pi/4$	$3\pi/2$	$7\pi/4$	2π
r	0	$\pi/4$	$\pi/2$	$3\pi/4$	π	$5\pi/4$	$3\pi/2$	$7\pi/4$	2π

We can plot these points on the plane, and then sketch a curve that fits the points. The resulting graph is a spiral.

Notice that the resulting graph cannot be the result of a function of the form $y = f(x)$, as it does not pass the vertical line test, even though it resulted from a function giving r in terms of θ.

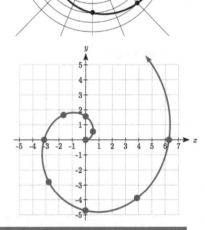

Although it is nice to see polar equations on polar grids, it is more common for polar graphs to be graphed on the Cartesian coordinate system, and so, the remainder of the polar equations will be graphed accordingly.

The spiral graph above on a Cartesian grid is shown here.

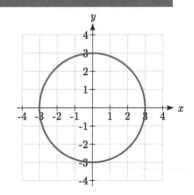

Example 6

Sketch a graph of the polar equation $r = 3$.

Recall that when a variable does not show up in the equation, it is saying that it does not matter what value that variable has; the output for the equation will remain the same. For example, the Cartesian equation $y = 3$ describes all the points where $y = 3$, no matter what the x values are, producing a horizontal line.

Likewise, this polar equation is describing all the points at a distance of 3 from the origin, no matter what the angle is, producing the graph of a **circle.**
The normal settings on graphing calculators and software graph on the Cartesian coordinate system with *y* being a function of *x*, where the graphing utility asks for *f(x)*, or simply *y* =.

To graph polar equations, you may need to change the mode of your calculator to Polar. You will know you have been successful in changing the mode if you now have *r* as a function of *θ*, where the graphing utility asks for *r(θ)*, or simply *r* =.

Example 7

Sketch a graph of the polar equation $r = 4\cos(\theta)$, and find an interval on which it completes one cycle.

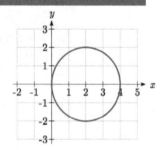

While we could again create a table, plot the corresponding points, and connect the dots, we can also turn to technology to directly graph it. Using technology, we produce the graph shown here, a circle passing through the origin.

Since this graph appears to close a loop and repeat itself, we might ask what interval of *θ* values yields the entire graph. At $\theta = 0$, $r = 4\cos(0) = 4$, yielding the point (4, 0). We want the next *θ* value when the graph returns to the point (4, 0). Solving for when $x = 4$ is equivalent to solving $r\cos(\theta) = 4$.

$r\cos(\theta) = 4$	Substituting the equation for *r* gives
$4\cos(\theta)\cos(\theta) = 4$	Dividing by 4 and simplifying
$\cos^2(\theta) = 1$	This has solutions when
$\cos(\theta) = 1$ or $\cos(\theta) = -1$	Solving these gives solutions
$\theta = 0$ or $\theta = \pi$	

This shows us at 0 radians we are at the point (0, 4), and again at *π* radians we are at the point (0, 4) having finished one complete revolution.

The interval $0 \le \theta < \pi$ yields one complete iteration of the circle.

Try it Now

3. Sketch a graph of the polar equation $r = 3\sin(\theta)$, and find an interval on which it completes one cycle.

The last few examples have all been circles. Next, we will consider two other "named" polar equations, **limaçons** and **roses**.

Example 8

Sketch a graph of the polar equation $r = 4\sin(\theta) + 2$. What interval of θ values corresponds to the inner loop?

This type of graph is called a **limaçon**.
Using technology, we can draw the graph. The inner loop begins and ends at the origin, where $r = 0$. We can solve for the θ values for which $r = 0$.

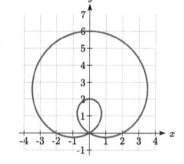

$0 = 4\sin(\theta) + 2$

$-2 = 4\sin(\theta)$

$\sin(\theta) = -\dfrac{1}{2}$

$\theta = \dfrac{7\pi}{6}$ or $\theta = \dfrac{11\pi}{6}$

This tells us that $r = 0$, so the graph passes through the origin, twice on the interval $[0, 2\pi)$.

The inner loop arises from the interval $\dfrac{7\pi}{6} \leq \theta \leq \dfrac{11\pi}{6}$.

This corresponds to where the function $r = 4\sin(\theta) + 2$ takes on negative values, as we could see if we graphed the function in the $r\theta$ plane.

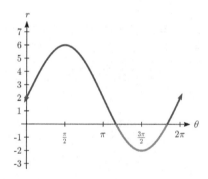

Example 9

Sketch a graph of the polar equation $r = \cos(3\theta)$. What interval of θ values describes one small loop of the graph?

This type of graph is called a 3 leaf **rose**.

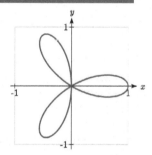

We can use technology to produce a graph. The interval $[0, \pi)$ yields one cycle of this function. As with the last problem, we can note that there is an interval on which one loop of this graph begins and ends at the origin, where $r = 0$. Solving for θ,

$0 = \cos(3\theta)$ Substitute $u = 3\theta$

$0 = \cos(u)$

$u = \dfrac{\pi}{2}$ or $u = \dfrac{3\pi}{2}$ or $u = \dfrac{5\pi}{2}$

Undo the substitution,

$$3\theta = \frac{\pi}{2} \quad \text{or} \quad 3\theta = \frac{3\pi}{2} \quad \text{or} \quad 3\theta = \frac{5\pi}{2}$$

$$\theta = \frac{\pi}{6} \quad \text{or} \quad \theta = \frac{\pi}{2} \quad \text{or} \quad \theta = \frac{5\pi}{6}$$

There are 3 solutions on $0 \le \theta < \pi$ which correspond to the 3 times the graph returns to the origin, but the first two solutions we solved for above are enough to conclude that one loop corresponds to the interval $\frac{\pi}{6} \le \theta < \frac{\pi}{2}$.

If we wanted to get an idea of how the computer drew this graph, consider when $\theta = 0$. $r = \cos(3\theta) = \cos(0) = 1$, so the graph starts at $(1,0)$. As we found above, at $\theta = \frac{\pi}{6}$ and $\theta = \frac{\pi}{2}$, the graph is at the origin. Looking at the equation, notice that any angle in between $\frac{\pi}{6}$ and $\frac{\pi}{2}$, for example at $\theta = \frac{\pi}{3}$, produces a negative r: $r = \cos\left(3 \cdot \frac{\pi}{3}\right) = \cos(\pi) = -1$.

θ	r	x	y
0	1	1	0
$\frac{\pi}{6}$	0	0	0
$\frac{\pi}{3}$	-1	$-\frac{1}{2}$	$-\frac{\sqrt{3}}{2}$
$\frac{\pi}{2}$	0	0	0

Notice that with a negative r value and an angle with terminal side in the first quadrant, the corresponding Cartesian point would be in the third quadrant. Since $r = \cos(3\theta)$ is negative on $\frac{\pi}{6} \le \theta < \frac{\pi}{2}$, this interval corresponds to the loop of the graph in the third quadrant.

Try it Now

4. Sketch a graph of the polar equation $r = \sin(2\theta)$. Would you call this function a **limaçon** or a **rose**?

Converting Equations

While many polar equations cannot be expressed nicely in Cartesian form (and vice versa), it can be beneficial to convert between the two forms, when possible. To do this we use the same relationships we used to convert points between coordinate systems.

Example 10

Rewrite the Cartesian equation $x^2 + y^2 = 6y$ as a polar equation.

We wish to eliminate x and y from the equation and introduce r and θ. Ideally, we would like to write the equation with r isolated, if possible, which represents r as a function of θ.

$x^2 + y^2 = 6y$	Remembering $x^2 + y^2 = r^2$ we substitute
$r^2 = 6y$	$y = r\sin(\theta)$ and so we substitute again
$r^2 = 6r\sin(\theta)$	Subtract $6r\sin(\theta)$ from both sides
$r^2 - 6r\sin(\theta) = 0$	Factor
$r(r - 6\sin(\theta)) = 0$	Use the zero factor theorem
$r = 6\sin(\theta)$ or $r = 0$	Since $r = 0$ is only a point, we reject that solution.

The solution $r = 6\sin(\theta)$ is fairly similar to the one we graphed in Example 7. In fact, this equation describes a circle with bottom at the origin and top at the point (0, 6).

Example 11

Rewrite the Cartesian equation $y = 3x + 2$ as a polar equation.

$y = 3x + 2$	Use $y = r\sin(\theta)$ and $x = r\cos(\theta)$
$r\sin(\theta) = 3r\cos(\theta) + 2$	Move all terms with r to one side
$r\sin(\theta) - 3r\cos(\theta) = 2$	Factor out r
$r(\sin(\theta) - 3\cos(\theta)) = 2$	Divide

$$r = \frac{2}{\sin(\theta) - 3\cos(\theta)}$$

In this case, the polar equation is more unwieldy than the Cartesian equation, but there are still times when this equation might be useful.

Example 12

Rewrite the polar equation $r = \dfrac{3}{1 - 2\cos(\theta)}$ as a Cartesian equation.

We want to eliminate θ and r and introduce x and y. It is usually easiest to start by clearing the fraction and looking to substitute values that will eliminate θ.

$r = \dfrac{3}{1 - 2\cos(\theta)}$ Clear the fraction

$r\left(1 - 2\cos(\theta)\right) = 3$ Use $\cos(\theta) = \dfrac{x}{r}$ to eliminate θ

$r\left(1 - 2\dfrac{x}{r}\right) = 3$ Distribute and simplify

$r - 2x = 3$ Isolate the r

$r = 3 + 2x$ Square both sides

$r^2 = (3 + 2x)^2$ Use $x^2 + y^2 = r^2$

$x^2 + y^2 = (3 + 2x)^2$

When our entire equation has been changed from r and θ to x and y we can stop unless asked to solve for y or simplify.

In this example, if desired, the right side of the equation could be expanded and the equation simplified further. However, the equation cannot be written as a function in Cartesian form.

Try it Now

5. a. Rewrite the Cartesian equation in polar form: $y = \pm\sqrt{3 - x^2}$

   b. Rewrite the polar equation in Cartesian form: $r = 2\sin(\theta)$

Example 13

Rewrite the polar equation $r = \sin(2\theta)$ in Cartesian form.

$r = \sin(2\theta)$ Use the double angle identity for sine

$r = 2\sin(\theta)\cos(\theta)$ Use $\cos(\theta) = \dfrac{x}{r}$ and $\sin(\theta) = \dfrac{y}{r}$

$r = 2 \cdot \dfrac{x}{r} \cdot \dfrac{y}{r}$ Simplify

$r = \dfrac{2xy}{r^2}$ Multiply by r^2

$r^3 = 2xy$ Since $x^2 + y^2 = r^2$, $r = \sqrt{x^2 + y^2}$

$\left(\sqrt{x^2 + y^2}\right)^3 = 2xy$

This equation could also be written as

$$\left(x^2 + y^2\right)^{3/2} = 2xy \quad \text{or} \quad x^2 + y^2 = \left(2xy\right)^{2/3}$$

Important Topics of This Section
Cartesian coordinate system
Polar coordinate system
Plotting points in polar coordinates
Converting coordinates between systems
Polar equations: Spirals, circles, limaçons and roses
Converting equations between systems

Try it Now Answers

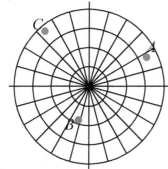

1.

2. a. $(r, \theta) = (2, \pi)$ converts to $(x, y) = (2\cos(\pi), 2\sin(\pi)) = (-2, 0)$

   b. $(x, y) = (0, -4)$ converts to $(r, \theta) = \left(4, \dfrac{3\pi}{2}\right) or \left(-4, \dfrac{\pi}{2}\right)$

3. $3\sin(\theta) = 0$ at $\theta = 0$ and $\theta = \pi$.
   It completes one cycle on the interval $0 \le \theta < \pi$.

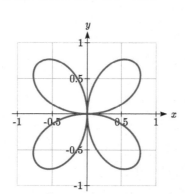

4. This is a 4-leaf rose.

5. a. $y = \pm\sqrt{3 - x^2}$ can be rewritten as $x^2 + y^2 = 3$, and becomes $r = \sqrt{3}$

b. $r = 2\sin(\theta)$. $r = 2\dfrac{y}{r}$. $r^2 = 2y$. $x^2 + y^2 = 2y$

Section 8.2 Exercises

Convert the given polar coordinates to Cartesian coordinates.

1. $\left(7, \dfrac{7\pi}{6}\right)$ 2. $\left(6, \dfrac{3\pi}{4}\right)$ 3. $\left(4, \dfrac{7\pi}{4}\right)$ 4. $\left(9, \dfrac{4\pi}{3}\right)$

5. $\left(6, -\dfrac{\pi}{4}\right)$ 6. $\left(12, -\dfrac{\pi}{3}\right)$ 7. $\left(3, \dfrac{\pi}{2}\right)$ 8. $(5, \pi)$

9. $\left(-3, \dfrac{\pi}{6}\right)$ 10. $\left(-2, \dfrac{2\pi}{3}\right)$ 11. $(3, 2)$ 12. $(7, 1)$

Convert the given Cartesian coordinates to polar coordinates.

13. $(4, 2)$ 14. $(8, 8)$ 15. $(-4, 6)$ 16. $(-5, 1)$

17. $(3, -5)$ 18. $(6, -5)$ 19. $(-10, -13)$ 20. $(-4, -7)$

Convert the given Cartesian equation to a polar equation.

21. $x = 3$ 22. $y = 4$ 23. $y = 4x^2$ 24. $y = 2x^4$

25. $x^2 + y^2 = 4y$ 26. $x^2 + y^2 = 3x$ 27. $x^2 - y^2 = x$ 28. $x^2 - y^2 = 3y$

Convert the given polar equation to a Cartesian equation.

29. $r = 3\sin(\theta)$ 30. $r = 4\cos(\theta)$

31. $r = \dfrac{4}{\sin(\theta) + 7\cos(\theta)}$ 32. $r = \dfrac{6}{\cos(\theta) + 3\sin(\theta)}$

33. $r = 2\sec(\theta)$ 34. $r = 3\csc(\theta)$

35. $r = \sqrt{r\cos(\theta) + 2}$ 36. $r^2 = 4\sec(\theta)\csc(\theta)$

Match each equation with one of the graphs shown.

37. $r = 2 + 2\cos(\theta)$ 38. $r = 2 + 2\sin(\theta)$ 39. $r = 4 + 3\cos(\theta)$

40. $r = 3 + 4\cos(\theta)$ 41. $r = 5$ 42. $r = 2\sin(\theta)$

A

B

C

D

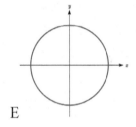

E

F

Match each equation with one of the graphs shown.

43. $r = \log(\theta)$ 44. $r = \theta\cos(\theta)$ 45. $r = \cos\left(\dfrac{\theta}{2}\right)$

46. $r = \sin(\theta)\cos^2(\theta)$ 47. $r = 1 + 2\sin(3\theta)$ 48. $r = 1 + \sin(2\theta)$

A

B

C

D

E

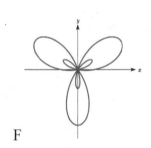

F

Sketch a graph of the polar equation.

49. $r = 3\cos(\theta)$ 　　　　　50. $r = 4\sin(\theta)$ 　　　　　51. $r = 3\sin(2\theta)$

52. $r = 4\sin(4\theta)$ 　　　　　53. $r = 5\sin(3\theta)$ 　　　　　54. $r = 4\sin(5\theta)$

55. $r = 3\cos(2\theta)$ 　　　　56. $r = 4\cos(4\theta)$ 　　　　57. $r = 2 + 2\cos(\theta)$

58. $r = 3 + 3\sin(\theta)$ 　　　59. $r = 1 + 3\sin(\theta)$ 　　　60. $r = 2 + 4\cos(\theta)$

61. $r = 2\theta$ 　　　　　　　62. $r = \dfrac{1}{\theta}$

63. $r = 3 + \sec(\theta)$, a conchoid 　　　　　64. $r = \dfrac{1}{\sqrt{\theta}}$, a lituus[1]

65. $r = 2\sin(\theta)\tan(\theta)$, a cissoid 　　　　66. $r = 2\sqrt{1 - \sin^2(\theta)}$, a hippopede

---

[1] This curve was the inspiration for the artwork featured on the cover of this book.

Section 8.3 Polar Form of Complex Numbers

From previous classes, you may have encountered "imaginary numbers" – the square roots of negative numbers – and, more generally, complex numbers which are the sum of a real number and an imaginary number. While these are useful for expressing the solutions to quadratic equations, they have much richer applications in electrical engineering, signal analysis, and other fields. Most of these more advanced applications rely on properties that arise from looking at complex numbers from the perspective of polar coordinates.

We will begin with a review of the definition of complex numbers.

Imaginary Number i

The most basic complex number is i, defined to be $i = \sqrt{-1}$, commonly called an **imaginary number**. Any real multiple of i is also an imaginary number.

Example 1

Simplify $\sqrt{-9}$.

We can separate $\sqrt{-9}$ as $\sqrt{9}\sqrt{-1}$. We can take the square root of 9, and write the square root of -1 as i.
$$\sqrt{-9} = \sqrt{9}\sqrt{-1} = 3i$$

A complex number is the sum of a real number and an imaginary number.

Complex Number

A **complex number** is a number $z = a + bi$, where a and b are real numbers
a is the real part of the complex number
b is the imaginary part of the complex number
$i = \sqrt{-1}$

Plotting a complex number

We can plot real numbers on a number line. For example, if we wanted to show the number 3, we plot a point:

To plot a complex number like $3 - 4i$, we need more than just a number line since there are two components to the number. To plot this number, we need two number lines, crossed to form a complex plane.

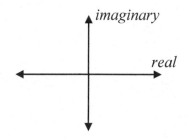

Complex Plane

In the **complex plane**, the horizontal axis is the real axis and the vertical axis is the imaginary axis.

Example 2

Plot the number $3 - 4i$ on the complex plane.

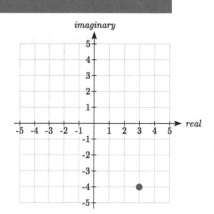

The real part of this number is 3, and the imaginary part is -4. To plot this, we draw a point 3 units to the right of the origin in the horizontal direction and 4 units down in the vertical direction.

Because this is analogous to the Cartesian coordinate system for plotting points, we can think about plotting our complex number $z = a + bi$ as if we were plotting the point (a, b) in Cartesian coordinates. Sometimes people write complex numbers as $z = x + yi$ to highlight this relation.

Arithmetic on Complex Numbers

Before we dive into the more complicated uses of complex numbers, let's make sure we remember the basic arithmetic involved. To add or subtract complex numbers, we simply add the like terms, combining the real parts and combining the imaginary parts.

Example 3

Add $3 - 4i$ and $2 + 5i$.

Adding $(3 - 4i) + (2 + 5i)$, we add the real parts and the imaginary parts
$3 + 2 - 4i + 5i$
$5 + i$

Try it Now
1. Subtract $2 + 5i$ from $3 - 4i$.

We can also multiply and divide complex numbers.

Example 4

Multiply: $4(2+5i)$.

To multiply the complex number by a real number, we simply distribute as we would when multiplying polynomials.

$$4(2+5i) \qquad \text{Distribute}$$
$$=4\cdot 2+4\cdot 5i \qquad \text{Simplify}$$
$$=8+20i$$

Example 5

Multiply: $(2-3i)(1+4i)$.

To multiply two complex numbers, we expand the product as we would with polynomials (the process commonly called FOIL – "first outer inner last").

$$(2-3i)(1+4i) \qquad \text{Expand}$$
$$=2+8i-3i-12i^2 \qquad \text{Since } i=\sqrt{-1},\ i^2=-1$$
$$=2+8i-3i-12(-1) \qquad \text{Simplify}$$
$$=14+5i$$

Example 6

Divide $\dfrac{(2+5i)}{(4-i)}$.

To divide two complex numbers, we have to devise a way to write this as a complex number with a real part and an imaginary part.

We start this process by eliminating the complex number in the denominator. To do this, we multiply the numerator and denominator by a special complex number so that the result in the denominator is a real number. The number we need to multiply by is called the **complex conjugate**, in which the sign of the imaginary part is changed. Here, $4+i$ is the complex conjugate of $4-i$. Of course, obeying our algebraic rules, we must multiply by $4+i$ on both the top and bottom.

$$\frac{(2+5i)}{(4-i)} \cdot \frac{(4+i)}{(4+i)}$$

In the numerator,

$(2+5i)(4+i)$ Expand

$= 8+20i+2i+5i^2$ Since $i = \sqrt{-1}$, $i^2 = -1$
$= 8+20i+2i+5(-1)$ Simplify
$= 3+22i$

Multiplying the denominator
$(4-i)(4+i)$ Expand

$(16-4i+4i-i^2)$ Since $i = \sqrt{-1}$, $i^2 = -1$
$(16-(-1))$
$= 17$

Combining this we get $\dfrac{3+22i}{17} = \dfrac{3}{17} + \dfrac{22i}{17}$

Try it Now
2. Multiply $3-4i$ and $2+3i$.

With the interpretation of complex numbers as points in a plane, which can be related to the Cartesian coordinate system, you might be starting to guess our next step – to refer to this point not by its horizontal and vertical components, but using its polar location, given by the distance from the origin and an angle.

Polar Form of Complex Numbers

Remember, because the complex plane is analogous to the Cartesian plane that we can think of a complex number $z = x + yi$ as analogous to the Cartesian point (x, y) and recall how we converted from (x, y) to polar (r, θ) coordinates in the last section.

Bringing in all of our old rules we remember the following:

$\cos(\theta) = \dfrac{x}{r}$ $x = r\cos(\theta)$

$\sin(\theta) = \dfrac{y}{r}$ $y = r\sin(\theta)$

$\tan(\theta) = \dfrac{y}{x}$ $x^2 + y^2 = r^2$

With this in mind, we can write $z = x + yi = r\cos(\theta) + ir\sin(\theta)$.

Example 7

Express the complex number $4i$ using polar coordinates.

On the complex plane, the number $4i$ is a distance of 4 from the origin at an angle of $\dfrac{\pi}{2}$, so $4i = 4\cos\left(\dfrac{\pi}{2}\right) + i4\sin\left(\dfrac{\pi}{2}\right)$

Note that the real part of this complex number is 0.

In the 18th century, Leonhard Euler demonstrated a relationship between exponential and trigonometric functions that allows the use of complex numbers to greatly simplify some trigonometric calculations. While the proof is beyond the scope of this class, you will likely see it in a later calculus class.

Polar Form of a Complex Number and Euler's Formula

The **polar form of a complex number** is $z = r\cos(\theta) + ir\sin(\theta)$.

An alternate form, which will be the primary one used, is $z = re^{i\theta}$

Euler's Formula states $re^{i\theta} = r\cos(\theta) + ir\sin(\theta)$

Similar to plotting a point in the polar coordinate system we need r and θ to find the polar form of a complex number.

Example 8

Find the polar form of the complex number -8.

Treating this is a complex number, we can write it as -8+0i.

Plotted in the complex plane, the number -8 is on the negative horizontal axis, a distance of 8 from the origin at an angle of π from the positive horizontal axis.

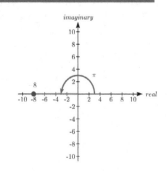

The polar form of the number -8 is $8e^{i\pi}$.

Plugging $r = 8$ and $\theta = \pi$ back into Euler's formula, we have:
$8e^{i\pi} = 8\cos(\pi) + 8i\sin(\pi) = -8 + 0i = -8$ as desired.

Example 9

Find the polar form of $-4+4i$.

On the complex plane, this complex number would correspond to the point (-4, 4) on a Cartesian plane. We can find the distance r and angle θ as we did in the last section.

$$r^2 = x^2 + y^2$$
$$r^2 = (-4)^2 + 4^2$$
$$r = \sqrt{32} = 4\sqrt{2}$$

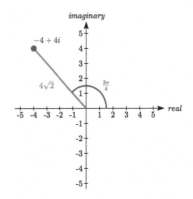

To find θ, we can use $\cos(\theta) = \dfrac{x}{r}$

$$\cos(\theta) = \frac{-4}{4\sqrt{2}} = -\frac{\sqrt{2}}{2}$$

This is one of known cosine values, and since the point is in the second quadrant, we can conclude that $\theta = \dfrac{3\pi}{4}$.

The polar form of this complex number is $4\sqrt{2}e^{\frac{3\pi}{4}i}$.

Example 10

Find the polar form of $-3-5i$.

On the complex plane, this complex number would correspond to the point (-3, -5) on a Cartesian plane. First, we find r.
$$r^2 = x^2 + y^2$$
$$r^2 = (-3)^2 + (-5)^2$$
$$r = \sqrt{34}$$

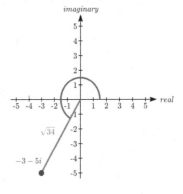

To find θ, we might use $\tan(\theta) = \dfrac{y}{x}$

$$\tan(\theta) = \frac{-5}{-3}$$

$$\theta = \tan^{-1}\left(\frac{5}{3}\right) = 1.0304$$

This angle is in the wrong quadrant, so we need to find a second solution. For tangent, we can find that by adding π.
$$\theta = 1.0304 + \pi = 4.1720$$

The polar form of this complex number is $\sqrt{34}e^{4.1720i}$.

Try it Now

3. Write $\sqrt{3} + i$ in polar form.

Example 11

Write $3e^{\frac{\pi}{6}i}$ in complex $a + bi$ form.

$$3e^{\frac{\pi}{6}i} = 3\cos\left(\frac{\pi}{6}\right) + i3\sin\left(\frac{\pi}{6}\right)$$ Evaluate the trig functions

$$= 3 \cdot \frac{\sqrt{3}}{2} + i3 \cdot \frac{1}{2}$$ Simplify

$$= \frac{3\sqrt{3}}{2} + i\frac{3}{2}$$

The polar form of a complex number provides a powerful way to compute powers and roots of complex numbers by using exponent rules you learned in algebra. To compute a power of a complex number, we:

1) Convert to polar form
2) Raise to the power, using exponent rules to simplify
3) Convert back to $a + bi$ form, if needed

Example 12

Evaluate $(-4 + 4i)^6$.

While we could multiply this number by itself five times, that would be very tedious. To compute this more efficiently, we can utilize the polar form of the complex number.

In an earlier example, we found that $-4 + 4i = 4\sqrt{2}e^{\frac{3\pi}{4}i}$. Using this,

$$(-4 + 4i)^6$$ Write the complex number in polar form

$$= \left(4\sqrt{2}e^{\frac{3\pi}{4}i}\right)^6$$ Utilize the exponent rule $(ab)^m = a^m b^m$

$$= \left(4\sqrt{2}\right)^6\left(e^{\frac{3\pi}{4}i}\right)^6$$ On the second factor, use the rule $(a^m)^n = a^{mn}$

$$= \left(4\sqrt{2}\right)^6 e^{\frac{3\pi}{4}i \cdot 6}$$ Simplify

$$= 32768e^{\frac{9\pi}{2}i}$$

At this point, we have found the power as a complex number in polar form. If we want the answer in standard $a + bi$ form, we can utilize Euler's formula.

$$32768e^{\frac{9\pi}{2}i} = 32768\cos\left(\frac{9\pi}{2}\right) + i32768\sin\left(\frac{9\pi}{2}\right)$$

Since $\dfrac{9\pi}{2}$ is coterminal with $\dfrac{\pi}{2}$, we can use our special angle knowledge to evaluate the sine and cosine.

$$32768\cos\left(\frac{9\pi}{2}\right) + i32768\sin\left(\frac{9\pi}{2}\right) = 32768(0) + i32768(1) = 32768i$$

We have found that $(-4+4i)^6 = 32768i$.

The result of the process can be summarized by DeMoivre's Theorem. This is a shorthand to using exponent rules.

DeMoivre's Theorem
If $z = r\left(\cos(\theta) + i\sin(\theta)\right)$, then for any integer n, $z^n = r^n\left(\cos(n\theta) + i\sin(n\theta)\right)$

We omit the proof, but note we can easily verify it holds in one case using Example 12:

$$(-4+4i)^6 = \left(4\sqrt{2}\right)^6\left(\cos\left(6\cdot\frac{3\pi}{4}\right) + i\sin\left(6\cdot\frac{3\pi}{4}\right)\right) = 32768\left(\cos\left(\frac{9\pi}{2}\right) + i\sin\left(\frac{9\pi}{2}\right)\right) = 32768i$$

Example 13

Evaluate $\sqrt{9i}$.

To evaluate the square root of a complex number, we can first note that the square root is the same as having an exponent of $\dfrac{1}{2}$: $\sqrt{9i} = (9i)^{1/2}$.

To evaluate the power, we first write the complex number in polar form. Since $9i$ has no real part, we know that this value would be plotted along the vertical axis, a distance of 9 from the origin at an angle of $\dfrac{\pi}{2}$. This gives the polar form: $9i = 9e^{\frac{\pi}{2}i}$.

$$\sqrt{9i} = (9i)^{1/2} \qquad\qquad \text{Use the polar form}$$

$$= \left(9 e^{\frac{\pi}{2} i} \right)^{1/2}$$ Use exponent rules to simplify

$$= 9^{1/2} \left(e^{\frac{\pi}{2} i} \right)^{1/2}$$

$$= 9^{1/2} e^{\frac{\pi}{2} i \cdot \frac{1}{2}}$$ Simplify

$$= 3 e^{\frac{\pi}{4} i}$$ Rewrite using Euler's formula if desired

$$= 3 \cos \left(\frac{\pi}{4} \right) + i 3 \sin \left(\frac{\pi}{4} \right)$$ Evaluate the sine and cosine

$$= 3 \frac{\sqrt{2}}{2} + i 3 \frac{\sqrt{2}}{2}$$

Using the polar form, we were able to find a square root of a complex number.

$$\sqrt{9i} = \frac{3\sqrt{2}}{2} + \frac{3\sqrt{2}}{2} i$$

Alternatively, using DeMoivre's Theorem we could write

$$\left(9 e^{\frac{\pi}{2} i} \right)^{1/2} = 9^{1/2} \left(\cos \left(\frac{1}{2} \cdot \frac{\pi}{2} \right) + i \sin \left(\frac{1}{2} \cdot \frac{\pi}{2} \right) \right) = 3 \left(\cos \left(\frac{\pi}{4} \right) + i \sin \left(\frac{\pi}{4} \right) \right) \text{ and simplify}$$

Try it Now

4. Evaluate $\left(\sqrt{3} + i \right)^6$ using polar form.

You may remember that equations like $x^2 = 4$ have two solutions, 2 and -2 in this case, though the square root $\sqrt{4}$ only gives one of those solutions. Likewise, the square root we found in Example 11 is only one of two complex numbers whose square is $9i$. Similarly, the equation $z^3 = 8$ would have three solutions where only one is given by the cube root. In this case, however, only one of those solutions, $z = 2$, is a real value. To find the others, we can use the fact that complex numbers have multiple representations in polar form.

Example 14

Find all complex solutions to $z^3 = 8$.

Since we are trying to solve $z^3 = 8$, we can solve for z as $z = 8^{1/3}$. Certainly one of these solutions is the basic cube root, giving $z = 2$. To find others, we can turn to the polar representation of 8.

Since 8 is a real number, is would sit in the complex plane on the horizontal axis at an angle of 0, giving the polar form $8e^{0i}$. Taking the 1/3 power of this gives the real solution:

$$\left(8e^{0i}\right)^{1/3} = 8^{1/3}\left(e^{0i}\right)^{1/3} = 2e^0 = 2\cos(0) + i2\sin(0) = 2$$

However, since the angle 2π is coterminal with the angle of 0, we could also represent the number 8 as $8e^{2\pi i}$. Taking the 1/3 power of this gives a first complex solution:

$$\left(8e^{2\pi i}\right)^{1/3} = 8^{1/3}\left(e^{2\pi i}\right)^{1/3} = 2e^{\frac{2\pi}{3}i} = 2\cos\left(\frac{2\pi}{3}\right) + i2\sin\left(\frac{2\pi}{3}\right) = 2\left(-\frac{1}{2}\right) + i2\left(\frac{\sqrt{3}}{2}\right) = -1 + \sqrt{3}i$$

For the third root, we use the angle of 4π, which is also coterminal with an angle of 0.

$$\left(8e^{4\pi i}\right)^{1/3} = 8^{1/3}\left(e^{4\pi i}\right)^{1/3} = 2e^{\frac{4\pi}{3}i} = 2\cos\left(\frac{4\pi}{3}\right) + i2\sin\left(\frac{4\pi}{3}\right) = 2\left(-\frac{1}{2}\right) + i2\left(-\frac{\sqrt{3}}{2}\right) = -1 - \sqrt{3}i$$

Altogether, we found all three complex solutions to $z^3 = 8$,
$$z = 2, \quad -1 + \sqrt{3}i, \quad -1 - \sqrt{3}i$$

Graphed, these three numbers would be equally spaced on a circle about the origin at a radius of 2.

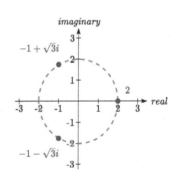

Important Topics of This Section
Complex numbers
Imaginary numbers
Plotting points in the complex coordinate system
Basic operations with complex numbers
Euler's Formula
DeMoivre's Theorem
Finding complex solutions to equations

Try it Now Answers

1. $(3-4i)-(2+5i)=1-9i$

2. $(3-4i)(2+3i)=18+i$

3. $\sqrt{3}+i$ would correspond with the point $\left(\sqrt{3},1\right)$ in the first quadrant.

   $r=\sqrt{\sqrt{3}^{2}+1^{2}}=\sqrt{4}=2$

   $\sin\left(\theta\right)=\dfrac{1}{2}$, so $\theta=\dfrac{\pi}{6}$

   $\sqrt{3}+i$ in polar form is $2e^{i\pi/6}$

4. $\left(\sqrt{3}+i\right)^{6}=\left(2e^{i\pi/6}\right)^{6}=2^{6}e^{i\pi}=64\cos(\pi)+i64\sin(\pi)=-64$

Section 8.3 Exercises

Simplify each expression to a single complex number.

1. $\sqrt{-9}$

2. $\sqrt{-16}$

3. $\sqrt{-6}\sqrt{-24}$

4. $\sqrt{-3}\sqrt{-75}$

5. $\dfrac{2+\sqrt{-12}}{2}$

6. $\dfrac{4+\sqrt{-20}}{2}$

Simplify each expression to a single complex number.

7. $(3+2i)+(5-3i)$

8. $(-2-4i)+(1+6i)$

9. $(-5+3i)-(6-i)$

10. $(2-3i)-(3+2i)$

11. $(2+3i)(4i)$

12. $(5-2i)(3i)$

13. $(6-2i)(5)$

14. $(-2+4i)(8)$

15. $(2+3i)(4-i)$

16. $(-1+2i)(-2+3i)$

17. $(4-2i)(4+2i)$

18. $(3+4i)(3-4i)$

19. $\dfrac{3+4i}{2}$

20. $\dfrac{6-2i}{3}$

21. $\dfrac{-5+3i}{2i}$

22. $\dfrac{6+4i}{i}$

23. $\dfrac{2-3i}{4+3i}$

24. $\dfrac{3+4i}{2-i}$

25. i^6

26. i^{11}

27. i^{17}

28. i^{24}

Rewrite each complex number from polar form into $a+bi$ form.

29. $3e^{2i}$

30. $4e^{4i}$

31. $6e^{\frac{\pi}{6}i}$

32. $8e^{\frac{\pi}{3}i}$

33. $3e^{\frac{5\pi}{4}i}$

34. $5e^{\frac{7\pi}{4}i}$

Rewrite each complex number into polar $re^{i\theta}$ form.

35. 6

36. -8

37. $-4i$

38. $6i$

39. $2+2i$

40. $4+4i$

41. $-3+3i$

42. $-4-4i$

43. $5+3i$

44. $4+7i$

45. $-3+i$

46. $-2+3i$

47. $-1-4i$

48. $-3-6i$

49. $5-i$

50. $1-3i$

Compute each of the following, leaving the result in polar $re^{i\theta}$ form.

51. $\left(3e^{\frac{\pi}{6}i}\right)\left(2e^{\frac{\pi}{4}i}\right)$

52. $\left(2e^{\frac{2\pi}{3}i}\right)\left(4e^{\frac{5\pi}{3}i}\right)$

53. $\dfrac{6e^{\frac{3\pi}{4}i}}{3e^{\frac{\pi}{6}i}}$

54. $\dfrac{24e^{\frac{4\pi}{3}i}}{6e^{\frac{\pi}{2}i}}$

55. $\left(2e^{\frac{\pi}{4}i}\right)^{10}$

56. $\left(3e^{\frac{\pi}{6}i}\right)^{4}$

57. $\sqrt{16e^{\frac{2\pi}{3}i}}$

58. $\sqrt{9e^{\frac{3\pi}{2}i}}$

Compute each of the following, simplifying the result into $a+bi$ form.

59. $(2+2i)^{8}$

60. $(4+4i)^{6}$

61. $\sqrt{-3+3i}$

62. $\sqrt{-4-4i}$

63. $\sqrt[3]{5+3i}$

64. $\sqrt[4]{4+7i}$

Solve each of the following equations for all complex solutions.

65. $z^{5}=2$

66. $z^{7}=3$

67. $z^{6}=1$

68. $z^{8}=1$

Section 8.4 Vectors

A woman leaves home, walks 3 miles north, then 2 miles southeast. How far is she from home, and in which direction would she need to walk to return home? How far has she walked by the time she gets home?

This question may seem familiar – indeed we did a similar problem with a boat in the first section of this chapter. In that section, we solved the problem using triangles. In this section, we will investigate another way to approach the problem using vectors, a geometric entity that indicates both a distance and a direction. We will begin our investigation using a purely geometric view of vectors.

A Geometric View of Vectors

Vector
A **vector** is an object that has both a length and a direction.   Geometrically, a vector can be represented by an arrow that has a fixed length and indicates a direction. If, starting at the point A, a vector, which means "carrier" in Latin, moves toward point B, we write $\overrightarrow{AB}$ to represent the vector.   A vector may also be indicated using a single letter in boldface type, like **u**, or by capping the letter representing the vector with an arrow, like $\vec{u}$.

Example 1

Draw a vector that represents the movement from the point $P(-1, 2)$ to the point $Q(3,3)$

By drawing an arrow from the first point to the second, we can construct a vector $\overrightarrow{PQ}$.

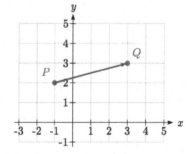

Try it Now

1. Draw a vector, $\vec{v}$, that travels from the origin to the point (3, 5).

Using this geometric representation of vectors, we can visualize the addition and scaling of vectors.

To add vectors, we envision a sum of two movements. To find $\vec{u} + \vec{v}$, we first draw the vector $\vec{u}$, then from the end of $\vec{u}$ we drawn the vector $\vec{v}$. This corresponds to the notion that first we move along the first vector, and then from that end position we move along the second vector. The sum $\vec{u} + \vec{v}$ is the new vector that travels directly from the beginning of $\vec{u}$ to the end of $\vec{v}$ in a straight path.

Adding Vectors Geometrically

To **add vectors geometrically**, draw $\vec{v}$ starting from the end of $\vec{u}$. The sum $\vec{u} + \vec{v}$ is the vector from the beginning of $\vec{u}$ to the end of $\vec{v}$.

Example 2

Given the two vectors shown below, draw $\vec{u} + \vec{v}$

We draw $\vec{v}$ starting from the end of $\vec{u}$, then draw in the sum $\vec{u} + \vec{v}$ from the beginning of $\vec{u}$ to the end of $\vec{v}$.

Notice that path of the walking woman from the beginning of the section could be visualized as the sum of two vectors. The resulting sum vector would indicate her end position relative to home.

Although vectors can exist anywhere in the plane, if we put the starting point at the origin it is easy to understand its size and direction relative to other vectors.

To scale vectors by a constant, such as $3\vec{u}$, we can imagine adding $\vec{u} + \vec{u} + \vec{u}$. The result will be a vector three times as long in the same direction as the original vector. If we were to scale a vector by a negative number, such as $-\vec{u}$, we can envision this as the opposite of $\vec{u}$; the vector so that $\vec{u} + (-\vec{u})$ returns us to the starting point. This vector $-\vec{u}$ would point in the opposite direction as $\vec{u}$ but have the same length.

Another way to think about scaling a vector is to maintain its direction and multiply its length by a constant, so that $3\vec{u}$ would point in the same direction but will be 3 times as long.

Scaling a Vector Geometrically

To **geometrically scale a vector** by a constant, scale the length of the vector by the constant.

Scaling a vector by a negative constant will reverse the direction of the vector.

Example 3

Given the vector shown, draw $3\vec{u}$, $-\vec{u}$, and $-2\vec{u}$.

The vector $3\vec{u}$ will be three times as long. The vector $-\vec{u}$ will have the same length but point in the opposite direction. The vector $-2\vec{u}$ will point in the opposite direction and be twice as long.

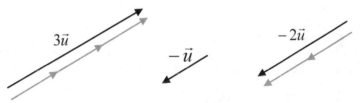

By combining scaling and addition, we can find the difference between vectors geometrically as well, since $\vec{u} - \vec{v} = \vec{u} + (-\vec{v})$.

Example 4

Given the vectors shown, draw $\vec{u} - \vec{v}$

From the end of $\vec{u}$ we draw $-\vec{v}$, then draw in the result.

Notice that the sum and difference of two vectors are the two diagonals of a parallelogram with the vectors $\vec{u}$ and $\vec{v}$ as edges.

Try it Now
2.   Using vector $\vec{v}$ from Try it Now #1, draw $-2\vec{v}$.

Component Form of Vectors

While the geometric interpretation of vectors gives us an intuitive understanding of vectors, it does not provide us a convenient way to do calculations.
For that, we need a handy way to represent vectors. Since a vector involves a length and direction, it would be logical to want to represent a vector using a length and an angle θ, usually measured from standard position.

Magnitude and Direction of a Vector
A vector $\vec{u}$ can be described by its magnitude, or length, $\|\vec{u}\|$, and an angle θ.
A vector with length 1 is called **unit vector**.

While this is very reasonable, and a common way to describe vectors, it is often more convenient for calculations to represent a vector by horizontal and vertical components.

Component Form of a Vector
The **component form** of a vector represents the vector using two components. $\vec{u} = \langle x, y \rangle$ indicates the vector represents a displacement of x units horizontally and y units vertically. 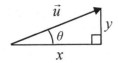 Notice how we can see the magnitude of the vector as the length of the hypotenuse of a right triangle, or in polar form as the radius, r.

Alternate Notation for Vector Components

Sometimes you may see vectors written as the combination of unit vectors $\vec{i}$ and $\vec{j}$, where $\vec{i}$ points the right and $\vec{j}$ points up. In other words, $\vec{i} = \langle 1,0 \rangle$ and $\vec{j} = \langle 0,1 \rangle$.

In this notation, the vector $\vec{u} = \langle 3,-4 \rangle$ would be written as $\vec{u} = 3\vec{i} - 4\vec{j}$ since both indicate a displacement of 3 units to the right, and 4 units down.

While it can be convenient to think of the vector $\vec{u} = \langle x, y \rangle$ as an arrow from the origin to the point (x, y), be sure to remember that most vectors can be situated anywhere in the plane, and simply indicate a displacement (change in position) rather than a position. It is common to need to convert from a magnitude and angle to the component form of the vector and vice versa. Happily, this process is identical to converting from polar coordinates to Cartesian coordinates, or from the polar form of complex numbers to the $a+bi$ form.

Example 5

Find the component form of a vector with length 7 at an angle of 135 degrees.

Using the conversion formulas $x = r\cos(\theta)$ and $y = r\sin(\theta)$, we can find the components

$$x = 7\cos(135°) = -\frac{7\sqrt{2}}{2}$$

$$y = 7\sin(135°) = \frac{7\sqrt{2}}{2}$$

This vector can be written in component form as $\left\langle -\frac{7\sqrt{2}}{2}, \frac{7\sqrt{2}}{2} \right\rangle$.

Example 6

Find the magnitude and angle θ representing the vector $\vec{u} = \langle 3,-2 \rangle$.

First we can find the magnitude by remembering the relationship between x, y and r:
$$r^2 = 3^2 + (-2)^2 = 13$$
$$r = \sqrt{13}$$

Second we can find the angle. Using the tangent,

$$\tan(\theta) = \frac{-2}{3}$$

$\theta = \tan^{-1}\left(-\frac{2}{3}\right) \approx -33.69°$, or written as a coterminal positive angle, $326.31°$. This angle is in the 4th quadrant as desired.

Try it Now

3. Using vector $\vec{v}$ from Try it Now #1, the vector that travels from the origin to the point (3, 5), find the components, magnitude and angle θ that represent this vector.

In addition to representing distance movements, vectors are commonly used in physics and engineering to represent any quantity that has both direction and magnitude, including velocities and forces.

Example 7

An object is launched with initial velocity 200 meters per second at an angle of 30 degrees. Find the initial horizontal and vertical velocities.

By viewing the initial velocity as a vector, we can resolve the vector into horizontal and vertical components.

$$x = 200\cos(30°) = 200 \cdot \frac{\sqrt{3}}{2} \approx 173.205 \text{ m/sec}$$

$$y = 200\sin(30°) = 200 \cdot \frac{1}{2} = 100 \text{ m/sec}$$

This tells us that, absent wind resistance, the object will travel horizontally at about 173 meters each second. Gravity will cause the vertical velocity to change over time – we'll leave a discussion of that to physics or calculus classes.

Adding and Scaling Vectors in Component Form

To add vectors in component form, we can simply add the corresponding components. To scale a vector by a constant, we scale each component by that constant.

> ### Combining Vectors in Component Form
>
> To **add, subtract, or scale vectors** in component form
>
> If $\vec{u} = \langle u_1, u_2 \rangle$, $\vec{v} = \langle v_1, v_2 \rangle$, and c is any constant, then
>
> $$\vec{u} + \vec{v} = \langle u_1 + v_1, u_2 + v_2 \rangle$$
>
> $$\vec{u} - \vec{v} = \langle u_1 - v_1, u_2 - v_2 \rangle$$
>
> $$c\vec{u} = \langle cu_1, cu_2 \rangle$$

Example 8

Given $\vec{u} = \langle 3, -2 \rangle$ and $\vec{v} = \langle -1, 4 \rangle$, find a new vector $\vec{w} = 3\vec{u} - 2\vec{v}$

Using the vectors given,
$\vec{w} = 3\vec{u} - 2\vec{v}$

$\quad = 3\langle 3, -2 \rangle - 2\langle -1, 4 \rangle$ Scale each vector

$\quad = \langle 9, -6 \rangle - \langle -2, 8 \rangle$ Subtract corresponding components

$\quad = \langle 11, -14 \rangle$

By representing vectors in component form, we can find the resulting displacement vector after a multitude of movements without needing to draw a lot of complicated non-right triangles. For a simple example, we revisit the problem from the opening of the section. The general procedure we will follow is:
1) Convert vectors to component form
2) Add the components of the vectors
3) Convert back to length and direction if needed to suit the context of the question

Example 9

A woman leaves home, walks 3 miles north, then 2 miles southeast. How far is she from home, and what direction would she need to walk to return home? How far has she walked by the time she gets home?

Let's begin by understanding the question in a little more depth. When we use vectors to describe a traveling direction, we often position things so north points in the upward direction, east points to the right, and so on, as pictured here.

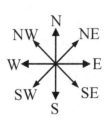

Consequently, travelling NW, SW, NE or SE, means we are travelling through the quadrant bordered by the given directions at a 45 degree angle.
With this in mind, we begin by converting each vector to components.

A walk 3 miles north would, in components, be $\langle 0,3 \rangle$.

A walk of 2 miles southeast would be at an angle of 45° South of East. Measuring from standard position the angle would be 315°.

Converting to components, we choose to use the standard position angle so that we do not have to worry about whether the signs are negative or positive; they will work out automatically.

$$\langle 2\cos(315°), 2\sin(315°) \rangle = \left\langle 2 \cdot \frac{\sqrt{2}}{2}, 2 \cdot \frac{-\sqrt{2}}{2} \right\rangle \approx \langle 1.414, -1.414 \rangle$$

Adding these vectors gives the sum of the movements in component form
$$\langle 0,3 \rangle + \langle 1.414, -1.414 \rangle = \langle 1.414, 1.586 \rangle$$

To find how far she is from home and the direction she would need to walk to return home, we could find the magnitude and angle of this vector.

Length $= \sqrt{1.414^2 + 1.586^2} = 2.125$

To find the angle, we can use the tangent
$$\tan(\theta) = \frac{1.586}{1.414}$$
$$\theta = \tan^{-1}\left(\frac{1.586}{1.414}\right) = 48.273° \text{ north of east}$$

Of course, this is the angle from her starting point to her ending point. To return home, she would need to head the opposite direction, which we could either describe as 180°+48.273° = 228.273° measured in standard position, or as 48.273° south of west (or 41.727° west of south).

She has walked a total distance of $3 + 2 + 2.125 = 7.125$ miles.

Keep in mind that total distance travelled is not the same as the length of the resulting displacement vector or the "return" vector.

Try it Now
4. In a scavenger hunt, directions are given to find a buried treasure. From a starting point at a flag pole you must walk 30 feet east, turn 30 degrees to the north and travel 50 feet, and then turn due south and travel 75 feet. Sketch a picture of these vectors, find their components, and calculate how far and in what direction you must travel to go directly to the treasure from the flag pole without following the map.

While using vectors is not much faster than using law of cosines with only two movements, when combining three or more movements, forces, or other vector quantities, using vectors quickly becomes much more efficient than trying to use triangles.

Example 10

Three forces are acting on an object as shown below, each measured in Newtons (N). What force must be exerted to keep the object in equilibrium (where the sum of the forces is zero)?

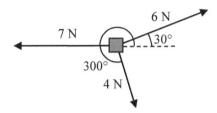

We start by resolving each vector into components.

The first vector with magnitude 6 Newtons at an angle of 30 degrees will have components

$$\langle 6\cos(30°), 6\sin(30°)\rangle = \left\langle 6 \cdot \frac{\sqrt{3}}{2}, 6 \cdot \frac{1}{2}\right\rangle = \langle 3\sqrt{3}, 3\rangle$$

The second vector is only in the horizontal direction, so can be written as $\langle -7, 0\rangle$.

The third vector with magnitude 4 Newtons at an angle of 300 degrees will have components

$$\langle 4\cos(300°), 4\sin(300°)\rangle = \left\langle 4 \cdot \frac{1}{2}, 4 \cdot \frac{-\sqrt{3}}{2}\right\rangle = \langle 2, -2\sqrt{3}\rangle$$

To keep the object in equilibrium, we need to find a force vector $\langle x, y\rangle$ so the sum of the four vectors is the zero vector, $\langle 0, 0\rangle$.

$\langle 3\sqrt{3}, 3\rangle + \langle -7, 0\rangle + \langle 2, -2\sqrt{3}\rangle + \langle x, y\rangle = \langle 0, 0\rangle$ Add component-wise

$\langle 3\sqrt{3} - 7 + 2, 3 + 0 - 2\sqrt{3}\rangle + \langle x, y\rangle = \langle 0, 0\rangle$ Simplify

$\langle 3\sqrt{3} - 5, 3 - 2\sqrt{3}\rangle + \langle x, y\rangle = \langle 0, 0\rangle$ Solve

$\langle x, y\rangle = \langle 0, 0\rangle - \langle 3\sqrt{3} - 5, 3 - 2\sqrt{3}\rangle$

$\langle x, y\rangle = \langle -3\sqrt{3} + 5, -3 + 2\sqrt{3}\rangle \approx \langle -0.196, 0.464\rangle$

This vector gives in components the force that would need to be applied to keep the object in equilibrium. If desired, we could find the magnitude of this force and direction it would need to be applied in.

Magnitude $= \sqrt{(-0.196)^2 + 0.464^2} = 0.504\,\text{N}$

Angle:

$$\tan(\theta) = \frac{0.464}{-0.196}$$

$$\theta = \tan^{-1}\left(\frac{0.464}{-0.196}\right) = -67.089°.$$

This is in the wrong quadrant, so we adjust by finding the next angle with the same tangent value by adding a full period of tangent:
$$\theta = -67.089° + 180° = 112.911°$$

To keep the object in equilibrium, a force of 0.504 Newtons would need to be applied at an angle of 112.911°.

Important Topics of This Section
Vectors, magnitude (length) & direction
Addition of vectors
Scaling of vectors
Components of vectors
Vectors as velocity
Vectors as forces
Adding & Scaling vectors in component form
Total distance travelled vs. total displacement

Try it Now Answers

1

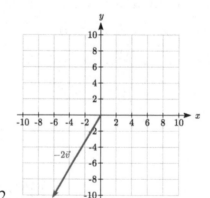

2.

2. $\vec{v} = \langle 3, 5 \rangle$ magnitude $= \sqrt{34}$ $\theta = \tan^{-1}\left(\dfrac{5}{3}\right) = 59.04°$

3.

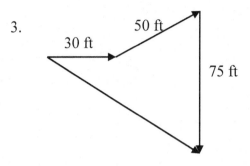

$\vec{v}_1 = \langle 30, 0 \rangle$ $\vec{v}_2 = \langle 50\cos(30°), 50\sin(30°) \rangle$ $\vec{v}_3 = \langle 0, -75 \rangle$

$\vec{v}_f = \langle 30 + 50\cos(30°), 50\sin(30°) - 75 \rangle = \langle 73.301, -50 \rangle$

Magnitude $= 88.73$ feet at an angle of $34.3°$ south of east.

Section 8.4 Exercises

Write the vector shown in component form.

1.

2.

Given the vectors shown, sketch $\vec{u} + \vec{v}$, $\vec{u} - \vec{v}$, and $2\vec{u}$.

3.

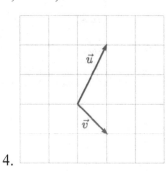

4.

Write each vector below as a combination of the vectors $\vec{u}$ and $\vec{v}$ from question #3.

5.

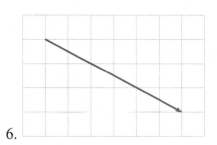

6.

From the given magnitude and direction in standard position, write the vector in component form.

7. Magnitude: 6, Direction: 45°

8. Magnitude: 10, Direction: 120°

9. Magnitude: 8, Direction: 220°

10. Magnitude: 7, Direction: 305°

Find the magnitude and direction of the vector.

11. $\langle 0,4 \rangle$ 12. $\langle -3,0 \rangle$ 13. $\langle 6,5 \rangle$ 14. $\langle 3,7 \rangle$

15. $\langle -2,1 \rangle$ 16. $\langle -10,13 \rangle$ 17. $\langle 2,-5 \rangle$ 18. $\langle 8,-4 \rangle$

19. $\langle -4,-6 \rangle$ 20. $\langle -1,9 \rangle$

Using the vectors given, compute $\vec{u} + \vec{v}$, $\vec{u} - \vec{v}$, and $2\vec{u} - 3\vec{v}$.

21. $\vec{u} = \langle 2,-3 \rangle$, $\vec{v} = \langle 1,5 \rangle$ 22. $\vec{u} = \langle -3,4 \rangle$, $\vec{v} = \langle -2,1 \rangle$

23. A woman leaves home and walks 3 miles west, then 2 miles southwest. How far from home is she, and in what direction must she walk to head directly home?

24. A boat leaves the marina and sails 6 miles north, then 2 miles northeast. How far from the marina is the boat, and in what direction must it sail to head directly back to the marina?

25. A person starts walking from home and walks 4 miles east, 2 miles southeast, 5 miles south, 4 miles southwest, and 2 miles east. How far have they walked? If they walked straight home, how far would they have to walk?

26. A person starts walking from home and walks 4 miles east, 7 miles southeast, 6 miles south, 5 miles southwest, and 3 miles east. How far have they walked? If they walked straight home, how far would they have to walk?

27. Three forces act on an object: $\vec{F_1} = \langle -8, -5 \rangle$, $\vec{F_2} = \langle 0, 1 \rangle$, $\vec{F_3} = \langle 4, -7 \rangle$. Find the net force acting on the object.

28. Three forces act on an object: $\vec{F_1} = \langle 2, 5 \rangle$, $\vec{F_2} = \langle 8, 3 \rangle$, $\vec{F_3} = \langle 0, -7 \rangle$. Find the net force acting on the object.

29. A person starts walking from home and walks 3 miles at 20° north of west, then 5 miles at 10° west of south, then 4 miles at 15° north of east. If they walked straight home, how far would they have to walk, and in what direction?

30. A person starts walking from home and walks 6 miles at 40° north of east, then 2 miles at 15° east of south, then 5 miles at 30° south of west. If they walked straight home, how far would they have to walk, and in what direction?

31. An airplane is heading north at an airspeed of 600 km/hr, but there is a wind blowing from the southwest at 80 km/hr. How many degrees off course will the plane end up flying, and what is the plane's speed relative to the ground?

32. An airplane is heading north at an airspeed of 500 km/hr, but there is a wind blowing from the northwest at 50 km/hr. How many degrees off course will the plane end up flying, and what is the plane's speed relative to the ground?

33. An airplane needs to head due north, but there is a wind blowing from the southwest at 60 km/hr. The plane flies with an airspeed of 550 km/hr. To end up flying due north, the pilot will need to fly the plane how many degrees west of north?

34. An airplane needs to head due north, but there is a wind blowing from the northwest at 80 km/hr. The plane flies with an airspeed of 500 km/hr. To end up flying due north, the pilot will need to fly the plane how many degrees west of north?

35. As part of a video game, the point (5, 7) is rotated counterclockwise about the origin through an angle of 35 degrees. Find the new coordinates of this point.

36. As part of a video game, the point (7, 3) is rotated counterclockwise about the origin through an angle of 40 degrees. Find the new coordinates of this point.

37. Two children are throwing a ball back and forth straight across the back seat of a car. The ball is being thrown 10 mph relative to the car, and the car is travelling 25 mph down the road. If one child doesn't catch the ball and it flies out the window, in what direction does the ball fly (ignoring wind resistance)?

38. Two children are throwing a ball back and forth straight across the back seat of a car. The ball is being thrown 8 mph relative to the car, and the car is travelling 45 mph down the road. If one child doesn't catch the ball and it flies out the window, in what direction does the ball fly (ignoring wind resistance)?

Section 8.5 Dot Product

Now that we can add, subtract, and scale vectors, you might be wondering whether we can multiply vectors. It turns out there are two different ways to multiply vectors, one which results in a number, and one which results in a vector. In this section, we'll focus on the first, called the **dot product** or **scalar product**, since it produces a single numeric value (a scalar). We'll begin with some motivation.

In physics, we often want to know how much of a force is acting in the direction of motion. To determine this, we need to know the angle between direction of force and the direction of motion. Likewise, in computer graphics, the lighting system determines how bright a triangle on the object should be based on the angle between object and the direction of the light. In both applications, we're interested in the angle between the vectors, so let's start there.

Suppose we have two vectors, $\vec{a} = \langle a_1, a_2 \rangle$ and $\vec{b} = \langle b_1, b_2 \rangle$. Using our polar coordinate conversions, we could write $\vec{a} = \langle |\vec{a}| \cos(\alpha), |\vec{a}| \sin(\alpha) \rangle$ and $\vec{b} = \langle |\vec{b}| \cos(\beta), |\vec{b}| \sin(\beta) \rangle$.

Now, if we knew the angles α and β, we wouldn't have much work to do - the angle between the vectors would be $\theta = \alpha - \beta$. While we certainly could use some inverse tangents to find the two angles, it would be great if we could find a way to determine the angle between the vector just from the vector components.

To help us manipulate $\theta = \alpha - \beta$, we might try introducing a trigonometric function:
$$\cos(\theta) = \cos(\alpha - \beta)$$

Now we can apply the difference of angles identity
$$\cos(\theta) = \cos(\alpha)\cos(\beta) + \sin(\alpha)\sin(\beta)$$

Now $a_1 = |\vec{a}| \cos(\alpha)$, so $\cos(\alpha) = \dfrac{a_1}{|\vec{a}|}$, and likewise for the other three components.

Making those substitutions,
$$\cos(\theta) = \frac{a_1}{|\vec{a}|}\frac{b_1}{|\vec{b}|} + \frac{a_2}{|\vec{a}|}\frac{b_2}{|\vec{b}|} = \frac{a_1 b_1 + a_2 b_2}{|\vec{a}||\vec{b}|}$$

$$|\vec{a}||\vec{b}| \cos(\theta) = a_1 b_1 + a_2 b_2$$

Notice the expression on the right is a very simple calculation based on the components of the vectors. It comes up so frequently we define it to be the **dot product** of the two vectors, notated by a dot. This gives us two definitions of the dot product.

Definitions of the Dot Product	
$\vec{a} \cdot \vec{b} = a_1 b_1 + a_2 b_2$	Component definition
$\vec{a} \cdot \vec{b} = \left\vert \vec{a} \right\vert \left\vert \vec{b} \right\vert \cos(\theta)$	Geometric definition

The first definition, $\vec{a} \cdot \vec{b} = a_1 b_1 + a_2 b_2$, gives us a simple way to calculate the dot product from components. The second definition, $\vec{a} \cdot \vec{b} = \left\vert \vec{a} \right\vert \left\vert \vec{b} \right\vert \cos(\theta)$, gives us a geometric interpretation of the dot product, and gives us a way to find the angle between two vectors, as we desired.

Example 1

Find the dot product $\langle 3, -2 \rangle \cdot \langle 5, 1 \rangle$.

Using the first definition, we can calculate the dot product by multiplying the x components and adding that to the product of the y components.

$$\langle 3, -2 \rangle \cdot \langle 5, 1 \rangle = (3)(5) + (-2)(1) = 15 - 2 = 13$$

Example 2

Find the dot product of the two vectors shown.

We can immediately see that the magnitudes of the two vectors are 7 and 6. We can quickly calculate that the angle between the vectors is 150°. Using the geometric definition of the dot product,

$$\vec{a} \cdot \vec{b} = \left\vert \vec{a} \right\vert \left\vert \vec{b} \right\vert \cos(\theta) = (6)(7)\cos(150°) = 42 \cdot \frac{\sqrt{3}}{2} = 21\sqrt{3}.$$

Try it Now

1. Calculate the dot product $\langle -7, 3 \rangle \cdot \langle -2, -6 \rangle$

Now we can return to our goal of finding the angle between vectors.

Example 3

An object is being pulled up a ramp in the direction $\langle 5,1 \rangle$ by a rope pulling in the direction $\langle 4,2 \rangle$. What is the angle between the rope and the ramp?

Using the component form, we can easily calculate the dot product.
$$\vec{a} \cdot \vec{b} = \langle 5,1 \rangle \cdot \langle 4,2 \rangle = (5)(4) + (1)(2) = 20 + 2 = 22$$

We can also calculate the magnitude of each vector.
$$|\vec{a}| = \sqrt{5^2 + 1^2} = \sqrt{26}, \quad |\vec{b}| = \sqrt{4^2 + 2^2} = \sqrt{20}$$

Substituting these values into the geometric definition, we can solve for the angle between the vectors.
$$\vec{a} \cdot \vec{b} = |\vec{a}||\vec{b}| \cos(\theta)$$
$$22 = \sqrt{26}\sqrt{20} \cos(\theta)$$
$$\theta = \cos^{-1}\left(\frac{22}{\sqrt{26}\sqrt{20}}\right) \approx 15.255°.$$

Example 4

Calculate the angle between the vectors $\langle 6,4 \rangle$ and $\langle -2,3 \rangle$.

Calculating the dot product, $\langle 6,4 \rangle \cdot \langle -2,3 \rangle = (6)(-2) + (4)(3) = -12 + 12 = 0$

We don't even need to calculate the magnitudes in this case since the dot product is 0.
$$\vec{a} \cdot \vec{b} = |\vec{a}||\vec{b}| \cos(\theta)$$
$$0 = |\vec{a}||\vec{b}| \cos(\theta)$$
$$\theta = \cos^{-1}\left(\frac{0}{|\vec{a}||\vec{b}|}\right) = \cos^{-1}(0) = 90°$$

With the dot product equaling zero, as in the last example, the angle between the vectors will always be 90°, indicating that the vectors are **orthogonal**, a more general way of saying perpendicular. This gives us a quick way to check if vectors are orthogonal. Also, if the dot product is positive, then the inside of the inverse cosine will be positive, giving an angle less than 90°. A negative dot product will then lead to an angle larger than 90°

Sign of the Dot Product	
If the dot product is:	
Zero	The vectors are **orthogonal** (perpendicular).
Positive	The angle between the vectors is less than 90°
Negative	The angle between the vectors is greater than 90°

Try it Now

2. Are the vectors $\langle -7, 3 \rangle$ and $\langle -2, -6 \rangle$ orthogonal? If not, find the angle between them.

Projections

In addition to finding the angle between vectors, sometimes we want to know how much one vector points in the direction of another. For example, when pulling an object up a ramp, we might want to know how much of the force is exerted in the direction of motion. To determine this we can use the idea of a **projection**.

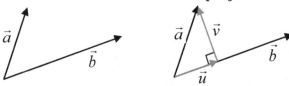

In the picture above, $\vec{u}$ is a projection of $\vec{a}$ onto $\vec{b}$. In other words, it is the portion of $\vec{a}$ that points in the same direction as $\vec{b}$.

To find the length of $\vec{u}$, we could notice that it is one side of a right triangle. If we define θ to be the angle between $\vec{a}$ and $\vec{u}$, then $\cos(\theta) = \dfrac{|\vec{u}|}{|\vec{a}|}$, so $|\vec{a}|\cos(\theta) = |\vec{u}|$.

While we could find the angle between the vectors to determine this magnitude, we could skip some steps by using the dot product directly. Since $\vec{a} \cdot \vec{b} = |\vec{a}||\vec{b}|\cos(\theta)$,

$|\vec{a}|\cos(\theta) = \dfrac{\vec{a} \cdot \vec{b}}{|\vec{b}|}$. Using this, we can rewrite $|\vec{u}| = |\vec{a}|\cos(\theta)$ as $|\vec{u}| = \dfrac{\vec{a} \cdot \vec{b}}{|\vec{b}|}$. This gives us

the length of the projection, sometimes denoted as $comp_{\vec{b}}\vec{a} = |\vec{u}| = \dfrac{\vec{a} \cdot \vec{b}}{|\vec{b}|}$.

To find the vector $\vec{u}$ itself, we could first scale $\vec{b}$ to a unit vector with length 1: $\dfrac{\vec{b}}{|\vec{b}|}$.

Multiplying this by the length of the projection will give a vector in the direction of $\vec{b}$ but with the correct length.

$$proj_{\vec{b}}\vec{a} = |\vec{u}|\frac{\vec{b}}{|\vec{b}|} = \left(\frac{\vec{a}\cdot\vec{b}}{|\vec{b}|}\right)\frac{\vec{b}}{|\vec{b}|} = \left(\frac{\vec{a}\cdot\vec{b}}{|\vec{b}|^2}\right)\vec{b}$$

Projection Vector

The projection of vector $\vec{a}$ onto $\vec{b}$ is $proj_{\vec{b}}\vec{a} = \left(\dfrac{\vec{a}\cdot\vec{b}}{|\vec{b}|^2}\right)\vec{b}$

The magnitude of the projection is $comp_{\vec{b}}\vec{a} = \dfrac{\vec{a}\cdot\vec{b}}{|\vec{b}|}$

Example 5

Find the projection of the vector $\langle 3,-2\rangle$ onto the vector $\langle 8,6\rangle$.

We will need to know the dot product of the vectors and the magnitude of the vector we are projecting onto.

$\langle 3,-2\rangle \cdot \langle 8,6\rangle = (3)(8) + (-2)(6) = 24 - 12 = 12$

$|\langle 8,6\rangle| = \sqrt{8^2 + 6^2} = \sqrt{64 + 36} = \sqrt{100} = 10$

The magnitude of the projection will be $\dfrac{\langle 3,-2\rangle \cdot \langle 8,6\rangle}{|\langle 8,6\rangle|} = \dfrac{12}{10} = \dfrac{6}{5}$.

To find the projection vector itself, we would multiply that magnitude by $\langle 8,6\rangle$ scaled to a unit vector.

$\dfrac{6}{5}\dfrac{\langle 8,6\rangle}{|\langle 8,6\rangle|} = \dfrac{6}{5}\dfrac{\langle 8,6\rangle}{10} = \dfrac{6}{50}\langle 8,6\rangle = \left\langle \dfrac{48}{50},\dfrac{36}{50}\right\rangle = \left\langle \dfrac{24}{25},\dfrac{18}{25}\right\rangle$.

Based on the sketch above, this answer seems reasonable.

3. Find the component of the vector $\langle -3, 4 \rangle$ that is *orthogonal* to the vector $\langle -8, 4 \rangle$

Work

In physics, when a constant force causes an object to move, the mechanical **work** done by that force is the product of the force and the distance the object is moved. However, we only consider the portion of force that is acting in the direction of motion.

This is simply the magnitude of the projection of the force vector onto the distance vector, $\dfrac{\vec{F} \cdot \vec{d}}{|\vec{d}|}$. The work done is the product of that component of force times the distance moved, the magnitude of the distance vector.

$$Work = \left(\frac{\vec{F} \cdot \vec{d}}{|\vec{d}|} \right) |\vec{d}| = \vec{F} \cdot \vec{d}$$

It turns out that work is simply the dot product of the force vector and the distance vector.

Work

When a force $\vec{F}$ causes an object to move some distance $\vec{d}$, the work done is

$$Work = \vec{F} \cdot \vec{d}$$

Example 6

A cart is pulled 20 feet by applying a force of 30 pounds on a rope held at a 30 degree angle. How much work is done?

Since work is simply the dot product, we can take advantage of the geometric definition of the dot product in this case.

$$Work = \vec{F} \cdot \vec{d} = |\vec{F}| \cdot |\vec{d}| \cos(\theta) = (30)(20) \cos(30°) \approx 519.615 \text{ ft-lbs.}$$

---

4. Find the work down moving an object from the point (1, 5) to (9, 14) by the force vector $\vec{F} = \langle 3, 2 \rangle$

Important Topics of This Section
Calculate Dot Product
Using component definition
Using geometric definition
Find the angle between two vectors
Sign of the dot product
Projections
Work

Try it Now Answers

1. $\langle -7,3 \rangle \cdot \langle -2,-6 \rangle = (-7)(-2) + (3)(-6) = 14 - 18 = -4$

2. In the previous Try it Now, we found the dot product was -4, so the vectors are not orthogonal. The magnitudes of the vectors are $\sqrt{(-7)^2 + 3^2} = \sqrt{58}$ and $\sqrt{(-2)^2 + 6^2} = \sqrt{40}$. The angle between the vectors will be

$$\theta = \cos^{-1}\left(\frac{-4}{\sqrt{58}\sqrt{40}} \right) \approx 94.764°$$

3. We want to find the component of $\langle -3,4 \rangle$ that is *orthogonal* to the vector $\langle -8,4 \rangle$. In the picture to the right, that component is vector $\vec{v}$. Notice that $\vec{u} + \vec{v} = \vec{a}$, so if we can find the projection vector, we can find $\vec{v}$.

$$\vec{u} = proj_{\vec{b}}\vec{a} = \left(\frac{\vec{a} \cdot \vec{b}}{|\vec{b}|^2} \right) \vec{b} = \left(\frac{\langle -3,4 \rangle \cdot \langle -8,4 \rangle}{\left(\sqrt{(-8)^2 + 4^2} \right)^2} \right) \langle -8,4 \rangle = \frac{40}{80} \langle -8,4 \rangle = \langle -4,2 \rangle.$$

Now we can solve $\vec{u} + \vec{v} = \vec{a}$ for $\vec{v}$.

$\vec{v} = \vec{a} - \vec{u} = \langle -3,4 \rangle - \langle -4,2 \rangle = \langle 1,2 \rangle$

4. The distance vector is $\langle 9 - 1, 14 - 5 \rangle = \langle 8,9 \rangle$.

The work is the dot product: $Work = \vec{F} \cdot \vec{d} = \langle 3,2 \rangle \cdot \langle 8,9 \rangle = 24 + 18 = 42$

Section 8.5 Exercises

Two vectors are described by their magnitude and direction in standard position. Find the dot product of the vectors.

1. Magnitude: 6, Direction: 45°; Magnitude: 10, Direction: 120°
2. Magnitude: 8, Direction: 220°; Magnitude: 7, Direction: 305°

Find the dot product of each pair of vectors.

3. $\langle 0,4 \rangle$; $\langle -3,0 \rangle$ 4. $\langle 6,5 \rangle$; $\langle 3,7 \rangle$

5. $\langle -2,1 \rangle$; $\langle -10,13 \rangle$ 6. $\langle 2,-5 \rangle$; $\langle 8,-4 \rangle$

Find the angle between the vectors

7. $\langle 0,4 \rangle$; $\langle -3,0 \rangle$ 8. $\langle 6,5 \rangle$; $\langle 3,7 \rangle$

9. $\langle 2,4 \rangle$; $\langle 1,-3 \rangle$ 10. $\langle -4,1 \rangle$; $\langle 8,-2 \rangle$

11. $\langle 4,2 \rangle$; $\langle 8,4 \rangle$ 12. $\langle 5,3 \rangle$; $\langle -6,10 \rangle$

13. Find a value for k so that $\langle 2,7 \rangle$ and $\langle k,4 \rangle$ will be orthogonal.

14. Find a value for k so that $\langle -3,5 \rangle$ and $\langle 2,k \rangle$ will be orthogonal.

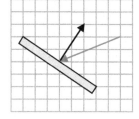

15. Find the magnitude of the projection of $\langle 8,-4 \rangle$ onto $\langle 1,-3 \rangle$.

16. Find the magnitude of the projection of $\langle 2,7 \rangle$ onto $\langle 4,5 \rangle$.

17. Find the projection of $\langle -6,10 \rangle$ onto $\langle 1,-3 \rangle$.

18. Find the projection of $\langle 0,4 \rangle$ onto $\langle 3,7 \rangle$.

19. A scientist needs to determine the angle of reflection when a laser hits a mirror. The picture shows the vector representing the laser beam, and a vector that is orthogonal to the mirror. Find the acute angle between these, the angle of reflection.

20. A triangle has coordinates at A: (1,4), B: (2,7), and C: (4,2). Find the angle at point B.

21. A boat is trapped behind a log lying parallel to the dock. It only requires 10 pounds of force to pull the boat directly towards you, but because of the log, you'll have to pull at a 45° angle. How much force will you have to pull with? (We're going to assume that the log is very slimy and doesn't contribute any additional resistance.)

22. A large boulder needs to be dragged to a new position. If pulled directly horizontally, the boulder would require 400 pounds of pulling force to move. We need to pull the boulder using a rope tied to the back of a large truck, forming a 15° angle from the ground. How much force will the truck need to pull with?

23. Find the work done against gravity by pushing a 20 pound cart 10 feet up a ramp that is 10° above horizontal. Assume there is no friction, so the only force is 20 pounds downwards due to gravity.

24. Find the work done against gravity by pushing a 30 pound cart 15 feet up a ramp that is 8° above horizontal. Assume there is no friction, so the only force is 30 pounds downwards due to gravity.

25. An object is pulled to the top of a 40 foot ramp that forms a 10° angle with the ground. It is pulled by rope exerting a force of 120 pounds at a 35° angle relative to the ground. Find the work done.

26. An object is pulled to the top of a 30 foot ramp that forms a 20° angle with the ground. It is pulled by rope exerting a force of 80 pounds at a 30° angle relative to the ground. Find the work done.

Section 8.6 Parametric Equations

Many shapes, even ones as simple as circles, cannot be represented as an equation where y is a function of x. Consider, for example, the path a moon follows as it orbits around a planet, which simultaneously rotates around a sun. In some cases, polar equations provide a way to represent such a path. In others, we need a more versatile approach that allows us to represent both the x and y coordinates in terms of a third variable, or parameter.

Parametric Equations
A system of **parametric equations** is a pair of functions $x(t)$ and $y(t)$ in which the x and y coordinates are the output, represented in terms of a third input parameter, t.

Example 1

Moving at a constant speed, an object moves at a steady rate along a straight path from coordinates (-5, 3) to the coordinates (3, -1) in 4 seconds, where the coordinates are measured in meters. Find parametric equations for the position of the object.

The x coordinate of the object starts at -5 meters, and goes to +3 meters, this means the x direction has changed by 8 meters in 4 seconds, giving us a rate of 2 meters per second. We can now write the x coordinate as a linear function with respect to time, t, $x(t) = -5 + 2t$. Similarly, the y value starts at 3 and goes to -1, giving a change in y value of 4 meters, meaning the y values have decreased by 4 meters in 4 seconds, for a rate of -1 meter per second, giving equation $y(t) = 3 - t$. Together, these are the parametric equations for the position of the object:

$x(t) = -5 + 2t$

$y(t) = 3 - t$

t	x	y
0	-5	3
1	-3	2
2	-1	1
3	1	0
4	3	-1

Using these equations, we can build a table of t, x, and y values. Because of the context, we limited ourselves to non-negative t values for this example, but in general you can use any values.

From this table, we could create three possible graphs: a graph of x vs. t, which would show the horizontal position over time, a graph of y vs. t, which would show the vertical position over time, or a graph of y vs. x, showing the position of the object in the plane.

Position of *x* as a function of time

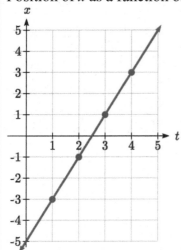

Position of *y* as a function of time

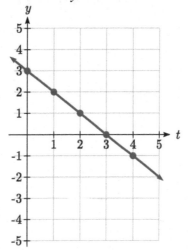

Position of *y* relative to *x*

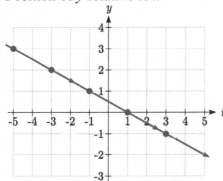

Notice that the parameter *t* does not explicitly show up in this third graph. Sometimes, when the parameter *t* does represent a quantity like time, we might indicate the direction of movement on the graph using an arrow, as shown above.

There is often no single parametric representation for a curve. In Example 1 we assumed the object was moving at a steady rate along a straight line. If we kept the assumption about the path (straight line) but did not assume the speed was constant, we might get a system like:

$$x(t) = -5 + 2t^2$$

$$y(t) = 3 - t^2$$

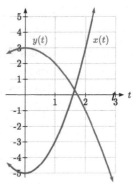

This starts at (-5, 3) when *t* = 0 and ends up at (3, -1) when *t* = 2. If we graph the *x(t)* and *y(t)* function separately, we can see that those are no longer linear, but if we graph *x* vs. *y* we will see that we still get a straight-line path.

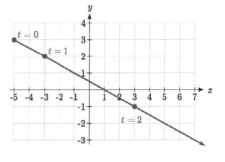

Example 2

Sketch a graph of

$$x(t) = t^2 + 1$$

$$y(t) = 2 + t$$

t	x	y
-3	10	-1
-2	5	0
-1	2	1
0	1	2
1	2	3
2	5	4

We can begin by creating a table of values. From this table, we can plot the (x, y) points in the plane, sketch in a rough graph of the curve, and indicate the direction of motion with respect to time by using arrows.

Notice that here the parametric equations describe a shape for which y is not a function of x. This is an example of why using parametric equations can be useful – since they can represent such a graph as a set of functions. This particular graph also appears to be a parabola where x is a function of y, which we will soon verify.

Example 3

Sketch a graph of

$$x(t) = 3\cos(t)$$

$$y(t) = 3\sin(t)$$

These equations should look familiar. Back when we first learned about sine and cosine we found that the coordinates of a point on a circle of radius r at an angle of θ will be $x = r\cos(\theta)$, $y = r\sin(\theta)$. The equations above are in the same form, with $r = 3$, and t used in place of θ.

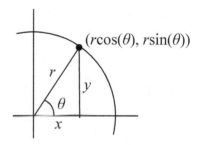

This suggests that for each value of t, these parametric equations give a point on a circle of radius 3 at the angle corresponding to t. At $t = 0$, the graph would be at $x = 3\cos(0)$, $y = 3\sin(0)$, the point (3,0). Indeed, these equations describe the equation of a circle, drawn counterclockwise.

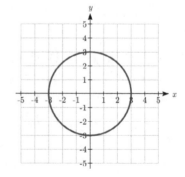

Interestingly, these similar parametric equations also describe the circle of radius 3:

$x(t) = 3\sin(t)$

$y(t) = 3\cos(t)$

The difference with these equations it the graph would start at $x = 3\sin(0)$, $y = 3\cos(0)$, the point (0,3). As t increases from 0, the x value will increase, indicating these equations would draw the graph in a clockwise direction.

While creating a t-x-y table, plotting points and connecting the dots with a smooth curve usually works to give us a rough idea of what the graph of a system of parametric equations looks like, it's generally easier to use technology to create these tables and (simultaneously) much nicer-looking graphs.

Example 4

Sketch a graph of $\begin{array}{l} x(t) = 2\cos(t) \\ y(t) = 3\sin(t) \end{array}$.

Notice first that this equation looks very similar to the ones from the previous example, except the coefficients are not equal. You might guess that the pairing of cos and sin will still produce rotation, but now x will vary from -2 to 2 while y will vary from -3 to 3, creating an ellipse.

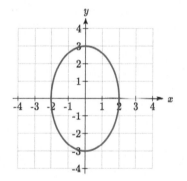

Using technology we can generate a graph of this equation, verifying it is indeed an ellipse.

Similar to graphing polar equations, you must change the MODE on your calculator (or select parametric equations on your graphing technology) before graphing a system of parametric equations. You will know you have successfully entered parametric mode when the equation input has changed to ask for a $x(t)=$ and $y(t)=$ pair of equations.

Try it Now

1. Sketch a graph of $\begin{array}{l} x(t) = 4\cos(3t) \\ y(t) = 3\sin(2t) \end{array}$. This is an example of a **Lissajous** figure.

Example 5

The populations of rabbits and wolves on an island over time are given by the graphs below. Use these graphs to sketch a graph in the *r-w* plane showing the relationship between the number of rabbits and number of wolves.

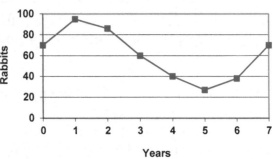

For each input *t*, we can determine the number of rabbits, *r*, and the number of wolves, *w*, from the respective graphs, and then plot the corresponding point in the *r-w* plane.

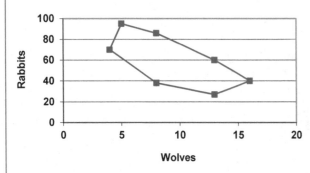

This graph helps reveal the cyclical interaction between the two populations.

Converting from Parametric to Cartesian

In some cases, it is possible to eliminate the parameter *t*, allowing you to write a pair of parametric equations as a Cartesian equation.

It is easiest to do this if one of the *x(t)* or *y(t)* functions can easily be solved for *t*, allowing you to then substitute the remaining expression into the second part.

Example 6

Write $\begin{array}{l} x(t) = t^2 + 1 \\ y(t) = 2 + t \end{array}$ as a Cartesian equation, if possible.

Here, the equation for *y* is linear, so is relatively easy to solve for *t*. Since the resulting Cartesian equation will likely not be a function, and for convenience, we drop the function notation.

$y = 2 + t$ Solve for t

$y - 2 = t$ Substitute this for t in the x equation

$x = (y - 2)^2 + 1$

Notice that this is the equation of a parabola with x as a function of y, with vertex at (1,2), opening to the right. Comparing this with the graph from Example 2, we see (unsurprisingly) that it yields the same graph in the x-y plane as did the original parametric equations.

Try it Now

2. Write $\begin{aligned} x(t) &= t^3 \\ y(t) &= t^6 \end{aligned}$ as a Cartesian equation, if possible.

Example 7

Write $\begin{aligned} x(t) &= \sqrt{t} + 2 \\ y(t) &= \log(t) \end{aligned}$ as a Cartesian equation, if possible.

We could solve either the first or second equation for t. Solving the first,

$x = \sqrt{t} + 2$

$x - 2 = \sqrt{t}$ Square both sides

$(x - 2)^2 = t$ Substitute into the y equation

$y = \log\left((x - 2)^2\right)$

Since the parametric equation is only defined for $t > 0$, this Cartesian equation is equivalent to the parametric equation on the corresponding domain. The parametric equations show that when $t > 0$, $x > 2$ and $y > 0$, so the domain of the Cartesian equation should be limited to $x > 2$.

To ensure that the Cartesian equation is as equivalent as possible to the original parametric equation, we try to avoid using domain-restricted inverse functions, such as the inverse trig functions, when possible. For equations involving trig functions, we often try to find an identity to utilize to avoid the inverse functions.

Example 8

Write $\begin{aligned} x(t) &= 2\cos(t) \\ y(t) &= 3\sin(t) \end{aligned}$ as a Cartesian equation, if possible.

To rewrite this, we can utilize the Pythagorean identity $\cos^2(t) + \sin^2(t) = 1$.

$x = 2\cos(t)$ so $\dfrac{x}{2} = \cos(t)$

$y = 3\sin(t)$ so $\dfrac{y}{3} = \sin(t)$

Starting with the Pythagorean Identity,

$\cos^2(t) + \sin^2(t) = 1$ Substitute in the expressions from the parametric form

$\left(\dfrac{x}{2}\right)^2 + \left(\dfrac{y}{3}\right)^2 = 1$ Simplify

$\dfrac{x^2}{4} + \dfrac{y^2}{9} = 1$

This is a Cartesian equation for the ellipse we graphed earlier.

Parameterizing Curves

While converting from parametric form to Cartesian can be useful, it is often more useful to parameterize a Cartesian equation – converting it into parametric form.

If the Cartesian equation gives one variable as a function of the other, then parameterization is trivial – the independent variable in the function can simply be defined as t.

Example 9

Parameterize the equation $x = y^3 - 2y$.

In this equation, x is expressed as a function of y. By defining $y = t$ we can then substitute that into the Cartesian equation, yielding $x = t^3 - 2t$. Together, this produces the parametric form:

$x(t) = t^3 - 2t$

$y(t) = t$

3. Write $x^2 + y^2 = 3$ in parametric form, if possible.

In addition to parameterizing Cartesian equations, we also can parameterize behaviors and movements.

Example 10

A robot follows the path shown. Create a table of values for the $x(t)$ and $y(t)$ functions, assuming the robot takes one second to make each movement.

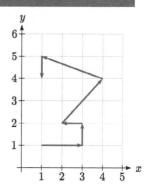

Since we know the direction of motion, we can introduce consecutive values for t along the path of the robot. Using these values with the x and y coordinates of the robot, we can create the tables. For example, we designate the starting point, at (1, 1), as the position at $t = 0$, the next point at (3, 1) as the position at $t = 1$, and so on.

t	0	1	2	3	4	5	6
x	1	3	3	2	4	1	1

t	0	1	2	3	4	5	6
y	1	1	2	2	4	5	4

Notice how this also ties back to vectors. The journey of the robot as it moves through the Cartesian plane could also be displayed as vectors and total distance traveled and displacement could be calculated.

Example 11

A light is placed on the edge of a bicycle tire as shown and the bicycle starts rolling down the street. Find a parametric equation for the position of the light after the wheel has rotated through an angle of θ.

Starting   Rotated by θ

Relative to the center of the wheel, the position of the light can be found as the coordinates of a point on a circle, but since the x coordinate begins at 0 and moves in the negative direction, while the y coordinate starts at the lowest value, the coordinates of the point will be given by:

$$x = -r\sin(\theta)$$
$$y = -r\cos(\theta)$$

The center of the wheel, meanwhile, is moving horizontally. It remains at a constant height of r, but the horizontal position will move a distance equivalent to the arclength of the circle drawn out by the angle, $s = r\theta$. The position of the center of the circle is then

$$x = r\theta$$

$$y = r$$

Combining the position of the center of the wheel with the position of the light on the wheel relative to the center, we get the following parametric equationw, with θ as the parameter:

$$x = r\theta - r\sin(\theta) = r(\theta - \sin(\theta))$$
$$y = r - r\cos(\theta) = r(1 - \cos(\theta))$$

The result graph is called a cycloid.

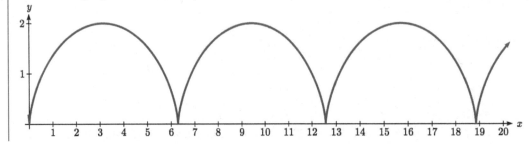

Example 12

A moon travels around a planet as shown, orbiting once every 10 days. The planet travels around a sun as shown, orbiting once every 100 days. Find a parametric equation for the position of the moon, relative to the center of the sun, after t days.

For this example, we'll assume the orbits are circular, though in real life they're actually elliptical.

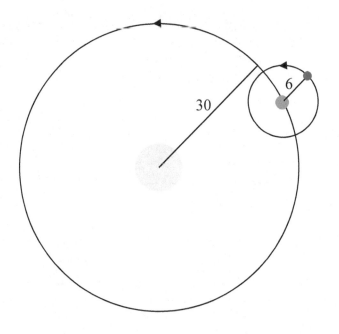

The coordinates of a point on a circle can always be written in the form
$$x = r\cos(\theta)$$
$$y = r\sin(\theta)$$

Since the orbit of the moon around the planet has a period of 10 days, the equation for the position of the *moon relative to the planet* will be
$$x(t) = 6\cos\left(\frac{2\pi}{10}t\right) = 6\cos\left(\frac{\pi}{5}t\right)$$
$$y(t) = 6\sin\left(\frac{2\pi}{10}t\right) = 6\sin\left(\frac{\pi}{5}t\right)$$

With a period of 100 days, the equation for the position of the *planet relative to the sun* will be
$$x(t) = 30\cos\left(\frac{2\pi}{100}t\right) = 30\cos\left(\frac{\pi}{50}t\right)$$
$$y(t) = 30\sin\left(\frac{2\pi}{100}t\right) = 30\sin\left(\frac{\pi}{50}t\right)$$

Combining these together, we can find the position of the *moon relative to the sun* as the sum of the components.
$$x(t) = 6\cos\left(\frac{\pi}{5}t\right) + 30\cos\left(\frac{\pi}{50}t\right)$$
$$y(t) = 6\sin\left(\frac{\pi}{5}t\right) + 30\sin\left(\frac{\pi}{50}t\right)$$

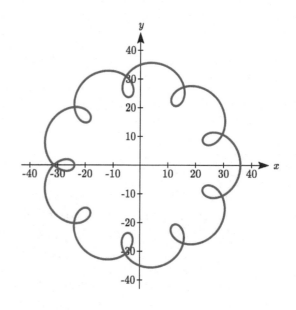

The resulting graph is shown here.

Try it Now

4. A wheel of radius 4 is rolled around the outside of a circle of radius 7. Find a parametric equation for the position of a point on the boundary of the smaller wheel. This shape is called an epicycloid.

Important Topics of This Section
Parametric equations
Graphing $x(t)$, $y(t)$ and the corresponding x-y graph
Sketching graphs and building a table of values
Converting parametric to Cartesian
Converting Cartesian to parametric (parameterizing curves)

Try it Now Answers

1.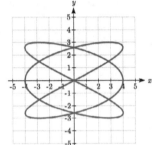

2. $y = \left(t^3\right)^2$, so $y = x^2$

3. $x(t) = 3\cos(t)$
   $y(t) = 3\sin(t)$

4. The center of the small wheel rotates in circle with radius 7+4=11.
   Since the circumference of the small circle is 8π and the circumference of the large circle is 22π, in the time it takes to roll around the large circle, the small circle will have rotated $\dfrac{22\pi}{8\pi} = \dfrac{11}{4}$ rotations. We use this as the stretch factor. The position of a point on the small circle will be the combination of the position of the center of the small wheel around the center of the large wheel, and the position of the point around the small wheel:

$$x(t) = 11\cos(t) - 4\cos\left(\frac{11}{4}t\right)$$

$$y(t) = 11\sin(t) - 4\sin\left(\frac{11}{4}t\right)$$

Section 8.6 Exercises

Match each set of equations with one of the graphs below.

1. $\begin{cases} x(t) = t \\ y(t) = t^2 - 1 \end{cases}$

2. $\begin{cases} x(t) = t - 1 \\ y(t) = t^2 \end{cases}$

3. $\begin{cases} x(t) = 4\sin(t) \\ y(t) = 2\cos(t) \end{cases}$

4. $\begin{cases} x(t) = 2\sin(t) \\ y(t) = 4\cos(t) \end{cases}$

5. $\begin{cases} x(t) = 2 + t \\ y(t) = 3 - 2t \end{cases}$

6. $\begin{cases} x(t) = -2 - 2t \\ y(t) = 3 + t \end{cases}$

A

B

C

D

E

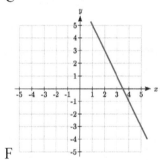
F

From each pair of graphs in the *t-x* and *t-y* planes shown, sketch a graph in the *x-y* plane.

7.

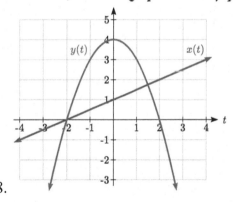
8.

From each graph in the *x-y* plane shown, sketch a graph of the parameter functions in the *t-x* and *t-y* planes.

9.

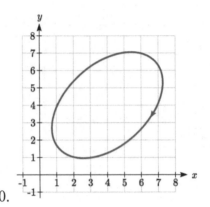

10.

Sketch the parametric equations for $-2 \le t \le 2$.

11. $\begin{cases} x(t)=1+2t \\ y(t)=t^2 \end{cases}$

12. $\begin{cases} x(t)=2t-2 \\ y(t)=t^3 \end{cases}$

Eliminate the parameter t to rewrite the parametric equation as a Cartesian equation

13. $\begin{cases} x(t)=5-t \\ y(t)=8-2t \end{cases}$

14. $\begin{cases} x(t)=6-3t \\ y(t)=10-t \end{cases}$

15. $\begin{cases} x(t)=2t+1 \\ y(t)=3\sqrt{t} \end{cases}$

16. $\begin{cases} x(t)=3t-1 \\ y(t)=2t^2 \end{cases}$

17. $\begin{cases} x(t)=2e^t \\ y(t)=1-5t \end{cases}$

18. $\begin{cases} x(t)=4\log(t) \\ y(t)=3+2t \end{cases}$

19. $\begin{cases} x(t)=t^3-t \\ y(t)=2t \end{cases}$

20. $\begin{cases} x(t)=t-t^4 \\ y(t)=t+2 \end{cases}$

21. $\begin{cases} x(t)=e^{2t} \\ y(t)=e^{6t} \end{cases}$

22. $\begin{cases} x(t)=t^5 \\ y(t)=t^{10} \end{cases}$

23. $\begin{cases} x(t)=4\cos(t) \\ y(t)=5\sin(t) \end{cases}$

24. $\begin{cases} x(t)=3\sin(t) \\ y(t)=6\cos(t) \end{cases}$

Parameterize (write a parametric equation for) each Cartesian equation

25. $y(x) = 3x^2 + 3$

26. $y(x) = 2\sin(x) + 1$

27. $x(y) = 3\log(y) + y$

28. $x(y) = \sqrt{y} + 2y$

29. $\dfrac{x^2}{4} + \dfrac{y^2}{9} = 1$

30. $\dfrac{x^2}{16} + \dfrac{y^2}{36} = 1$

Parameterize the graphs shown.

31.

32.

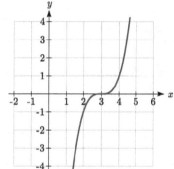

33.

34.

35. Parameterize the line from $(-1, 5)$ to $(2, 3)$ so that the line is at $(-1, 5)$ at $t = 0$, and at $(2, 3)$ at $t = 1$.

36. Parameterize the line from $(4, 1)$ to $(6, -2)$ so that the line is at $(4, 1)$ at $t = 0$, and at $(6, -2)$ at $t = 1$.

The graphs below are created by parameteric equations of the form $\begin{cases} x(t) = a\cos(bt) \\ y(t) = c\sin(dt) \end{cases}$.

Find the values of a, b, c, and d to achieve each graph.

37.

38.

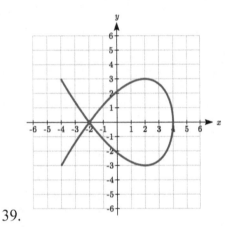

39.

40.

41. An object is thrown in the air with vertical velocity 20 ft/s and horizontal velocity 15 ft/s. The object's height can be described by the equation $y(t) = -16t^2 + 20t$, while the object moves horizontally with constant velocity 15 ft/s. Write parametric equations for the object's position, then eliminate time to write height as a function of horizontal position.

42. A skateboarder riding on a level surface at a constant speed of 9 ft/s throws a ball in the air, the height of which can be described by the equation $y(t) = -16t^2 + 10t + 5$. Write parametric equations for the ball's position, then eliminate time to write height as a function of horizontal position.

43. A carnival ride has a large rotating arm with diameter 40 feet centered 35 feet off the ground. At each end of the large arm are two smaller rotating arms with diameter 16 feet each. The larger arm rotates once every 5 seconds, while the smaller arms rotate once every 2 seconds. If you board the ride when the point P is closest to the ground, find parametric equations for your position over time.

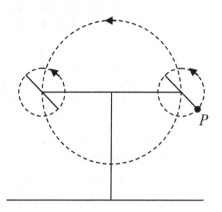

44. A hypocycloid is a shape generated by tracking a fixed point on a small circle as it rolls around the inside of a larger circle. If the smaller circle has radius 1 and the large circle has radius 6, find parametric equations for the position of the point P as the smaller wheel rolls in the direction indicated.

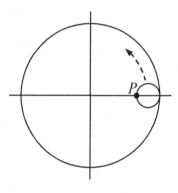

Chapter 9: Conics

In this chapter, we will explore a set of shapes defined by a common characteristic: they can all be formed by slicing a cone with a plane. These families of curves have a broad range of applications in physics and astronomy, from describing the shape of your car headlight reflectors to describing the orbits of planets and comets.

Section 9.1 Ellipses

The National Statuary Hall[1] in Washington, D.C. is an oval-shaped room called a whispering chamber because the shape makes it possible for sound to reflect from the walls in a special way. Two people standing in specific places are able to hear each other whispering even though they are far apart. To determine where they should stand, we will need to better understand ellipses.

An ellipse is a type of **conic section**, a shape resulting from intersecting a plane with a cone and looking at the curve where they intersect. They were discovered by the Greek mathematician Menaechmus over two millennia ago.

The figure below[2] shows two types of conic sections. When a plane is perpendicular to the axis of the cone, the shape of the intersection is a circle. A slightly titled plane creates an oval-shaped conic section called an **ellipse**.

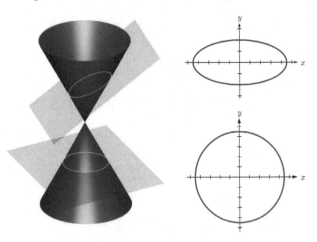

---

[1] Photo by Gary Palmer, Flickr, CC-BY, https://www.flickr.com/photos/gregpalmer/2157517950

[2] Pbroks13 (https://commons.wikimedia.org/wiki/File:Conic_sections_with_plane.svg), "Conic sections with plane", cropped to show only ellipse and circle by L Michaels, CC BY 3.0

An ellipse can be drawn by placing two thumbtacks in a piece of cardboard then cutting a piece of string longer than the distance between the thumbtacks. Tack each end of the string to the cardboard, and trace a curve with a pencil held taught against the string. An ellipse is the set of all points where the sum of the distances from two fixed points is constant. The length of the string is the constant, and the two thumbtacks are the fixed points, called foci.

Ellipse Definition and Vocabulary

An **ellipse** is the set of all points $Q(x, y)$ for which the sum of the distance to two fixed points $F_1(x_1, y_1)$ and $F_2(x_2, y_2)$, called the **foci** (plural of focus), is a constant k: $d(Q, F_1) + d(Q, F_2) = k$.

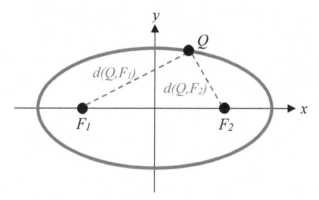

The **major axis** is the line passing through the foci.

Vertices are the points on the ellipse which intersect the major axis.

The **major axis length** is the length of the line segment between the vertices.

The **center** is the midpoint between the vertices (or the midpoint between the foci).

The **minor axis** is the line perpendicular to the minor axis passing through the center.

Minor axis endpoints are the points on the ellipse which intersect the minor axis.

The minor axis endpoints are also sometimes called **co-vertices**.

The **minor axis length** is the length of the line segment between minor axis endpoints.

Note that which axis is major and which is minor will depend on the orientation of the ellipse. In the ellipse shown at right, the foci lie on the y axis, so that is the major axis, and the x axis is the minor axis. Because of this, the vertices are the endpoints of the ellipse on the y axis, and the minor axis endpoints (co-vertices) are the endpoints on the x axis.

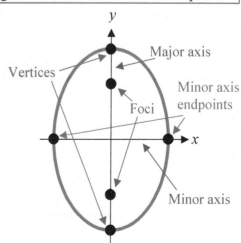

Ellipses Centered at the Origin

From the definition above we can find an equation for an ellipse. We will find it for a ellipse centered at the origin $C(0,0)$ with foci at $F_1(c,0)$ and $F_2(-c,0)$ where $c > 0$.

Suppose $Q(x, y)$ is some point on the ellipse. The distance from F_1 to Q is
$$d(Q, F_1) = \sqrt{(x-c)^2 + (y-0)^2} = \sqrt{(x-c)^2 + y^2}$$

Likewise, the distance from F_2 to Q is
$$d(Q, F_2) = \sqrt{(x-(-c))^2 + (y-0)^2} = \sqrt{(x+c)^2 + y^2}$$

From the definition of the ellipse, the sum of these distances should be constant:
$d(Q, F_1) + d(Q, F_2) = k$ so that
$$\sqrt{(x-c)^2 + y^2} + \sqrt{(x+c)^2 + y^2} = k$$

If we label one of the vertices $(a,0)$, it should satisfy the equation above since it is a point on the ellipse. This allows us to write k in terms of a.
$$\sqrt{(a-c)^2 + 0^2} + \sqrt{(a+c)^2 + 0^2} = k$$
$|a-c| + |a+c| = k$ Since $a > c$, these will be positive
$(a-c) + (a+c) = k$
$2a = k$

Substituting that into our equation, we will now try to rewrite the equation in a friendlier form.

$\sqrt{(x-c)^2 + y^2} + \sqrt{(x+c)^2 + y^2} = 2a$	Move one radical
$\sqrt{(x-c)^2 + y^2} = 2a - \sqrt{(x+c)^2 + y^2}$	Square both sides
$\left(\sqrt{(x-c)^2 + y^2}\right)^2 = \left(2a - \sqrt{(x+c)^2 + y^2}\right)^2$	Expand
$(x-c)^2 + y^2 = 4a^2 - 4a\sqrt{(x+c)^2 + y^2} + (x+c)^2 + y^2$	Expand more
$x^2 - 2xc + c^2 + y^2 = 4a^2 - 4a\sqrt{(x+c)^2 + y^2} + x^2 + 2xc + c^2 + y^2$	

Combining like terms and isolating the radical leaves

$4a\sqrt{(x+c)^2 + y^2} = 4a^2 + 4xc$	Divide by 4
$a\sqrt{(x+c)^2 + y^2} = a^2 + xc$	Square both sides again
$a^2\left((x+c)^2 + y^2\right) = a^4 + 2a^2xc + x^2c^2$	Expand
$a^2\left(x^2 + 2xc + c^2 + y^2\right) = a^4 + 2a^2xc + x^2c^2$	Distribute
$a^2x^2 + 2a^2xc + a^2c^2 + a^2y^2 = a^4 + 2a^2xc + x^2c^2$	Combine like terms

$$a^2x^2 - x^2c^2 + a^2y^2 = a^4 - a^2c^2 \qquad \text{Factor common terms}$$
$$(a^2 - c^2)x^2 + a^2y^2 = a^2(a^2 - c^2)$$

Let $b^2 = a^2 - c^2$. Since $a > c$, we know $b > 0$. Substituting b^2 for $a^2 - c^2$ leaves
$$b^2x^2 + a^2y^2 = a^2b^2 \qquad \text{Divide both sides by } a^2b^2$$
$$\frac{x^2}{a^2} + \frac{y^2}{b^2} = 1$$

This is the standard equation for an ellipse. We typically swap a and b when the major axis of the ellipse is vertical.

Equation of an Ellipse Centered at the Origin in Standard Form

The standard form of an equation of an ellipse centered at the origin $C(0,0)$ depends on whether the major axis is horizontal or vertical. The table below gives the standard equation, vertices, minor axis endpoints, foci, and graph for each.

Major Axis	Horizontal	Vertical
Standard Equation	$\dfrac{x^2}{a^2} + \dfrac{y^2}{b^2} = 1$	$\dfrac{x^2}{b^2} + \dfrac{y^2}{a^2} = 1$
Vertices	$(-a, 0)$ and $(a, 0)$	$(0, -a)$ and $(0, a)$
Minor Axis Endpoints	$(0, -b)$ and $(0, b)$	$(-b, 0)$ and $(b, 0)$
Foci	$(-c, 0)$ and $(c, 0)$ where $b^2 = a^2 - c^2$	$(0, -c)$ and $(0, c)$ where $b^2 = a^2 - c^2$
Graph		

Example 1

Put the equation of the ellipse $9x^2 + y^2 = 9$ in standard form. Find the vertices, minor axis endpoints, length of the major axis, and length of the minor axis. Sketch the graph, then check using a graphing utility.

The standard equation has a 1 on the right side, so this equation can be put in standard form by dividing by 9:

$$\frac{x^2}{1} + \frac{y^2}{9} = 1$$

Since the y-denominator is greater than the x-denominator, the ellipse has a vertical major axis. Comparing to the general standard form equation $\frac{x^2}{b^2} + \frac{y^2}{a^2} = 1$, we see the value of $a = \sqrt{9} = 3$ and the value of $b = \sqrt{1} = 1$.

The vertices lie on the y-axis at $(0, \pm a) = (0, \pm 3)$.
The minor axis endpoints lie on the x-axis at $(\pm b, 0) = (\pm 1, 0)$.
The length of the major axis is $2(a) = 2(3) = 6$.
The length of the minor axis is $2(b) = 2(1) = 2$.

To sketch the graph we plot the vertices and the minor axis endpoints. Then we sketch the ellipse, rounding at the vertices and the minor axis endpoints.

 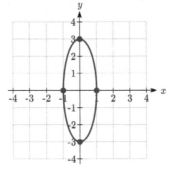

To check on a graphing utility, we must solve the equation for y. Isolating y^2 gives us
$$y^2 = 9(1 - x^2)$$

Taking the square root of both sides we get
$$y = \pm 3\sqrt{1 - x^2}$$

Under **Y=** on your graphing utility enter the two halves of the ellipse as $y = 3\sqrt{1 - x^2}$ and $y = -3\sqrt{1 - x^2}$. Set the window to a comparable scale to the sketch with xmin = -5, xmax = 5, ymin= -5, and ymax = 5.
Here's an example output on a TI-84 calculator:

Plot1 Plot2 Plot3 ■\Y₁■3√1-X² ■\Y₂■-3√1-X²	WINDOW Xmin= -5 Xmax=5 Xscl=1 Ymin= -5 Ymax=5 Yscl=1 Xres=1	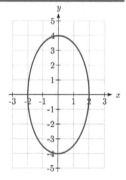

Sometimes we are given the equation. Sometimes we need to find the equation from a graph or other information.

Example 2

Find the standard form of the equation for an ellipse centered at (0,0) with horizontal major axis length 28 and minor axis length 16.

Since the center is at (0,0) and the major axis is horizontal, the ellipse equation has the standard form $\dfrac{x^2}{a^2}+\dfrac{y^2}{b^2}=1$. The major axis has length $2a = 28$ or $a = 14$. The minor axis has length $2b = 16$ or $b = 8$. Substituting gives $\dfrac{x^2}{16^2}+\dfrac{y^2}{8^2}=1$ or $\dfrac{x^2}{256}+\dfrac{y^2}{64}=1$.

Try it Now

1. Find the standard form of the equation for an ellipse with horizontal major axis length 20 and minor axis length 6.

Example 3

Find the standard form of the equation for the ellipse graphed here.

The center is at (0,0) and the major axis is vertical, so the standard form of the equation will be $\dfrac{x^2}{b^2}+\dfrac{y^2}{a^2}=1$.

From the graph we can see the vertices are (0,4) and (0,-4), giving $a = 4$.
The minor-axis endpoints are (2,0) and (-2,0), giving $b = 2$.

The equation will be $\dfrac{x^2}{2^2}+\dfrac{y^2}{4^2}=1$ or $\dfrac{x^2}{4}+\dfrac{y^2}{16}=1$.

Ellipses Not Centered at the Origin

Not all ellipses are centered at the origin. The graph of such an ellipse is a shift of the graph centered at the origin, so the standard equation for one centered at (h, k) is slightly different. We can shift the graph right h units and up k units by replacing x with $x - h$ and y with $y - k$, similar to what we did when we learned transformations.

Equation of an Ellipse Centered at (h, k) in Standard Form

The standard form of an equation of an ellipse centered at the point $C(h,k)$ depends on whether the major axis is horizontal or vertical. The table below gives the standard equation, vertices, minor axis endpoints, foci, and graph for each.

Major Axis	Horizontal	Vertical
Standard Equation	$\dfrac{(x-h)^2}{a^2}+\dfrac{(y-k)^2}{b^2}=1$	$\dfrac{(x-h)^2}{b^2}+\dfrac{(y-k)^2}{a^2}=1$
Vertices	$(h \pm a,\, k)$	$(h,\, k \pm a)$
Minor Axis Endpoints	$(h,\, k \pm b)$	$(h \pm b,\, k)$
Foci	$(h \pm c,\, k)$ where $b^2 = a^2 - c^2$	$(h,\, k \pm c)$ where $b^2 = a^2 - c^2$
Graph		

Example 4

Put the equation of the ellipse $x^2 + 2x + 4y^2 - 24y = -33$ in standard form. Find the vertices, minor axis endpoints, length of the major axis, and length of the minor axis. Sketch the graph.

To rewrite this in standard form, we will need to complete the square, twice.

Looking at the x terms, $x^2 + 2x$, we like to have something of the form $(x+n)^2$. Notice that if we were to expand this, we'd get $x^2 + 2nx + n^2$, so in order for the coefficient on x to match, we'll need $(x+1)^2 = x^2 + 2x + 1$. However, we don't have a +1 on the left side of the equation to allow this factoring. To accommodate this, we will add 1 to both sides of the equation, which then allows us to factor the left side as a perfect square:

$x^2 + 2x + 1 + 4y^2 - 24y = -33 + 1$

$(x+1)^2 + 4y^2 - 24y = -32$

Repeating the same approach with the y terms, first we'll factor out the 4.

$4y^2 - 24y = 4(y^2 - 6y)$

Now we want to be able to write $4(y^2 - 6y)$ as $4(y+n)^2 = 4(y^2 + 2ny + n^2)$. For the coefficient of y to match, n will have to -3, giving

$4(y-3)^2 = 4(y^2 - 6y + 9) = 4y^2 - 24y + 36$.

To allow this factoring, we can add 36 to both sides of the equation.

$(x+1)^2 + 4y^2 - 24y + 36 = -32 + 36$

$(x+1)^2 + 4(y^2 - 6y + 9) = 4$

$(x+1)^2 + 4(y-3)^2 = 4$

Dividing by 4 gives the standard form of the equation for the ellipse

$$\frac{(x+1)^2}{4} + \frac{(y-3)^2}{1} = 1$$

Since the x-denominator is greater than the y-denominator, the ellipse has a horizontal major axis. From the general standard equation $\frac{(x-h)^2}{a^2} + \frac{(h-k)^2}{b^2} = 1$ we see the value of $a = \sqrt{4} = 2$ and the value of $b = \sqrt{1} = 1$.

The center is at $(h, k) = (-1, 3)$.
The vertices are at $(h \pm a, k)$ or $(-3, 3)$ and $(1, 3)$.
The minor axis endpoints are at $(h, k \pm b)$ or $(-1, 2)$ and $(-1, 4)$.

The length of the major axis is $2(a) = 2(2) = 4$.
The length of the minor axis is $2(b) = 2(1) = 2$.

To sketch the graph we plot the vertices and the minor axis endpoints. Then we sketch the ellipse, rounding at the vertices and the minor axis endpoints.

 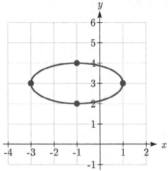

Example 5

Find the standard form of the equation for an ellipse centered at (-2,1), a vertex at (-2,4) and passing through the point (0,1).

The center at (-2,1) and vertex at (-2,4) means the major axis is vertical since the *x*-values are the same. The ellipse equation has the standard form $\dfrac{(x-h)^2}{b^2} + \dfrac{(y-k)^2}{a^2} = 1$.

The value of $a = 4-1=3$. Substituting $a = 3$, $h = -2$, and $k = 1$ gives $\dfrac{(x+2)^2}{b^2} + \dfrac{(y-1)^2}{3^2} = 1$. Substituting for x and y using the point (0,1) gives

$$\frac{(0+2)^2}{b^2} + \frac{(1-1)^2}{3^2} = 1.$$

Solving for b gives $b=2$.

The equation of the ellipse in standard form is $\dfrac{(x+2)^2}{2^2} + \dfrac{(y-1)^2}{3^2} = 1$ or

$$\frac{(x+2)^2}{4} + \frac{(y-1)^2}{9} = 1.$$

Try it Now

2. Find the center, vertices, minor axis endpoints, length of the major axis, and length of the minor axis for the ellipse $(x-4)^2 + \dfrac{(y+2)^2}{4} = 1$.

Bridges with Semielliptical Arches

Arches have been used to build bridges for centuries, like in the Skerton Bridge in England which uses five semielliptical arches for support[3]. Semielliptical arches can have engineering benefits such as allowing for longer spans between supports.

Example 6

A bridge over a river is supported by a single semielliptical arch. The river is 50 feet wide. At the center, the arch rises 20 feet above the river. The roadway is 4 feet above the center of the arch. What is the vertical distance between the roadway and the arch 15 feet from the center?

Put the center of the ellipse at (0,0) and make the span of the river the major axis.

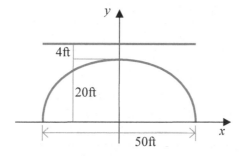

Since the major axis is horizontal, the equation has the form $\dfrac{x^2}{a^2}+\dfrac{y^2}{b^2}=1$.

The value of $a=\dfrac{1}{2}(50)=25$ and the value of $b=20$, giving $\dfrac{x^2}{25^2}+\dfrac{y^2}{15^2}=1$.

Substituting $x=15$ gives $\dfrac{15^2}{25^2}+\dfrac{y^2}{20^2}=1$. Solving for y, $y=20\sqrt{1-\dfrac{225}{625}}=16$.

The roadway is $20+4=24$ feet above the river. The vertical distance between the roadway and the arch 15 feet from the center is $24-16=8$ feet.

---

[3] Maxine Armstrong
(https://commons.wikimedia.org/wiki/File:Skerton_Bridge,_Lancaster,_England.JPG), "Skerton Bridge, Lancaster, England", CC BY-SA

Ellipse Foci

The location of the foci can play a key role in ellipse application problems. Standing on a focus in a whispering gallery allows you to hear someone whispering at the other focus. To find the foci, we need to find the length from the center to the foci, c, using the equation $b^2 = a^2 - c^2$. It looks similar to, but is not the same as, the Pythagorean Theorem.

Example 7

The National Statuary Hall whispering chamber is an elliptical room 46 feet wide and 96 feet long. To hear each other whispering, two people need to stand at the foci of the ellipse. Where should they stand?

We could represent the hall with a horizontal ellipse centered at the origin. The major axis length would be 96 feet, so $a = \dfrac{1}{2}(96) = 48$, and the minor axis length would be 46 feet, so $b = \dfrac{1}{2}(46) = 23$. To find the foci, we can use the equation $b^2 = a^2 - c^2$.

$23^2 = 48^2 - c^2$
$c^2 = 48^2 - 23^2$
$c = \sqrt{1775} \approx \pm 42$ ft.

To hear each other whisper, two people would need to stand $2(42) = 84$ feet apart along the major axis, each about $48 - 42 = 6$ feet from the wall.

Example 8

Find the foci of the ellipse $\dfrac{(x-2)^2}{4} + \dfrac{(y+3)^2}{29} = 1$.

The ellipse is vertical with an equation of the form $\dfrac{(x-h)^2}{b^2} + \dfrac{(y-k)^2}{a^2} = 1$.

The center is at $(h, k) = (2, -3)$. The foci are at $(h, k \pm c)$.

To find length c we use $b^2 = a^2 - c^2$.
Substituting gives $4 = 29 - c^2$ or $c = \sqrt{25} = 5$.

The ellipse has foci $(2, -3 \pm 5)$, or $(2, -8)$ and $(2, 2)$.

Example 9

Find the standard form of the equation for an ellipse with foci (-1,4) and (3,4) and major axis length 10.

Since the foci differ in the x-coordinates, the ellipse is horizontal with an equation of the form $\dfrac{(x-h)^2}{a^2}+\dfrac{(h-k)^2}{b^2}=1$.

The center is at the midpoint of the foci $\left(\dfrac{x_1+x_2}{2},\dfrac{y_1+y_2}{2}\right)=\left(\dfrac{(-1)+3}{2},\dfrac{4+4}{2}\right)=(1,4)$.

The value of a is half the major axis length: $a=\dfrac{1}{2}(10)=5$.

The value of c is half the distance between the foci: $c=\dfrac{1}{2}(3-(-1))=\dfrac{1}{2}(4)=2$.

To find length b we use $b^2=a^2-c^2$. Substituting a and c gives $b^2=5^2-2^2=21$.

The equation of the ellipse in standard form is $\dfrac{(x-1)^2}{5^2}+\dfrac{(y-4)^2}{21}=1$ or

$\dfrac{(x-1)^2}{25}+\dfrac{(y-4)^2}{21}=1$.

Try it Now

3. Find the standard form of the equation for an ellipse with focus (2,4), vertex (2,6), and center (2,1).

Planetary Orbits

It was long thought that planetary orbits around the sun were circular. Around 1600, Johannes Kepler discovered they were actually elliptical[4]. His first law of planetary motion says that planets travel around the sun in an elliptical orbit with the sun as one of the foci.

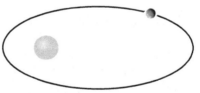

The length of the major axis can be found by measuring the planet's *aphelion*, its greatest distance from the sun, and *perihelion*, its shortest distance from the sun, and summing them together.

---

[4] Technically, they're approximately elliptical. The orbits of the planets are not exactly elliptical because of interactions with each other and other celestial bodies.

Example 10

Mercury's aphelion is 35.98 million miles and its perihelion is 28.58 million miles. Write an equation for Mercury's orbit.

Let the center of the ellipse be (0,0) and its major axis be horizontal so the equation will have form $\dfrac{x^2}{a^2} + \dfrac{y^2}{b^2} = 1$.

The length of the major axis is $2a = 35.98 + 28.58 = 64.56$ giving $a = 32.28$ and $a^2 = 1041.9984$.

Since the perihelion is the distance from the focus to one vertex, we can find the distance between the foci by subtracting twice the perihelion from the major axis length: $2c = 64.56 - 2(28.58) = 7.4$ giving $c = 3.7$.

Substitution of a and c into $b^2 = a^2 - c^2$ yields $b^2 = 32.28^2 - 3.7^2 = 1028.3084$.

The equation is $\dfrac{x^2}{1041.9984} + \dfrac{y^2}{1028.3084} = 1$.

Important Topics of This Section
Ellipse Definition
Ellipse Equations in Standard Form
Ellipse Foci
Applications of Ellipses

Try it Now Answers

1. $2a = 20$, so $a = 10$. $2b = 6$, so $b = 3$. $\dfrac{x^2}{100} + \dfrac{y^2}{9} = 1$

2. Center (4, -2). Vertical ellipse with $a = 2$, $b = 1$.
   Vertices at (4, -2±2) = (4,0) and (4,-4),
   minor axis endpoints at (4±1, -2) = (3,-2) and (5,-2),
   major axis length 4, minor axis length 2

3. Vertex, center, and focus have the same x-value, so it's a vertical ellipse.
   Using the vertex and center, $a = 6 - 1 = 5$
   Using the center and focus, $c = 4 - 1 = 3$
   $b^2 = 5^2 - 3^2$. $b = 4$.
   $\dfrac{(x-2)^2}{16} + \dfrac{(y-1)^2}{25} = 1$

Section 9.1 Exercises

In problems 1–4, match each graph with one of the equations A–D.

A. $\dfrac{x^2}{4}+\dfrac{y^2}{9}=1$ B. $\dfrac{x^2}{9}+\dfrac{y^2}{4}=1$ C. $\dfrac{x^2}{9}+y^2=1$ D. $x^2+\dfrac{y^2}{9}=1$

1.   2.   3.   4.

        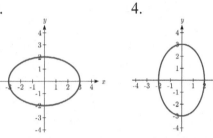

In problems 5–14, find the vertices, the minor axis endpoints, length of the major axis, and length of the minor axis. Sketch the graph. Check using a graphing utility.

5. $\dfrac{x^2}{4}+\dfrac{y^2}{25}=1$ 6. $\dfrac{x^2}{16}+\dfrac{y^2}{4}=1$ 7. $\dfrac{x^2}{4}+y^2=1$ 8. $x^2+\dfrac{y^2}{25}=1$

9. $x^2+25y^2=25$ 10. $16x^2+y^2=16$ 11. $16x^2+9y^2=144$

12. $16x^2+25y^2=400$ 13. $9x^2+y^2=18$ 14. $x^2+4y^2=12$

In problems 15–16, write an equation for the graph.
15.   16.

    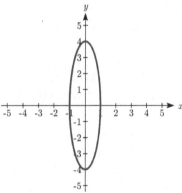

In problems 17–20, find the standard form of the equation for an ellipse satisfying the given conditions.
17. Center (0,0), horizontal major axis length 64, minor axis length 14

18. Center (0,0), vertical major axis length 36, minor axis length 18

19. Center (0,0), vertex (0,3), $b=2$

20. Center (0,0), vertex (4,0), $b=3$

In problems 21–28, match each graph to equations A-H.

A. $\dfrac{(x-2)^2}{4}+\dfrac{(y-1)^2}{9}=1$

E. $\dfrac{(x+2)^2}{4}+\dfrac{(y+1)^2}{9}=1$

B. $\dfrac{(x-2)^2}{4}+\dfrac{(y-1)^2}{16}=1$

F. $\dfrac{(x+2)^2}{4}+\dfrac{(y+1)^2}{16}=1$

C. $\dfrac{(x-2)^2}{16}+\dfrac{(y-1)^2}{4}=1$

G. $\dfrac{(x+2)^2}{16}+\dfrac{(y+1)^2}{4}=1$

D. $\dfrac{(x-2)^2}{9}+\dfrac{(y-1)^2}{4}=1$

H. $\dfrac{(x+2)^2}{9}+\dfrac{(y+1)^2}{4}=1$

21.

22.

23.

24.

25.

26.

27.

28.

In problems 29–38, find the vertices, the minor axis endpoints, length of the major axis, and length of the minor axis. Sketch the graph. Check using a graphing utility.

29. $\dfrac{(x-1)^2}{25}+\dfrac{(y+2)^2}{4}=1$

30. $\dfrac{(x+5)^2}{16}+\dfrac{(y-3)^2}{36}=1$

31. $(x+2)^2+\dfrac{(y-3)^2}{25}=1$

32. $\dfrac{(x-1)^2}{25}+(y-6)^2=1$

33. $4x^2+8x+4+y^2=16$

34. $x^2+4y^2+16y+16=36$

35. $x^2+2x+4y^2+16y=-1$

36. $4x^2+16x+y^2-8y=4$

37. $9x^2-36x+4y^2+8y=104$

38. $4x^2+8x+9y^2+36y=-4$

In problems 39–40, write an equation for the graph.

39.   40.

      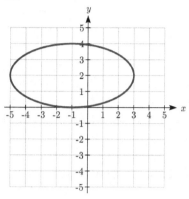

In problems 41–42, find the standard form of the equation for an ellipse satisfying the given conditions.

41. Center (-4,3), vertex(-4,8), point on the graph (0,3)

42. Center (1,-2), vertex(-5,-2), point on the graph (1,0)

43. **Window** A window in the shape of a semiellipse is 12 feet wide and 4 feet high. What is the height of the window above the base 5 feet from the center ?

44. **Window** A window in the shape of a semiellipse is 16 feet wide and 7 feet high. What is the height of the window above the base 4 feet from the center?

45. **Bridge** A bridge over a river is supported by a semielliptical arch. The river is 150 feet wide. At the center, the arch rises 60 feet above the river. The roadway is 5 feet above the center of the arch. What is the vertical distance between the roadway and the arch 45 feet from the center?

46. **Bridge** A bridge over a river is supported by a semielliptical arch. The river is 1250 feet wide. At the center, the arch rises 175 feet above the river. The roadway is 3 feet above the center of the arch. What is the vertical distance between the roadway and the arch 600 feet from the center?

47. **Racetrack** An elliptical racetrack is 100 feet long and 90 feet wide. What is the width of the racetrack 20 feet from a vertex on the major axis?

48. **Racetrack** An elliptical racetrack is 250 feet long and 150 feet wide. What is the width of the racetrack 25 feet from a vertex on the major axis?

In problems 49-52, find the foci.

49. $\dfrac{x^2}{19} + \dfrac{y^2}{3} = 1$

50. $\dfrac{x^2}{2} + \dfrac{y^2}{38} = 1$

51. $(x+6)^2 + \dfrac{(y-1)^2}{26} + = 1$

52. $\dfrac{(x-3)^2}{10} + (y+5)^2 = 1$

In problems 53-72, find the standard form of the equation for an ellipse satisfying the given conditions.

53. Major axis vertices (±3,0), c=2

54. Major axis vertices (0,±7), c=4

55. Foci (0,±5) and major axis length 12

56. Foci (±3,0) and major axis length 8

57. Foci (±5,0), vertices (±7,0)

58. Foci (0,±2), vertices (0,±3)

59. Foci (0,±4) and x-intercepts (±2,0)

60. Foci (±3,0) and y-intercepts (0,±1)

61. Center (0,0), major axis length 8, foci on x-axis, passes through point $\left(2, \sqrt{6}\right)$

62. Center (0,0), major axis length 12, foci on y-axis, passes through point $\left(\sqrt{10},4\right)$

63. Center (-2,1), vertex (-2,5), focus (-2,3)

64. Center (-1,-3), vertex (-7,-3), focus (-4,-3)

65. Foci (8,2) and (-2,2), major axis length 12

66. Foci (-1,5) and (-1,-3), major axis length 14

67. Vertices (3,4) and (3,-6), c= 2

68. Vertices (2,2) and (-4,2), c= 2

69. Center (1,3), focus (0,3), passes through point (1,5)

70. Center (-1,-2), focus (1,-2), passes through point (2,-2)

71. Focus (-15,-1), vertices (-19,-1) and (15,-1)

72. Focus (-3,2), vertices (-3,4) and (-3,-8)

73. **Whispering Gallery** If an elliptical whispering gallery is 80 feet long and 25 feet wide, how far from the center of room should someone stand on the major axis of the ellipse to experience the whispering effect? Round to two decimal places.

74. **Billiards** Some billiards tables are elliptical and have the foci marked on the table. If such a one is 8 feet long and 6 feet wide, how far are the foci from the center of the ellipse? Round to two decimal places.

75. **Planetary Orbits** The orbits of planets around the sun are approximately elliptical with the sun as a focus. The *aphelion* is a planet's greatest distance from the sun and the *perihelion* is its shortest. The length of the major axis is the sum of the aphelion and the perihelion. Earth's aphelion is 94.51 million miles and its perihelion is 91.40 million miles. Write an equation for Earth's orbit.

76. **Satellite Orbits** The orbit of a satellite around Earth is elliptical with Earth's center as a focus. The satellite's maximum height above the Earth is 170 miles and its minimum height above the Earth is 90 miles. Write an equation for the satellite's orbit. Assume Earth is spherical and has a radius of 3960 miles.

77. **Eccentricity** e of an ellipse is the ratio $\dfrac{c}{a}$ where c is the distance of a focus from the center and a is the distance of a vertex from the center. Write an equation for an ellipse with eccentricity 0.8 and foci at (-4,0) and (4,0).

78. **Confocal** ellipses have the same foci. Show that, for $k > 0$, all ellipses of the form $\dfrac{x^2}{6+k} + \dfrac{y^2}{k} = 1$ are confocal.

79. The **latus rectum** of an ellipse is a line segment with endpoints on the ellipse that passes through a focus and is perpendicular to the major axis. Show that $\dfrac{2b^2}{a}$ is the length of the latus rectum of $\dfrac{x^2}{a^2} + \dfrac{y^2}{b^2} = 1$ where $a > b$.

Section 9.2 Hyperbolas

In the last section, we learned that planets have approximately elliptical orbits around the sun. When an object like a comet is moving quickly, it is able to escape the gravitational pull of the sun and follows a path with the shape of a **hyperbola**. Hyperbolas are curves that can help us find the location of a ship, describe the shape of cooling towers, or calibrate seismological equipment.

The hyperbola is another type of conic section created by intersecting a plane with a double cone, as shown below[5].

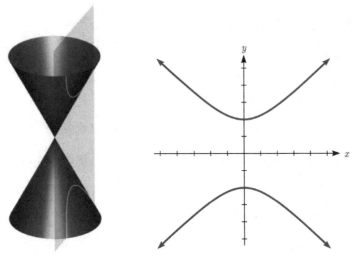

The word "hyperbola" derives from a Greek word meaning "excess." The English word "hyperbole" means exaggeration. We can think of a hyperbola as an excessive or exaggerated ellipse, one turned inside out.

We defined an ellipse as the set of all points where the sum of the distances from that point to two fixed points is a constant. A hyperbola is the set of all points where the absolute value of the difference of the distances from the point to two fixed points is a constant.

---

[5] Pbroks13 (https://commons.wikimedia.org/wiki/File:Conic_sections_with_plane.svg), "Conic sections with plane", cropped to show only a hyperbola by L Michaels, CC BY 3.0

Hyperbola Definition

A **hyperbola** is the set of all points $Q(x, y)$ for which the absolute value of the difference of the distances to two fixed points $F_1(x_1, y_1)$ and $F_2(x_2, y_2)$ called the **foci** (plural for focus) is a constant k: $\left| d(Q, F_1) - d(Q, F_2) \right| = k$.

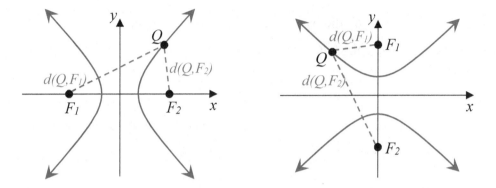

The **transverse axis** is the line passing through the foci.

Vertices are the points on the hyperbola which intersect the transverse axis.

The **transverse axis length** is the length of the line segment between the vertices.

The **center** is the midpoint between the vertices (or the midpoint between the foci).

The other axis of symmetry through the center is the **conjugate axis**.

The two disjoint pieces of the curve are called **branches**.

A hyperbola has two **asymptotes**.

Which axis is the transverse axis will depend on the orientation of the hyperbola. As a helpful tool for graphing hyperbolas, it is common to draw a **central rectangle** as a guide. This is a rectangle drawn around the center with sides parallel to the coordinate axes that pass through each vertex and co-vertex. The asymptotes will follow the diagonals of this rectangle.

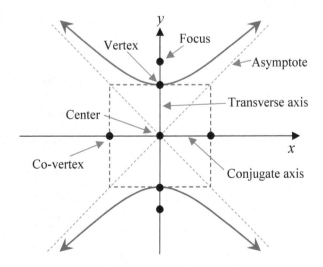

Hyperbolas Centered at the Origin

From the definition above we can find an equation of a hyperbola. We will find it for a hyperbola centered at the origin $C(0,0)$ opening horizontally with foci at $F_1(c,0)$ and $F_2(-c,0)$ where $c > 0$.

Suppose $Q(x, y)$ is a point on the hyperbola. The distances from Q to F_1 and Q to F_2 are:

$$d(Q,F_1) = \sqrt{(x-c)^2 + (y-0)^2} = \sqrt{(x-c)^2 + y^2}$$
$$d(Q,F_2) = \sqrt{(x-(-c))^2 + (y-0)^2} = \sqrt{(x+c)^2 + y^2}\,.$$

From the definition, the absolute value of the difference should be constant:

$$\left| d(Q,F_1) - d(Q,F_2) \right| = \left| \sqrt{(x-c)^2 + y^2} - \sqrt{(x+c)^2 + y^2} \right| = k$$

Substituting in one of the vertices $(a,0)$, we can determine k in terms of a:

$$\left| \sqrt{(a-c)^2 + 0^2} - \sqrt{(a+c)^2 + 0^2} \right| = k$$

$$\left\| a-c \right| - \left| a+c \right\| = k \qquad\qquad \text{Since } c > a,\ |a-c| = c-a$$

$$\left| (c-a) - (a+c) \right| = k$$

$$k = \left| -2a \right| = \left| 2a \right|$$

Using $k = 2a$ and removing the absolute values,

$$\sqrt{(x-c)^2 + y^2} - \sqrt{(x+c)^2 + y^2} = \pm 2a \qquad\qquad \text{Move one radical}$$

$$\sqrt{(x-c)^2 + y^2} = \pm 2a + \sqrt{(x+c)^2 + y^2} \qquad\qquad \text{Square both sides}$$

$$(x-c)^2 + y^2 = 4a^2 \pm 4a\sqrt{(x+c)^2 + y^2} + (x+c)^2 + y^2 \quad \text{Expand}$$

$$x^2 - 2xc + c^2 + y^2 = 4a^2 \pm 4a\sqrt{(x+c)^2 + y^2} + x^2 + 2xc + c^2 + y^2$$

Combining like terms leaves

$$-4xc = 4a^2 \pm 4a\sqrt{(x+c)^2 + y^2} \qquad\qquad \text{Divide by 4}$$

$$-xc = a^2 \pm a\sqrt{(x+c)^2 + y^2} \qquad\qquad \text{Isolate the radical}$$

$$\pm a\sqrt{(x+c)^2 + y^2} = -a^2 - xc \qquad\qquad \text{Square both sides again}$$

$$a^2\left((x+c)^2 + y^2\right) = a^4 + 2a^2 xc + x^2 c^2 \qquad\qquad \text{Expand and distribute}$$

$$a^2 x^2 + 2a^2 xc + a^2 c^2 + a^2 y^2 = a^4 + 2a^2 xc + x^2 c^2 \qquad \text{Combine like terms}$$

$$a^2 y^2 + a^2 c^2 - a^4 = x^2 c^2 - a^2 x^2 \qquad\qquad \text{Factor common terms}$$

$$a^2 y^2 + a^2\left(c^2 - a^2\right) = \left(c^2 - a^2\right)x^2$$

Let $b^2 = c^2 - a^2$. Since $c > a$, $b > 0$. Substituting b^2 for $c^2 - a^2$ leaves

$a^2 y^2 + a^2 b^2 = b^2 x^2$ Divide both sides by $a^2 b^2$

$\dfrac{y^2}{b^2} + 1 = \dfrac{x^2}{a^2}$ Rewrite

$\dfrac{x^2}{a^2} - \dfrac{y^2}{b^2} = 1$

We can see from the graphs of the hyperbolas that the branches appear to approach asymptotes as x gets large in the negative or positive direction. The equations of the horizontal hyperbola asymptotes can be derived from its standard equation.

$\dfrac{x^2}{a^2} - \dfrac{y^2}{b^2} = 1$ Solve for y

$y^2 = b^2 \left(\dfrac{x^2}{a^2} - 1 \right)$ Rewrite 1 as $\dfrac{x^2}{a^2} \dfrac{a^2}{x^2}$

$y^2 = b^2 \left(\dfrac{x^2}{a^2} - \dfrac{x^2}{a^2} \dfrac{a^2}{x^2} \right)$ Factor out $\dfrac{x^2}{a^2}$

$y^2 = b^2 \dfrac{x^2}{a^2} \left(1 - \dfrac{a^2}{x^2} \right)$ Take the square root

$y = \pm \dfrac{b}{a} x \sqrt{1 - \dfrac{a^2}{x^2}}$

As $x \to \pm\infty$ the quantity $\dfrac{a^2}{x^2} \to 0$ and $\sqrt{1 - \dfrac{a^2}{x^2}} \to 1$, so the asymptotes are $y = \pm \dfrac{b}{a} x$.

Similarly, for vertical hyperbolas the asymptotes are $y = \pm \dfrac{a}{b} x$.

The standard form of an equation of a hyperbola centered at the origin $C(0,0)$ depends on whether it opens horizontally or vertically. The following table gives the standard equation, vertices, foci, asymptotes, construction rectangle vertices, and graph for each.

Equation of a Hyperbola Centered at the Origin in Standard Form

Opens	Horizontally	Vertically
Standard Equation	$\dfrac{x^2}{a^2} - \dfrac{y^2}{b^2} = 1$	$\dfrac{y^2}{a^2} - \dfrac{x^2}{b^2} = 1$
Vertices	$(-a, 0)$ and $(a, 0)$	$(0, -a)$ and $(0, a)$
Foci	$(-c, 0)$ and $(c, 0)$ where $b^2 = c^2 - a^2$	$(0, -c)$ and $(0, c)$ Where $b^2 = c^2 - a^2$
Asymptotes	$y = \pm \dfrac{b}{a} x$	$y = \pm \dfrac{a}{b} x$
Construction Rectangle Vertices	$(a, b), (-a, b), (a, -b), (-a, -b)$	$(b, a), (-b, a), (b, -a), (-b, -a)$
Graph		

Example 1

Put the equation of the hyperbola $y^2 - 4x^2 = 4$ in standard form. Find the vertices, length of the transverse axis, and the equations of the asymptotes. Sketch the graph. Check using a graphing utility.

The equation can be put in standard form $\dfrac{y^2}{4} - \dfrac{x^2}{1} = 1$ by dividing by 4.

Comparing to the general standard equation $\dfrac{y^2}{a^2} - \dfrac{x^2}{b^2} = 1$ we see that $a = \sqrt{4} = 2$ and $b = \sqrt{1} = 1$.

Since the x term is subtracted, the hyperbola opens vertically and the vertices lie on the y-axis at $(0, \pm a) = (0, \pm 2)$.

The length of the transverse axis is $2(a) = 2(2) = 4$.

Equations of the asymptotes are $y = \pm \dfrac{a}{b} x$ or $= \pm 2x$.

To sketch the graph we plot the vertices of the construction rectangle at $(\pm b, \pm a)$ or $(-1,-2)$, $(-1,2)$, $(1,-2)$, and $(1,2)$. The asymptotes are drawn through the diagonals of the rectangle and the vertices plotted. Then we sketch in the hyperbola, rounded at the vertices and approaching the asymptotes.

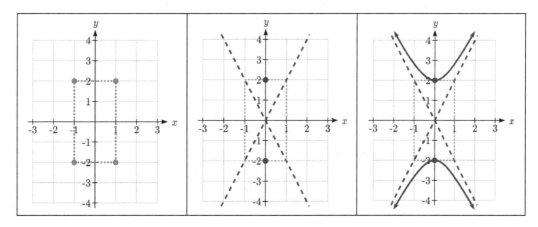

To check on a graphing utility, we must solve the equation for y. Isolating y^2 gives us $y^2 = 4(1 + x^2)$.

Taking the square root of both sides we find $y = \pm 2\sqrt{1 + x^2}$.

Under **Y=** enter the two halves of the hyperbola and the two asymptotes as $y = 2\sqrt{1 + x^2}$, $y = -2\sqrt{1 + x^2}$, $y = 2x$, and $y = -2x$. Set the window to a comparable scale to the sketch with xmin = -4, xmax = 4, ymin= -3, and ymax = 3.

Plot1 Plot2 Plot3	WINDOW	
\Y₁■2√1+X²	Xmin=■4	
\Y₂■-2√1+X²	Xmax=4	
\Y₃■2X	Xscl=1	
\Y₄■-2X	Ymin=-3	
	Ymax=3	
	Yscl=1	
	Xres=1	

Sometimes we are given the equation. Sometimes we need to find the equation from a graph or other information.

Example 2

Find the standard form of the equation for a hyperbola with vertices at (-6,0) and (6,0) and asymptote $y = \dfrac{4}{3}x$.

Since the vertices lie on the x-axis with a midpoint at the origin, the hyperbola is horizontal with an equation of the form $\dfrac{x^2}{a^2} - \dfrac{y^2}{b^2} = 1$. The value of a is the distance from the center to a vertex. The distance from (6,0) to (0,0) is 6, so $a = 6$.

The asymptotes follow the form $y = \pm \dfrac{b}{a}x$. From $y = \dfrac{3}{4}x$ we see $\dfrac{3}{4} = \dfrac{b}{a}$ and substituting $a = 6$ give us $\dfrac{3}{4} = \dfrac{b}{6}$. Solving yields $b = 8$.

The equation of the hyperbola in standard form is $\dfrac{x^2}{6^2} - \dfrac{y^2}{8^2} = 1$ or $\dfrac{x^2}{36} - \dfrac{y^2}{64} = 1$.

Try it Now

1. Find the standard form of the equation for a hyperbola with vertices at (0,-8) and (0,8) and asymptote $y = 2x$

Example 3

Find the standard form of the equation for a hyperbola with vertices at (0, 9) and (0,-9) and passing through the point (8,15).

Since the vertices lie on the y-axis with a midpoint at the origin, the hyperbola is vertical with an equation of the form $\dfrac{y^2}{a^2} - \dfrac{x^2}{b^2} = 1$. The value of a is the distance from the center to a vertex. The distance from (0,9) to (0,0) is 9, so $a = 9$.

Substituting $a = 9$ and the point (8,15) gives $\dfrac{15^2}{9^2} - \dfrac{8^2}{b^2} = 1$. Solving for b ycilds

$b = \sqrt{\dfrac{9^2 \left(8^2\right)}{15^2 - 9^2}} = 6$.

The standard equation for the hyperbola is $\dfrac{y^2}{9^2} - \dfrac{x^2}{6^2} = 1$ or $\dfrac{y^2}{81} - \dfrac{x^2}{36} = 1$.

Hyperbolas Not Centered at the Origin

Not all hyperbolas are centered at the origin. The standard equation for one centered at (h, k) is slightly different.

Equation of a Hyperbola Centered at (h, k) in Standard Form

The standard form of an equation of a hyperbola centered at $C(h,k)$ depends on whether it opens horizontally or vertically. The table below gives the standard equation, vertices, foci, asymptotes, construction rectangle vertices, and graph for each.

Opens	Horizontally	Vertically
Standard Equation	$\dfrac{(x-h)^2}{a^2} - \dfrac{(y-k)^2}{b^2} = 1$	$\dfrac{(y-k)^2}{a^2} - \dfrac{(x-h)^2}{b^2} = 1$
Vertices	$(h \pm a, k)$	$(h, k \pm a)$
Foci	$(h \pm c, k)$ where $b^2 = c^2 - a^2$	$(h, k \pm c)$ where $b^2 = c^2 - a^2$
Asymptotes	$y - k = \pm\dfrac{b}{a}(x-h)$	$y - k = \pm\dfrac{a}{b}(x-h)$
Construction Rectangle Vertices	$(h \pm a, k \pm b)$	$(h \pm b, k \pm a)$
Graph		

Example 4

Write an equation for the hyperbola in the graph shown.

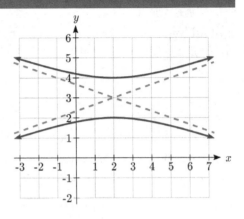

The center is at (2,3), where the asymptotes cross. It opens vertically, so the equation will look like

$$\frac{(y-3)^2}{a^2} - \frac{(x-2)^2}{b^2} = 1.$$

The vertices are at (2,2) and (2,4). The distance from the center to a vertex is $a = 4 - 3 = 1$.

If we were to draw in the construction rectangle, it would extend from $x = -1$ to $x = 5$. The distance from the center to the right side of the rectangle gives $b = 5 - 2 = 3$.

The standard equation of this hyperbola is $\frac{(y-3)^2}{1^2} - \frac{(x-2)^2}{3^2} = 1$, or

$$(y-3)^2 - \frac{(x-2)^2}{9} = 1.$$

Example 5

Put the equation of the hyperbola $9x^2 + 18x - 4y^2 + 16y = 43$ in standard form. Find the center, vertices, length of the transverse axis, and the equations of the asymptotes. Sketch the graph, then check on a graphing utility.

To rewrite the equation, we complete the square for both variables to get
$$9(x^2 + 2x + 1) - 4(y^2 - 4y + 4) = 43 + 9 - 16$$
$$9(x+1)^2 - 4(y-2)^2 = 36$$

Dividing by 36 gives the standard form of the equation, $\frac{(x+1)^2}{4} - \frac{(y-2)^2}{9} = 1$

Comparing to the general standard equation $\frac{(x-h)^2}{a^2} - \frac{(h-k)^2}{b^2} = 1$ we see that

$a = \sqrt{4} = 2$ and $b = \sqrt{9} = 3$.

Since the y term is subtracted, the hyperbola opens horizontally.
The center is at $(h, k) = (-1, 2)$.
The vertices are at $(h \pm a, k)$ or (-3, 2) and (1,2).
The length of the transverse axis is $2(a) = 2(2) = 4$.

Equations of the asymptotes are $y - k = \pm \frac{b}{a}(x - h)$ or $y - 2 = \pm \frac{3}{2}(x+1)$.

To sketch the graph we plot the corners of the construction rectangle at $(h\pm a, k\pm b)$ or $(1, 5)$, $(1, -1)$, $(-3,5)$, and $(-3,-1)$.   The asymptotes are drawn through the diagonals of the rectangle and the vertices plotted. Then we sketch in the hyperbola rounded at the vertices and approaching the asymptotes.

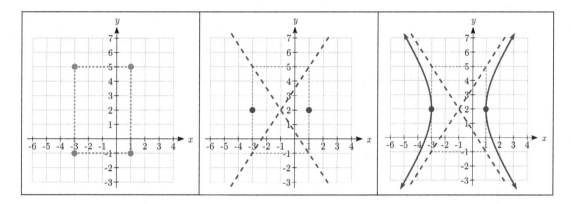

To check on a graphing utility, we must solve the equation for y.

$$y = 2 \pm \sqrt{9\left(\frac{(x+1)^2}{4} - 1\right)}.$$

Under **Y=** enter the two halves of the hyperbola and the two asymptotes as

$$y = 2 + \sqrt{9\left(\frac{(x+1)^2}{4} - 1\right)}, \quad y = 2 - \sqrt{9\left(\frac{(x+1)^2}{4} - 1\right)}, \quad y = \frac{3}{2}(x+1) + 2, \text{ and}$$

$$y = -\frac{3}{2}(x+1) + 2.$$ Set the window to a comparable scale to the sketch, then graph.

Note that the gaps you see on the calculator are not really there; they're a limitation of the technology.

```
Plot1  Plot2  Plot3              WINDOW
■\Y₁■2+√(9((X+1)²/4-1))          Xmin=-6
                                 Xmax=4
■\Y₂■2-√(9((X+1)²/4-1))          Xscl=1
■\Y₃■3/2(X+1)+2                  Ymin=-3
■\Y₄■-3/2(X+1)+2                 Ymax=7
                                 Yscl=1
                                 Xres=1
```

Example 6

Find the standard form of the equation for a hyperbola with vertices at $(-2,-5)$ and $(-2,7)$, and asymptote $y = \frac{3}{2}x + 4$.

Since the vertices differ in the y-coordinates, the hyperbola opens vertically with an equation of the form $\dfrac{(y-k)^2}{a^2} - \dfrac{(x-h)^2}{b^2} = 1$ and asymptote equations of the form $y - k = \pm \dfrac{a}{b}(x-h)$.

The center will be halfway between the vertices, at $\left(-2, \dfrac{-5+7}{2}\right) = (-2,1)$.

The value of a is the distance from the center to a vertex. The distance from $(-2,1)$ to $(-2,-5)$ is 6, so $a = 6$.

While our asymptote is not given in the form $y - k = \pm \dfrac{a}{b}(x-h)$, notice this equation would have slope $\dfrac{a}{b}$. We can compare that to the slope of the given asymptote equation to find b. Setting $\dfrac{3}{2} = \dfrac{a}{b}$ and substituting $a = 6$ gives us $b = 4$.

The equation of the hyperbola in standard form is $\dfrac{(y-1)^2}{6^2} - \dfrac{(x+2)^2}{4^2} = 1$ or $\dfrac{(y-1)^2}{36} - \dfrac{(x+2)^2}{16} = 1$.

Try it Now

2. Find the center, vertices, length of the transverse axis, and equations of the asymptotes for the hyperbola $\dfrac{(x+5)^2}{9} - \dfrac{(y-2)^2}{36} = 1$.

Hyperbola Foci

The location of the foci can play a key role in hyperbola application problems. To find them, we need to find the length from the center to the foci, c, using the equation $b^2 = c^2 - a^2$. It looks similar to, but is not the same as, the Pythagorean Theorem.

Compare this with the equation to find length c for ellipses, which is $b^2 = a^2 - c^2$. If you remember that for the foci to be inside the ellipse they have to come before the vertices $(c < a)$, it's clear why we would calculate a^2 minus c^2. To be inside a hyperbola, the foci have to go beyond the vertices $(c > a)$, so we can see for hyperbolas we need c^2 minus a^2, the opposite.

Example 7

Find the foci of the hyperbola $\dfrac{(y+1)^2}{4} - \dfrac{(x-3)^2}{5} = 1$.

The hyperbola is vertical with an equation of the form $\dfrac{(y-k)^2}{a^2} - \dfrac{(x-h)^2}{b^2} = 1$.
The center is at $(h, k) = (3, -1)$. The foci are at $(h, k \pm c)$.

To find length c we use $b^2 = c^2 - a^2$. Substituting gives $5 = c^2 - 4$ or $c = \sqrt{9} = 3$.

The hyperbola has foci $(3, -4)$ and $(3, 2)$.

Example 8

Find the standard form of the equation for a hyperbola with foci $(5, -8)$ and $(-3, -8)$ and vertices $(4, -8)$ and $(-2, -8)$.

Since the vertices differ in the x-coordinates, the hyperbola opens horizontally with an equation of the form $\dfrac{(x-h)^2}{a^2} - \dfrac{(y-k)^2}{b^2} = 1$.

The center is at the midpoint of the vertices
$\left(\dfrac{x_1 + x_2}{2}, \dfrac{y_1 + y_2}{2} \right) = \left(\dfrac{4 + (-2)}{2}, \dfrac{-8 + (-8)}{2} \right) = (1, -8)$.

The value of a is the horizontal length from the center to a vertex, or $a = 4 - 1 = 3$.
The value of c is the horizontal length from the center to a focus, or $= 5 - 1 = 4$.
To find length b we use $b^2 = c^2 - a^2$. Substituting gives $b^2 = 16 - 9 = 7$.

The equation of the hyperbola in standard form is $\dfrac{(x-1)^2}{3^2} - \dfrac{(y-(-8))^2}{7} = 1$ or

$\dfrac{(x-1)^2}{9} - \dfrac{(y+8)^2}{7} = 1$.

Try it Now

3. Find the standard form of the equation for a hyperbola with focus $(1,9)$, vertex $(1,8)$, center $(1,4)$.

LORAN

Before GPS, the Long Range Navigation (LORAN) system was used to determine a ship's location. Two radio stations A and B simultaneously sent out a signal to a ship. The difference in time it took to receive the signal was computed as a distance locating the ship on the hyperbola with the A and B radio stations as the foci. A second pair of radio stations C and D sent simultaneous signals to the ship and computed its location on the hyperbola with C and D as the foci. The point P where the two hyperbolas intersected gave the location of the ship.

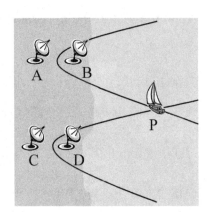

Example 9

Stations A and B are 150 kilometers apart and send a simultaneous radio signal to the ship. The signal from B arrives 0.0003 seconds before the signal from A. If the signal travels 300,000 kilometers per second, find the equation of the hyperbola on which the ship is positioned.

Stations A and B are at the foci, so the distance from the center to one focus is half the distance between them, giving $c = \frac{1}{2}(150) = 75$ km.

By letting the center of the hyperbola be at (0,0) and placing the foci at (±75,0), the equation $\frac{x^2}{a^2} - \frac{y^2}{b^2} = 1$ for a hyperbola centered at the origin can be used.

The difference of the distances of the ship from the two stations is
$k = 300,000\dfrac{\text{km}}{\text{s}} \cdot (0.0003\,\text{s}) = 90\,\text{km}$. From our derivation of the hyperbola equation we determined $k = 2a$, so $a = \frac{1}{2}(90) = 45$.

Substituting a and c into $b^2 = c^2 - a^2$ yields $b^2 = 75^2 - 45^2 = 3600$.

The equation of the hyperbola in standard form is $\dfrac{x^2}{45^2} - \dfrac{y^2}{3600} = 1$ or $\dfrac{x^2}{2025} - \dfrac{y^2}{3600} = 1$.

To determine the position of a ship using LORAN, we would need an equation for the second hyperbola and would solve for the intersection. We will explore how to do that in the next section.

Important Topics of This Section
Hyperbola Definition
Hyperbola Equations in Standard Form
Hyperbola Foci
Applications of Hyperbolas
Intersections of Hyperbolas and Other Curves

Try it Now Answers

1. The vertices are on the y axis so this is a vertical hyperbola.
   The center is at the origin.
   $a = 8$

   Using the asymptote slope, $\dfrac{8}{b} = 2$, so $b = 4$.

   $$\dfrac{y^2}{64} - \dfrac{x^2}{16} = 1$$

2. Center (-5, 2). This is a horizontal hyperbola. $a = 3$. $b = 6$.
   transverse axis length 6,
   Vertices will be at (-5±3,2) = (-2,2) and (-8,2),

   Asymptote slope will be $\dfrac{6}{3} = 2$. Asymptotes: $y - 2 = \pm 2(x + 5)$

3. Focus, vertex, and center have the same x value so this is a vertical hyperbola.
   Using the vertex and center, $a = 9 - 4 = 5$
   Using the focus and center, $c = 8 - 4 = 4$
   $b^2 = 5^2 - 4^2$. $b = 3$.

   $$\dfrac{(y-4)^2}{16} - \dfrac{(x-1)^2}{9} = 1$$

Section 9.2 Exercises

In problems 1–4, match each graph to equations A–D.

A. $\dfrac{x^2}{4} - \dfrac{y^2}{9} = 1$ B. $\dfrac{x^2}{9} - \dfrac{y^2}{4} = 1$ C. $y^2 - \dfrac{x^2}{9} = 1$ D. $\dfrac{y^2}{9} - x^2 = 1$

1.   2.   3.   4.

      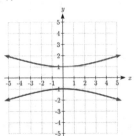

In problems 5–14, find the vertices, length of the transverse axis, and equations of the asymptotes. Sketch the graph. Check using a graphing utility.

5. $\dfrac{x^2}{4} - \dfrac{y^2}{25} = 1$ 6. $\dfrac{y^2}{16} - \dfrac{x^2}{9} = 1$ 7. $y^2 - \dfrac{x^2}{4} = 1$ 8. $x^2 - \dfrac{y^2}{25} = 1$

9. $x^2 - 9y^2 = 9$ 10. $y^2 - 4x^2 = 4$ 11. $9y^2 - 16x^2 = 144$

12. $16x^2 - 25y^2 = 400$ 13. $9x^2 - y^2 = 18$ 14. $4y^2 - x^2 = 12$

In problems 15–16, write an equation for the graph.

15.   16.

                     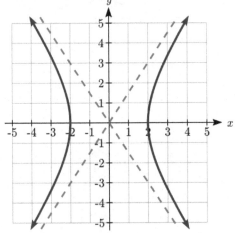

In problems 17–22, find the standard form of the equation for a hyperbola satisfying the given conditions.

17. Vertices at (0,4) and (0, -4); asymptote $y = \dfrac{1}{2}x$

18. Vertices at (-6,0) and (6,0); asymptote $y = 3x$

19. Vertices at (-3,0) and (3,0); passes through (5,8)

20. Vertices at (0, 4) and (0, -4); passes through (6, 5)

21. Asymptote y = x; passes through (5, 3)

22. Asymptote y = x; passes through (12, 13)

In problems 23–30, match each graph to equations A–H.

A. $\dfrac{(x-1)^2}{9} - \dfrac{(y-2)^2}{4} = 1$

E. $\dfrac{(y-2)^2}{4} - \dfrac{(x-1)^2}{9} = 1$

B. $\dfrac{(x+1)^2}{9} - \dfrac{(y+2)^2}{4} = 1$

F. $\dfrac{(y+2)^2}{4} - \dfrac{(x+1)^2}{9} = 1$

C. $\dfrac{(x+1)^2}{9} - \dfrac{(y+2)^2}{16} = 1$

G. $\dfrac{(y+2)^2}{4} - \dfrac{(x+1)^2}{16} = 1$

D. $\dfrac{(x-1)^2}{9} - \dfrac{(y-2)^2}{16} = 1$

H. $\dfrac{(y-2)^2}{4} - \dfrac{(x-1)^2}{16} = 1$

23.

24.

25.

26.

27.

28.

29.

30.

In problems 31–40, find the center, vertices, length of the transverse axis, and equations of the asymptotes. Sketch the graph. Check using a graphing utility.

31. $\dfrac{(x-1)^2}{25} - \dfrac{(y+2)^2}{4} = 1$

32. $\dfrac{(y-3)^2}{16} - \dfrac{(x+5)^2}{36} = 1$

33. $\dfrac{(y-1)^2}{9} - (x+2)^2 = 1$

34. $\dfrac{(x-1)^2}{25} - (y-6)^2 = 1$

35. $4x^2 - 8x - y^2 = 12$

36. $4y^2 + 16y - 9x^2 = 20$

37. $4y^2 - 16y - x^2 - 2x = 1$

38. $4x^2 - 16x - y^2 + 6y = 29$

39. $9x^2 + 36x - 4y^2 + 8y = 4$

40. $9y^2 + 36y - 16x^2 - 96x = -36$

In problems 41–42, write an equation for the graph.

41.

42.

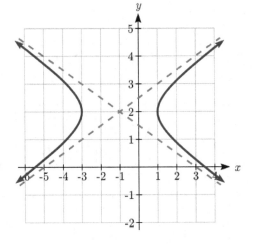

In problems 43–44, find the standard form of the equation for a hyperbola satisfying the given conditions.

43. Vertices (-1,-2) and (-1,6); asymptote $y - 2 = 2(x+1)$

44. Vertices (-3,-3) and (5,-3); asymptote $y + 3 = \dfrac{1}{2}(x-1)$

In problems 45–48, find the center, vertices, length of the transverse axis, and equations of the asymptotes. Sketch the graph. Check using a graphing utility.

45. $y = \pm 4\sqrt{9x^2 - 1}$

46. $y = \pm \dfrac{1}{4}\sqrt{9x^2 + 1}$

47. $y = 1 \pm \dfrac{1}{2}\sqrt{9x^2 + 18x + 10}$

48. $= -1 \pm 2\sqrt{9x^2 - 18x + 8}$

In problems 49–54, find the foci.

49. $\dfrac{y^2}{6} - \dfrac{x^2}{19} = 1$

50. $x^2 - \dfrac{y^2}{35} = 1$

51. $\dfrac{(x-1)^2}{15} - (y-6)^2 = 1$

52. $\dfrac{(y-3)^2}{47} - \dfrac{(x+5)^2}{2} = 1$

53. $y = 1 \pm \dfrac{4}{3}\sqrt{x^2 + 8x + 25}$

54. $y = -3 \pm \dfrac{12}{5}\sqrt{x^2 - 4x - 21}$

In problems 55–66, find the standard form of the equation for a hyperbola satisfying the given conditions.

55. Foci (5,0) and (-5,0), vertices (4,0) and (4,0)

56. Foci (0,26) and (0, -26), vertices (0,10) and (0,-10)

57. Focus (0, 13), vertex (0,12), center (0,0)

58. Focus (15, 0), vertex (12, 0), center (0,0)

59. Focus (17, 0) and (-17,0), asymptotes $y = \dfrac{8}{15}x$ and $y = -\dfrac{8}{15}x$

60. Focus (0, 25) and (0, 25), asymptotes $y = \dfrac{24}{7}x$ and $y = -\dfrac{24}{7}x$

61. Focus (10, 0) and (-10, 0), transverse axis length 16

62. Focus (0, 34) and (0, -34), transverse axis length 32

63. Foci (1, 7) and (1, -3), vertices (1, 6) and (1,-2)

64. Foci (4, -2) and (-6, -2), vertices (2, -2) and (-4, -2)

65. Focus (12, 3), vertex (4, 3), center (-1, 3)

66. Focus (-3, 15), vertex (-3, 13), center (-3, -2)

67. **LORAN** Stations A and B are 100 kilometers apart and send a simultaneous radio signal to a ship. The signal from A arrives 0.0002 seconds before the signal from B. If the signal travels 300,000 kilometers per second, find an equation of the hyperbola on which the ship is positioned if the foci are located at A and B.

68. **Thunder and Lightning** Anita and Samir are standing 3050 feet apart when they see a bolt of light strike the ground. Anita hears the thunder 0.5 seconds before Samir does. Sound travels at 1100 feet per second. Find an equation of the hyperbola on which the lighting strike is positioned if Anita and Samir are located at the foci.

69. **Cooling Tower** The cooling tower for a power plant has sides in the shape of a hyperbola. The tower stands 179.6 meters tall. The diameter at the top is 72 meters. At their closest, the sides of the tower are 60 meters apart. Find an equation that models the sides of the cooling tower.

70. **Calibration** A seismologist positions two recording devices 340 feet apart at points A and B. To check the calibration, an explosive is detonated between the devices 90 feet from point A. The time the explosions register on the devices is noted and the difference calculated. A second explosion will be detonated east of point A. How far east should the second explosion be positioned so that the measured time difference is the same as for the first explosion?

71. **Target Practice** A gun at point A and a target at point B are 200 feet apart. A person at point C hears the gun fire and hit the target at exactly the same time. Find an equation of the hyperbola on which the person is standing if the foci are located at A and B. A fired bullet has a velocity of 2000 feet per second. The speed of sound is 1100 feet per second.

72. **Comet Trajectories** A comet passes through the solar system following a hyperbolic trajectory with the sun as a focus. The closest it gets to the sun is 3×10^8 miles. The figure shows the trajectory of the comet, whose path of entry is at a right angle to its path of departure. Find an equation for the comet's trajectory. Round to two decimal places.

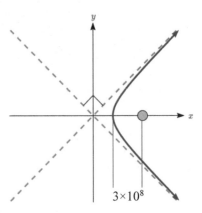

73. The **conjugate** of the hyperbola $\dfrac{x^2}{a^2} - \dfrac{y^2}{b^2} = 1$ is $\dfrac{x^2}{a^2} - \dfrac{y^2}{b^2} = -1$. Show that $5y^2 - x^2 + 25 = 0$ is the conjugate of $x^2 - 5y^2 + 25 = 0$.

74. The **eccentricity** e of a hyperbola is the ratio $\dfrac{c}{a}$, where c is the distance of a focus from the center and a is the distance of a vertex from the center. Find the eccentricity of $\dfrac{x^2}{9} - \dfrac{y^2}{16} = 1$.

75. An **equilateral hyperbola** is one for which $a = b$. Find the eccentricity of an equilateral hyperbola.

76. The **latus rectum** of a hyperbola is a line segment with endpoints on the hyperbola that passes through a focus and is perpendicular to the transverse axis. Show that $\dfrac{2b^2}{a}$ is the length of the latus rectum of $\dfrac{x^2}{a^2} - \dfrac{y^2}{b^2} = 1$.

77. **Confocal** hyperbolas have the same foci. Show that, for $0 < k < 6$, all hyperbolas of the form $\dfrac{x^2}{k} - \dfrac{y^2}{6-k} = 1$ are confocal.

Section 9.3 Parabolas and Non-Linear Systems

To listen for signals from space, a radio telescope uses a dish in the shape of a parabola to focus and collect the signals in the receiver.

While we studied parabolas earlier when we explored quadratics, at the time we didn't discuss them as a conic section. A parabola is the shape resulting from when a plane parallel to the side of the cone intersects the cone[6].

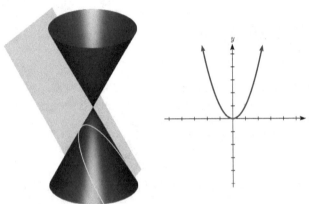

Parabola Definition and Vocabulary

A **parabola** with vertex at the origin can be defined by placing a fixed point at $F(0, p)$ called the **focus**, and drawing a line at $y = -p$, called the **directrix**. The parabola is the set of all points $Q(x, y)$ that are an equal distance between the fixed point and the directrix.

For general parabolas,

The **axis of symmetry** is the line passing through the foci, perpendicular to the directrix.

The **vertex** is the point where the parabola crosses the axis of symmetry.

The distance from the vertex to the focus, p, is the **focal length**.

---

[6] Pbroks13 (https://commons.wikimedia.org/wiki/File:Conic_sections_with_plane.svg), "Conic sections with plane", cropped to show only parabola, CC BY 3.0

Equations for Parabolas with Vertex at the Origin

From the definition above we can find an equation of a parabola. We will find it for a parabola with vertex at the origin, $C(0,0)$, opening upward with focus at $F(0,p)$ and directrix at $y = -p$.

Suppose $Q(x,y)$ is some point on the parabola. The distance from Q to the focus is

$$d(Q,F) = \sqrt{(x-0)^2 + (y-p)^2} = \sqrt{x^2 + (y-p)^2}$$

The distance from the point Q to the directrix is the difference of the y-values:
$$d = y - (-p) = y + p$$

From the definition of the parabola, these distances should be equal:

$\sqrt{x^2 + (y-p)^2} = y + p$ Square both sides

$x^2 + (y-p)^2 = (y+p)^2$ Expand

$x^2 + y^2 - 2py + p^2 = y^2 + 2py + p^2$ Combine like terms

$x^2 = 4py$

This is the standard conic form of a parabola that opens up or down (vertical axis of symmetry), centered at the origin. Note that if we divided by $4p$, we would get a more familiar equation for the parabola, $y = \dfrac{x^2}{4p}$. We can recognize this as a transformation of the parabola $y = x^2$, vertically compressed or stretched by $\dfrac{1}{4p}$.

Using a similar process, we could find an equation of a parabola with vertex at the origin opening left or right. The focus will be at $(p,0)$ and the graph will have a horizontal axis of symmetry and a vertical directrix. The standard conic form of its equation will be

$y^2 = 4px$, which we could also write as $x = \dfrac{y^2}{4p}$.

Example 1

Write the standard conic equation for a parabola with vertex at the origin and focus at (0, -2).

With focus at (0, -2), the axis of symmetry is vertical, so the standard conic equation is $x^2 = 4py$. Since the focus is (0, -2), $p = -2$.

The standard conic equation for the parabola is $x^2 = 4(-2)y$, or

$x^2 = -8y$

For parabolas with vertex not at the origin, we can shift these equations, leading to the equations summarized next.

Equation of a Parabola with Vertex at (*h*, *k*) in Standard Conic Form

The standard conic form of an equation of a parabola with vertex at the point (h,k) depends on whether the axis of symmetry is horizontal or vertical. The table below gives the standard equation, vertex, axis of symmetry, directrix, focus, and graph for each.

	Horizontal	Vertical
Standard Equation	$(y-k)^2 = 4p(x-h)$	$(x-h)^2 = 4p(y-k)$
Vertex	(*h*, *k*)	(*h*, *k*)
Axis of symmetry	$y = k$	$x = h$
Directrix	$x = h - p$	$y = k - p$
Focus	$(h + p, k)$	$(h, k + p)$
Graph	An example with $p < 0$	An example with $p > 0$

Since you already studied quadratics in some depth earlier, we will primarily explore the new concepts associated with parabolas, particularly the focus.

Example 2

Put the equation of the parabola $y = 8(x-1)^2 + 2$ in standard conic form. Find the vertex, focus, and axis of symmetry.

From your earlier work with quadratics, you may already be able to identify the vertex as (1,2), but we'll go ahead and put the parabola in the standard conic form. To do so, we need to isolate the squared factor.

$y = 8(x-1)^2 + 2$ Subtract 2 from both sides

$y - 2 = 8(x-1)^2$ Divide by 8

$\dfrac{(y-2)}{8} = (x-1)^2$

This matches the general form for a vertical parabola, $(x-h)^2 = 4p(y-k)$, where $4p = \dfrac{1}{8}$. Solving this tells us $p = \dfrac{1}{32}$. The standard conic form of the equation is

$$(x-1)^2 = 4\left(\frac{1}{32}\right)(y-2).$$

The vertex is at (1,2). The axis of symmetry is at $x = 1$.

The directrix is at $y = 2 - \dfrac{1}{32} = \dfrac{63}{32}$.

The focus is at $\left(1, 2 + \dfrac{1}{32}\right) = \left(1, \dfrac{65}{32}\right)$.

Example 3

A parabola has its vertex at (1,5) and focus at (3,5). Find an equation for the parabola.

Since the vertex and focus lie on the line $y = 5$, that is our axis of symmetry.

The vertex (1,5) tells us $h = 1$ and $k = 5$.

Looking at the distance from the vertex to the focus, $p = 3 - 1 = 2$.

Substituting these values into the standard conic form of an equation for a horizontal parabola gives the equation

$(y-5)^2 = 4(2)(x-1)$

$(y-5)^2 = 8(x-1)$

Note this could also be rewritten by solving for x, resulting in

$x = \dfrac{1}{8}(y-5)^2 + 1$

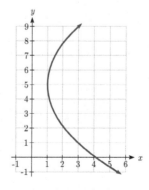

Try it Now
1. A parabola has its vertex at (-2,3) and focus at (-2,2). Find an equation for this parabola.

Applications of Parabolas

In an earlier section, we learned that ellipses have a special property that a ray eminating from one focus will be reflected back to the other focus, the property that enables the whispering chamber to work. Parabolas also have a special property, that any ray eminating from the focus will be reflected parallel to the axis of symmetry. Reflectors in flashlights take advantage of this property to focus the light from the bulb into a collimated beam. The same property can be used in reverse, taking parallel rays of sunlight or radio signals and directing them all to the focus.

Example 4

A solar cooker is a parabolic dish that reflects the sun's rays to a central point allowing you to cook food. If a solar cooker has a parabolic dish 16 inches in diameter and 4 inches tall, where should the food be placed?

We need to determine the location of the focus, since that's where the food should be placed. Positioning the base of the dish at the origin, the shape from the side looks like:

The standard conic form of an equation for the parabola would be $x^2 = 4py$. The parabola passes through (4, 8), so substituting that into the equation, we can solve for p:

$8^2 = 4(p)(4)$

$p = \dfrac{8^2}{16} = 4$

The focus is 4 inches above the vertex. This makes for a very convenient design, since then a grate could be placed on top of the dish to hold the food.

Try it Now
2. A radio telescope is 100 meters in diameter and 20 meters deep. Where should the receiver be placed?

Non-Linear Systems of Equations

In many applications, it is necessary to solve for the intersection of two curves. Many of the techniques you may have used before to solve systems of linear equations will work for non-linear equations as well, particularly substitution. You have already solved some examples of non-linear systems when you found the intersection of a parabola and line while studying quadratics, and when you found the intersection of a circle and line while studying circles.

Example 4

Find the points where the ellipse $\dfrac{x^2}{4} + \dfrac{y^2}{25} = 1$ intersects the circle $x^2 + y^2 = 9$.

To start, we might multiply the ellipse equation by 100 on both sides to clear the fractions, giving $25x^2 + 4y^2 = 100$.

A common approach for finding intersections is substitution. With these equations, rather than solving for x or y, it might be easier to solve for x^2 or y^2. Solving the circle equation for x^2 gives $x^2 = 9 - y^2$. We can then substitute that expression for x^2 into the ellipse equation.

$25x^2 + 4y^2 = 100$ Substitute $x^2 = 9 - y^2$

$25\left(9 - y^2\right) + 4y^2 = 100$ Distribute

$225 - 25y^2 + 4y^2 = 100$ Combine like terms

$-21y^2 = -125$ Divide by -21

$y^2 = \dfrac{125}{21}$ Use the square root to solve

$y = \pm\sqrt{\dfrac{125}{21}} = \pm\dfrac{5\sqrt{5}}{\sqrt{21}}$

We can substitute each of these y values back in to $x^2 = 9 - y^2$ to find x

$x^2 = 9 - \left(\sqrt{\dfrac{125}{21}}\right)^2 = 9 - \dfrac{125}{21} = \dfrac{189}{21} - \dfrac{125}{21} = \dfrac{64}{21}$

$x = \pm\sqrt{\dfrac{64}{21}} = \pm\dfrac{8}{\sqrt{21}}$

There are four points of intersection: $\left(\pm\dfrac{8}{\sqrt{21}}, \pm\dfrac{5\sqrt{5}}{\sqrt{21}}\right)$.

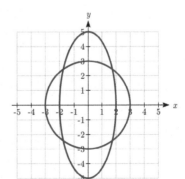

It's worth noting there is a second technique we could have used in the previous example, called elimination. If we multiplied the circle equation by -4 to get $-4x^2 - 4y^2 = -36$, we can then add it to the ellipse equation, eliminating the variable y.

$25x^2 + 4y^2 = 100$

$-4x^2 - 4y^2 = -36$ Add the left sides, and add the right sides

$21x^2 = 64$ Solve for x

$x = \pm\sqrt{\dfrac{64}{21}} = \pm\dfrac{8}{\sqrt{21}}$

Example 5

Find the points where the hyperbola $\dfrac{y^2}{4} - \dfrac{x^2}{9} = 1$ intersects the parabola $y = 2x^2$.

We can solve this system of equations by substituting $y = 2x^2$ into the hyperbola equation.

$\dfrac{(2x^2)^2}{4} - \dfrac{x^2}{9} = 1$ Simplify

$\dfrac{4x^4}{4} - \dfrac{x^2}{9} = 1$ Simplify, and multiply by 9

$9x^4 - x^2 = 9$ Move the 9 to the left

$9x^4 - x^2 - 9 = 0$

While this looks challenging to solve, we can think of it as a "quadratic in disguise," since $x^4 = (x^2)^2$. Letting $u = x^2$, the equation becomes

$9u^2 - u^2 - 9 = 0$ Solve using the quadratic formula

$u = \dfrac{-(-1) \pm \sqrt{(-1)^2 - 4(9)(-9)}}{2(9)} = \dfrac{1 \pm \sqrt{325}}{18}$ Solve for x

$x^2 = \dfrac{1 \pm \sqrt{325}}{18}$ But $1 - \sqrt{325} < 0$, so

$x = \pm\sqrt{\dfrac{1 + \sqrt{325}}{18}}$ This leads to two real solutions

$x \approx 1.028, -1.028$

Substituting these into $y = 2x^2$, we can find the corresponding y values. The curves intersect at the points (1.028, 2.114) and (-1.028, 2.114).

3. Find the points where the line $y = 4x$ intersect the ellipse $\dfrac{y^2}{4} - \dfrac{x^2}{16} = 1$

Solving for the intersection of two hyperbolas allows us to utilize the LORAN navigation approach described in the last section.

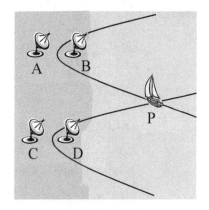

In our example, stations A and B are 150 kilometers apart and send a simultaneous radio signal to the ship. The signal from B arrives 0.0003 seconds before the signal from A. We found the equation of the hyperbola in standard form would be

$$\frac{x^2}{2025} - \frac{y^2}{3600} = 1$$

Example 10

Continuing the situation from the last section, suppose stations C and D are located 200 km due south of stations A and B and 100 km apart. The signal from D arrives 0.0001 seconds before the signal from C, leading to the equation $\dfrac{x^2}{225} - \dfrac{(y+200)^2}{2275} = 1$. Find the position of the ship.

To solve for the position of the boat, we need to find where the hyperbolas intersect. This means solving the system of equations. To do this, we could start by solving both equations for x^2. With the first equation from the previous example,

$\dfrac{x^2}{2025} - \dfrac{y^2}{3600} = 1$ Move the y term to the right

$\dfrac{x^2}{2025} = 1 + \dfrac{y^2}{3600}$ Multiply both sides by 2025

$x^2 = 2025 + \dfrac{2025y^2}{3600}$ Simplify

$x^2 = 2025 + \dfrac{9y^2}{16}$

With the second equation, we repeat the same process

$\dfrac{x^2}{225} - \dfrac{(y+200)^2}{2275} = 1$ Move the y term to the right and multiply by 225

$x^2 = 225 + \dfrac{225(y+200)^2}{2275}$ Simplify

$$x^2 = 225 + \frac{9(y+200)^2}{91}$$

Now set these two expressions for x^2 equal to each other and solve.

$$2025 + \frac{9y^2}{16} = 225 + \frac{9(y+200)^2}{91}$$ Subtract 225 from both sides

$$1800 + \frac{9y^2}{16} = \frac{9(y+200)^2}{91}$$ Divide by 9

$$200 + \frac{y^2}{16} = \frac{(y+200)^2}{91}$$ Multiply both sides by $16 \cdot 91 = 1456$

$$291200 + 91y^2 = 16(y+200)^2$$ Expand and distribute

$$291200 + 91y^2 = 16y^2 + 6400y + 640000$$ Combine like terms on one side

$$75y^2 - 6400y - 348800 = 0$$ Solve using the quadratic formula

$$y = \frac{-(-6400) \pm \sqrt{(-6400)^2 - 4(75)(-348800)}}{2(75)} \approx 123.11 \text{ km or } -37.78 \text{ km}$$

We can find the associated x values by substituting these y-values into either hyperbola equation. When $y \approx 123.11$,

$$x^2 \approx 2025 + \frac{9(123.11)^2}{16}$$

$$x \approx \pm 102.71$$

When $y \approx -37.78$km,

$$x^2 \approx 2025 + \frac{9(-37.78)^2}{16}$$

$$x \approx \pm 53.18$$

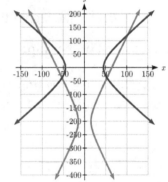

This provides 4 possible locations for the ship. Two can be immediately discarded, as they're on land. Navigators would use other navigational techniques to decide between the two remaining locations.

Important Topics of This Section
Parabola Definition
Parabola Equations in Standard Form
Applications of Parabolas
Solving Non-Linear Systems of Equations

Try it Now Answers

1. Axis of symmetry is vertical, and the focus is below the vertex.

$p = 2 - 3 = -1$.

$(x - (-2))^2 = 4(-1)(y - 3)$, or $(x + 2)^2 = -4(y - 3)$.

2. The standard conic form of the equation is $x^2 = 4py$.

Using (50,20), we can find that $50^2 = 4p(20)$, so $p = 31.25$ meters.

The receiver should be placed 31.25 meters above the vertex.

3. Substituting $y = 4x$ gives $\dfrac{(4x)^2}{4} - \dfrac{x^2}{16} = 1$. Simplify

$\dfrac{16x^2}{4} - \dfrac{x^2}{16} = 1$. Multiply by 16 to get

$64x^2 - x^2 = 16$

$x = \pm\sqrt{\dfrac{16}{63}} = \pm 0.504$

Substituting those into $y = 4x$ gives the corresponding y values.

The curves intersect at (0.504, 2.016) and (-0.504, -2.016).

Section 9.3 Exercises

In problems 1–4, match each graph with one of the equations A–D.

A. $y^2 = 4x$ B. $x^2 = 4y$ C. $x^2 = 8y$ D. $y^2 + 4x = 0$

1.

2.

3.

4.

In problems 5–14, find the vertex, axis of symmetry, directrix, and focus of the parabola.

5. $y^2 = 16x$ 6. $x^2 = 12y$ 7. $y = 2x^2$ 8. $x = -\dfrac{y^2}{8}$

9. $x + 4y^2 = 0$ 10. $8y + x^2 = 0$ 11. $(x-2)^2 = 8(y+1)$

12. $(y+3)^2 = 4(x-2)$ 13. $y = \dfrac{1}{4}(x+1)^2 + 4$ 14. $x = -\dfrac{1}{12}(y+1)^2 + 1$

In problems 15–16, write an equation for the graph.

15.

16.

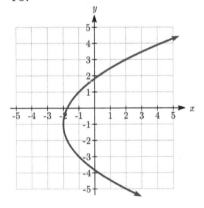

In problems 17-20, find the standard form of the equation for a parabola satisfying the given conditions.

17. Vertex at (2,3), opening to the right, focal length 3

18. Vertex at (-1,2), opening down, focal length 1

19. Vertex at (0,3), focus at (0,4)

20. Vertex at (1,3), focus at (0,3)

21. The mirror in an automobile headlight has a parabolic cross-section with the light bulb at the focus. On a schematic, the equation of the parabola is given as $x^2 = 4y^2$. At what coordinates should you place the light bulb?

22. If we want to construct the mirror from the previous exercise so that the focus is located at $(0,0.25)$, what should the equation of the parabola be?

23. A satellite dish is shaped like a paraboloid of revolution. This means that it can be formed by rotating a parabola around its axis of symmetry. The receiver is to be located at the focus. If the dish is 12 feet across at its opening and 4 feet deep at its center, where should the receiver be placed?

24. Consider the satellite dish from the previous exercise. If the dish is 8 feet across at the opening and 2 feet deep, where should we place the receiver?

25. A searchlight is shaped like a paraboloid of revolution. A light source is located 1 foot from the base along the axis of symmetry. If the opening of the searchlight is 2 feet across, find the depth.

26. If the searchlight from the previous exercise has the light source located 6 inches from the base along the axis of symmetry and the opening is 4 feet wide, find the depth.

In problems 27–34, solve each system of equations for the intersections of the two curves.

27. $y = 2x$
$y^2 - x^2 = 1$

28. $y = x + 1$
$2x^2 + y^2 = 1$

29. $x^2 + y^2 = 11$
$x^2 - 4y^2 = 1$

30. $2x^2 + y^2 = 4$
$y^2 - x^2 = 1$

31. $y = x^2$
$y^2 - 6x^2 = 16$

32. $x = y^2$
$\dfrac{x^2}{4} + \dfrac{y^2}{9} = 1$

33. $x^2 - y^2 = 1$
$4y^2 - x^2 = 1$

34. $x^2 = 4(y - 2)$
$x^2 = 8(y + 1)$

35. A LORAN system has transmitter stations A, B, C, and D at (-125,0), (125,0), (0, 250), and (0,-250), respectively. A ship in quadrant two computes the difference of its distances from A and B as 100 miles and the difference of its distances from C and D as 180 miles. Find the *x*- and *y*-coordinates of the ship's location. Round to two decimal places.

36. A LORAN system has transmitter stations A, B, C, and D at (-100,0), (100,0), (-100, -300), and (100,-300), respectively. A ship in quadrant one computes the difference of its distances from A and B as 80 miles and the difference of its distances from C and D as 120 miles. Find the *x*- and *y*-coordinates of the ship's location. Round to two decimal places.

Section 9.4 Conics in Polar Coordinates

In the preceding sections, we defined each conic in a different way, but each involved the distance between a point on the curve and the focus. In the previous section, the parabola was defined using the focus and a line called the directrix. It turns out that all conic sections (circles, ellipses, hyperbolas, and parabolas) can be defined using a single relationship.

Conic Sections General Definition

A **conic section** can be defined by placing a fixed point at the origin, $F(0,0)$, called the **focus**, and drawing a line L called the **directrix** at $x = \pm p$ or $y = \pm p$. The conic section is the set of all points $Q(x, y)$ for which the ratio of the distance from Q to F to the distance from Q to the directrix is some positive constant e, called the **eccentricity**. In other words, $\dfrac{d(Q,F)}{d(Q,L)} = e$.

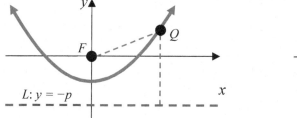

Warning: the eccentricity, e, is **not** the Euler constant $e \approx 2.71828$ we studied with exponentials

---

The Polar Form of a Conic

To create a general equation for a conic section using the definition above, we will use polar coordinates. Represent $Q(x, y)$ in polar coordinates so $(x, y) = (r\cos(\theta), r\sin(\theta))$. For now, we'll focus on the case of a horizontal directrix at $y = -p$, as in the picture above on the left.

The distance from the focus to the point Q in polar is just r.
The distance from the point Q to the directrix $y = -p$ is $r\sin(\theta) - (-p) = p + r\sin(\theta)$

The ratio of these should be the constant eccentricity e, so

$$\frac{d(Q,F)}{d(Q,L)} = e$$ Substituting in the expressions for the distances,

$$\frac{r}{p + r\sin(\theta)} = e$$

To have a standard polar equation, we need to solve for r. Start by clearing the fraction.

$r = e(p + r\sin(\theta))$ Distribute

$r = ep + er\sin(\theta)$ Move terms with r to the left

$r - er\sin(\theta) = ep$ Factor the r

$r(1 - e\sin(\theta)) = ep$ Divide

$r = \dfrac{ep}{1 - e\sin(\theta)}$

We could repeat the same approach for a directrix at $y = p$ and for vertical directrices to obtain the polar equations below.

Polar Equation for a Conic Section

A **conic section** with a focus at the origin, **eccentricity** e, and **directrix** at $x = \pm p$ or $y = \pm p$ will have polar equation:

$r = \dfrac{ep}{1 \pm e\sin(\theta)}$ when the directrix is $y = \pm p$

$r = \dfrac{ep}{1 \pm e\cos(\theta)}$ when the directrix is $x = \pm p$

Example 1

Write the polar equation for a conic section with eccentricity 3 and directrix at $x = 2$.

We are given $e = 3$ and $p - 2$. Since the directrix is vertical and at a positive x value, we use the equation involving cos with the positive sign.

$r = \dfrac{(3)(2)}{1 + 3\cos(\theta)} = \dfrac{6}{1 + 3\cos(\theta)}$

Graphing that using technology reveals it's an equation for a hyperbola.

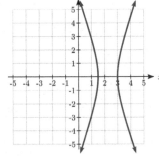

Try it Now

1. Write a polar equation for a conic with eccentricity 1 and directrix at $y = -3$.

Relating the Polar Equation to the Shape

It was probably not obvious to you that the polar equation in the last example would give the graph of a hyperbola. To explore the relationship between the polar equation and the shape, we will try to convert the polar equation into a Cartesian one. For simplicity, we will consider the case where the directrix is $x = 1$.

$r = \dfrac{e}{1 + e\cos(\theta)}$ Multiply by the denominator

$r(1 + e\cos(\theta)) = e$ Rewrite $\cos(\theta) = \dfrac{x}{r}$

$r\left(1 + e\dfrac{x}{r}\right) = e$ Distribute

$r + ex = e$ Isolate r

$r = e - ex$ Square both sides

$r^2 = (e - ex)^2$ Rewrite $r^2 = x^2 + y^2$ and expand

$x^2 + y^2 = e^2 - 2e^2x + e^2x^2$ Move variable terms to the left

$x^2 + 2e^2x - e^2x^2 + y^2 = e^2$ Combine like terms

$x^2(1 - e^2) + 2e^2x + y^2 = e^2$

We could continue, by completing the square with the x terms, to eventually rewrite this in the standard form as $\left(\dfrac{(1-e^2)^2}{e^2}\right)\left(x - \dfrac{e^2}{1-e^2}\right)^2 + \left(\dfrac{1-e^2}{e^2}\right)y^2 = 1$, but happily there's no need for us to do that.

In the equation $x^2(1 - e^2) + 2e^2x + y^2 = e^2$, we can see that:

When $e < 1$, the coefficients of both x^2 and y^2 are positive, resulting in ellipse.

When $e > 1$, the coefficient of x^2 is negative while the coefficient of y^2 is positive, resulting in a hyperbola.

When $e = 1$, the x^2 will drop out of the equation, resulting in a parabola.

Relation Between the Polar Equation of a Conic and its Shape
For a **conic section** with a focus at the origin, **eccentricity** e, and **directrix** at $x = \pm p$ or $y = \pm p$,   when $0 < e < 1$, the graph is an ellipse   when $e = 1$, the graph is a parabola   when $e > 1$, the graph is a hyperbola

Taking a more intuitive approach, notice that if $e < 1$, the denominator $1 + e\cos(\theta)$ will always be positive and so r will always be positive. This means that the radial distance r is defined and finite for every value of θ, including $\dfrac{\pi}{2}$, with no breaks. The only conic with this characteristic is an ellipse.

If $e = 1$, the denominator will be positive for all values of θ, except $-\pi$ where the denominator is 0 and r is undefined. This fits with a parabola, which has a point at every angle except at the angle pointing along the axis of symmetry away from the vertex.

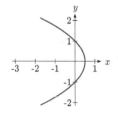

If $e > 1$, then the denominator will be zero at two angles other than $\pm\dfrac{\pi}{2}$, and r will be negative for a set of θ values. This division of positive and negative radius values would result in two distinct branches of the graph, fitting with a hyperbola.

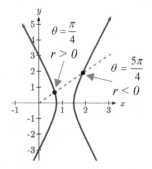

Example 2

For each of the following conics with focus at the origin, identify the shape, the directrix, and the eccentricity.

a. $r = \dfrac{8}{1 - 2\sin(\theta)}$ b. $r = \dfrac{6}{3 - 2\cos(\theta)}$ c. $r = \dfrac{8}{5 + 5\sin(\theta)}$

a. This equation is already in standard form $r = \dfrac{ep}{1 \pm e\sin(\theta)}$ for a conic with horizontal directrix at $y = -p$.

The eccentricity is the coefficient of $\sin(\theta)$, so $e = 2$.

Since $e = 2 > 1$, the shape will be a hyperbola.

Looking at the numerator, $ep = 8$, and substituting $e = 2$ gives $p = 4$. The directrix is $y = -4$.

b. This equation is not in standard form, since the constant in the denominator is not 1. To put it into standard form, we can multiply the numerator and denominator by 1/3.

$$r = \frac{6}{3-2\cos(\theta)} \cdot \frac{\frac{1}{3}}{\frac{1}{3}} = \frac{6\left(\frac{1}{3}\right)}{(3-2\cos(\theta))\left(\frac{1}{3}\right)} = \frac{2}{1-\frac{2}{3}\cos(\theta)}$$

This is the standard form for a conic with vertical directrix $x = -p$. The eccentricity is the coefficient on $\cos(\theta)$, so $e = \frac{2}{3}$.

Since $0 < e < 1$, the shape is an ellipse.

Looking at the numerator, $ep = 2$, so $\frac{2}{3}p = 2$, giving $p = 3$. The directrix is $x = -3$.

c. This equation is also not in standard form. Multiplying the numerator and denominator by 1/5 will put it in standard form.

$$r = \frac{8}{5+5\sin(\theta)} \cdot \frac{\frac{1}{5}}{\frac{1}{5}} = \frac{8\left(\frac{1}{5}\right)}{(5+5\sin(\theta))\left(\frac{1}{5}\right)} = \frac{\frac{8}{5}}{1+\sin(\theta)}$$

This is the standard form for a conic with horizontal directrix at $y = p$.
The eccentricity is the coefficient on $\sin(\theta)$, so $e = 1$. The shape will be a parabola.

Looking at the numerator, $ep = \frac{8}{5}$. Since $e = 1$, $p = \frac{8}{5}$. The directrix is $y = \frac{8}{5}$.

Notice that since the directrix is above the focus at the origin, the parabola will open downward.

Try it Now

2. Identify the shape, the directrix, and the eccentricity of $r = \dfrac{9}{4+2\cos(\theta)}$

Graphing Conics from the Polar Form

Identifying additional features of a conic in polar form can be challenging, which makes graphing without technology likewise challenging. We can utilize our understanding of the conic shapes from earlier sections to aid us.

Example 3

Sketch a graph of $r = \dfrac{3}{1 - 0.5\sin(\theta)}$ and write its Cartesian equation.

This is in standard form, and we can identify that $e = 0.5$, so the shape is an ellipse. From the numerator, $ep = 3$, so $0.5p = 3$, giving $p = 6$. The directrix is $y = -6$.

To sketch a graph, we can start by evaluating the function at a few convenient θ values, and finding the corresponding Cartesian coordinates.

$\theta = 0 \qquad r = \dfrac{3}{1 - 0.5\sin(0)} = \dfrac{3}{1} = 3 \qquad\qquad (x, y) = (3, 0)$

$\theta = \dfrac{\pi}{2} \qquad r = \dfrac{3}{1 - 0.5\sin\left(\dfrac{\pi}{2}\right)} = \dfrac{3}{1 - 0.5} = 6 \qquad (x, y) = (0, 6)$

$\theta = \pi \qquad r = \dfrac{3}{1 - 0.5\sin(\pi)} = \dfrac{3}{1} = 3 \qquad\qquad (x, y) = (-3, 0)$

$\theta = \dfrac{3\pi}{2} \qquad r = \dfrac{3}{1 - 0.5\sin\left(\dfrac{3\pi}{2}\right)} = \dfrac{3}{1 + 0.5} = 2 \qquad (x, y) = (0, -2)$

Plotting these points and remembering the origin is one of the foci gives an idea of the shape, which we could sketch in. To get a better understanding of the shape, we could use these features to find more.

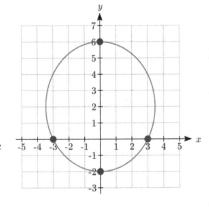

The vertices are at (0, -2) and (0, 6), so the center must be halfway between, at $\left(0, \dfrac{-2 + 6}{2}\right) = (0, 2)$. Since the vertices are a distance a from the center, $a = 6 - 2 = 4$.

One focus is at (0, 0), a distance of 2 from the center, so $c = 2$, and the other focus must be 2 above the center, at (0, 4).

We can now solve for b: $b^2 = a^2 - c^2$, so $b^2 = 4^2 - 2^2 = 10$, hence $b = \pm\sqrt{10}$. The minor axis endpoints would be at $\left(-\sqrt{10}, 2\right)$ and $\left(\sqrt{10}, 2\right)$.

We can now use the center, a, and b to write the Cartesian equation for this curve:

$\dfrac{x^2}{10} + \dfrac{(y - 2)^2}{16} = 1$

Try it Now

3. Sketch a graph of $r = \dfrac{6}{1 + 2\cos(\theta)}$ and identify the important features.

Important Topics of This Section
Polar equations for Conic Sections
Eccentricity and Directrix
Determining the shape of a polar conic section

Try it Now Answers

1. $r = \dfrac{(1)(3)}{1 - \sin(\theta)}$. $\quad r = \dfrac{3}{1 - \sin(\theta)}$

2. We can convert to standard form by multiplying the top and bottom by $\dfrac{1}{4}$.

$r = \dfrac{\dfrac{9}{4}}{1 + \dfrac{1}{2}\cos(\theta)}$. Eccentricity $= \dfrac{1}{2}$, so the shape is an ellipse.

The numerator is $ep = \dfrac{1}{2}p = \dfrac{9}{4}$. The directrix is $x = \dfrac{9}{2}$.

3. The eccentricity is $e = 2$, so the graph of the equation is a hyperbola. The directrix is $x = 3$. Since the directrix is a vertical line and the focus is at the origin, the hyperbola is horizontal.

$\theta = 0 \qquad r = \dfrac{6}{1 + 2\cos(0)} = \dfrac{6}{1 + 2} = 2 \qquad (x, y) = (2, 0)$

$\theta = \dfrac{\pi}{2} \qquad r = \dfrac{6}{1 + 2\cos\left(\dfrac{\pi}{2}\right)} = \dfrac{6}{1} = 6 \qquad (x, y) = (0, 6)$

$\theta = \pi \qquad r = \dfrac{6}{1 + 2\cos(\pi)} = \dfrac{6}{1 - 2} = -6 \qquad (x, y) = (6, 0)$

$\theta = \dfrac{3\pi}{2} \qquad r = \dfrac{6}{1 + 2\cos\left(\dfrac{3\pi}{2}\right)} = \dfrac{6}{1} = 6 \qquad (x, y) = (0, -6)$

Plotting those points, we can connect the three on the left with a smooth curve to form one branch of the hyperbola, and the other branch will be a mirror image passing through the last point.

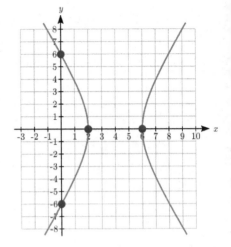

The vertices are at (2,0) and (6,0).

The center of the hyperbola would be at the midpoint of the vertices, at (4,0).
The vertices are a distance $a = 2$ from the center.
The focus at the origin is a distance $c = 4$ from the center.
Solving for b, $b^2 = 4^2 - 2^2 = 12$. $b = \pm\sqrt{12} = \pm 2\sqrt{3}$.

The asymptotes would be $y = \pm\sqrt{3}(x-4)$.

The Cartesian equation of the hyperbola would be:
$$\frac{(x-4)^2}{4} - \frac{y^2}{12} = 1$$

Section 9.4 Exercises

In problems 1–8, find the eccentricity and directrix, then identify the shape of the conic.

1. $r = \dfrac{12}{1 + 3\cos(\theta)}$

2. $r = \dfrac{4}{1 - \sin(\theta)}$

3. $r = \dfrac{2}{4 - 3\sin(\theta)}$

4. $r = \dfrac{7}{2 - \cos(\theta)}$

5. $r = \dfrac{1}{5 - 5\cos(\theta)}$

6. $r = \dfrac{6}{3 + 8\cos(\theta)}$

7. $r = \dfrac{4}{7 + 2\cos(\theta)}$

8. $r = \dfrac{16}{4 + 3\sin(\theta)}$

In problems 9–14, find a polar equation for a conic having a focus at the origin with the given characteristics.

9. Directrix $x = -4$, eccentricity $e = 5$.

10. Directrix $y = -2$, eccentricity $e = 3$.

11. Directrix $y = 3$, eccentricity $e = \dfrac{1}{3}$.

12. Directrix $x = 5$, eccentricity $e = \dfrac{3}{4}$.

13. Directrix $y = -2$, eccentricity $e = 1$.

14. Directrix $x = -3$, eccentricity $e = 1$.

In problems 15–20, sketch a graph of the conic. Use the graph to help you find important features and write a Cartesian equation for the conic.

15. $r = \dfrac{9}{1 - 2\cos(\theta)}$

16. $r = \dfrac{4}{1 + 3\sin(\theta)}$

17. $r = \dfrac{12}{3 + \sin(\theta)}$

18. $r = \dfrac{15}{3 - 2\cos(\theta)}$

19. $r = \dfrac{6}{1 + \cos(\theta)}$

20. $r = \dfrac{4}{1 - \sin(\theta)}$

21. At the beginning of the chapter, we defined an ellipse as the set of all points Q for which the sum of the distance from each focus to Q is constant. Mathematically, $d(Q, F_1) + d(Q, F_2) = k$. It is not obvious that this definition and the one provided in this section involving the directrix are related. In this exercise, we will start with the definition from this section and attempt to derive the earlier formula from it.

   a. Draw an ellipse with foci at $(c, 0)$ and $(-c, 0)$, vertices at $(a, 0)$ and $(-a, 0)$, and directrixes at $x = p$ and $x = -p$. Label the foci F_1 and F_2. Label the directrixes L_1 and L_2. Label some point (x, y) on the ellipse Q.

   b. Find formulas for $d(Q, L_1)$ and $D(Q, L_2)$ in terms of x and p.

   c. From the definition of a conic in this section, $\dfrac{d(Q, F_1)}{d(Q, L_1)} = e$. Likewise, $\dfrac{d(Q, F_2)}{d(Q, L_2)} = e$ as well. Use these ratios, with your answers from part (b) above, to find formulas for $d(Q, F_1)$ and $D(Q, F_2)$ in terms of e, x, and p.

   d. Show that the sum, $d(Q, F_1) + d(Q, F_2)$, is constant. This establishes that the definitions are connected.

   e. Let Q be a vertex. Find the distances $d(Q, F_1)$ and $D(Q, F_2)$ in terms of a and c. Then combine this with your result from part (d) to find a formula for p in terms of a and e.

   f. Let Q be a vertex. Find the distances $D(Q, L_2)$ and $D(Q, F_2)$ in terms of a, p, and c. Use the relationship $\dfrac{d(Q, F_2)}{d(Q, L_2)} = e$, along with your result from part (e), to find a formula for e in terms of a and c.

22. When we first looked at hyperbolas, we defined them as the set of all points Q for which the absolute value of the difference of the distances to two fixed points is constant. Mathematically, $|d(Q, F_1) - d(Q, F_2)| = k$. Use a similar approach to the one in the last exercise to obtain this formula from the definition given in this section. Find a formula for e in terms of a and c.

Solutions to Selected Exercises

Chapter 1

Section 1.1

1. a. $f(40) = 13$

   b. 2 Tons of garbage per week is produced by a city with a population of 5,000.

3. a. In 1995 there are 30 ducks in the lake

   b. In 2000 there are 40 ducks in the late

5. a ,b, d, e   7. a, b   9. a, b, d

11. b   13. b, c, e, f   15. $f(1) = 1$, $f(3) = 1$

17. $g(2) = 4$, $g(-3) = 2$ 19. $f(3) = 53$, $f(2) = 1$

	$f(-2)$	$f(-1)$	$f(0)$	$f(1)$	$f(2)$
21.	8	6	4	2	0
23.	49	18	3	4	21
25.	4	-1	0	1	-4
27.	4	4.414	4.732	5	5.236
29.	-4	-6	-6	-4	0
31.	5	DNE	-3	-1	-1/3
33.	1/4	1/2	1	2	4

35. a. -6   b.-16   37. a. 5   b. $-\dfrac{5}{3}$

39. a. iii   b. viii c. I   d. ii   e. vi   f. iv   g. v   h. vii

41. a. iv   b. ii   c. v   d. I   e. vi   f. iii

43. $(x-3)^2 + (y+9)^2 = 36$

45. (a)   (b)   (c)

  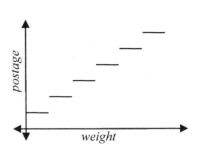

47a. t b. a c. r d. L: (c, t) and K: (a, p)

642

Section 1.2

1. D: [-5, 3) R: [0,2]   3. D: $2 < t \le 8$ R: $6 \le g(t) < 8$

5. D: [0,4] R: [-3, 0]   7. $[2,\infty)$ 9. $(-\infty,3]$

11. $(-\infty,6)\cup(6,\infty)$ 13. $(-\infty,-\frac{1}{2})\cup\left(-\frac{1}{2},\infty\right)$

15. $[-4,4)\cup(4,\infty)$ 17. $(-\infty,-11)\cup(-11,2)\cup(2,\infty)$

	$f(-1)$	$f(0)$	$f(2)$	$f(4)$
19.	-4	6	20	34
21.	-1	-2	7	5
23.	-5	3	3	16

25. $f(x)=\begin{cases} 2 & if & -6 \le x \le -1 \\ -2 & if & -1 < x \le 2 \\ -4 & if & 2 < x \le 4 \end{cases}$ 27. $f(x)=\begin{cases} 3 & if & x \le 0 \\ x^2 & if & x > 0 \end{cases}$

29. $f(x)=\begin{cases} \dfrac{1}{x} & if & x < 0 \\ \sqrt{x} & if & x \ge 0 \end{cases}$

31.

33.

35.

643

Section 1.3

1. a) 6 million dollars per year   b) 2 million dollars per year

3. $\dfrac{4-5}{4-1} = -\dfrac{1}{3}$ 5. 6   7. 27

9. $\dfrac{352}{27}$ 11. $4b+4$ 13. 3

15. $-\dfrac{1}{13h+169}$ 17. $9+9h+3h^2$ 19. $4x+2h$

21. Increasing: $(-1.5, 2)$. Decreasing: $(-\infty, -1.5) \cup (2, \infty)$

23. Increasing: $(-\infty, 1) \cup (3, 4)$. Decreasing: $(1,3) \cup (4, \infty)$

25. Increasing, concave up   27. Decreasing, concave down

29. Decreasing, concave up   31. Increasing, concave down

33. Concave up $(-\infty, 1)$. Concave down $(1, \infty)$. Inflection point at (1, 2)

35. Concave down $(-\infty, 3) \cup (3, \infty)$

37. Local minimum at (3, -22).
    Inflection points at (0,5) and (2, -11).
    Increasing on $(3, \infty)$. Decreasing $(-\infty, 3)$
    Concave up $(-\infty, 0) \cup (2, \infty)$. Concave down $(0, 2)$

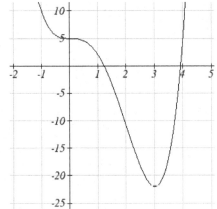

39. Local minimum at (-2, -2)
    Decreasing $(-3, -2)$
    Increasing $(-2, \infty)$
    Concave up $(-3, \infty)$

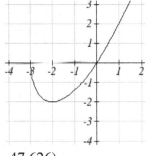

41. Local minimums at (-3.152, -47.626)
    and (2.041, -32.041)
    Local maximum at (-0.389, 5.979)
    Inflection points at (-2, -24) and (1, -15)
    Increasing $(-3.152, -0.389) \cup (2.041, \infty)$
    Decreasing $(-\infty, -3.152) \cup (-0.389, 2.041)$
    Concave up $(-\infty, -2) \cup (1, \infty)$
    Concave down $(-2, 1)$

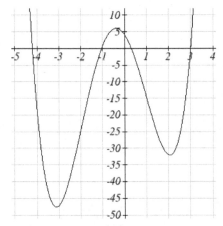

Section 1.4

1. $f(g(0)) = 36$. $g(f(0)) = -57$

3. $f(g(0)) = 4$. $g(f(0)) = 4$

5. 4   7. 9   9. 4   11. 7   13. 0   15. 4   17. 3   19. 2

21. $f(g(x)) = \dfrac{x}{7}$ $\qquad$ $g(f(x)) = 7x - 36$

23. $f(g(x)) = x + 3$ $\qquad$ $g(f(x)) = \sqrt{x^2 + 3}$

25. $f(g(x)) = |5x + 1|$ $\qquad$ $g(f(x)) = 5|x| + 1$

27. $f(g(h(x))) = \left(\sqrt{x} - 6\right)^4 + 6$

29. b $\qquad\qquad$ 31a. $r(V(t)) = \sqrt[3]{\dfrac{3(10 + 20t)}{4\pi}}$ b. 4.609in

33. $(0, \infty)$ $\qquad$ 35. $\left(-\infty, \dfrac{1}{3}\right) \cup \left(\dfrac{1}{3}, 1\right) \cup (1, \infty)$ $\qquad$ 37. $[2, 5) \cup (5, \infty)$

39. $g(x) = x + 2, f(x) = x^2$ $\qquad$ 41. $f(x) = \dfrac{3}{x}, g(x) = x - 5$

43. $f(x) = 3 + \sqrt{x}, g(x) = x - 2$, or $f(x) = 3 + x, g(x) = \sqrt{x - 2}$

45a. $f(f(x)) = a(ax + b) + b = \left(a^2\right)x + (ab + b)$

b. $g(x) = \sqrt{6}x - \dfrac{8}{\sqrt{6} + 1}$ or $g(x) = -\sqrt{6}x - \dfrac{8}{1 - \sqrt{6}}$

47a. $C(f(s)) = \dfrac{70\left(\dfrac{s}{60}\right)^2}{10 + \left(\dfrac{s}{60}\right)^2}$ $\qquad$ b. $C(g(h)) = \dfrac{70(60h)^2}{10 + (60h)^2}$

c. $v(C(m)) = \dfrac{5280}{3600}\left(\dfrac{70m^2}{10 + m^2}\right)$

645

Section 1.5

1. Horizontal shift right 49 units

3. Horizontal shift left 3 units

5. Vertical shift up 5 units

7. Vertical shift down 2 units

9. Horizontal shift right 2 units, Vertical shift up 3 units

11. $f(x+2)+1 = \sqrt{x+2}+1$

13. $f(x-3)-4 = \dfrac{1}{x-3} - 4$

15. $g(x) = f(x-1), \quad h(x) = f(x)+1$

17.

19.

21.

23.

25. $y = |x-3| - 2$

27. $y = \sqrt{x+3} - 1$

29. $y = -\sqrt{x}$

31.

33a. $-f(-x) = -6^{-x}$ b. $-f(x+2)-3 = -6^{x+2} - 3$

35. $y = -(x+1)^2 + 2$

37. $y = \sqrt{-x} + 1$

39a. Even   b. Neither   c. Odd

646

41. Reflect *f(x)* about the *x*-axis

43. Vertically stretch *y* values by 4

45. Horizontally compress *x* values by 1/5

47. Horizontally stretch *x* values by 3

49. Reflect *f(x)* about the *y*-axis and vertically stretch *y* values by 3

51. $f(-4x) = |-4x|$

53. $\dfrac{1}{3}f(x+2) - 3 = \dfrac{1}{3(x+2)^2} - 3$

55. $f(2(x-5)) + 1 = (2(x-5))^2 + 1$

57. Horizontal shift left 1 unit, vertical stretch y values by 4, vertical shift down 5 units

    becomes

59. Horizontal shift right 4 units, vertical stretch y values by 2, reflect over x axis, vertically shift up 3 units.

    becomes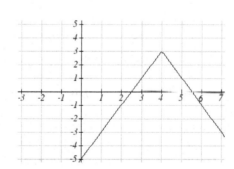

647

61. Vertically compress y values by ½

 becomes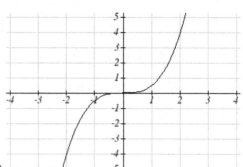

63. Horizontally stretch x values by 3, vertical shift down 3 units

 becomes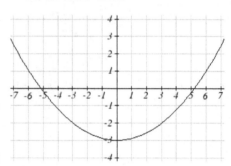

65. Reflected over the y axis, horizontally shift right 4 units $a(x) = \sqrt{-(x-4)}$

 becomes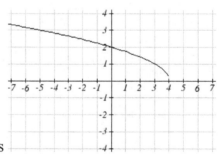

67. This function is increasing on $(-1, \infty)$ and decreasing on $(-\infty, -1)$

69. This function is decreasing on $(-\infty, 4)$

71. This function is concave down on $(-3, \infty)$ and concave up on $(-\infty, -3)$

73. This function is concave up everywhere

75. $f(-x)$ 77. $3f(x)$ 79. $2f(-x)$

81. $2f\left(\dfrac{1}{2}x\right)$ 83. $2f(x) - 2$ 85. $-f(x+1) + 3$

87. $y = -2(x+2)^2 + 3$

89. $y = \left(\dfrac{1}{2}(x-1)\right)^3 + 2$

91. $y = \sqrt{2(x+2)} + 1$

93. $y = \dfrac{-1}{(x-2)^2} + 3$

95. $y = -2|x+1| + 3$

97. $y = \sqrt[3]{-\dfrac{1}{2}(x-2)} + 1$

99. $f(x) = \begin{cases} (x+3)^2 + 1 & \text{if } x \le -2 \\ \dfrac{1}{2}|x-2| + 3 & \text{if } x > -2 \end{cases}$

101. $f(x) = \begin{cases} 1 & \text{if } x < -2 \\ -2(x+1)^2 + 4 & \text{if } -2 \le x \le 1 \\ \sqrt[3]{x-2} + 1 & \text{if } x > 1 \end{cases}$

103a. *Domain*: $3.5 \le x \le 6$ d. *Range*: $-9 \le y \le 7$

Section 1.6

1. 6   3. -4   5. ½

7a. 3   b. 2   c. 2   d. 3

9a. 0   b. 7   c. 1   d. 3

11.

x	1	4	7	12	16
$f^{-1}(x)$	3	6	9	13	14

13. $f^{-1}(x) = x - 3$

15. $f^{-1}(x) = -x + 2$

17. $f^{-1}(x) = \dfrac{x-7}{11}$

19. Restricted domain $x \ge -7$, $f^{-1}(x) = \sqrt{x} - 7$

21. Restricted domain $x \ge 0$, $f^{-1}(x) = \sqrt{x+5}$

23a. $f(g(x)) = \left(\sqrt[3]{x+5}\right)^3 - 5 = x$ b. $g(f(x)) = \sqrt[3]{x^3 - 5 + 5} = x$

c. This means that they are inverse functions (of each other)

Chapter 2

Section 2.1

1. $P(t) = 1700t + 45000$ 3. $D(t) = 10 + 2t$ 5. $M(n) = 40 - 2n$

7. Increasing   9. Decreasing   11. Decreasing

13. Increasing   15. Decreasing   17. 3

19. $-\dfrac{1}{3}$ 21. $\dfrac{4}{5}$ 23. $\dfrac{2}{3}$

25. -0.05 mph   (or 0.05 miles per hour toward her home)

27. Population is decreasing by 400 people per year

29. Monthly charge in dollars has an initial base charge of \$24, and increases by \$0.10 for each minute talked

31. Terry started at an elevation of 3,000 ft and is descending by 70ft per second.

33. $y = \dfrac{3}{5}x - 1$ 35. $y = 3x - 2$ 37. $y = -\dfrac{1}{3}x + \dfrac{11}{3}$

39. $y = -1.5x - 3$ 41. $y = \dfrac{2}{3}x + 1$ 43. $y = -2x + 3$

45. $P(n) = -0.004n + 34$

47. The 1st, 3rd & 4th tables are linear: respectively

   1. $g(x) = -3x + 5$ 3. $f(x) = 5x - 5$ 4. $k(x) = 3x - 2$

49a. $C = \dfrac{5}{9}F - \dfrac{160}{9}$ b. $F = \dfrac{9}{5}C + 32$ c. $-9.4°F$

Section 2.2

1. E   3. D   5. B

7.

9.

11.

13.

650

15.

17.

19.

21.

23. a. $g(x) = \dfrac{3}{4}(x+2) - 4$ b. ¾   c. -5/2

25. $y = 3$

27. $x = -3$

	Vertical Intercept	Horizontal Intercept
29.	(0,2)	(2,0)
31.	(0,-5)	(5/3, 0)
33.	(0,4)	(-10,0)

35. Line 1: $m = -10$ Line 2: $m = -10$ Parallel

37. Line 1: $m = -2$ Line 2: $m = 1$ Neither

39. Line 1: $m = -\dfrac{2}{3}$ Line 2: $m = \dfrac{3}{2}$ Perpendicular

41. $y = -5x - 2$ 43. $y = \dfrac{1}{2}t + 1$ 45. (-1,1)

47. (1.2, 10)   49. Plan B saves money if the miles are $> 111\dfrac{1}{9}$

51. $f(x) = \begin{cases} 2x+3 & if & -3 \le x < -1 \\ x-1 & if & -1 \le x \le 2 \\ -2 & if & 2 < x \le 5 \end{cases}$

Section 2.3

1a. 696 people   b. 4 years   c. 174 people per year

  d. 305 people   e. $P(t) = 305 + 174t$ f. 2219 people.

3a. $C(x) = 0.15x + 10$

  b. The flat monthly fee is $10 and there is an additional $0.15 fee for each additional minute used

  c. $113.05

5a. $P(t) = 190t + 4170$ b. 6640 moose

7a. $R(t) = 16 - 2.1t$ b. 5.5 billion cubic feet   c. During the year 2017

9. More than 133 minutes   11. More than $42,857.14 worth of jewelry

13. 20.012 square units   15. 6 square units

17. $A = -\dfrac{b^2}{2m}$

19a. Hawaii   b. $80,640   c. During the year 1933

21. 26.225 miles

Section 2.4

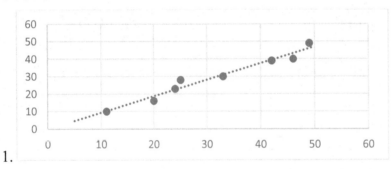

1.

3. $y = 1.971x - 3.519$, $r = 0.967$ 5. $y = -0.901x + 26.04$, $r = -0.968$

7. $17.483 \approx 17 \, situps$ 9. D   11. A

13. Yes, trend appears linear because $r = 0.994$ and will exceed 35% near the end of the year 2019.

Section 2.5

1. $y = \dfrac{1}{2}|x+2|+1$

3. $y = -3|x-3|+3$

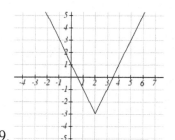

5.

7.

9.

11. $x = -\dfrac{9}{5}$ or $x = \dfrac{13}{5}$

13. $x = \dfrac{1}{2}$ or $x = \dfrac{15}{2}$

15. $x = -\dfrac{5}{3}$ or $x = -\dfrac{1}{3}$

	Horizontal Intercepts	Vertical Intercept
17.	(-6, 0) and (4, 0)	(0, -8)
19.	none	(0, -7)

21. $-11 < x < 1$ or $(-11, 1)$

23. $x \geq 5$, $x \leq -1$ or $(-\infty, -1] \cup [5, \infty)$

25. $-\dfrac{13}{3} < x < -\dfrac{5}{3}$ or $(-\dfrac{13}{3}, -\dfrac{5}{3})$

Chapter 3

Section 3.1

1. As $x \to \infty$, $f(x) \to \infty$ As $x \to -\infty$, $f(x) \to \infty$
3. As $x \to \infty$, $f(x) \to \infty$ As $x \to -\infty$, $f(x) \to -\infty$
5. As $x \to \infty$, $f(x) \to -\infty$ As $x \to -\infty$, $f(x) \to -\infty$
7. As $x \to \infty$, $f(x) \to -\infty$ As $x \to -\infty$, $f(x) \to \infty$
9. 7th Degree, Leading coefficient 4
11. 2nd Degree, Leading coefficient -1
13. 4th Degree, Leading coefficient -2
15. 3rd Degree, Leading coefficient 6
17. As $x \to \infty$, $f(x) \to -\infty$ As $x \to -\infty$, $f(x) \to -\infty$
19. As $x \to \infty$, $f(x) \to \infty$ As $x \to -\infty$, $f(x) \to \infty$
21. intercepts: 5, turning points: 4   23. 3
25. 5   27. 3   29. 5
31. Horizontal Intercepts (1,0), (-2, 0), (3, 0)   Vertical Intercept (0, 12)
33. Horizontal Intercepts (1/3, 0) (-1/2, 0)   Vertical Intercept (0, 2)

653

Section 3.2

1. $f(x) = (x-2)^2 - 3$ 3. $f(x) = -2(x-2)^2 + 7$ 5. $f(x) = \frac{1}{2}(x-3)^2 - 1$

	Vertex	Vertical Intercept	Horizontal Intercepts
7.	$(-2.5, -0.5)$	$(0,12)$	$(-2, 0)$ $(-3, 0)$
9.	$(2.5, -8.5)$	$(0,4)$	$(0.438, 0)$ $(4.562, 0)$
11.	$(0.75, 1.25)$	$(0,-1)$	$(0.191, 0)$ $(1.309, 0)$

13. $f(x) = (x-6)^2 - 4$ 15. $f(x) = 2(x+2)^2 - 18$ 17. $b = 32$ and $c = -39$

19. $f(x) = -\frac{2}{3}(x+3)(x-1)$ 21. $f(x) = \frac{3}{5}(x-2)(x-5)$

23. $f(x) = -\frac{1}{4}(x-4)^2$ 25. $f(x) = -\frac{1}{9}(x+3)^2 + 2$

27a. 234m   b. 2909.561 ft   c. 47.735 seconds
29a. 3 ft   b. 111 ft   c. 72.497 ft
31. 24.91 in by 24.91 in

33. $125\, ft$ by $83\frac{1}{3}\, ft$

35. 24.6344 cm
37. $10.70

Section 3.3

$C(t)$	C, intercepts	t, intercepts
1.	$(0,48)$	$(4,0), (-1,0), (6,0)$
3.	$(0,0)$	$(0,0), (2,0), (-1,0)$
5.	$(0,0)$	$(0,0), (1,0), (3,0)$

7. (-1.646, 0) (3.646, 0) (5,0)
9. As $t \to \infty$, $h(t) \to \infty$ $t \to -\infty$, $h(t) \to -\infty$
11. As $t \to \infty$, $p(t) \to -\infty$ $t \to -\infty$, $p(t) \to -\infty$

13.

15.

654

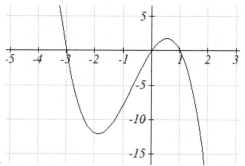

17.

19. $(3, \infty)$

21. $(-\infty, -2) \cup (1, 3)$

23. $[3.5, 6]$

25. $(-\infty, 1] \cup [4, \infty)$

27. $[-2, -2] \cup [3, \infty)$

29. $(-\infty, -4) \cup (-4, 2) \cup (2, \infty)$

31. $y = -\frac{2}{3}(x+2)(x-1)(x-3)$

33. $y = \frac{1}{3}(x-1)^2(x-3)^2(x+3)$

35. $y = -15(x-1)^2(x-3)^3$

37. $y = \frac{1}{2}(x+2)(x-1)(x-3)$

39. $y = -(x+1)^2(x-2)$

41. $y = -\frac{1}{24}(x+3)(x+2)(x-2)(x-4)$

43. $y = \frac{1}{24}(x+4)(x+2)(x-3)^2$

45. $y = \frac{1}{12}(x+2)^2(x-3)^2$

47. $y = \frac{1}{6}(x+3)(x+2)(x-1)^3$

49. $y = -\frac{1}{16}(x+3)(x+1)(x-2)^2(x-4)$

51. Base 2.58, Height 3.336

Section 3.4

1. $4x^2 + 3x - 1 = (x-3)(4x+15) + 44$

3. $5x^4 - 3x^3 + 2x^2 - 1 = (x^2+4)(5x^2-3x-18) + (12x+71)$

5. $9x^3 + 5 = (2x-3)\left(\frac{9}{2}x^2 + \frac{27}{4}x + \frac{81}{8}\right) + \frac{283}{8}$

7. $(3x^2 - 2x + 1) = (x-1)(3x+1) + 2$

9. $(3 - 4x - 2x^2) = (x+1)(-2x-2) + 5$

11. $(x^3 + 8) = (x+2)(x^2 - 2x + 4) + 0$

13. $(18x^2 - 15x - 25) = \left(x - \frac{5}{3}\right)(18x + 15) + 0$

15. $(2x^3 + x^2 + 2x + 1) = \left(x + \frac{1}{2}\right)(2x^2 + 2) + 0$

17. $(2x^3 - 3x + 1) = \left(x - \frac{1}{2}\right)\left(2x^2 + x - \frac{5}{2}\right) - \frac{1}{4}$

19. $(x^4 - 6x^2 + 9) = (x - \sqrt{3})(x^3 + \sqrt{3}x^2 - 3x - 3\sqrt{3}) + 0$

21. $x^3 - 6x^2 + 11x - 6 = (x-1)(x-2)(x-3)$

23. $3x^3 + 4x^2 - x - 2 = 3\left(x - \dfrac{2}{3}\right)(x+1)^2$

25. $x^3 + 2x^2 - 3x - 6 = (x+2)(x+\sqrt{3})(x-\sqrt{3})$

27. $4x^4 - 28x^3 + 61x^2 - 42x + 9 = 4\left(x - \dfrac{1}{2}\right)^2 (x-3)^2$

Section 3.5

1. All of the real zeros lie in the interval $[-7,7]$

- Possible rational zeros are ± 1, ± 2, ± 3

3. All of the real zeros lie in the interval $[-13,13]$

- Possible rational zeros are ± 1, ± 2, ± 3, ± 4, ± 6, ± 12

5. All of the real zeros lie in the interval $[-8,8]$

- Possible rational zeros are ± 1, ± 7

7. All of the real zeros lie in the interval $[-3,3]$

- Possible rational zeros are $\pm\dfrac{1}{17}$, $\pm\dfrac{2}{17}$, $\pm\dfrac{5}{17}$, $\pm\dfrac{10}{17}$, ± 1, ± 2, ± 5, ± 10

9. All of the real zeros lie in the interval $\left[-\dfrac{14}{3}, \dfrac{14}{3}\right]$

- Possible rational zeros are $\pm\dfrac{1}{3}$, $\pm\dfrac{2}{3}$, $\pm\dfrac{5}{3}$, $\pm\dfrac{10}{3}$, ± 1, ± 2, ± 5, ± 10

11. $x = -2$, $x = 1$, $x = 3$ (each has mult. 1)

13. $x = -2$ (mult. 2), $x = 1$ (mult. 1), $x = 3$ (mult. 1)

15. $x = 7$ (mult. 1)

17. $x = \dfrac{5}{17}$, $x = \pm\sqrt{2}$ (each has mult. 1)

19. $x = -2$, $x = \dfrac{3 \pm \sqrt{69}}{6}$ (each has mult. 1)

21. $x = 0$, $x = \dfrac{5 \pm \sqrt{61}}{18}$ (each has mult. 1)

23. $x = \pm\sqrt{3}$ (each has mult. 1)

25. $x = \pm\sqrt{5}$ (each has mult. 1)

27. $x = \sqrt[3]{-2} = -\sqrt[3]{2}$, $x = \sqrt[3]{5}$ (each has mult. 1)

29. $x = 2$, $x = \pm\sqrt{2}$ (each has mult. 1)

31. $x = -4$ (mult. 3), $x = 6$ (mult. 2)

Section 3.6

1. $3i$

3. -12

5. $1+\sqrt{3}i$

7. $8-i$

9. $-11+4i$

11. $-12+8i$

13. $30-10i$

15. $11+10i$

17. 20

19. $\dfrac{3}{2}+2i$

21. $\dfrac{3}{2}+\dfrac{5}{2}i$

23. $-\dfrac{1}{25}-\dfrac{18}{25}i$

25. $f(x)=x^2-4x+13=(x-(2+3i))(x-(2-3i))$.   Zeros: $x=2\pm3i$

27. $f(x)=3x^2+2x+10=3\left(x-\left(-\dfrac{1}{3}+\dfrac{\sqrt{29}}{3}i\right)\right)\left(x-\left(-\dfrac{1}{3}-\dfrac{\sqrt{29}}{3}i\right)\right)$.   Zeros: $x=-\dfrac{1}{3}\pm\dfrac{\sqrt{29}}{3}i$

29. $f(x)=x^3+6x^2+6x+5=(x+5)(x^2+x+1)=(x+5)\left(x-\left(-\dfrac{1}{2}+\dfrac{\sqrt{3}}{2}i\right)\right)\left(x-\left(-\dfrac{1}{2}-\dfrac{\sqrt{3}}{2}i\right)\right)$

  Zeros: $x=-5,\ x=-\dfrac{1}{2}\pm\dfrac{\sqrt{3}}{2}i$

31. $f(x)=x^3+3x^2+4x+12=(x+3)\left(x^2+4\right)=(x+3)(x+2i)(x-2i)$.   Zeros: $x=-3,\pm2i$

33. $f(x)=x^3+7x^2+9x-2=(x+2)\left(x-\left(-\dfrac{5}{2}+\dfrac{\sqrt{29}}{2}\right)\right)\left(x-\left(-\dfrac{5}{2}-\dfrac{\sqrt{29}}{2}\right)\right)$

  Zeros: $x=-2,\ x=-\dfrac{5}{2}\pm\dfrac{\sqrt{29}}{2}$

35. $f(x)=4x^4-4x^3+13x^2-12x+3=\left(x-\dfrac{1}{2}\right)^2\left(4x^2+12\right)=4\left(x-\dfrac{1}{2}\right)^2(x+i\sqrt{3})(x-i\sqrt{3})$

  Zeros: $x=\dfrac{1}{2},\ x=\pm\sqrt{3}i$

37. $f(x)=x^4+x^3+7x^2+9x-18=(x+2)(x-1)\left(x^2+9\right)=(x+2)(x-1)(x+3i)(x-3i)$

  Zeros: $x=-2,1,\pm3i$

39.

$f(x)=-3x^4-8x^3-12x^2-12x-5=(x+1)^2\left(-3x^2-2x-5\right)=-3(x+1)^2\left(x-\left(-\dfrac{1}{3}+\dfrac{\sqrt{14}}{3}i\right)\right)\left(x-\left(-\dfrac{1}{3}-\dfrac{\sqrt{14}}{3}i\right)\right)$

  Zeros: $x=-1,\ x=-\dfrac{1}{3}\pm\dfrac{\sqrt{14}}{3}i$

41. $f(x)=x^4+9x^2+20=\left(x^2+4\right)\left(x^2+5\right)=(x-2i)(x+2i)\left(x-i\sqrt{5}\right)\left(x+i\sqrt{5}\right)$

  Zeros: $x=\pm2i,\pm i\sqrt{5}$

Section 3.7

1. D 3. A

	Vertical Asymptotes	Horizontal Asymptote	Vertical y-Intercept	Horizontal x-intercept
5.	$x = -4$	$y = 2$	$(0, -3/4)$	$(3/2, 0)$
7.	$x = 2$	$y = 0$	$(0, 1)$	DNE
9.	$x = -4, 1\frac{1}{3}$	$y = 1$	$(0, 5/16)$	$(-1/3, 0), (5, 0)$
11.	$x = -1$, hole at $x = 1$	$y = 1$	$(0, 3)$	$(-3, 0)$
13.	$x = 4$	none $y = 2x$ (oblique)	$(0, \frac{1}{4})$	$(-1, 0), (1/2, 0)$
15.	$x = 0, 4$	$y = 0$	DNE	$(-2, 0), (2/3, 0)$
17.	$x = -2, 4$	$y = 1$	$(0, -15/16)$	$(1, 0), (-3, 0), (5, 0)$

5.

7.

9.

11.

13.

15.

17.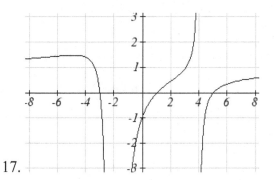

19. $y = \dfrac{50(x-2)(x+1)}{(x+5)(x-5)}$

21. $y = \dfrac{7(x-4)(x+6)}{(x+4)(x+5)}$

23. $y = \dfrac{1(x-2)^2}{2(x+1)}$

25. $y = \dfrac{4(x-3)}{(x+3)(x-4)}$

27. $y = \dfrac{27(x-2)}{(x+3)(x-3)^2}$

29. $y = \dfrac{1(x+3)(x-2)}{3(x-1)}$

31. $y = \dfrac{-6(x-1)^2}{(x+3)(x-2)^2}$

33. $y = -\dfrac{2(x)(x-3)}{(x+3)(x-4)}$

35. $y = \dfrac{2(x-1)^3}{(x+1)(x-2)^2}$

37. $y = \dfrac{(x-4)(x-2)}{(x-4)(x+1)}$

39. $y = 3x - 2$

41. $y = \dfrac{1}{2}x + 1$

43. $y = -2x + 1$

45. a. $C(n) = \dfrac{4}{20+n}$ b. $C(10) \approx 13.33\%$ c. 80 mL d. as $n \to \infty, C \to 0$

Section 3.8

1. Domain $(4, \infty)$ Inverse $f^{-1}(x) = \sqrt{x} + 4$

3. Domain $(-\infty, 0)$ Inverse $f^{-1}(x) = -\sqrt{12 - x}$

5. Domain $(-\infty, \infty)$ Inverse $f^{-1}(x) = \sqrt[3]{\dfrac{x-1}{3}}$

7. $f^{-1}(x) = \dfrac{(x-9)^2}{4} + 1$

9. $f^{-1}(x) = \left(\dfrac{x-9}{2}\right)^3$

11. $f^{-1}(x) = \dfrac{2-8x}{x}$

13. $f^{-1}(x) = \dfrac{3-7x}{x-1}$

15. $f^{-1}(x) = \dfrac{5x-4}{3+4x}$

17. 65.574 mph

19. 34.073 mph

21. 14.142 feet

Chapter 4

Section 4.1

1. Linear   3. Exponential   5. Neither

7. $P(t) = 11,000(1.085)^t$ 9. 47622 Fox

11. $17561.70   13. $y = 6(5)^x$ 15. $y = 2000(0.1)^x$

17. $y = 3(2)^x$ 19. $y = \left(\frac{1}{6}\right)^{-\frac{3}{5}}\left(\frac{1}{6}\right)^{\frac{x}{5}} = 2.93(0.699)^x$ 21. $y = \frac{1}{8}(2)^x$

23. $34.32\,mg$ 25. 1.39%; $155,368.09   27. $4,813.55

29. Annual $\approx$ $7353.84   Quarterly $\approx$ $7,469.63   Monthly $\approx$ $7,496.71
     Continuously $\approx$ $7,510.44

31. 3.03%   33. 7.4 years

35a. $w(t) = (1.113)(1.046)^t$ b. $1.11   c. Below what the model predicts $\approx$ $5.70

Section 4.2

1. B   3. A   5. E   7. D   9. C

11.   13.   15.

17. $y = 4^x + 4$ 19. $y = 4^{x+2}$ 21. $y = -4^x$

23. As $x \to \infty$ $f(x) \to -\infty$. As $x \to -\infty$ $f(x) \to -1$

25. As $x \to \infty$ $f(x) \to -2$ As $x \to -\infty$ $f(x) \to \infty$

27. As $x \to \infty$ $f(x) \to 2$ As $x \to -\infty$ $f(x) \to \infty$

29. $y = -2^{x+2} + 1 = -4(2)^x + 1$ 31. $y = -2(2)^{-x} + 3$

33. $y = -2(3)^x + 7$ 35. $y = 2\left(\frac{1}{2}\right)^x - 4$

Section 4.3

1. $4^m = q$ 3. $a^c = b$ 5. $10^t = v$

7. $e^n = w$ 9. $\log_4(y) = x$ 11. $\log_c(k) = d$

13. $\log(b) = a$ 15. $\ln(h) = k$ 17. 9

19. 1/8   21. 1000   23. e^2

25. 2   27. -3   29. ½

31. 4

33. -3

35. -2

37. -1.398

39. 2.708

41. $\dfrac{\log(14)}{\log(5)} \approx 1.6397$

43. $\dfrac{\log\left(\dfrac{1}{15}\right)}{\log(7)} \approx -1.392$

45. $\dfrac{\ln(17)}{5} \approx 0.567$

47. $\dfrac{\dfrac{\log(38)}{\log(3)}+5}{4} \approx 2.078$

49. $\dfrac{\log(5)}{\log(1.03)} \approx 54.449$

51. $\dfrac{\log\left(\dfrac{8}{3}\right)}{3\log(1.04)} \approx 8.335$

53. $\dfrac{\ln\left(\dfrac{1}{5}\right)}{-0.12} \approx 13.412$

55. $\dfrac{\log\left(\dfrac{5}{8}\right)}{\log\left(\dfrac{1}{2}\right)} \approx 0.678$

57. $f(t) = 300e^{-0.0943t}$

59. $f(t) = 10e^{0.03922t}$

61. $f(t) = 150(1.0618)^t$

63. $f(t) = 50(0.98807)^t$

65. During the year 2013

67. During the year 2074

69. $\approx 34\,hours$

71. 13.532 years

Section 4.4

1. $\log_3(4)$ 3. $\log_3(7)$ 5. $\log_3(5)$ 7. $\log_7(2)$ 9. $\log(6x^9)$

11. $\ln(2x^7)$

13. $\log\left(x^2(x+1)^3\right)$

15. $\log\left(\dfrac{xz^3}{\sqrt{y}}\right)$

17. $15\log(x)+13\log(y)-19\log(z)$

19. $-2\ln(a)+4\ln(b)-5\ln(c)$

21. $\dfrac{3}{2}\log(x)-2\log(y)$

23. $\ln(y)+\dfrac{1}{2}\left(\ln(y)-\ln(1-y)\right)$

25. $\dfrac{8}{3}\log(x)+\dfrac{14}{3}\log(y)$

27. $x \approx -0.717$

29. $x \approx -6.395$

31. $t \approx 17.329$

33. $x = \dfrac{2}{7}$

35. $x \approx 0.123$

37. $x \approx 4.642$

39. $x \approx 30.158$

41. $x \approx -2.889$.

43. $x \approx 6.873\,or\;x \approx -0.873$

45. $x = \dfrac{12}{11} \approx 1.091$

47. $x = 10$

661

Section 4.5

1. Domain: $: x > 5$ V. A. @ $x = 5$

3. Domain: $x < 3$ V.A. @ $x = 3$

5. Domain: $x > -\dfrac{1}{3}$ V.A. @ $x = -\dfrac{1}{3}$

7. Domain: $x < 0$ V.A. @ $x = 0$

9.

11.

13.

15.

17. $y = \dfrac{1}{\log(2)}\log\big(-(x-1)\big)$ 19. $y = -\dfrac{3}{\log(3)}\log(x+4)$

21. $y = \dfrac{3}{\log(4)}\log(x+2)$ 23. $y = -\dfrac{2}{\log(5)}\log\big(-(x-5)\big)$

Section 4.6

1. $f(t) = 13(0.9195)^t$.   2 mg will remain after 22.3098 minutes

3. $f(t) = 200(0.999564)^t$. $f(1000) = 129.3311$ mg

5. $r = -0.06448$.   Initial mass: 9.9018 mg.   After 3 days: 0.01648 mg

7. $f(t) = 250(0.9909)^t$.   Half-life = 75.8653 minutes

9. $f(t) = a(0.999879)^t$.   60% ($0.60a$) would remain after 4222.813 years

11. $P(t) = 1500(1.02337)^t$ (t in minutes). After 2 hours = 24000.

    After 100 minutes = 15119

13. a) 610.5143 (about 611)   b) 25.6427 minutes   c) 10431.21   d) 106.9642 minutes

15. 23.1914 years

17. 53.319 hours

19. $T(t) = 90(0.99166)^t + 75$.   a) 134.212 deg   b) 112.743 minutes

662

21. a) b) 100 c) 269.487 d) 7.324 years

23. $\log(x) = -0.5$. $x = 0.3162$ 25. $\log(x) = 1.5$. $x = 31.623$

27.

| Whisper | | | | | | | | Vacuum | | | | | | | | Jet |

10^{-10} 10^{-9} 10^{-8} 10^{-7} 10^{-6} 10^{-5} 10^{-4} 10^{-3} 10^{-2} 10^{-1} 10^{0} 10^{1} 10^{2}

29. 63095.7 times more intense   31. MMS magnitude 5.817

33. a) about 1640671 b) 1.4 hours c) No, because $(2.042727)^{0.693147} \approx e^{0.495105}$

d) Anja's data predicts a continuous growth rate of 0.4116, which is much smaller than the rate 0.495105 you calculated. Our model would overestimate the number of cells.

35. a) The curve that increases rapidly at first is $M(p)$

b) $H(100) = 0.9775$

c) Myoglobin: $M(20) = 0.9524$. Hemoglobin: $H(20) = 0.3242$

d) At 20 torrs: 0.6282. At 40 torrs: 0.2060. At 60 torrs: 0.0714

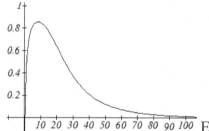 Efficiency seems to be maximized at about 8 torr

37. a) $C(t) = 1.03526^t$, or $C(t) = e^{0.03466t}$

b) Volume of one cell: $\frac{4}{3}\pi(50 \times 10^{-4})^3 \approx 5.236 \times 10^{-7}$ cm^3, so will need about

1.9099 × 10^6 cells for a volume of 1cm^3. $C(t) = 1.9099 \times 10^6$ after 417.3 hours

39. 31.699 days

663

Section 4.7

1. $\log(f(x)) = \log(1.3)x + \log(4)$

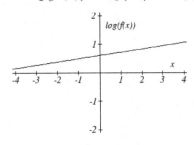

3. $\log(f(x)) = \log(0.2)x + 1$

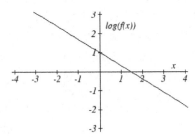

5. $y = e^{\frac{1}{2}x-1} = e^{-1}e^{\frac{1}{2}x} \approx 0.368(1.6487)^x$

7. $y = 10^{-x-2} = 10^{-2}10^{-1x} = 0.01(0.1)^x$

9. $y = 776.682(1.426)^x$　　　　　11. $y = 731.92(0.738)^x$

13. Expenditures are approximately \$205

15. $y = 7.599(1.016)^x$ $r = 0.83064$, $y = 0.1493x + 7.4893$, $r = 0.81713$. Using the better function, we predict electricity will be 11.157 cents per kwh

Chapter 5

Section 5.1

1. 10

3. $(x-8)^2 + (y+10)^2 = 8^2$

5. $(x-7)^2 + (y+2)^2 = 293$

7. $(x-5)^2 + (y-8)^2 = 13$

9.

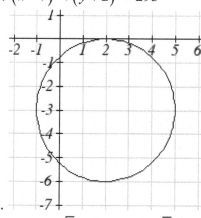

11. $(0, 3+\sqrt{5})$ and $(0, 3-\sqrt{5})$

13. (1.3416407865, 7.683281573)

15. (-1.07335, 2.8533)

17. 29.87 miles

Section 5.2

1.

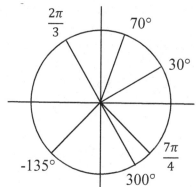

3. π

5. 150°

7. 325°

9. 54°

11. $\dfrac{8\pi}{9}$

13. $\dfrac{\pi}{2}$

15. 35 miles

17. 8π cm

19. 5.7596 miles

21. 28.6479°

23. 14.1372 cm²

25. 3960 rad/min   630.254 RPM

27. 2.094 in/sec, π/12 rad/sec, 2.5 RPM

29. 75,398.22 mm/min = 1.257 m/sec

31. Angular speed: π/12 rad/hr. Linear speed: 1036.73 miles/hr

Section 5.3

1. a. III   b. II

3. $-\dfrac{4}{5}$

5. $-\dfrac{4\sqrt{3}}{7}$

7. $-\dfrac{\sqrt{55}}{8}$

9.   a. reference: 45°. Quadrant III. $\sin(225°) = -\dfrac{\sqrt{2}}{2}$. $\cos(225°) = -\dfrac{\sqrt{2}}{2}$

   b. reference: 60°. Quadrant IV. $\sin(300°) = -\dfrac{\sqrt{3}}{2}$. $\cos(300°) = \dfrac{1}{2}$

   c. reference: 45°. Quadrant II. $\sin(135°) = \dfrac{\sqrt{2}}{2}$. $\cos(135°) = -\dfrac{\sqrt{2}}{2}$

   d. reference: 30°. Quadrant III. $\sin(210°) = -\dfrac{1}{2}$. $\cos(210°) = -\dfrac{\sqrt{3}}{2}$

11.   a. reference: $\dfrac{\pi}{4}$. Quadrant III. $\sin\left(\dfrac{5\pi}{4}\right) = -\dfrac{\sqrt{2}}{2}$. $\cos\left(\dfrac{5\pi}{4}\right) = -\dfrac{\sqrt{2}}{2}$

   b. reference: $\dfrac{\pi}{6}$. Quadrant III. $\sin\left(\dfrac{7\pi}{6}\right) = -\dfrac{1}{2}$. $\cos\left(\dfrac{7\pi}{6}\right) = -\dfrac{\sqrt{3}}{2}$

   c. reference: $\dfrac{\pi}{3}$. Quadrant IV. $\sin\left(\dfrac{5\pi}{3}\right) = -\dfrac{\sqrt{3}}{2}$. $\cos\left(\dfrac{5\pi}{3}\right) = \dfrac{1}{2}$

   d. reference: $\dfrac{\pi}{4}$. Quadrant II. $\sin\left(\dfrac{3\pi}{4}\right) = \dfrac{\sqrt{2}}{2}$. $\cos\left(\dfrac{3\pi}{4}\right) = -\dfrac{\sqrt{2}}{2}$

13.   a. $\sin\left(-\dfrac{3\pi}{4}\right) = -\dfrac{\sqrt{2}}{2}$ $\cos\left(-\dfrac{3\pi}{4}\right) = -\dfrac{\sqrt{2}}{2}$

   b. $\sin\left(\dfrac{23\pi}{6}\right) = -\dfrac{1}{2}$ $\cos\left(\dfrac{23\pi}{6}\right) = \dfrac{\sqrt{3}}{2}$

   c. $\sin\left(-\dfrac{\pi}{2}\right) = -1$ $\cos\left(-\dfrac{\pi}{2}\right) = 0$

   d. $\sin(5\pi) = 0$ $\cos(5\pi) = -1$

15. a. $\dfrac{2\pi}{3}$ b. 100°   c. 40°   d. $\dfrac{5\pi}{3}$ e. 235°

17. a. $\dfrac{5\pi}{3}$ b. 280°   c. 220°   d. $\dfrac{2\pi}{3}$ e. 55°

19. (-11.491, -9.642)

Section 5.4

1. $\sec(\theta) = \sqrt{2}$, $\csc(\theta) = \sqrt{2}$, $\tan(\theta) = 1$, $\cot(\theta) = 1$

3. $\sec(\theta) = -\dfrac{2\sqrt{3}}{3}$, $\csc(\theta) = 2$, $\tan(\theta) = -\dfrac{\sqrt{3}}{3}$, $\cot(\theta) = -\sqrt{3}$

5. $\sec(\theta) = -2$, $\csc(\theta) = \dfrac{2\sqrt{3}}{3}$, $\tan(\theta) = -\sqrt{3}$, $\cot(\theta) = -\dfrac{\sqrt{3}}{3}$

7. a. $\sec(135°) = -\sqrt{2}$ b. $\csc(210°) = -2$ c. $\tan(60°) = \sqrt{3}$.   d. $\cot(225°) = 1$

9. $\cos(\theta) = -\dfrac{\sqrt{7}}{4}$, $\sec(\theta) = -\dfrac{4\sqrt{7}}{7}$, $\csc(\theta) = \dfrac{4}{3}$, $\tan(\theta) = -\dfrac{3\sqrt{7}}{7}$, $\cot(\theta) = -\dfrac{\sqrt{7}}{3}$

11. $\sin(\theta) = -\dfrac{2\sqrt{2}}{3}$, $\csc(\theta) = -\dfrac{3\sqrt{2}}{4}$, $\sec(\theta) = -3$, $\tan(\theta) = 2\sqrt{2}$, $\cot(\theta) = \dfrac{\sqrt{2}}{4}$

13. $\sin(\theta) = \dfrac{12}{13}$, $\cos(\theta) = \dfrac{5}{13}$, $\sec(\theta) = \dfrac{13}{5}$, $\csc(\theta) = \dfrac{13}{12}$, $\cot(\theta) = \dfrac{5}{12}$

15. a. $\sin(0.15) = 0.1494$ $\cos(0.15) = 0.9888$ $\tan(0.15) = 0.1511$
    b. $\sin(4) = -0.7568$ $\cos(4) = -0.6536$ $\tan(4) = 1.1578$
    c. $\sin(70°) = 0.9397$ $\cos(70°) = 0.3420$ $\tan(70°) = 2.7475$
    d. $\sin(283°) = -0.9744$ $\cos(283°) = 0.2250$ $\tan(283°) = -4.3315$

17. $\sec(t)$ 19. $\tan(t)$ 21. $\tan(t)$ 23. $\cot(t)$ 25. $\left(\sec(t)\right)^2$

Section 5.5

1. $\sin(A) = \dfrac{5\sqrt{41}}{41}$, $\cos(A) = \dfrac{4\sqrt{41}}{41}$, $\tan(A) = \dfrac{5}{4}$

   $\sec(A) = \dfrac{\sqrt{41}}{4}$, $\csc(A) = \dfrac{\sqrt{41}}{5}$, $\cot(A) = \dfrac{4}{5}$

3. $c = 14$, $b = 7\sqrt{3}$, $B = 60°$ 5. $a = 5.3171$, $c = 11.3257$, $A = 28°$

7. $a = 9.0631$, $b = 4.2262$, $B = 25°$ 9. 32.4987 ft

11. 836.2698 ft   13. 460.4069 ft

15. 660.35 feet   17. 28.025 ft

19. 143.0427   21. 86.6685

Chapter 6

Section 6.1

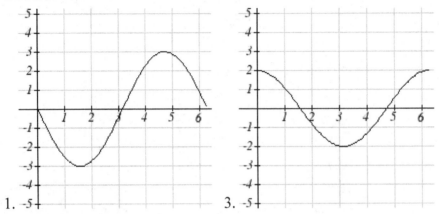

1.

3.

5. Amp: 3. Period= 2. Midline: y= -4. $f(t) = 3\sin(\pi t) - 4$

7. Amp: 2. Period= 4π. Midline: y= 1. $f(t) = 2\cos\left(\frac{1}{2}t\right) + 1$

9. Amp: 2. Period= 5. Midline: y= 3. $f(t) = -2\cos\left(\frac{2\pi}{5}t\right) + 3$

11. Amp: 3, Period $= \frac{\pi}{4}$, Shift: 4 left, Midline: $y = 5$

13. Amp: 2, Period $= \frac{2\pi}{3}$, Shift: 7 right, Midline: $y = 4$

15. Amp: 1, Period $= 12$, Shift: 6 left, Midline: $y = -3$

17. $f(x) = 4\sin\left(\frac{\pi}{5}(x+1)\right)$

19. $f(x) = \cos\left(\frac{\pi}{5}(x+2)\right)$

21. $D(t) = 50 - 7\sin\left(\frac{\pi}{12}t\right)$

23 a. Amp: 12.5. Midline: $y = 13.5$. Period: 10

   b. $h(t) = -12.5\cos\left(\frac{\pi}{5}t\right) + 13.5$

   c. $h(5) = 26$ meters

Section 6.2

1. II

3. I

5. Period: $\dfrac{\pi}{4}$. Horizontal shift: 8 right

7. Period: 8. Horizontal shift: 1 left

9. Period: 6. Horizontal shift: 3 left

11.

13.

15.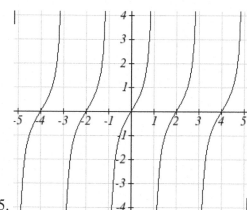

17. $f(x) = 2\sec\left(\dfrac{\pi}{2}x\right) - 1$

19. $f(x) = 2\csc\left(\dfrac{\pi}{4}x\right) + 1$

21. $\tan(-x) = 1.5$

23. $\sec(-x) = 2$

25. $\csc(-x) = 5$

27. $-\csc(x)$

Section 6.3

1. $\dfrac{\pi}{4}$

3. $-\dfrac{\pi}{6}$

5. $\dfrac{\pi}{3}$

7. $\dfrac{3\pi}{4}$

9. $\dfrac{\pi}{4}$

11. $-\dfrac{\pi}{3}$

13. 1.9823

15. -0.9273

17. 44.427°

19. $\dfrac{\pi}{4}$

21. $-\dfrac{\pi}{6}$

23. $\dfrac{2\sqrt{10}}{7}$

25. $\dfrac{1}{\sqrt{17}}$

27. $\dfrac{\sqrt{25-x^2}}{5}$

29. $\dfrac{3x}{\sqrt{9x^2+1}}$

Section 6.4

1. $\dfrac{5\pi}{4},\dfrac{7\pi}{4}$

3. $\dfrac{\pi}{3},\dfrac{5\pi}{3}$

5. $\dfrac{\pi}{2}$

7. $\dfrac{\pi}{2},\dfrac{3\pi}{2}$

9. $\dfrac{\pi}{4}+2\pi k,\dfrac{7\pi}{4}+2\pi k$, where k is an integer

11. $\dfrac{7\pi}{6}+2\pi k,\dfrac{11\pi}{6}+2\pi k$, where k is an integer

13. $\dfrac{\pi}{18}+\dfrac{2\pi}{3}k,\dfrac{5\pi}{18}+\dfrac{2\pi}{3}k$, where k is an integer

15. $\dfrac{5\pi}{12}+\dfrac{2\pi}{3}k,\dfrac{7\pi}{12}+\dfrac{2\pi}{3}k$, where k is an integer

17. $\dfrac{\pi}{6}+\pi k,\dfrac{5\pi}{6}+\pi k$, where k is an integer

19. $\dfrac{\pi}{4}+\dfrac{2\pi}{3}k,\dfrac{5\pi}{12}+\dfrac{2\pi}{3}k$, where k is an integer

21. $4+8k$, where k is an integer

23. $\dfrac{1}{6}+2k,\dfrac{5}{6}+2k$, where k is an integer

25. 0.2734, 2.8682

27. 3.7603, 5.6645

29. 2.1532, 4.1300

31. 0.7813, 5.5019

33. 0.04829, 0.47531

35. 0.7381, 1.3563

37. 0.9291, 3.0709

39. 1.3077, 4.6923

Section 6.5

1. $c = \sqrt{89}$, $A = 57.9946°$, $B = 32.0054°$

3. $b = \sqrt{176}$, $A = 27.8181°$, $B = 62.1819°$

5. $y(x) = 6\sin\left(\dfrac{\pi}{2}(x-1)\right) + 4$

7. $D(t) = 50 - 13\cos\left(\dfrac{\pi}{12}(t-5)\right)$

9. a. $P(t) = 129 - 25\cos\left(\dfrac{\pi}{6}t\right)$ b. $P(t) = 129 - 25\cos\left(\dfrac{\pi}{6}(t-3)\right)$

11. 75 degrees   13. 8

15. 2.80869431742   17. 5.035 months

Chapter 7

Section 7.1

1. $\dfrac{7\pi}{6}, \dfrac{11\pi}{6}$ 3. $\dfrac{\pi}{3}, \dfrac{5\pi}{3}$

5. $\dfrac{2}{3} + 8k$, and $\dfrac{10}{3} + 8k$, where k is an integer

7. $\dfrac{5\pi}{12} + k\pi$ and $\dfrac{7\pi}{12} + k\pi$, where k is an integer

9. $0.1339 + 10k$ and $8.6614 + 10k$, where k is an integer

11. $1.1438 + \dfrac{2\pi}{3}k$ and $1.9978 + \dfrac{2\pi}{3}k$, where k is an integer

13. $\dfrac{\pi}{2}, \dfrac{3\pi}{2}, 0.644, 2.498$ 15. 0.056, 1.515, 3.197, 4.647

17. $0, \pi, \dfrac{\pi}{3}, \dfrac{5\pi}{3}$ 19. $\dfrac{\pi}{6}, \dfrac{5\pi}{6}, \dfrac{7\pi}{6}, \dfrac{11\pi}{6}$

21. 1.183, 1.958, 4.325, 5.100   23. $\dfrac{3\pi}{2}, \dfrac{7\pi}{6}, \dfrac{11\pi}{6}$

25. $\pi, \dfrac{\pi}{3}, \dfrac{5\pi}{3}$ 27. 1.823, 4.460

29. 2.301, 3.983, 0.723, 5.560   31. 3.305, 6.120

33. $0, \dfrac{\pi}{3}, \dfrac{2\pi}{3}, \pi, \dfrac{4\pi}{3}, \dfrac{5\pi}{3}$

35. $0, \dfrac{\pi}{4}, \dfrac{3\pi}{4}, \pi, \dfrac{5\pi}{4}, \dfrac{7\pi}{4}$

37. $\dfrac{\pi}{6}, \dfrac{2\pi}{3}, \dfrac{5\pi}{6}, \dfrac{4\pi}{3}$

39. $0, \pi, 1.231, 5.052$

41. $\dfrac{\pi}{3}, \dfrac{5\pi}{3}$

Section 7.2

1. $\dfrac{\sqrt{2}+\sqrt{6}}{4}$

3. $\dfrac{-\sqrt{2}-\sqrt{6}}{4}$

5. $\dfrac{\sqrt{2}-\sqrt{6}}{4}$

7. $\dfrac{\sqrt{2}+\sqrt{6}}{4}$

9. $\dfrac{\sqrt{3}}{2}\sin(x)-\dfrac{1}{2}\cos(x)$

11. $-\dfrac{\sqrt{3}}{2}\cos(x)+\dfrac{1}{2}\sin(x)$

13. $\sec(t)$

15. $\tan(x)$

17. $8\big(\cos(5x)-\cos(27x)\big)$

19. $\sin(8x)+\sin(2x)$

21. $2\cos(5t)\cos(t)$

23. $2\sin(5x)\cos(2x)$

25. a. $\left(\dfrac{2}{3}\right)\left(-\dfrac{1}{4}\right)+\left(-\dfrac{\sqrt{5}}{3}\right)\left(\dfrac{\sqrt{15}}{4}\right)=\dfrac{-2-5\sqrt{3}}{12}$

b. $\left(-\dfrac{\sqrt{5}}{3}\right)\left(-\dfrac{1}{4}\right)+\left(\dfrac{2}{3}\right)\left(\dfrac{\sqrt{15}}{4}\right)=\dfrac{\sqrt{5}+2\sqrt{15}}{12}$

27. $0.373+\dfrac{2\pi}{3}k$ and $0.674+\dfrac{2\pi}{3}k$, where k is an integer

29. $2\pi k$, where k is an integer

31. $\dfrac{\pi}{7}+\dfrac{4\pi}{7}k$, $\dfrac{3\pi}{7}+\dfrac{4\pi}{7}k$, $\dfrac{\pi}{3}+\dfrac{4\pi}{3}k$, and $\pi+\dfrac{4\pi}{3}k$, where k is an integer

33. $\dfrac{7\pi}{12}+\pi k$, $\dfrac{11\pi}{12}+\pi k$, and $\dfrac{\pi}{4}k$, where k is an integer

35. $2\sqrt{13}\sin(x+5.3004)$ or $2\sqrt{13}\sin(x-0.9828)$

37. $\sqrt{29}\sin(3x+0.3805)$

39. $0.3681, 3.8544$

41. $0.7854, 1.8158$

43. $\tan(6t)$

Section 7.3

1. a. $\dfrac{3\sqrt{7}}{32}$ b. $\dfrac{31}{32}$ c. $\dfrac{3\sqrt{7}}{31}$ 3. $\cos(56°)$

5. $\cos(34°)$ 7. $\cos(18x)$

9. $2\sin(16x)$ 11. $0, \pi, 2.4189, 3.8643$

13. $0.7297, 2.4119, 3.8713, 5.5535$

15. $\dfrac{\pi}{6}, \dfrac{\pi}{2}, \dfrac{5\pi}{6}, \dfrac{3\pi}{2}$

17. a. $\dfrac{2\pi}{9}, \dfrac{4\pi}{9}, \dfrac{8\pi}{9}, \dfrac{10\pi}{9}, \dfrac{14\pi}{9}, \dfrac{16\pi}{9}, 0, \dfrac{2\pi}{3}, \dfrac{4\pi}{3}$

19. $\dfrac{1+\cos(10x)}{2}$

21. $\dfrac{3}{8} - \dfrac{1}{2}\cos(16x) + \dfrac{1}{8}\cos(32x)$

23. $\dfrac{1}{16} - \dfrac{1}{16}\cos(2x) + \dfrac{1}{16}\cos(4x) - \dfrac{1}{16}\cos(2x)\cos(4x)$

25. a. $\sqrt{\dfrac{1}{2} + \dfrac{2\sqrt{3}}{7}}$ b. $\sqrt{\dfrac{1}{2} - \dfrac{2\sqrt{3}}{7}}$ c. $\dfrac{1}{7 - 4\sqrt{3}}$

Section 7.4

1. $y = 3\sin\left(\dfrac{\pi}{6}(x-3)\right) - 1$

3. Amplitude: 8, Period: $\dfrac{1}{3}$ second, Frequency: 3 Hz (cycles per second)

5. $P(t) = -19\cos\left(\dfrac{\pi}{6}t\right) + \dfrac{40}{3}t + 650$ 7. $P(t) = -33\cos\left(\dfrac{\pi}{6}t\right) + 900(1.07)^t$

9. $D(t)=10(0.85)^t \cos(36\pi t)$

11. $D(t)=17(0.9145)^t \cos(28\pi t)$

13. a. IV   b. III

15. $y=6(4)^x +5\sin\left(\dfrac{\pi}{2}x\right)$

17. $y=-3\sin\left(\dfrac{\pi}{2}\right)+2x+7$

19. $y=8\left(\dfrac{1}{2}\right)^x \cos\left(\dfrac{\pi}{2}x\right)+3$

Chapter 8

Section 8.1

1.

3.

5.

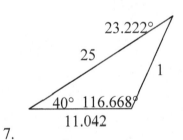

7.

9. $\beta=68°, a=14.711, c=20.138$

11. $\beta=28.096°, \gamma=32.904°, c=16.149$

13. Not possible.

15. $\beta=64.243°, \gamma=72.657°, c=257.328$ OR $\beta=115.757°, \gamma=21.143°, c=97.238$

17.

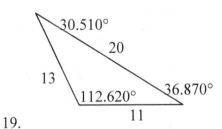

19.

21. $c=2.066, \alpha=52.545°, \beta=86.255°$

23. $a = 11.269, \beta = 27.457°, \gamma = 32.543°$

25. 177.562   27. 978.515 ft

29. Distance to A: 565.258 ft. Distance to shore: 531.169 ft

31. 529.014 m   33. 173.877 feet

35. 4.642 km, 2.794 km   37. 757.963 ft

39. 2371.129 miles   41. 65.375 cm^2

43. 7.72

Section 8.2

1. $\left(-\dfrac{7\sqrt{3}}{2}, -\dfrac{7}{2}\right)$ 3. $\left(2\sqrt{2}, -2\sqrt{2}\right)$

5. $\left(3\sqrt{2}, -3\sqrt{2}\right)$ 7. $(0,3)$

9. $\left(-\dfrac{3\sqrt{3}}{2}, -\dfrac{3}{2}\right)$ 11. $(-1.248, 2.728)$

13. $\left(2\sqrt{5}, 0.464\right)$ 15. $\left(2\sqrt{13}, 2.159\right)$

17. $\left(\sqrt{34}, 5.253\right)$ 19. $\left(\sqrt{269}, 4.057\right)$

21. $r = 3\sec(\theta)$ 23. $r = \dfrac{\sin(\theta)}{4\cos^2(\theta)}$

25. $r = 4\sin(\theta)$ 27. $r = \dfrac{\cos(\theta)}{\left(\cos^2(\theta) - \sin^2(\theta)\right)}$

29. $x^2 + y^2 = 3y$ 31. $y + 7x = 4$

33. $x = 2$ 35. $x^2 + y^2 = x + 2$

37. A   39. C   41. E   43. C   45. D   47. F

49.   51.   53.

675

55.

57.

59.

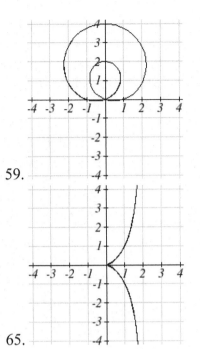

61.

63.

65.

Section 8.3

1. $3i$

3. -12

5. $1+\sqrt{3}i$

7. $8-i$

9. $-11+4i$

11. $-12+8i$

13. $30-10i$

15. $11+10i$

17. 20

19. $\dfrac{3}{2}+2i$

21. $\dfrac{3}{2}+\dfrac{5}{2}i$

23. $-\dfrac{1}{25}-\dfrac{18}{25}i$

25. -1

27. i

29. $3\cos(2)+3\sin(2)i=-1.248+2.728i$

31. $3\sqrt{3}+3i$

33. $-\dfrac{3\sqrt{2}}{2}-\dfrac{3\sqrt{2}}{2}i$

35. $6e^{0i}$

37. $4e^{\frac{3\pi}{2}i}$

39. $2\sqrt{2}e^{\frac{\pi}{4}i}$

41. $3\sqrt{2}e^{\frac{3\pi}{4}i}$

43. $\sqrt{34}e^{0.540i}$

45. $\sqrt{10}e^{2.820i}$

47. $\sqrt{17}e^{4.467i}$

49. $\sqrt{26}e^{6.086i}$

51. $6e^{\frac{5\pi}{12}i}$

53. $2e^{\frac{7\pi}{12}i}$

55. $1024e^{\frac{5\pi}{2}i}$

57. $4e^{\frac{\pi}{3}i}$

59. 4096

61. $0.788 + 1.903i$

63. $1.771 + 0.322i$

65. $\sqrt[5]{2} \approx 1.149, 0.355 + 1.092i, -0.929 + 0.675i, -0.929 - 0.675i, 0.355 - 1.092i$

67. $1, \dfrac{1}{2} + \dfrac{\sqrt{3}}{2}i, -\dfrac{1}{2} + \dfrac{\sqrt{3}}{2}i, -1, -\dfrac{1}{2} - \dfrac{\sqrt{3}}{2}i, \dfrac{1}{2} - \dfrac{\sqrt{3}}{2}i$

Section 8.4

1. $-4, 2$

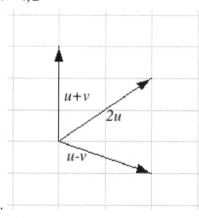

3. The vectors do not need to start at the same point

5. $3\vec{v} - \vec{u}$

7. $3\sqrt{2}, 3\sqrt{2}$

9. $-6.128, -5.142$

11. Magnitude: 4, Direction: 90°

13. Magnitude: 7.810, Direction: 39.806°

15. Magnitude: 2.236, Direction: 153.435°

17. Magnitude: 5.385, Direction: 291.801°

19. Magnitude: 7.211, Direction: 236.310°

21. $\vec{u} + \vec{v} = \langle 3,2 \rangle$, $\vec{u} - \vec{v} = \langle 1,-8 \rangle$, $2\vec{u} - 3\vec{v} = \langle 1,-21 \rangle$

23. 4.635 miles, 17.764 deg N of E

25. 17 miles. 10.318 miles

27. $\overrightarrow{F_{net}} = -4, -11$

29. Distance: 2.868. Direction: 86.474° North of West, or 3.526° West of North

31. 4.924 degrees. 659 km/hr

33. 4.424 degrees

35. (0.081, 8.602)

37. 21.801 degrees, relative to the car's forward direction

677

Section 8.5

1. $6 \cdot 10 \cdot \cos(75°) = 15.529$ 3. $(0)(-3) + (4)(0) = 0$ 5. $(-2)(-10) + (1)(13) = 33$

7. $\cos^{-1}\left(\dfrac{0}{\sqrt{4}\sqrt{3}}\right) = 90°$ 9. $\cos^{-1}\left(\dfrac{(2)(1) + (4)(-3)}{\sqrt{2^2 + 4^2}\sqrt{1^2 + (-3)^2}}\right) = 135°$

11. $\cos^{-1}\left(\dfrac{(4)(8) + (2)(4)}{\sqrt{4^2 + 8^2}\sqrt{2^2 + 4^2}}\right) = 0°$ 13. $(2)(k) + (7)(4) = 0$, $k = -14$

15. $\dfrac{(8)(1) + (-4)(-3)}{\sqrt{1^2 + (-3)^2}} = 6.325$ 17. $\left(\dfrac{(-6)(1) + (10)(-3)}{\sqrt{1^2 + (-3)^2}^2}\right)\langle 1, -3 \rangle = \langle -3.6, 10.8 \rangle$

19. The vectors are $\langle 2,3 \rangle$ and $\langle -5,-2 \rangle$. The acute angle between the vectors is $34.509°$

21. 14.142 pounds   23. $\langle 10\cos(10°), 10\sin(10°) \rangle \cdot \langle 0, -20 \rangle$, so 34.7296 ft-lbs

25. $40 \cdot 120 \cdot \cos(25°) = 4350.277$ ft-lbs

Section 8.6

1. C   3. E   5. F

7.

9. x(t)

y(t)

11.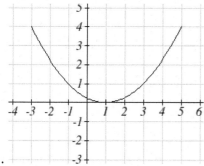

13. $y = -2 + 2x$

15. $y = 3\sqrt{\dfrac{x-1}{2}}$

17. $x = 2e^{\frac{1-y}{5}}$ or $y = 1 - 5\ln\left(\dfrac{x}{2}\right)$

19. $x = \left(\dfrac{y}{2}\right)^3 - \dfrac{y}{2}$

21. $y = x^3$

23. $\left(\dfrac{x}{4}\right)^2 + \left(\dfrac{y}{5}\right)^2 = 1$

25. $\begin{cases} x(t) = t \\ y(t) = 3t^2 + 3 \end{cases}$

27. $\begin{cases} x(t) = 3\log(t) + t \\ y(t) = t \end{cases}$

29. $\begin{cases} x(t) = 2\cos(t) \\ y(t) = 3\sin(t) \end{cases}$

31. $\begin{cases} x(t) = t^3 \\ y(t) = t + 2 \end{cases}$

33. $\begin{cases} x(t) = t - 1 \\ y(t) = -t^2 \end{cases}$

35. $\begin{cases} x(t) = -1 + 3t \\ y(t) = 5 - 2t \end{cases}$

37. $\begin{cases} x(t) = 4\cos(3t) \\ y(t) = 6\sin(t) \end{cases}$

39. $\begin{cases} x(t) = 4\cos(2t) \\ y(t) = 3\sin(3t) \end{cases}$

41. $y(x) = -16\left(\dfrac{x}{15}\right)^2 + 20\left(\dfrac{x}{15}\right)$

43. $\begin{cases} x(t) = 20\sin\left(\dfrac{2\pi}{5}t\right) + 8\sin(\pi t) \\ y(t) = 35 - 20\cos\left(\dfrac{2\pi}{5}t\right) - 8\cos(\pi t) \end{cases}$

Chapter 9

Section 9.1

1. D   3. B

5. Vertices (0,±5), minor axis endpoints (±2,0), major length = 10, minor length = 4

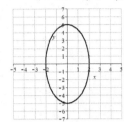

7. Vertices (±2,0), minor axis endpoints (0,±1), major length = 4, minor length = 2

9. Vertices (±5,0), minor axis endpoints (0,±1), major length = 10, minor length = 2

11. Vertices (0,±4), minor axis endpoints (±3,0), major length = 8, minor length = 6

13. Vertices $\left(0,\pm3\sqrt{2}\right)$, minor axis endpoints $\left(\pm\sqrt{2},0\right)$, major length = $6\sqrt{2}$, minor length = $2\sqrt{2}$

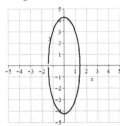

680

15. $\dfrac{x^2}{16}+\dfrac{y^2}{4}=1$ 17. $\dfrac{x^2}{1024}+\dfrac{y^2}{49}=1$ 19. $\dfrac{x^2}{4}+\dfrac{y^2}{9}=1$

21. B 23. C 25. F 27. G

29. Center (1,-2), vertices (6,-2) and (-4,-2), minor axis endpoints (1,0) and (1,-4), major length= 10, minor length = 4

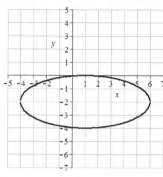

31. Center (-2,3), vertices (-2,8) and (-2,-2), minor axis endpoints (-1,3) and (-3,3), major length = 10, minor length = 2

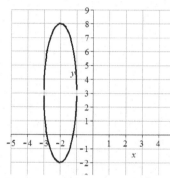

33. Center (-1,0), vertices (-1,4) and (-1,-4), minor axis endpoints (-1,0) and (3,0), major length = 8, minor length = 4

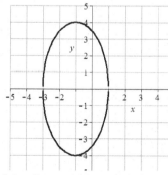

35. Center (-1,-2), vertices (3,-2) and (-5,-2), minor axis endpoints (-1,0) and (-1,-4), major length = 8, minor length = 4

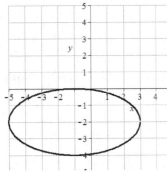

37. Center (2,-1), vertices (2,5) and (2,-7), minor axis endpoints (6,-1) and (-2,-1), major length = 12, minor length = 8

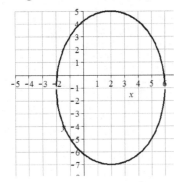

39. $(x-3)^2 + \dfrac{(y+1)^2}{16} = 1$ 41. $\dfrac{(x+4)^2}{16} + \dfrac{(y-3)^2}{25} = 1$

43. 2.211083 feet   45. 17 feet   47. 64 feet   49. (±4,0)   51. (-6,6) and (-6,-4)

53. $\dfrac{x^2}{9} + \dfrac{y^2}{5} = 1$ 55. $\dfrac{x^2}{11} + \dfrac{y^2}{36} = 1$ 57. $\dfrac{x^2}{49} + \dfrac{y^2}{24} = 1$ 59. $\dfrac{x^2}{4} + \dfrac{y^2}{20} = 1$

61. $\dfrac{x^2}{16} + \dfrac{y^2}{8} = 1$ 63. $\dfrac{(x+2)^2}{12} + \dfrac{(y-1)^2}{16} = 1$ 65. $\dfrac{(x-3)^2}{36} + \dfrac{(y-2)^2}{11} = 1$

67. $\dfrac{(x-3)^2}{21} + \dfrac{(y+1)^2}{25} = 1$ 69. $\dfrac{(x-1)^2}{4} + \dfrac{(y-3)^2}{5} = 1$ 71. $\dfrac{(x+2)^2}{289} + \dfrac{(y+1)^2}{120} = 1$

73. 31.22 feet   75. $\dfrac{x^2}{8640.632025} + \dfrac{y^2}{8638.214} = 1$ 77. $\dfrac{x^2}{25} + \dfrac{y^2}{9} = 1$

79. The center is at (0,0). Since $a > b$, the ellipse is horizontal. Let (c,0) be the focus on the positive x-axis. Let (c, h) be the endpoint in Quadrant 1 of the latus rectum passing through (c,0).

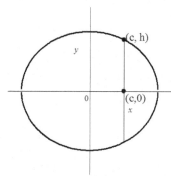

The distance between the focus and latus rectum endpoint can be found by substituting (c,0) and (c,h) into the distance formula $h = \sqrt{(x_1 - x_2)^2 + (y_1 - y_2)^2}$ which yields $h = \sqrt{(c-c)^2 + (h-0)^2} = h$. So h is half the latus rectum distance. Substituting (c,h) into the ellipse equation to find h gives $\dfrac{c^2}{a^2} + \dfrac{h^2}{b^2} = 1$. Solve for h yields

$$h^2 = b^2\left(1 - \frac{c^2}{a^2}\right) = b^2\left(\frac{a^2}{a^2} - \frac{c^2}{a^2}\right) = b^2\left(\frac{a^2 - c^2}{a^2}\right) = b^2\left(\frac{b^2}{a^2}\right) = \frac{b^4}{a^2}. \text{ so } h = \sqrt{\frac{b^4}{a^2}} = \frac{b^2}{a}. \text{ The}$$

distance of the latus rectum is $2h = \dfrac{2b^2}{a}$.

Section 9.2

1. B   3. D

5. Vertices (± 2,0), transverse length = 4, asymptotes y = $\pm 5/2$x,

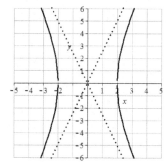

7. Vertices $(0, \pm 1)$, transverse length = 2, asymptotes $y = \pm 1/2x$,

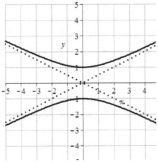

9. Vertices $(\pm 3, 0)$, transverse length = 6, asymptotes $y = \pm 1/3x$,

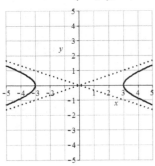

11. Vertices $(0, \pm 4)$, transverse length = 8, asymptotes $y = \pm 4/3x$

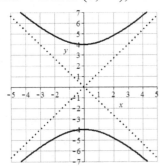

13. Vertices $(\pm \sqrt{2}, 0)$, transverse length = $2\sqrt{2}$, asymptotes $y = \pm 3x$,

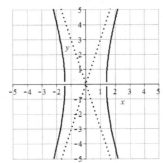

15. $\dfrac{y^2}{4} - \dfrac{x^2}{9} = 1$　　17. $\dfrac{y^2}{16} - \dfrac{x^2}{64} = 1$　　19. $\dfrac{x^2}{9} - \dfrac{y^2}{36} = 1$　　21. $\dfrac{x^2}{16} - \dfrac{y^2}{16} = 1$

23. C　　　25. H　　　27. B　　　29. A

31. Center $(1, -2)$, vertices $(6, -2)$ and $(-4, -2)$, transverse length = 10, asymptotes $y = \pm 2/5(x-1)-2$

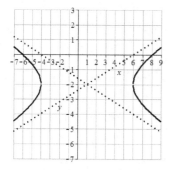

33. Center (-2,1), vertices (-2,4) and (-2,-2), transverse length = 6, asymptotes y =±3(x+2)+1

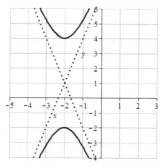

35. Center (1,0), vertices (3,0) and (-1,0), transverse length = 4, asymptotes y =±2(x-1)

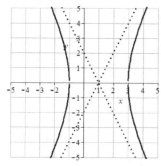

37. Center (-1,2), vertices (-1,4) and (-1,0), transverse length = 4, asymptotes y =±1/2(x+1)+2

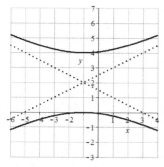

39. Center (-2,1), vertices (0,1) and (-4,1), transverse length = 4, asymptotes y =±3/2(x+2)+1

685

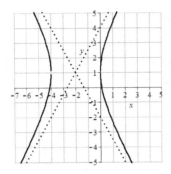

41. $\dfrac{(y+1)^2}{9} - \dfrac{(x-4)^2}{4} = 1$ 43. $\dfrac{(y-2)^2}{16} - \dfrac{(x+1)^2}{4} = 1$

45. Center (0,0), vertices (±1/3,0), transverse length = 2/3, asymptotes y = ±12x

47. Center (-1,1), vertices (-1,3/2) and (-1,1/2), transverse length = 1, asymptotes y = ± 3/2 (x + 1) +1

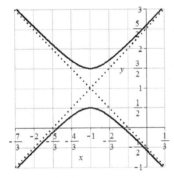

49. Foci (0,±5) 51. Foci (5,6) and (-3,6) 53. Foci (-4,6) and (-4,-4)

55. $\dfrac{x^2}{16} - \dfrac{y^2}{9} = 1$ 57. $\dfrac{y^2}{144} - \dfrac{x^2}{25} = 1$ 59. $\dfrac{x^2}{225} - \dfrac{y^2}{64} = 1$ 61. $\dfrac{x^2}{64} - \dfrac{y^2}{36} = 1$

63. $\dfrac{(y-2)^2}{16} - \dfrac{(x-1)^2}{9} = 1$ 65. $\dfrac{(x+1)^2}{25} - \dfrac{(y-3)^2}{144} = 1$ 67. $\dfrac{x^2}{900} - \dfrac{y^2}{1600} = 1$

69. $\dfrac{x^2}{900} - \dfrac{y^2}{14400.3636} = 1$

71. $\dfrac{x^2}{3025} - \dfrac{y^2}{6975} = 1$

73. $5y^2 - x^2 + 25 = 0$ can be put in the form $\dfrac{y^2}{5} - \dfrac{x^2}{25} = -1$. $x^2 - 5y^2 + 25 = 0$ can be put

in the form $\dfrac{y^2}{5} - \dfrac{x^2}{25} = 1$ showing they are conjugate.

75. $\sqrt{2}$ 77. No matter the value of k, the foci are at $\left(\pm\sqrt{6}, 0\right)$

Section 9.3

1. C   3. A

5. Vertex: (0,0). Axis of symmetry: $y = 0$. Directrix: $x = -4$. Focus: (4,0)

7. Vertex: (0,0). Axis of symmetry: $x = 0$. Directrix: $y = -1/8$. Focus: (0,1/8)

9. Vertex: (0,0). Axis of symmetry: $y = 0$. Directrix: $x = 1/16$. Focus: (-1/16,0)

11. Vertex: (2,-1). Axis of symmetry: $x = 2$. Directrix: $y = -3$. Focus: (2,1)

13. Vertex: (-1,4). Axis of symmetry: $x = -1$. Directrix: $y = 3$. Focus: (-1,5)

15. $(y-1)^2 = -(x-3)$ 17. $(y-3)^2 = 12(x-2)$ 19. $x^2 = 4(y-3)$

21. At the focus, (0,1)   23. 2.25 feet above the vertex.   25. 0.25 ft

27. $\left(\dfrac{1}{\sqrt{3}}, \dfrac{2}{\sqrt{3}}\right)\left(\dfrac{-1}{\sqrt{3}}, \dfrac{-2}{\sqrt{3}}\right)$ 29. $\left(3, \sqrt{2}\right)\left(3, -\sqrt{2}\right)\left(-3, \sqrt{2}\right)\left(-3, -\sqrt{2}\right)$

31. $\left(2\sqrt{2}, 8\right)\left(-2\sqrt{2}, 8\right)$

33. $\left(\sqrt{\dfrac{5}{3}}, \sqrt{\dfrac{2}{3}}\right)\left(-\sqrt{\dfrac{5}{3}}, \sqrt{\dfrac{2}{3}}\right)\left(\sqrt{\dfrac{5}{3}}, -\sqrt{\dfrac{2}{3}}\right)\left(-\sqrt{\dfrac{5}{3}}, -\sqrt{\dfrac{2}{3}}\right)$

35. (-64.50476622, 93.37848007) $\approx$ (-64.50, 93.38)

687

Section 9.4

1. $e = 3$. Directrix: $x = 4$. Hyperbola.   3. $e = 3/4$. Directrix: $y = -2/3$. Ellipse.

5. $e = 1$. Directrix: $x = -1/5$. Parabola.   7. $e = 2/7$. Directrix: $x = 2$. Ellipse.

9. $r = \dfrac{20}{1 - 5\cos(\theta)}$ 11. $r = \dfrac{1}{1 + \dfrac{1}{3}\sin(\theta)}$, or $r = \dfrac{3}{3 + \sin(\theta)}$

13. $r = \dfrac{2}{1 - \sin(\theta)}$

15. Hyperbola. Vertices at $(-9,0)$ and $(-3,0)$
    Center at $(-6,0)$. $a = 3$. $c = 6$, so $b = \sqrt{27}$
    $\dfrac{(x+6)^2}{9} - \dfrac{y^2}{27} = 1$

17. Ellipse. Vertices at $(0,3)$ and $(0,-6)$
    Center at $(0,-1.5)$. $a = 4.5$, $c = 1.5$, $b = \sqrt{18}$
    $\dfrac{x^2}{18} + \dfrac{(y+1.5)^2}{20.25} = 1$

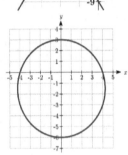

19. Parabola. Vertex at $(3,0)$. $p = 3$.
    $y^2 = -12(x - 3)$

21. a)

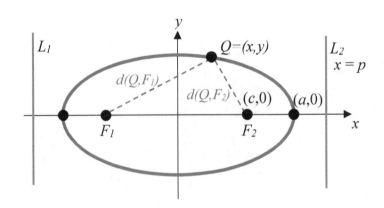

b) $d(Q,L_1) = x - (-p) = x + p$, $d(Q,L_2) = p - x$

c) $d(Q,F_1) = ed(Q,L_1) = e(x+p)$. $d(Q,F_2) = ed(Q,L_2) = e(p-x)$

d) $d(Q,F_1) + d(Q,F_2) = e(x+p) + e(p-x) = 2ep$, a constant.

e) At $Q = (a, 0)$, $d(Q,F_1) = a - (-c) = a + c$, and $d(Q,F_2) = a - c$, so
$$d(Q,F_1) + d(Q,F_2) = (a+c) + (a-c) = 2a$$

Combining with the result above, $2ep = 2a$, so $p = \dfrac{a}{e}$.

f) $d(Q,F_2) = a - c$, and $d(Q,L_2) = p - a$

$$\frac{d(Q,F_2)}{d(Q,L_2)} = e, \text{ so } \frac{a-c}{p-a} = e.$$

$a - c = e(p-a)$. Using the result from (e),

$$a - c = e\left(\frac{a}{e} - a\right)$$

$$a - c = a - ea$$

$$e = \frac{c}{a}$$

Index